Reinforced Concrete Design

Reinforced Concrete Design (RC) is performed mostly by the limit state method throughout the world. This book covers the fundamental concepts and principles of RC design developing the topics from the basic theories and assumptions. Building upon the possible revisions to the mother code of concrete in India, IS:456-2000, it explains the RC design provisions of IRC:112-2020, which are in line with international standards. In addition to strength design, serviceability and ductility design are also covered.

Features:

- Highlights the basic philosophy of RC design and behaviour of the sections up to and beyond limit state.
- Clarifies limit state theory from the basic assumptions provided in relevant Indian and international standards, IS:456, IRC:112 and Eurocode:2.
- Includes design aids or tools for standard and high strength concrete up to M90 grade as per different codes of practice.
- Explains the concept of ductility of reinforced concrete sections subjected to flexure with or without axial loads from fundamental principles.
- Covers fundamentals on serviceability requirements in reinforced concrete structures.
- Illustrates the design methodology of shear walls and includes design aids developed using basic principles as per relevant codes of practice.
- Explains reinforced concrete design provisions as per latest national and international standards and these are expected to be in line with those to be included in the forthcoming revision of IS:456.

This book is aimed at graduate students, researchers and professionals in civil engineering, construction engineering and concrete.

Reinforced Concrete Design

Limit State Method and Beyond

Santanu Bhanja

CRC Press is an imprint of the
Taylor & Francis Group, an **informa** business

Designed cover image: © Santanu Bhanja

First edition published 2024
by CRC Press
2385 NW Executive Center Drive, Suite 320, Boca Raton FL 33431

and by CRC Press
4 Park Square, Milton Park, Abingdon, Oxon, OX14 4RN

CRC Press is an imprint of Taylor & Francis Group, LLC

ISBN: 9781032458892 (hbk)
ISBN: 9781032541457 (pbk)
ISBN: 9781003415398 (ebk)

DOI: 10.1201/9781003415398

Typeset in Times
by Newgen Publishing UK

Dedicated with deep love and heartfelt gratitude to my most Revered Dadu, Beloved Mother Rai Kishori, and Dearest Father Benoyendra Kumar without whose blessings this book would not have seen the light of day.

Contents

Preface ix
About the Author xi
Acknowledgements xiii

Chapter 1 Introduction 1

Chapter 2 Basic Material – Concrete 5

Chapter 3 General Overview of IS:456-2000 and IRC:112-2020 – General Provisions of Indian Standards on Concrete 22

Chapter 4 Provisions of Standards or Codes of Practice on Reinforcement: IS:1786-2008, IS:456-2000, SP:16-1980, IRC:112-2020 and Eurocode-2 28

Chapter 5 General Considerations for Analysis of RC Structures 41

Chapter 6 General Considerations for Design of RC Structures 49

Chapter 7 Philosophy of Reinforced Concrete Design 53

Chapter 8 Working Stress Method 60

Chapter 9 Limit State Method as per IS:456 74

Chapter 10 Limit State of Collapse in Flexure as per IS:456 80

Chapter 11 Limit State of Collapse in Compression as per IS:456 114

Chapter 12 Limit State of Collapse in Shear 181

Chapter 13 Development Length of Reinforcement in RC Sections 189

Chapter 14 Limit State of Serviceability 201

Chapter 15 Limit State Method as per IRC:112-2020 for RC Sections 214

Chapter 16 Design beyond Limit State 270

Chapter 17 Ductility Design 277

Chapter 18 Earthquake-Resistant Design of Structures 306

Chapter 19 Design of Shear Walls Following the Fundamental Principles of Limit State Method as per Indian and International Standards 312

Chapter 20 Design of Staircases 367

Chapter 21 Design of Foundations 376

Index 391

Preface

Reinforced concrete design is performed mostly by Limit State Method throughout the globe. The design principles are similar with certain modifications in different parts of the world. The design philosophies are outlined in their standards or codes of practice which follow the prescriptive method of design. The mother code of concrete in India, IS-456 – "Indian Standard, Plain and Reinforced Concrete – Code of Practice (Fourth Revision)" was last revised in 2000. The design parameters given in the standard are not valid for high-strength concrete from M65 onwards and the stress–strain relationships of HYSD steels have become outdated. The current philosophy based on a load-based prescriptive method of design also has several shortcomings and disadvantages.

The author has been associated with concrete over the last three and a half decades as a student, researcher, designer and then as an academician involved in teaching at the under-graduate, post-graduate levels and beyond and as a trainer of teachers. In addition, the author has had the privilege of working in the industry as a design and site engineer for 7 years before switching to academics. From past experience it is felt that reinforced concrete design is not a mere process or operation or method, but a philosophy. Unfortunately, it is highly underrated, especially amongst students and practicing engineers. Also, the design philosophies of RC design which are evolving with time lack proper interpretation and implications. Limitations of the design method and standards need to be properly understood by designers. Any structural design is based on the stress–strain relationship of the materials involved. These basic requirements are not properly highlighted in our standards or in the commonly available books. This point has been given prime importance in interpreting the design philosophy. The present Indian Standard does not have any explanatory handbook nor design aids or tools necessary for design. The design aids which were available for the previous version of the Standard have been withdrawn. Though the limit state method of design has significant improvements over the previous design philosophies, performing design of reinforced concrete elements following the first principles can be quite cumbersome and design aids are almost invariably necessary.

RC design for earthquake-resistant structures is growing as an important branch in design but, unfortunately, it is not properly covered in the present standards. The standard on Ductile Design and Detailing of Reinforced Concrete Structures – IS:13920-2016, only provides some do's and don'ts for incorporating ductility in design without going into the fundamental ductility design considerations. The design methodology of shear walls, which are almost routinely used nowadays in multistorey buildings, as per fundamental principles of the limit state method, is neither properly furnished in our standards nor in the books currently available. Shear wall design should be based on interaction charts developed following the fundamental principles of the limit state method. However, these are not commonly available for use in India. The closed-form equations provided in IS:13920-2016 do not seem to be adequate for proper design of shear walls. The concept of ductility, and its evaluation and application to RC sections are yet to be incorporated into the codes of practice. This book is a humble effort to address these areas.

The commonly available books mostly deal with RCC design in a systematic and stereotypical manner covering the different aspects of design mostly elaborated by numerical examples. However, this subject is not like physics or mathematics which are governed by strict laws and hypotheses. RCC design is something creative, and the same problem can be handled by different designers in different ways. Hence the subject needs to be treated freely with an open mind. This book makes a humble attempt to cover the fundamental concepts and principles of RC design, developing the topics from the basic theories and assumptions. The objective or aim is not to clarify the methodology by numerical examples (which are profusely available nowadays) but to develop or initiate interest in the subject.

The mother code of concrete in India, IS:456-2000, is on the verge of revision whereby significant changes are expected to be introduced. It is most likely that the modified standard will have reinforced concrete design provisions similar to IRC:112-2020, Code of Practice for Concrete Road Bridges, which is in line with international standards like Eurocode 2: Design of Concrete Structures – Part 1-1: General Rules and Rules for Buildings. The important RC design provisions of these standards have been covered in this book. The book focuses on fundamental concepts and principles with necessary explanations, and hence the number of numerical examples is quite limited. An attempt has been made to address the fundamental assumptions and considerations of design, which are not commonly covered in the standard books available. To accommodate these, some of the common topics available in standard books have not been elaborately dealt with, for which the author sincerely apologizes.

After a brief introduction in Chapter 1, Chapters 2, 3 and 4 deal with materials and provisions in Indian Standards regarding concrete and reinforcement. Chapter 5 deals with analysis, whereas Chapters 6 and 7 cover design considerations and philosophy. Chapter 8 is dedicated to the working stress method of design and the reasons for its deletion from the current national and international standards. Chapters 9–14 deal with the limit state method, highlighting the design in flexure, compression, shear, bond and serviceability. Chapter 15 deals with important design provisions of IRC 112 and Eurocode 2. In Chapter 16, design beyond the limit state has been highlighted. In Chapter 17, ductility design has been covered in detail. Earthquake-resistant design features in Chapter 18. Detail deliberation on shear wall design, along with the development of interaction charts, are covered in Chapter 19. Chapter 20 covers the basic aspects of staircase design, while the design of isolated footings has been covered in Chapter 21.

Knowledge is eternal, its horizon has no boundaries and it is ever expanding. According to Swami Vivekananda, the source of all knowledge is the human brain. As human civilization progresses, knowledge will expand. If the present endeavour can ignite interest in this subject, the author, who considers himself a student in the field of civil engineering, would consider that his efforts have been truly rewarded.

This book is designed to serve students, researchers and practicing engineers associated with the design of reinforced concrete structures as the basic design philosophy up to and beyond the limit state has been highlighted. The author sincerely requests that readers, students and practicing engineers convey their feedback, suggestions and criticisms for further improvement of this humble effort.

Santanu Bhanja

About the Author

Santanu Bhanja is currently Professor and former Head-of-the-Department of Civil Engineering, National Institute of Technical Teachers' Training and Research, Kolkata, under Ministry of Education, Govt of India. He graduated with a degree in civil engineering from Jadavpur University, Kolkata, received M Tech degree from IIT, Kharagpur, and was awarded PhD from Jadavpur University, Kolkata. He has served at Bengal Engineering College (DU), Shibpur, as a Lecturer (presently Assistant Professor level) and then as an Assistant Professor (presently Associate Professor level) at the Bengal Engineering and Science University, Shibpur, Howrah, West Bengal (presently IIEST, Shibpur), for about a decade before taking up his present assignment. Prof. Bhanja has also worked in the industry for about 7 years, mostly as a Design Engineer. Besides teaching, he has provided technical consultancy as a Proof-Consultant for the private and government sectors regarding the design of several prestigious projects. He has authored several research papers which have been published in journals of international repute and has presented numerous papers in national and international conferences and seminars. Prof. Bhanja has more than three decades of experience with concrete as a researcher, teacher of students at the under-graduate, post-graduate levels and beyond and as a trainer of the technical faculty members of degree and diploma level engineering institutions. He has organized several workshops, seminars, webinars and training programmes in the field of concrete.

Acknowledgements

I express my Sastanga Pranams and deep regards at the lotus feet of my most Revered Dadu and my beloved parents, mother Rai Kishori and father Benoyendra Kumar, and earnestly request them to shower their choicest blessings and good wishes from RAMAKRISHNALOKA. They have been and will remain a constant source of love, care, affection and encouragement till I breathe my last.

With deep gratitude and regards, I pay my respectful pranams to my teacher, Professor Bratish Sengupta, ex-Professor-in-Charge of Civil Engineering Department, Jadavpur University, Kolkata, for igniting my passion for concrete and being a constant source of inspiration as long as he was physically present in this world. I sincerely pray for his blessings from his heavenly abode.

I express my appreciation to the Director, National Institute of Technical Teachers' Training and Research, Kolkata, for providing the necessary help and cooperation for completion of the book.

This book has seen the light of the day due to the untiring help and cooperation of my numerous students, who have been solely responsible for bringing it into the current form and shape. The names of only a few have been mentioned here. I am highly obliged to him for his dedicated support.

I acknowledge with sincere thanks the untiring support provided by my ex-student Sri Bibhas Mondal for extending all forms of technical support, typing the manuscript, drawing the figures and carrying out the necessary composing work.

The constant support of the research fellows working under my supervision, Mrs Anasuya Mondal and Jayanta Nath Chowdhury, deserve special mention. They were always eager to walk an extra mile for me and have always extended their selfless help happily. I am deeply indebted to them for their untiring support during the preparation and proof checking of the manuscript. Thanks are due to all my students for their tireless help and constant support. In fact, the idea of authoring a book came from them.

The cooperation, love, encouragement and warmth of heart received from all of my family members cannot be expressed by mere good choice of words. I wish to express my deep regards and heartfelt love for their untiring patience and sacrifice during the entire period of preparing this book when I was heavily preoccupied with my work.

I also express my heartfelt thanks to Dr Gagandeep Singh and Ms Jahnavi Vaid of CRC Press, Taylor and Francis Group, for their unconditional support and cooperation at all the stages during the process of the preparation of the manuscript. No words of appreciation can do justice to their cooperation. I also wish to put on record my appreciation for the excellent cooperation received at the editorial, proof checking and production stages.

10 December, 2023
Kolkata
Santanu Bhanja

1 Introduction

1.1 REINFORCED CONCRETE – DEFINITION

Concrete literally forms the basis of our modern society. There is scarcely any aspect of life that does not depend directly or indirectly on concrete. We play, work, study and live in concrete structures to which we drive over concrete roads and bridges. Our goods are transported by trucks travelling on concrete highways, by trains which run on rails supported on concrete sleepers, by ships which moor at concrete piers in harbours protected by concrete breakwaters and by aeroplanes which land and take off on concrete runways. Water for drinking purposes and for growing crops is stored in large concrete dams and electricity is generated by burning coal in thermal power plants made of concrete. Hence, we can see that concrete forms a part of our everyday life.

Concrete is very strong in compression but very weak in tension. In real-life structures tension and compression go hand to hand and the structural elements must be capable of resisting both. Hence engineers have been reinforcing concrete by suitable means so that the sections can resist both tension and compression. This has paved the way for reinforced concrete, which is a composite material in which reinforcements are embedded into the elements in such a manner that concrete and reinforcements act together. One common form of reinforcement is steel reinforcing bars which have very good bonding properties with concrete and very similar coefficient of thermal expansion values. The success of reinforced concrete lies in the perfect bonding between these two materials. The journey of reinforced concrete started about one and half centuries ago, and during the initial stages people were very doubtful and suspicious about this material. Today, concrete is the second largest per capita consumed material in the world, coming only after water, which explains the importance of this construction material throughout the globe. Conventionally, concrete was made with cement and the term cement-concrete became analogous with concrete, but nowadays other types of concrete have also emerged, such as geo-polymer concrete, which is made without cement. Similarly, steel reinforcements have been commonly used in concrete, which are now paving the way for other types of reinforcements like fibre-reinforced polymers. However, till date concrete commonly refers to cement concrete and reinforcement with steel.

1.2 PLAIN AND STRUCTURAL CONCRETE

Plain concrete refers to concrete without reinforcements. Some applications of plain concrete are in the form of levelling and water proofing courses for substructures, for slabs directly resting on bearings, etc.

Structural concrete refers to elements of concrete which are provided with reinforcements, and these elements take part in the load path of the structures. Structural concrete, which is reinforced

DOI: 10.1201/9781003415398-1

concrete, plays a very important role in the modern construction industries throughout the globe. It is because of this material that different shapes and sizes of structural elements can be used in structures, thereby significantly improving their aesthetics, functionality and application.

1.3 ADVANTAGES AND DISADVANTAGES OF REINFORCED CONCRETE

1.3.1 Advantages of concrete

All over the world concrete and structural steel are considered as the two major construction materials. At times they complement each other and at times they compete with one another. The advantages of concrete are manyfold, and a list would be almost endless. Among the major advantages are the following:

(i) Economical – economic and widespread availability of raw materials;
(ii) Ability to be cast into any shape – its versatility and adaptability;
(iii) Durability – nominal maintenance is required during service;
(iv) Fire resistance – concrete is non-combustible and does not emit toxic fumes when subjected to fire. Compared to structural steel, the fire resistance of concrete is significantly high;
(v) On-site production;
(vi) Aesthetic properties;
(vii) Ecological construction – concrete is the safest home for tons of pozzolanic and industrial by-products (fly ash, blast furnace slag, etc.). Stockpiling of these waste materials can cause air pollution and dumping into ponds or streams results in the release of toxic metals that are normally present in them only in small amounts. Hence concrete may be described as Nilkantha, or similar to Lord Shiva, who drank poison to save the world.

1.3.2 Disadvantages of concrete

The disadvantages of concrete include the following:

(i) Low tensile strength;
(ii) Low ductility;
(iii) Volume instability;
(iv) Low strength to weight ratio – high load capacity requires comparatively large masses of concrete;
(v) Corrosion of reinforcements – throughout the world the major reason for failure of concrete structures is corrosion of reinforcements, which is a major concern for engineers.

1.4 ROLE OF CONCRETE AND REINFORCEMENT ON THE PROPERTIES OF REINFORCED CONCRETE

Concrete would not have gained its present status as a principal construction material, but for the invention of reinforced concrete. The idea of reinforcing concrete with steel has resulted in a new composite material, having the ability to resist significant tensile stresses, which was otherwise impossible. Thus, the construction of load-bearing members became viable with this new material. Concrete is strong, robust, durable and has high compressive strength. The ability of concrete to be cast into any size and shape is unique amongst all the construction materials, which has given engineers ample freedom that has resulted in many splendid structures throughout the world. The steel reinforcements embedded in concrete compensate its inability of carrying tension. Due to the perfect bond between steel and concrete, strain compatibility is maintained and the two materials act harmoniously. Reinforcements which are highly ductile impart ductility to reinforced concrete even though the material concrete is highly brittle. Our demands from concrete are constantly increasing and new-generation concrete needs to be designed for strength, serviceability and ductility along with the requirements of stability, economy and aesthetics. Reinforcements can impart this very important attribute of ductility to concrete provided the correct design and detailing have been performed. Reinforced concrete can be made ductile in regard to flexural failure only, as for other modes of failure the material behaves predominantly in a brittle manner. Reinforced concrete design was evaluated as good or excellent even a few decades back on the basis of its strength, serviceability, economy and aesthetics, whereas the present definition of good design refers to enhanced performance of the reinforced concrete structures up to and beyond the design loads. Reinforcements cater to the effects of temperature, creep and shrinkage if properly designed and detailed. They have the capacity to resist tension along with compression.

1.5 GRADES OF STRUCTURAL CONCRETE PERMITTED BY IS:456 AND GRADES OF STEEL PERMITTED IN IS:1786

IS:456-2000 classifies concrete into the following classes:

(a) Ordinary concrete having grades M10 to M20;
(b) Standard concrete ranging from M25 to M60;
(c) High-strength concrete, M65 to M100.

Here M refers to the concrete mix and the number indicates the specified characteristic compressive strength. The design parameters given in the Indian standard are valid for grades of concrete up to M60 only. IS:456 refers to IS:1786 – Indian Standard for High-Strength Deformed Steel Bars and Wires for Concrete Reinforcements – Specifications. The present version of this standard, published in 2008, specifies the requirements of deformed bars and wires for use as reinforcements in concrete in the following strength grades:

(a) Fe 415, Fe 415D, Fe 415S;
(b) Fe 500, Fe 500D, Fe 500S;
(c) Fe 550, Fe 550D;
(d) Fe 600;
(e) Fe 650;
(f) Fe 700.

Here the figure following the symbol Fe indicates the specified minimum 0.2% proof stress or yield stress in N/mm^2 and D and S indicate enhanced and additional requirements.

However, in IS:456-2000, design parameters are specified for HYSD bars for grades Fe 415 and Fe 500 only. IS: 13920:2016 – Ductile Design and Detailing of Reinforced Concrete Structures Subjected to Seismic Forces – Code of Practice, allows Fe 550 as the maximum grade of HYSD steel to be used in concrete.

1.6 SUMMARY

This chapter defines reinforced concrete and lists its advantages and disadvantages. The roles of concrete and reinforcements have also been discussed. Grades of concrete and steel permitted in the Indian standards have been covered also.

BIBLIOGRAPHY

IS 456:2000 (2007). *Plain and Reinforced Concrete – Code of Practice*, Fourth Revision, Tenth Reprint April 2007, Bureau of Indian Standards, New Delhi.

IS 1786:2008. (2008). *Indian Standard for High Strength Deformed Steel Bars and Wires for Concrete Reinforcements – Specifications*, Fourth Revision, Bureau of Indian Standards, New Delhi.

Mindess, S. and Young, J. Francis. (1981). *Concrete*, Prentice Hall, Upper Saddle River, New Jersey.

Nadim Hassoun, M. and Al-Manaseer, A. (2015). *Structural Concrete Theory and Design*, Sixth Edition, Wiley, Hoboken, New Jersey.

Neville, A. M. (1996). *Properties of Concrete*, Fourth Edition, ELBS with Addison Wesley Longman, England.

Park, R. and Paulay, T. (1975). *Reinforced Concrete Structures*, Wiley, New York.

Pillai, S. U. and Menon, D. (2012). *Reinforced Concrete Design*, Third Edition, Reprint 2012, McGraw Hill, New Delhi.

Shah, V. L. and Karve, S. R. (2016). *Limit State Theory and Design of Reinforced Concrete*, Eighth Edition, Structures Publications, Pune, India.

Varghese, P. C. (2008). *Limit State Design of Reinforced Concrete*, PHI Learning, New Delhi.

Varghese, P. C. (2009). *Advanced Reinforced Concrete Design*, PHI Learning, New Delhi.

2 Basic Material
Concrete

2.1 INTRODUCTION

Concrete is unique among all the construction materials in that it is designed specifically for a particular project from the materials available at the site. The site engineer has full control and responsibility in the design of concrete. If it is not properly designed, it will result in sub-standard performance. Hence the civil engineer must have a thorough knowledge of the properties of concrete and the procedures that lead to the production of good concrete with the required properties. In order to prepare a good dish, having only good-quality ingredients will not be sufficient; the knowledge and skill of the chef play a vital role. In a similar manner it can be stated that only having good-quality ingredients will not result in good-quality concrete until and unless there is a skilled and knowledgeable engineer to design the entire process.

2.2 INGREDIENTS OF CONCRETE – CEMENT, AGGREGATES, WATER AND ADMIXTURES

2.2.1 Cement

Cement is a finely pulverized material which by itself is not a binder but which develops the binding property as a result of hydration (chemical reaction between cement and water). A cement is called hydraulic when the hydration products are stable in an aqueous environment. The most commonly used hydraulic cement for making concrete is Portland cement, which essentially consists of hydraulic calcium silicates.

Portland cement is a hydraulic cement produced by pulverizing clinkers consisting primarily of hydraulic calcium silicates, usually containing one or more forms of calcium sulphate as an interground addition. It is a cement obtained by intimately mixing together mainly calcareous and argillaceous materials, burning them at a clinkering temperature and grinding the resulting clinker. The name Portland cement was given due to the resemblance of the colour and quality of the hardened cement to Portland stone.

2.2.1.1 Manufacture

Since calcium silicates are the primary constituents of Portland cement, the raw materials for cement production must provide calcium and silica in suitable forms and proportions.

Sources of calcium include naturally occurring calcium carbonate materials such as limestone, chalk, marl and seashells.

DOI: 10.1201/9781003415398-2

Sources of silica include clay and shale (rather than quartz and sandstone).

The following chemical reactions take place in a cement kiln:

$$\left.\begin{array}{l}\text{Limestone} \rightarrow CaO + CO_2 \\ \text{Clay} \rightarrow SiO_2 + Al_2O_3 + Fe_2O_3 + H_2O\end{array}\right\} \Rightarrow \left\{\begin{array}{l}3CaO.SiO_2 \\ 2CaO.SiO_2 \\ 3CaO.Al_2O_3 \\ 4CaO.\ Al_2O_3.\ Fe_2O_3\end{array}\right.$$

2.2.1.2 Chemical composition

The basic raw materials used in the manufacture of cement are (i) lime, (ii) silica, (iii) alumina and (iv) iron oxide. These oxides interact with each other in a rotary kiln to form complex compounds. During manufacture the rate of cooling and fineness of grinding affect the properties of the cement. The relative proportions of the different oxides influence the properties of cement. The approximate oxide composition limits of OPC are shown in Table 2.1.

Determination of compound composition from chemical analysis

The oxides present in the raw materials when subjected to high clinkering temperature combine with each other to form complex compounds. The identification of the major compounds has been done by R. H. Bogue and they are called "Bogue's compounds". These are described in Table 2.2.

Calcium silicates form a major part of the compound composition of cement; together they constitute 70–80% of cement. The remainder is provided by calcium aluminates, alumino ferrites and some minor compounds such as alkalis. Even small changes in oxide composition can cause large

TABLE 2.1
Oxide composition of cement

Oxide	Content (%)
CaO	60–67
SiO_2	17–25
Al_2O_3	3–8
Fe_2O_3	0.5–6
MgO	0.1–4
Alkalis (such as K_2O, Na_2O)	0.4–1.3
SO_3	1.0–3.0

TABLE 2.2
Compound compositions of cement

Name of compound	Formula	Abbreviation of formula
Tricalcium silicate	$3CaO.SiO_2$	C_3S
Dicalcium silicate	$2CaO.SiO_2$	C_2S
Tricalcium aluminate	$3CaO.Al_2O_3$	C_3A
Tetra-calcium alumino ferrite	$4CaO.Al_2O_3.Fe_2O_3$	C_4AF

C—CaO, S—SiO_2, A—Al_2O_3, F—Fe_2O_3, H—H_2O

changes in the compound composition. The properties of cement and hence of the resultant concrete made with it are influenced by the type and proportion of the compounds present.

2.2.1.3 Hydration of cement

Anhydrous Portland cement does not bind sand and gravel, it acquires its adhesive property only when mixed with water. This is because of the chemical reactions of cement with water, commonly referred to as the hydration of cement, and the products are collectively termed products of hydration. These reactions yield products that possess setting and hardening characteristics.

Hydration of Cement Compounds

(a) The silicates (C_3S and C_2S) react with water to form calcium silicate hydrate (CSH) and calcium hydroxide (CH). CSH is responsible for strength, but CH is highly undesirable and causes durability problems.
(b) C_3S readily reacts with water, producing more heat of hydration and is responsible for the early strength of concrete.
(c) C_2S hydrates more slowly, produces less heat of hydration and is responsible for the later strength of concrete. The CSH produced by C_2S is of better quality than that by C_3S.
(d) Reaction of C_3A with water is very fast and may lead to flash set. Gypsum is added at the time of grinding to prevent this flash set. The hydrated calcium aluminates do not contribute to the strength of cement paste, rather they pose durability problems.
(e) The hydrated product of C_4AF also does not contribute to the strength but is more stable than hydrated C_3A.

2.2.1.4 Fineness

In addition to the compound composition, the fineness of a cement affects its reactivity with water. The finer the cement, the more rapidly it will react. For a given compound composition, the rate of reactivity and hence the strength development can be enhanced by finer grinding of cement.

2.2.1.5 Structure of hydrated cement paste

The mechanical properties of hardened cement and concrete depend on:

(i) Chemical composition of the hydrated cement;
(ii) Physical structure of the hydration products.

Fresh cement paste is a plastic mass consisting of water and cement but once it has set its gross volume remains approximately constant.

At any stage of hydration, the hardened paste consists of very poorly crystallized hydrates of various compounds, collectively termed as gel, unhydrated cement, calcium hydroxide, some minor compounds and residue of water-filled spaces in the fresh paste. These voids are called capillary pores. Within the gel there exist interstitial voids, called gel pores. Capillary pores are one to two orders of magnitude larger than gel pores.

Water added to cement is:

(i) Partly used up in chemical reactions;
(ii) Partly occupies gel pores;
(iii) The remainder, if any, causes capillary pores.

Water required for complete hydration of cement in a sealed container is 38% of the weight of cement. Out of this, 23% is required for the chemical reactions and 15% for filling up the gel pores.

2.2.2 Aggregates

Aggregates consist of granular material. They are classed as:

(i) Coarse aggregates (CA) – aggregate particles mostly >4.75 mm;
(ii) Fine aggregates (FA) – aggregate particles mostly <4.75 mm.

Aggregates occupy about 70–75% of the volume of concrete and influence the different characteristics and properties of concrete. Hence the properties of the aggregates need to be studied to assess their contribution to concrete.

Source
The natural aggregates owe their origin to the three bedrocks: igneous, sedimentary and metamorphic:

- Igneous rocks result from cooling of molten magma;
- Sedimentary rocks are formed from deposition, cementation and consolidation;
- Metamorphic rocks are formed from igneous and sedimentary rocks subjected to high temperature and pressure which change their strength and texture.

Size
FA – aggregate sample most of which passes through 4.75 mm IS sieve.
CA – aggregate sample most of which is retained on 4.75 mm IS sieve.

Shape
Aggregates can be classified as rounded, irregular or partly rounded, angular and flaky.

Surface texture
The surface texture of aggregates can be termed as glassy, smooth, crystalline, granular, honeycombed and porous.

Strength
Generally, this is not very important but it plays an important role in high-strength concrete (HSC).

Quality of aggregate
Hard, strong, dense, durable, free from reins (foliations) are desirable qualities. Flaky, elongated ones are to be avoided.

2.2.2.1 Influence of aggregates on the properties of concrete

The size, shape and surface texture govern the specific surface area of the aggregates, i.e., surface area per unit weight. The greater the specific surface area, the greater will be the bonding between the paste matrix and aggregates and the greater will be the strength of the concrete. However, a higher specific surface area will result in reduced workability of concrete. Generally, the strength of aggregates does not affect the strength of concrete (for normal- to medium-strength concrete) as the paste phase is weaker than the aggregate phase and failure generally occurs through the paste or transition phase.

Grading
A well-graded aggregate results in strong and durable concrete.

2.2.3 Water

Potable water is fit for concreting.

2.2.4 Admixtures

Admixtures can be defined as the fourth ingredient added to concrete, in addition to cement, aggregates and water, immediately before or during mixing, to modify one or more specific properties of concrete either in the fresh or hardened state.

Historically, admixtures are almost as old as concrete itself. It is known that the Romans used animal fat, milk, and blood to improve their concrete. Although these were probably added to improve the workability; blood (due to haemoglobin) is a very effective air-entraining agent. The Romans were certainly unaware of this, but the blood might well have improved the durability of their concrete. In more recent times, calcium chloride was often used as an accelerator. However, the systematic study of admixtures began with the introduction of air-entraining agents in the 1930s. This was also accidental – one of the grinding aids used in cement production was beef tallow, and it was found that where tallow was used, the resulting concrete was much more resistant to freezing and thawing. The leakage of oil into the cement in older grinding units had a similar effect. From these beginnings, it is unusual now to find concrete to which some admixture has not been added.

2.2.4.1 Benefits of admixtures

The reason for the large growth in the use of admixtures is that they are capable of imparting considerable physical and economic benefits to concrete. These benefits include the use of concrete under circumstances where previously there existed considerable difficulties.

Admixtures, although not always cheap, do not necessarily represent additional expenditure because their use can result in concomitant savings, for example, in the cost of labour required to effect compaction, in the cement content which would otherwise be necessary, or in improving durability without the use of additional measures.

It should be stressed that, while properly used admixtures are beneficial to concrete, they is no remedy for poor-quality mix ingredients, for the use of incorrect mix proportions, or for poor workmanship in transporting, placing and compaction.

2.2.4.2 Reasons for using admixtures

To modify properties of fresh concrete, mortar and grout:

(a) Increase workability without increasing water content or decrease water content at the same workability;
(b) Retard or accelerate time of initial setting;
(c) Reduce segregation;
(d) Improve pumpability;
(e) Reduce the rate of slump loss.

To modify proportions of hardened concrete, mortar or grout:

(a) Accelerate/decelerate the rate of strength development at early ages;
(b) Increase the strengths indirectly;
(c) Increase the durability of concrete.

2.2.4.3 Different types of admixtures as per Indian Standard IS:9103

(a) Accelerating admixture;

(b) Retarding admixture;
(c) Water-reducing admixture;
(d) Air-entraining admixture;
(e) Superplasticizing admixture.

Admixtures are commonly classified by their function in concrete but often they exhibit some additional action. The classification of ASTM C 494-92 is as follows:

Type A	Water-reducing
Type B	Retarding
Type C	Accelerating
Type D	Water-reducing and retarding
Type E	Water-reducing and accelerating
Type F	High-range water-reducing or superplasticizing
Type G	High-range water-reducing and retarding, or superplasticizing and retarding

Water-reducing admixtures

When water is added to cement, cement particles floc together entrapping water particles. The mode of action of water-reducing admixtures is to deflocculate or to disperse the cement agglomerates into smaller fragments, thus releasing the entrapped water. The principal active components of water-reducing admixtures are surface active agents which are adsorbed on the cement particles giving them a negative charge which leads to repulsion leading to deflocculation.

Superplasticizers (SP)

These act in the same way as water-reducing admixtures but are significantly more active. SPs do not fundamentally affect the strength of hydrated cement paste, the main effect being a better distribution of cement particles leading to their better hydration.

Use of superplasticizers

The availability of superplasticizers has revolutionized the use of concrete in a number of ways, making it possible to place it where it was not possible to do so before. Superplasticizers also make it possible to produce concrete with significantly superior strength and other properties, now termed as high-performance concrete.

They can be successfully used in concrete containing fly ash and are particularly valuable when silica fume is present in the mix because that material significantly increases the water demand of the mix.

Superplasticizers do not influence shrinkage, creep, modulus of elasticity or resistance to freezing and thawing.

2.3 PROPERTIES OF CONCRETE

2.3.1 Properties of fresh concrete

From the stage of mixing till it is transported, placed in the formwork and compacted, fresh concrete should satisfy a number of requirements which may be summarized as follows:

(a) The mix should be stable, in that it should not segregate during transportation and placing. The tendency of bleeding should be minimized.
(b) The mix should be cohesive and mobile enough to be placed in the form around the reinforcement and should be able to be cast into the required shape.

(c) The mix should be as amenable to proper and thorough compaction as possible in the situation of placing and with the facilities of compaction.
(d) It should be possible to obtain a satisfactory surface finish.

The diverse requirements of stability, mobility, compactability, placeability and finishability of fresh concrete mentioned above are collectively referred to as "workability". Workability of fresh concrete is thus a composite property.

It is also clear that the optimum workability of concrete varies from situation to situation, and concrete which can be described as workable for pouring into large sections with minimum reinforcement may not be equally workable for pouring in thin sections with a heavier concentration of reinforcement. A concrete may not be workable when compacted by hand but may be satisfactory when mechanical vibration is used.

2.3.2 Properties of hardened concrete

The most important properties of concrete in the hardened state are strength and durability.

2.3.2.1 Strength of concrete

Strength of concrete is commonly considered as its most important property, although durability, impermeability and volume stability may be equally significant. It is, perhaps, unfortunate that it has become the custom to assess concrete quality only in terms of its compressive strength. The general assumption is that an improvement in concrete strength will improve all other properties as well. Although this is often the case, there are many important exceptions. For example, an increase in cement content that may be intended to increase the strength may also increase the amount of shrinkage and creep. Nevertheless, strength usually gives an overall picture of the quality of concrete because strength is directly related to the structure of the hardened cement paste.

The factors that can affect the strength of concrete can be classified into four categories: constituent materials, methods of preparation, curing procedures and testing conditions.

The strength of concrete is its resistance to rupture. It may be measured in a number of ways, such as strength in compression, in tension, in shear or in flexure.

Effect of porosity on strength

Concrete may be considered to be a brittle material as fracturing takes place at a moderately low strain. The primary factor which governs the strength of brittle materials is porosity. The water–cement ratio determines the porosity of the hardened cement paste at any stage of hydration.

2.3.2.2 Water–cement ratio

The strength of concrete at a given age and cured at a prescribed temperature is assumed to depend primarily on two factors only: the water–cement (w/c) ratio and the degree of compaction. When concrete is fully compacted (in practice this is taken to mean that the hardened concrete contains about 1% of air voids) its strength is taken to be inversely proportional to the w/c ratio. A law was established by Duff Abrams' in 1919. According to this law:

$$f_C = \frac{k_1}{k_2^{w/c}}$$

where w/c represents the water–cement ratio of the mix (originally taken by volume) and k_1 and k_2 are empirical constants. This is presented in Figure 2.1.

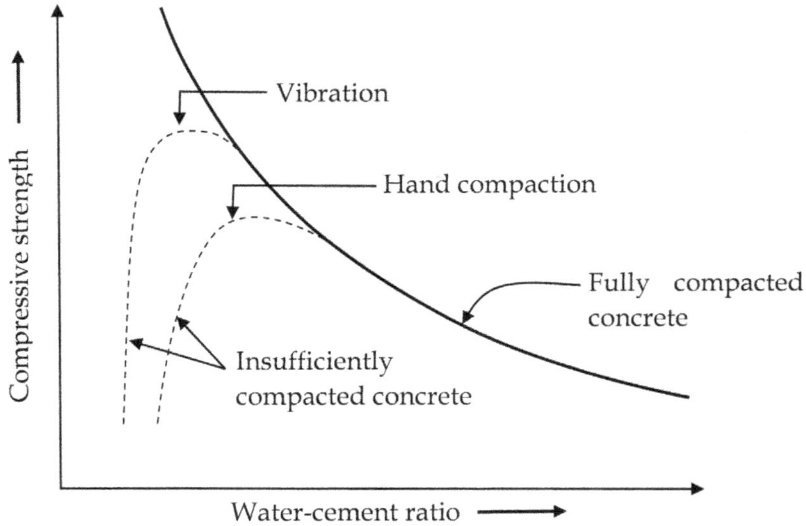

FIGURE 2.1 Relationship between compressive Strength and w/c (Abrams law).

The limitations of Abram's law are that this law does not consider the type of cement, its physical and chemical properties, degree of hydration, temperature of hydration, the air content of concrete or the effects of aggregates.

One example of the limitations of the w/c law is illustrated in Figure 2.1. Concrete that is not properly compacted will contain large voids, which contribute to its porosity. Thus at low w/c ratios, where full compaction is difficult to achieve, Abram's law ceases to be valid. Modern methods of compaction and the use of superplasticizers have made it possible to achieve good compaction even at very low w/c ratios and hence Abram's law may be considered to be valid.

Nevertheless, in practice, the w/c ratio is the largest single factor affecting the strength of fully compacted concrete.

2.3.2.3 Factors affecting strength of concrete

1) W/c ratio
   Even though the strength of concrete is dependent on capillary porosity, it is not an easy quantity to measure or predict. The capillary porosity of a properly compacted concrete at any degree of hydration is determined by the w/c ratio. Hence, one can ensure the strength of properly compacted concrete by specifying the w/c ratio.

2) Time
   The strength of concrete increases with time but the rate of strength gain depends upon the type of cement, curing conditions, initial w/c ratio, etc.

3) Cement
   The chemical composition and fineness of cement influence concrete strength. Strength gain depends on the chemical composition of the cement. The rate of hydration depends on the fineness of the cement.

4) Aggregate
   The strength of concrete also depends on the shape, texture and size of aggregates. The smoother and more rounded the aggregates are, the lower is the strength, while the rougher

the texture and angular, the greater is the concrete strength, since this results in better bond between aggregate and paste. The larger is the aggregate size, the lower is the surface area per unit weight, i.e., specific surface area. Hence, for constant workability water requirements decrease. W/c can thus be reduced, which in turn will increase the strength of the concrete. On the other hand, increasing the size of aggregate will reduce the surface area and hence the bond between paste and aggregate will be reduced resulting in lower strength.

5) Admixture
   Admixtures (chemical) have very little effect on the ultimate strength of concrete as long as porosity is not affected. Admixture affects the rate of strength gain either by accelerating or decelerating the hydration of cement.

2.3.2.4 Stress–strain relationship

To calculate the deformation and deflection of structural members, one must know the relationship between stress and strain. Concrete behaves nearly elastically at low stress levels, however at higher stress levels the behaviour is inelastic.

The definition of pure elasticity is that strains appear and disappear immediately on application and removal of stress.

Sometimes, a major assumption is made in approximating the stress–strain relationship to be a straight line, i.e., treating concrete as a linearly elastic material. This is used in the working stress method of the design of structural concrete. Here permissible stress is limited to one-third of the ultimate load.

For all practical purposes, concrete may be considered to be a linear elastic material when it is loaded up to 50% of the ultimate load. The modulus of elasticity of concrete can be determined from the stress–strain relationship. For concrete, different moduli of elasticity are in vogue, as shown in Figure 2.2.

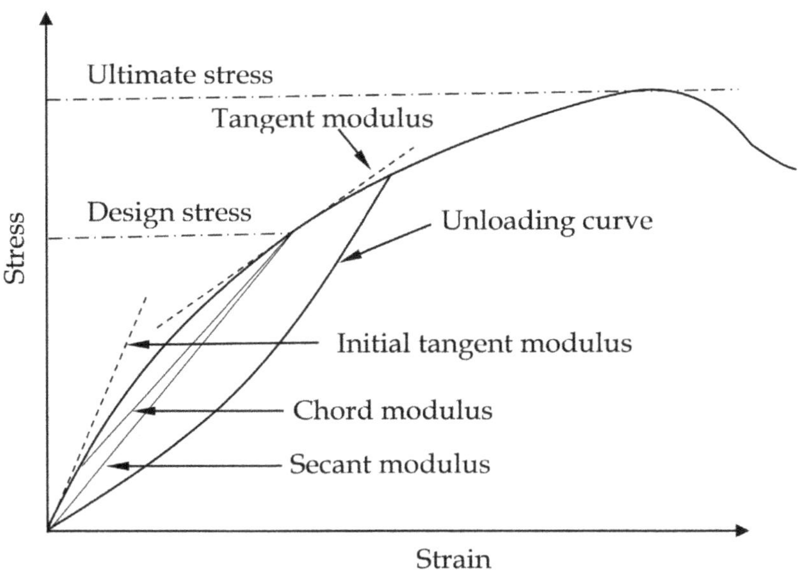

FIGURE 2.2 Typical stress–strain diagram for concrete, showing the different elastic moduli.

2.3.2.5 Durability of concrete

Durability of concrete can be defined as its resistance to deteriorating influences residing within the concrete or present in the environment to which it is exposed. The durability of concrete is one of its most important properties because it is essential that concrete be capable of withstanding the conditions for which it has been designed throughout the life of the structure.

2.4 BASICS OF CONCRETE MIX DESIGN

The proportioning of concrete mixtures, more commonly referred to as "mix design", is a process that consists of two interrelated steps:

1) Selection of the suitable ingredients (cement, aggregates, water and admixtures, if any) of concrete; and
2) Determining their relative quantities (i.e., "proportioning") to produce, as economically as possible, concrete of the appropriate workability, strength and durability. Mix design remains an empirical procedure. The properties of concrete in fresh state are workability and in hardened state are compressive strength and durability.

A design in the strict sense of the word is not possible. The materials used are essentially variable and many of their properties cannot be assessed truly quantitatively. One only tries to make an intelligent guess at the economical combinations of the ingredients. Hence, in order to obtain a satisfactory mix one should not only calculate or estimate the proportions of the materials but also perform trial mixes. The properties of these mixes are checked and adjustments in the mix proportions are made until a fully satisfactory mix is obtained.

2.4.1 Economy

The cost of concrete includes cost of materials, labour and equipment. Labour and equipment costs are independent of the type and quality of concrete. Materials costs are most important; cement is the most expensive ingredient, and a lower cement content results in an economical mix with less shrinkage and creep. Economy is also related to the degree of quality control available at a site.

2.4.2 Workability

A properly designed mix must be capable of being placed and compacted properly with the equipment available. Segregation and bleeding should be avoided. Concrete should be supplied at the minimum workability that will permit adequate placement.

2.4.3 Strength and durability

Concrete specifications require a minimum compressive strength. Specifications may also require that the concrete meet certain durability requirements such as resistance to freezing and thawing or chemical attack.

The process of mix design involves the satisfactory resolution of a host of requirements. Since these requirements cannot all be optimized simultaneously, some compromises (between strength and workability) will be necessary. Again, it must be remembered that even a perfect mix will not perform properly, unless the proper placing, finishing and curing procedures are carried out.

The proportioning of concrete mixes is based on the use of certain relationships established from experimental data which act as a guide for the selection of the best combination of ingredients to achieve the desirable properties.

2.4.4 Parameters considered in concrete mix design

The design of concrete mix is based on the following parameters:

(a) Grade designation;
(b) Type of cement;
(c) Maximum nominal size of aggregate;
(d) Minimum cement content;
(e) Maximum w/c ratio;
(f) Workability.

2.4.5 Methods of concrete mix design

Mix design methods are primarily based on empirical relations (expressed through graphs, charts, tables, etc.) which have been developed from extensive experimental investigations. These relationships merely act as a guide and should be followed, invariably, by the making of trial mixes. Hence concrete mix design is still very much a problem of trial and error, and it remains largely an empirical procedure. Basically, all methods follow the same basic principles with some minor variations.

2.4.6 Purpose/objectives of mix design

The purpose of concrete mix design is to ensure the most optimum proportions of the constituent materials to meet the requirements. Mix design should ensure that the concrete:

(i) Complies with the specifications of structural strength (as laid down in the standards);
(ii) Complies with the durability requirements to resist the environment to which it will be exposed;
(iii) Is capable of being mixed, transported, compacted as efficiently as possible;
(iv) And, lastly, is as economical as possible.

2.4.7 Approach to concrete mix design

Rational proportioning of the ingredients of concrete is the essence of concrete mix design. The proportions are controlled by factors governing the plastic state and hardened states of concrete. If the condition of the plastic concrete is not satisfactory the performance of the hardened concrete will be considerably hampered. The essential functional properties of hardened concrete such as strength, durability and surface finish can only be ensured if the workability and cohesiveness of the fresh concrete are as specified for the working conditions.

2.4.8 Mix design of ordinary and standard grades of concrete as per Indian Standard IS: 10262-2019

The basic steps of mix design may be listed as follows:

1) Determination of target mean compressive strength at 28 days. Since the compressive strength of concrete is highly variable, quality control measures on site and statistical control measures are adopted for reducing the variability. Hence, in order to achieve the characteristic strength of concrete, mix design is performed for a mean strength higher than the characteristic compressive strength using statistical concepts;
2) Estimation of air content;

3) Selection of the water–cement ratio;
4) Selection of water content based on workability requirements;
5) Calculation of cement/cementitious materials content;
6) Estimation of coarse and fine aggregates content;
7) Water correction on calculated water content for moist/dry aggregates;
8) Performing trial mixes in the laboratory to finalize the mix proportions.

For performing the mix design at least four trial mixes are to be performed in the laboratory: trial mixes 1, 2, 3 and 4.

2.4.9 Concept of Trial Mixes

The concrete to be designed should have the desired levels of economy, durability, workability and strength.

The process of mix design is based on the following assumptions:

1. The compressive strength of concrete is primarily governed by the water–cement ratio.
2. For given aggregate characteristics, the workability of concrete is governed by the water content.

Trial mixes are performed to satisfy workability first and then the strength of concrete by following these two assumptions.

The proportions of trial mix 1 are determined following the guidelines as specified in the different codes of practice or specialized literature of different countries.

Trial mix 1 satisfies durability requirements (as per the standard or code of practice followed) and economy (as a specific guideline or standard methodology has been followed). This mix is cast and checked for workability at the laboratory. Mere calculations using empirical relationships can generally never yield the desired workability or strength of concrete. By using the concept of trial mixes first workability is achieved and then strength. Necessary adjustments are made in the water and/or admixture content to achieve the desired level of workability.

With this changed water content but the same w/c as in trial mix 1, mix proportions are calculated – this constitutes trial mix 2. This mix now satisfies durability, economy and workability. The final consideration is compressive strength. The w/c is varied within a range of 10% of a preselected value. It has been observed that within this range the desired strength is achieved. Hence two more mixes, trial mix 3 and trial mix 4 are calculated, by varying the water cement ratio by 5% less and 5% more than that considered in trial mix 2.

The compressive strengths of trial mixes 2, 3 and 4 are determined at 28 days. A plot of c/w versus compressive strength is plotted as shown in Figure 2.3. The c/w (and thus w/c) corresponding to the desired strength is read off. The mix proportions corresponding to this w/c and water content as per trial mix 2 are calculated. This is the final mix having the desired levels of workability, durability, economy and compressive strength.

The constituents of the trial mixes are as follows:

Trail mix 1 → $(W/C)_1$, $(W)_1$, $(C)_1$, $(FA)_1$, $(CA)_1$, (ChA) (may be used)
Trail mix 2 → $(W/C)_1$, $(W)_2$, $(C)_2$, $(FA)_2$, $(CA)_2$, $(ChA)_2$ (may be kept constant or varied)
Trail mix 3 → $(W/C)_1 + 0.05$, $(W)_2$, $(C)_3$, $(FA)_3$, $(CA)_3$, $(ChA)_2$
Trail mix 4 → $(W/C)_1 - 0.05$, $(W)_2$, $(C)_4$, $(FA)_4$, $(CA)_4$, $(ChA)_2$
Final mix → $(W/C)_{\text{from graph}}$, $(W)_2$, $(C)_f$, $(FA)_f$, $(CA)_f$, $(ChA)_2$

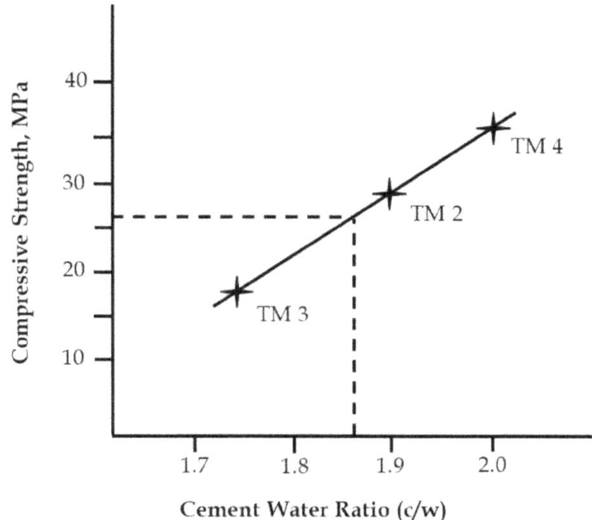

FIGURE 2.3 Relationship between the cement–water ratio and 28-day compressive strength for trial mixes.

where W, C, FA, CA and ChA represent the contents (kg/m^3) of water, cement, fine aggregates, coarse aggregates and chemical admixtures, respectively. Where mineral admixtures are used, the methodology is similar, only the content is either preassigned or determined by trial and error to achieve the desired result.

Since the relationship of strength versus water–cement ratio is rectangular hyperbolic in nature, the plot between strength and cement–water ratio is a straight line. (valid up to a c/w ratio of 2.6).

2.5 HIGH-PERFORMANCE CONCRETE (HPC) – DEFINITION, MATERIALS AND PROPERTIES

The concept of HPC has certainly evolved with time. What exactly is "high-performance"? Various parameters have been attached to HPC, high strength being a popular descriptor. While equating HPC with high strength certainly has some merit, it does not present a complete or accurate picture. Other properties of the concrete must also be considered and may even override the strength issue.

HPC is not a revolutionary material. It is a development of normal concrete. Previously it was termed high-strength concrete but, in many cases, high durability was also required and hence the term HPC was coined. With the advancement of concrete technology, the meaning of the term high strength has changed significantly over the years. According to Neville, HPC contains the following ingredients: common good-quality aggregates, OPC (although rapid hardening Portland cement may be used when high early strength is required) at a very high content, 450–550 kg/m^3, silica fume, generally 5–15% by mass of the total cementitious material, sometimes other cementitious materials such as fly ash or ggbf slag and always a superplasticizer. The dosage of SP is high 5–15 litres per m^3 of concrete, allowing a reduction of water content of about 45–75 kg/m^3 of concrete. It is essential that HPC be capable of being placed in the structure by conventional methods and that it is cured in the usual manner.

The concrete is very dense, it has a minimal volume of capillary pores, and these pores become segmented upon curing. At the same time, a significant proportion of Portland cement remains unhydrated, even when the concrete is in contact with water because water cannot penetrate through the pore system so as to reach the unhydrated remnants of Portland cement. These remnants can be viewed as very fine "aggregate" particles which are extremely well bonded to the products of hydration.

However, it can be easily observed that this high performance need not be limited to only the high-strength concretes as mentioned earlier. In general, a "high-performance concrete" can be defined as that concrete which has the highest durability for any given strength. Any concrete that satisfies certain criteria aimed to overcome limitations of conventional concrete can be called high-performance concrete (HPC). Some of the requirements of HPC are as follows:

(a) A much improved resistance to environmental influences (durability in service): high-durability concrete;
(b) A substantially increased structural capacity while maintaining adequate durability: high-strength concrete;
(c) A significantly reduced construction time, e.g., to permit rapid opening or reopening of roads to traffic, without compromising long-term serviceability: high early-strength concrete;
(d) The ability of the fresh concrete to flow easily to fill form work and encapsulate reinforcing bars while maintaining homogeneity: high workability.

It is, therefore, not possible to provide a unique definition of HPC without considering the performance requirements of the intended use of the concrete.

The American Concrete Institute (ACI) defines high-performance concrete as concrete meeting special combinations of performance and uniformity requirements that cannot always be achieved routinely when using conventional constituents and normal mixing, placing and curing practices. A commentary on the definition states that a high-performance concrete is one in which certain characteristics are developed for a particular application and environment. Examples of characteristics that may be considered critical for an application are:

1) Ease of placement;
2) Compaction without segregation;
3) Early age strength;
4) Long-term mechanical properties;
5) Permeability;
6) Density;
7) Heat of hydration;
8) Toughness;
9) Volume stability;
10) Long life in severe environments.

A high-strength concrete is always a high-performance concrete, but a high-performance concrete may not always be a high-strength concrete.

Most high-performance concretes produced today contain materials in addition to Portland cement to help achieve the compressive strength or durability performance. These materials include fly ash, silica fume and ground-granulated blast furnace slag used separately or in combination. At the same time, chemical admixtures such as high-range water-reducers are needed to ensure that the concrete is easy to transport, place and finish. For high-strength concretes, a combination of mineral and chemical admixtures is almost always essential to ensure achievement of the required strength.

Most high-performance concretes have a high cementitious content and a water–cementitious material ratio of 0.40 or less. However, the proportions of the individual constituents vary depending on local preferences and local materials. Mix proportions developed in one part of the country do not necessarily work in a different location. Many trial batches are usually necessary before a successful mix is developed.

High-performance concretes are also more sensitive to changes in constituent material properties than conventional concretes. Variations in the chemical and physical properties of the cementitious materials and chemical admixtures need to be carefully monitored. This means that a greater degree of quality control is required for the successful production of high-performance concrete.

2.5.1 High-performance concrete principle

Making high-performance concrete is more complicated than producing typical concrete. The reason for this is that as the compressive strength increases, the concrete properties are no longer related only to the water/binder ratio, the fundamental parameter governing the properties of typical concrete by virtue of the hydrated cement paste.

Concrete can be considered as a non-homogeneous material composed of three separate phases:

- The hydrated cement paste;
- The transition zone between aggregate and hydrated cement paste;
- The aggregates.

Improvements in all these phases are required for HPC.

2.5.2 High-performance concrete: requirements for constituent materials and mix proportioning

The selection of constituent materials is the first step in producing concrete. Sometimes, in normal concrete technology not enough attention is paid to this selection, which is the origin of problems on many occasions. Selection of materials to produce HPC can be considered critical if success is desired.

The necessary requirements of constituent materials for normal concrete may not be adequate to produce HPC. General rules on necessary requirements for component materials of high-performance concrete include crushed, good-quality, maximum size 12–14 mm coarse aggregate, river sand, high-strength cements; compatibility between the superplasticizer and the cement chosen, etc. Some of these rules are not absolute, as there are still some disagreements and uncertainties as to how to select the best materials. On the other hand, these general suggestions are not sufficient and preliminary studies in the laboratory have to be extended to select materials for the mix.

Tests to evaluate the suitability of a material to produce HPC must be centred on its capability of reaching maximum strength with minimum water demand.

2.5.3 Selection of materials

The main ingredients of HPC are almost the same as that of conventional concrete. These are:

(a) Cement;
(b) Fine aggregate;
(c) Coarse aggregate;
(d) Water;
(e) Mineral admixtures (fine filler and/or pozzolanic supplementary cementitious materials);
(f) Chemical admixtures (plasticizers, superplasticizers, retarders, air-entraining agents).

Effective production of high-strength concrete is achieved by carefully selecting, controlling, and proportioning all of the ingredients. To achieve higher strength concretes, optimum proportions must be selected, considering the cement and mineral admixture characteristics, aggregate quality, paste proportion, aggregate paste interaction, admixture type and dosage rate, and mixing. Evaluating cement, mineral and chemical admixtures, and aggregate from various potential sources in varying proportions will indicate the optimum combination of materials.

2.5.4 Ratio of Coarse to Fine Aggregate

For normal-strength concretes, the ratio of coarse to fine aggregate (for a 14 mm maximum size of aggregate) is in the range of 0.9–1.4. However, for high-strength concrete, the coarse/fine ratio is much higher. For instance, Peterman and Carrasquillo recommend a coarse/fine ratio used in practice vary in the range of 1.5–1.8.

2.5.5 Admixtures

With the advent of admixtures, both chemical and mineral (pozzolanic), it is now possible to make concretes tailored to a specific application, particularly in the development of high-strength concretes. Adjustments are required in the mix proportions when the admixtures are used to achieve optimum benefits in the fresh and hardened states. At this stage, it is important to note that an improvement in certain characteristics of concrete (green or hardened state) may lead to changes in the other characteristics.

Chemical admixtures

The selection of a good and efficient superplasticizer is also crucial when making HPC, because not all superplasticizer types and brands react in the same way with a particular cement. Experience has shown that not all commercial superplasticizers have the same efficiency in dispersing the cement particles within the mix, in reducing the amount of mixing water, and in controlling the rheology of very low water/binder ratio mixtures during the first hour after contact between the cement and water.

Mineral admixtures

Mineral admixtures form an essential part of the HPC mix. These are used for various purposes, depending upon their properties. More than the chemical composition, mineralogical and granulometric characteristics determine the influence of the mineral admixture's role in enhancing the properties of concrete.

Mix proportions for HPC are influenced by many factors, including specified performance properties, locally available materials, local experience, personal preferences and cost. With today's technology, there are many products available for use in concrete to enhance its properties. Consequently, there are many alternatives for mix proportions that will result in concrete with the desired properties.

2.6 SUMMARY

The properties of ingredient materials of concrete have been discussed in this chapter along with the properties of fresh and hardened concrete. The effects of the concrete ingredients on the properties of fresh and hardened concrete have been discussed. The properties of concrete and its ingredients are so highly variable that mere calculations on the basis of charts and tables developed from past experience can never yield the desirable proportions needed for performing its mix design. Hence, the process of mix proportioning is almost invariably followed by the making of trial mixes. The fundamental principles of performing four trial mixes have been clearly highlighted in this chapter. The properties of concrete in the fresh and hardened states have also been covered in brief. An introduction to the basis of high-performance concrete has also been included.

BIBLIOGRAPHY

Gambhir, M. L. (2013). *Concrete Technology: Theory and Practice*, McGraw Hill, New Delhi.

Ghosh, S. N. (1993). *Progress in Cement and Concrete Mineral Admixtures in Cement and Concrete*, First Edition, Akademia Books International, New Delhi.

IS 383:2016 (2016). *Coarse and Fine Aggregate for Concrete – Specification*, Third Revision, Bureau of Indian Standards, New Delhi.

IS 456:2000 (2007). *Plain and Reinforced Concrete – Code of Practice*, Fourth Revision, Tenth Reprint April 2007, Bureau of Indian Standards, New Delhi.

IS 10262:2019 (2019). *Indian Standard Concrete Mix Proportioning – Guidelines*, First Revision, Bureau of Indian Standards, New Delhi.

Neville, A. M. (1996). *Properties of Concrete*, Fourth Edition, ELBS with Adhison Wesley Longman, England.

Papadakis, V. G., Antiohos, S. and Tsimas, S. (2002). "Supplementary cementing materials in concrete Part-II: A fundamental estimation of the efficiency factor", *Cement and Concrete Research*, Vol. 32, 1533–1538.

Pillai, S. U. and Menon, D. (2012). *Reinforced Concrete Design*, Third Edition, Reprint 2012, McGraw Hill, New Delhi.

3 General Overview of IS:456-2000 and IRC:112-2020

General Provisions of Indian Standards on Concrete

3.1 INTRODUCTION

IS:456 is considered as the umbrella code for concrete in India. This code was first published in 1953 under the title "Code of Practice for Plain and Reinforced Concrete for General Building Construction". The first revision was published in 1957 and the second revision was released in 1964 when the title was changed to "Code of Practice for Plain and Reinforced Concrete", thereby enlarging the scope of this standard to cover other types of structures. The third and fourth revisions were published in 1978 and 2000, respectively. This code, being the mother code for concrete in India, covers the basic requirements of plain and reinforced concrete. Apart from making plain concrete, it covers general design considerations, special design requirements for structural members and systems, philosophy of design, stress–strain relationships of the materials, assumptions and methodology of strength and serviceability design. etc. This code was published in July, 2000, and since then five amendments have been published till 2019, whereby some significant modifications in the codal clauses have been incorporated. Presently, this standard has been taken up for revision.

The standard has five sections and nine annexures as follows:

- **Sections**
  1. General;
  2. Materials, Workmanship, Inspection and Testing;
  3. General Design Consideration;
  4. Special Design Requirements for Structural Members and Systems;
  5. Structural Design (Limit State Method).

- **Annexures**
  A. List of Referred Indian Standards;
  B. Structural Design (Working Stress Method);
  C. Calculation of Deflection;
  D. Slabs Spanning in Two Directions;
  E. Effective Length of Columns;
  F. Calculation of Crack Width;
  G. Moments of Resistance for Rectangular and T-sections;
  H. Committee Composition;
  J. Self-Compacting Concrete (Amendment 3).

DOI: 10.1201/9781003415398-3

The standards or codes of practice deal with the prescriptive method of design based on loads and provide the guidelines of design in the form of do's and don'ts, ensure consistency among designers and provide them with legal validity. This document may be considered as a mine of information based on the fruits of collective wisdom.

IS:456 is the basic code for concrete, and separate codes for special structures such as shells, folded plates, arches, bridges, chimneys, hydraulic, blast-resistant and liquid-resistant structures should be read in conjunction with IS:456. This standard deals with strength and serviceability design. However, nowadays, ductility design is gaining importance for aseismic design and hence an additional standard, IS:13920-2016 – Ductile Design and Detailing of Reinforced Concrete Structures Subjected to Seismic Forces – code of practice, is to be used along with IS:456 for earthquake zones III, IV and V as specified in the relevant code on earthquake-resistant design, IS:1893 (Part 1) – 2016.

3.2 PROVISIONS FOR PLAIN CONCRETE AND REINFORCEMENTS

In the fourth revision of IS:456, which was published in 2000, more emphasis was placed on material and durability aspects, rather than on analysis, design and detailing considerations. The coverage of materials and durability aspects were comprehensive, whereas the modifications on analysis, design and detailing aspects were not very significant. The standard deals with both plain and reinforced concrete and section 2 covers materials, workmanship, inspection and testing.

3.2.1 Materials and general considerations

Regarding the primary ingredient of concrete, i.e., cement, the standard along with its amendments allow all three grades of OPC along with other blended cements for use in concrete. The code permits the use of different types of mineral admixtures. The standard specifies that chemical admixtures, if used, should comply with the requirements of IS:9103 (Specifications for Admixture for Concrete). It is mentioned that the aggregates should conform to the requirements of IS:383 (Specifications for Coarse and Fine Aggregates from Natural Sources and Concrete). Aggregates are specified with respect to their nominal maximum size, which can be taken as the size of the sieve through which, if the aggregate sample is sieved, only a small percentage of the sample will be retained to the tune of 5–10%. It has been mentioned in the standard that 20 mm aggregate is suitable for most work. The method of testing water for concrete and the permissible limits of chloride and sulphates have also been described. Potable water is considered satisfactory for making concrete. The pH value of water should not be less than 6.

The standard permits different types of reinforcement such as mild steel and medium tensile bars, high-strength deformed steel bars, hard drawn steel wire fabric and structural steel. Nowadays, for making concrete, mostly high-strength deformed bars are used. IS 456:2000 refers to IS:1786 – Specification for High-Strength Deformed Steel Bars and Wires for Concrete Reinforcement for use of high-strength deformed steel bars for use in concrete in which the following steel grades have been permitted: Fe 415, Fe 415D, Fe 415S, Fe 500, Fe 500D, Fe 500S, Fe 550, Fe 550D, Fe 600, Fe 650 and Fe 700. However, IS:456 refers to mild steel of grade Fe 250 and two HYSD steel grades of Fe415 and Fe500 only in all strength and serviceability calculations. Higher grades of steel result in a reduction in ductility. Since ductility plays a vital role in aseismic design, very high strengths of steel are not desirable in reinforced concrete. A capping on the maximum grades of steel has been provided in IS:13920 – Code of Practice for Ductile Design and Detailing of Reinforced Concrete Structures Subjected to Seismic Forces. The maximum permitted grade of steel has been specified as Fe 550 for seismic zones other than zone II. It may be mentioned that

India is divided into four seismic zones starting with zone II, which is least seismically vulnerable zone, to zone V, which is the most vulnerable zone. Hence it appears that the mother code of concrete is lagging behind the specifications for concrete reinforcement as laid out in the other allied standards.

Concrete has been divided into three groups based on the grade designation: ordinary concrete from M10 to M20, standard concrete from M20 to M60 and high-strength concrete from M65 to M100. However, it has been categorically mentioned that for grades of concrete higher than M60, design parameters given in the standard may not be applicable and designers should refer to specialist literature and experimental results. The workability of concrete has been specified with respect to the slump test. Regarding durability considerations, the standard provides a comprehensive guideline in the form of environmental exposure condition, shape and size of member, exposure to chemical attack, constituent materials, workmanship, etc. Five different environmental exposure conditions have been specified in qualitative terms and it is at the discretion of designers to choose the most suitable exposure condition for their structures. The least severe exposure condition is mild and the most severe one has been designated as extreme. The grade of concrete along with the cover to reinforcement, minimum cement content, maximum water cement ratio, etc. have been defined with respect to these exposure conditions. The maximum cement content has been specified as 450 kg/m^3 from a durability aspect. The minimum grades of structural concrete have been specified as M20 for superstructure and M25 for substructure or foundations. The general guidelines for concrete mix proportioning have been specified. Concrete mixes have been classified as nominal mix and design mix concrete. It has been specified that nominal mix may be performed for grades of concrete M20 or lower. Though the general aspect of design mix concrete has been specified, the methodology of mix design has not been included into the standard. Details of the mix proportions for nominal mix concrete along with the relative proportions of all the ingredients including water have been clearly described in the standard.

3.2.2 Workmanship

Regarding the production of concrete, the quality assurance measures have been clearly defined. The different stages of concrete production, namely batching, mixing, transporting, placing, compacting, curing and finishing, have been elaborated in detail. Regarding batching, it has been specified that all ingredients of concrete shall be batched by mass except water and chemical admixtures. Volume batching may be allowed under special conditions, and with certain limitations. Concrete should be properly and quickly transported and placed with care such that the maximum free fall is restricted to 1.5 metres. It has been specified that concrete should be thoroughly compacted using mechanical vibrators. Regarding curing, the minimum and recommended times for different types of concrete have been specified in the standard and moist curing methods have been preferred to membrane curing. The code specifies that constant and strict supervision at all stages of concrete constructions need to be properly ensured.

It has been specified that concrete should be mixed in a mechanical mixer. Regarding assembly of reinforcement, it has been specified that all reinforcements should be properly placed and held in position as per the relevant drawings.

Requirements of sampling and compressive strength assessment of concrete specimens have been clearly specified in the standard. Concrete is a material whose properties are highly variable and in order to reduce the variability quality control at the site and statistical quality control need to be enforced. The acceptance and rejection of concrete shall be based on the 28-day compressive strength and the acceptance criteria have been stipulated in the code with respect to both average and individual strength of samples. The test result of each sample is the average strength of three specimens which should not vary by more than 15 percent of the average. Each specimen consists of a 150 × 150 × 150 mm cube cast, cured and tested as per standard conditions. Test result of each sample is the mean strength of three 150 mm cubes (or specimens) which constitute the sample. Hence, for meeting acceptance criteria, the test results of at least four consecutive samples are necessary. Test sets should consist of samples 1, 2, 3, 4 then 5, 6, 7, 8 and so on. As per the standard, concrete of M15 and above shall be deemed to comply with the strength requirements when both the following requirements are met:

1. Mean of the group of four non-overlapping consecutive test results should have a value (N/mm^2) ≥ f_{ck} + 0.825 × established standard deviation
   or f_{ck} +3, whichever is greater.
2. Individual test result should have a value (N/mm^2) ≥ f_{ck}-3,
   where f_{ck} = characteristic compressive strength of concrete.

3.2.3 Inspection and Testing

The standard specifies a detailed inspection procedure covering materials, workmanship and construction. In case of doubt regarding the grade of concrete used, either due to poor workmanship or based on the result of cube strength test, core test and/or load test is specified.

From the above discussions it can be concluded that IS:456-2000 provides significant information and guidelines for producing good-quality concrete.

3.3 PROVISIONS OF IS:456-2000 ON STRUCTURAL CONCRETE

In section 3 of the code, general design considerations are mentioned which include the basics of loading, analysis, overall stability, durability requirements, design considerations and reinforcement detailing. This section covers the general requirements for slabs, beams and columns. It is specifically mentioned that structures and structural elements shall normally be designed by the limit state method and only if the limit state method cannot be conveniently adopted, the working stress method may be used.

Section 4 of the standard deals with special design requirements for structural members and systems covering the requirements for corbels, deep beams, different types of flooring systems, walls, stairs and footings.

Section 5 is on structural design covering the basic aspects of the limit state method of design. This method deals with strength and serviceability design only. New-generation concrete structures need to possess another attribute, which is ductility, so that the performance of the structures even under overloading is desirable and the risk of loss of life and property can be reduced. This aspect is not covered in this standard and some basic requirements of ductility design are included in an allied standard – IS:13920-2016, Code of Practice for Ductile Design and Detailing of Reinforced Concrete Structure Subjected to Seismic Forces.

Structural design by the working stress method has been stripped from the status of a separate section and has been included in Annexure B, to highlight the fact that this method is losing its importance and acceptability for present-day structures.

The annexures deal with referred Indian standards, working stress method, deflection calculations, two-way slabs, effective length calculations, crack width calculations, expression for moments of resistance for rectangular and T sections, committee composition and self-compacting concrete.

3.4 PROVISIONS OF IRC:112-2020 – CODE OF PRACTICE FOR CONCRETE ROAD BRIDGES

This standard deals with the structural use of plain cement concrete, reinforced concrete, prestressed concrete and composite construction using concrete elements in bridges. It covers design principles, detailed design criteria and practical rules, material specifications, workmanship, etc. which aim for safe, serviceable, durable and economical bridges.

This standard has been completely based on the limit state method of design superseding both the working loads and allowable stress methods permitted in the previous version. The code has 18 sections and 10 annexures. Only the general provisions for the design of reinforced concrete sections will be discussed as design provisions for bridges are beyond the scope of this book.

The provisions for designing with reinforced concrete are on a par with the requirements of RC design as laid out in current international standards (Eurocode 2) and it is expected that the new revision of IS:456 will be in line with these requirements.

3.4.1 Provisions for Plain Concrete and Reinforcements (Untensioned)

Section 6 covers material properties and their design values. Regarding untensioned steel reinforcements, the various grades of steel permitted by the code are as follows: mild steel (MS) grade 1 and high-strength deformed steel (HSD) Fe 415, Fe 415D, Fe 415S, Fe 500, Fe 500D, Fe 500S, Fe 550, Fe 550D and Fe 600. The relevant BIS standards for mild steel and HSD steel are IS:432 and IS:1786, respectively. A bilinear stress–strain diagram of reinforcing steel for design has been prescribed in the form of two different types: a simplified bilinear diagram and an idealized bilinear diagram. The idealized bilinear diagram includes the effect of strain hardening and has a capping on the maximum strain. Meanwhile, the simplified bilinear diagram has a horizontal plastic branch at constant stress without a capping on the maximum strain. Concrete has been classified into three categories: ordinary concrete of M15 and M20, standard concrete of M15 to M50 and high-performance concrete from M30 to M90. The stress and deformation characteristics for normal concrete specifying the strains along with corresponding stresses, modulus of elasticity, etc., are specified in this section. The stress–strain relationships of unconfined concrete for the design of sections have been specified in the form of parabolic-rectangular type and other equivalent stress–strain relations.

In section 18 of the code, materials, quality control and workmanship are specified. The specifications for different grades of untensioned steel have been provided in this section. Regarding materials in concrete, all three grades of OPC and blended Portland cements have been permitted, and chemical admixture conforming to IS:9103 has been allowed. Regarding mineral admixtures, fly ash conforming to grade 1 of IS:3812, ground granulated blast furnace slag (GGBS) conforming to IS:12089 and silica fume conforming to IS:15388 have been permitted. All coarse and fine aggregates conforming to IS:383 have been allowed for use in concrete. The requirements of water for concrete making have been specified in this standard. Relative proportions of ingredients of ordinary concrete are also specified. The mix design of standard and high-performance concrete should be performed as per the national code or established procedures. Sampling, testing and acceptance criteria have been specified in the code. Acceptance criteria with respect to compressive strength require compliance with both conditions mentioned below:

a) The mean strength determined from any group of four consecutive non-overlapping samples shall exceed the specified characteristic compressive strength by 3 MPa.
b) The strength of any sample is not less than the specified characteristic compressive strength minus 3 MPa.

Quality control and workmanship requirements are also specified in the standard. Along with the quality assurance measures, the stages for the production of concrete are also discussed, and guidelines for inspection and testing of structures are also mentioned.

The basis of design in the form of limit state philosophy has been provided in section 5 of the code. Two basic groups of limit state have been prescribed: ultimate limit state (ULS) covering static equilibrium and failure of structural elements under ultimate design load and serviceability limit state (SLS) dealing with conditions of the structure under serviceability design loads.

Section 7 prescribes the analysis part.

In section 8, the ultimate limit state of linear elements for bending and axial forces are specified. The assumptions for the ultimate limit state method of design along with strain and stress distributions have been enumerated in this section in detail. In contrary to IS:456, this section deals with RC sections subjected to flexure and axial forces and the possible domains of strain distribution have been highlighted.

Section 9 deals with three-dimensional elements, whereas in section 10 ultimate limit state of shear, punching shear and torsion are specified. Detailing requirements of structural members are covered in section 16.

IRC:112-2020 deals specifically with concrete bridges. However, the general requirements for the design of reinforced concrete sections are quite similar to the requirements provided in international standards like Eurocode 2 (EN1992-1-1, 2004, Design of Concrete Structures – part-1-1: General Rules and Rules for Buildings). This standard, which has been published very recently, incorporates specifications for the present-day materials of reinforced concrete including their stress–strain relationships. The design philosophy is also on a par with international guidelines. This standard has prescribed significant improvements in the material properties such as the design parameters for higher strength concrete up to M90 and stress–strain relationships of HSD bars made by current techniques and reinforced concrete design philosophies with respect to IS:456-2000. The material specifications and design philosophy of the revised version of IS:456 are expected to be revised in line with these requirements.

3.5 SUMMARY

In this chapter the basic aspects of plain and reinforced concrete as envisaged in IS:456-2000 have been discussed. It is observed that the code provides very useful information in the field of materials and plain and reinforced concrete making though these portions require immediate revision based on the requirements of current international standards. Recently, the code of practice for concrete bridges has been revised in the form of IRC:112-2020, whereby it is observed that a significant amount of improvement in the field of concrete and reinforcement has been incorporated. The design philosophy for RC sections has also been improved. These improvements are similar to those available in international standards and it is expected that the forthcoming revision of IS:456 will be in line with these provisions.

BIBLIOGRAPHY

EN 1992-1-1 (2004). *Eurocode 2: Design of Concrete Structures – Part 1-1: General Rules and Rules for Buildings*, European Committee for Standardization, Brussels.

IS 456:2000 (2007). *Plain and Reinforced Concrete – Code of Practice*, Fourth Revision, Tenth Reprint April 2007, Bureau of Indian Standards, New Delhi.

IS 1893 (Part 1):2016 (2016). *Criteria for Earthquake Resistant Design of Structures, Part 1, General Provisions and Buildings*, Sixth Revision, Bureau of Indian Standards, New Delhi.

IS 13920-2016 (July 2016). *Ductile Design and Detailing of Reinforced Concrete Structures Subjected to Seismic Forces – Code of Practice*, First Revision, Bureau of Indian Standards, New Delhi.

IRC:112-2020 (2020). *Code of Practice For Concrete Road Bridges*, First Revision, Indian Road Congress, New Delhi.

IRC: SP-105-2015 (January 2015). *Explanatory Handbook to IRC:112-2011 Code of Practice for Concrete Road Bridges*, Indian Road Congress, New Delhi.

SP: 24 (S & T)-1983 (1984). *Explanatory Handbook on Indian Standard Code of Practice for Plain and Reinforced Concrete (IS: 456-1978)*, Indian Standard Institution, New Delhi.

4 Provisions of Standards or Codes of Practice on Reinforcement

IS:1786-2008, IS:456-2000, SP:16-1980, IRC:112-2020 and Eurocode-2

4.1 INTRODUCTION

Concrete is a material which is very strong in compression but very weak in tension. Since in real-life structures tension and compression almost always act together, any structural element made of concrete needs to be reinforced by providing reinforcements. Steel and concrete are compatible materials, having similar coefficients of thermal expansion. Steel reinforcements are embedded in concrete to carry flexural and direct tension, compression if necessary and for internal actions such as the effects of temperature, shrinkage, cracking, etc. In any reinforced concrete structural element, the amount of reinforcement in the cross-section is significantly small. Steel is a highly ductile material, whereas concrete is a highly brittle one. The duty of the designer is to design and detail the element in such a manner that the mode of failure is ductile in nature so that risk of loss of life and property can be significantly reduced. While the production of concrete falls within the domain of civil engineering, the making of steel falls within the purview of multiple engineering disciplines such as mechanical, metallurgy, etc. Civil engineers are the major end-users of structural steel.

4.2 GENERAL REQUIREMENTS AS PER INDIAN STANDARDS: IS:456-2000, IS:1786-2008, IRC:112-2020, AND EUROCODE 2

As per IS:456-2000 reinforcement in concrete shall be any of the following:

(a) Mild steel and medium tensile steel bars;
(b) High-strength deformed steel bars;
(c) Hard-drawn steel wire fabric;
(d) Structural steel.

As per IS:456 design parameters are specified only for mild steel having grade Fe 250 and for high yield strength deformed bars of grades Fe 415 and Fe 500 only. The stress–strain relationship for HYSD bars as prescribed by the standards are for reinforcements, made by the CTD process, which do not have a definite yield point.

As per IRC:112-2020, reinforcement shall consist of hot-rolled, thermo-mechanical or heat-treated rods, decoiled rods or cold-worked steel for various grades of untensioned steel reinforcements as described in Table 4.1.

As per Eurocode 2, general principles and rules for reinforcing steel are provided which can be in the form of bars, decoiled rods, welded fabric and lattice girders. The required properties of

DOI: 10.1201/9781003415398-4

TABLE 4.1
Grades of reinforcing steel as per IRC:112-2020

Types of steel	Grades or designation
Mild steel	Grade 1
High-strength deformed steel (HSD)	Fe 415
	Fe415D
	Fe 415S
	Fe 500
	Fe 500D
	Fe 500S
	Fe 550
	Fe 550D
	Fe 600

reinforcing steels shall be verified using the testing procedure as prescribed in the relevant European standard. The properties of reinforcement suitable for use with Eurocode 2 indicate three types of bars or decoiled rods: Class A, Class B and Class C, with a characteristic yield strength f_{yk} or $f_{0.2k}$ ranging from 400 to 600 MPa, where $f_{0.2k}$ refers to characteristic 0.2% proof stress of reinforcement. Other relevant properties are also specified in the standard.

As per IS:1786-2008 the different grades of steel which can be used as reinforcement in concrete are as follows:

(a) Fe 415, Fe 415D, Fe 415S;
(b) Fe 500, Fe 500D, Fe 500S;
(c) Fe 550, Fe 550D;
(d) Fe 600;
(e) Fe 650;
(f) Fe 700.

Notes:

i. The figure following the symbol Fe indicates the specified minimum 0.2% proof stress or yield stress in N/mm^2;
ii. The letter D or S following the strength grade indicates the categories with the same specified minimum 0.2% proof stress or yield stress but with enhanced and additional requirements.

The yield strength of reinforcing bars can be increased by different methods. In India, even a couple of decades ago, high-strength bars were generally produced by the cold twisting process, as a result of which IS:456-2000 had prescribed the stress–strain relationship of high-strength deformed bars made using the CTD process.

Nowadays, steel is produced generally by thermo-mechanical treatment because reinforcements produced by this method have significant advantages over those produced by the CTD process. Basic steel is made from virgin iron ore through a blast furnace. The rebars are hot rolled from steel billets and subjected to thermo-mechanical treatment (see Figure 4.1) in three successive stages:

a. Quenching – The hot rolled bar is rapidly quenched by a special water spray system which hardens the surface of the bar to an optimal depth through the formation of a martensitic rim while the core remains hot and austenitic.

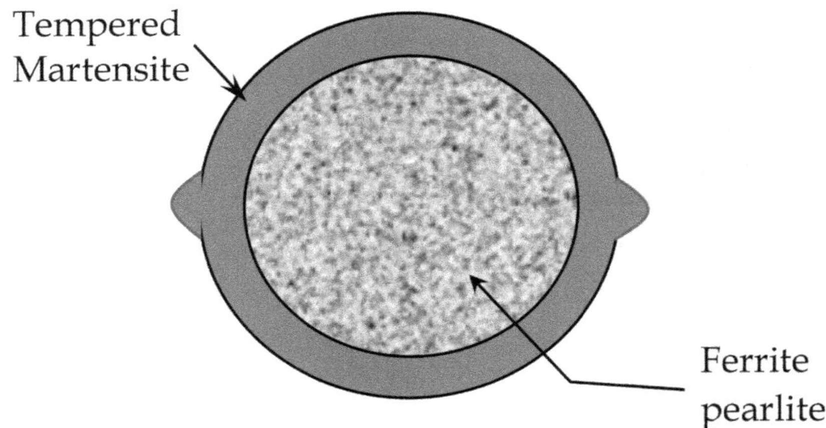

FIGURE 4.1 Typical cross-section of rebar made by the TMT process.

b. Self-tempering – When the bar leaves the quenching box the core remains hot compared to the surface, allowing heat to flow from the core to the surface causing tempering of the outer martensitic layer into a structure called tempered martensite. The core still remains austenitic at this stage.
c. Atmospheric cooling – This takes place on the cooling bed where the austenitic core is transformed into a ductile ferrite–pearlite structure. Thus, the final form consists of an optimum combination of a strong outer layer (tempered martensite) with a ductile ferrite–pearlite core. The outer shell is responsible for imparting strength to the reinforcement, while the inner core provides ductility.

4.3 MECHANICAL PROPERTIES

The mechanical properties of high-strength deformed bars and wires as specified by IS:1786-2008 are shown in Table 4.2. The bars should satisfy the requirements of the bend test and rebend test, and bond as specified in the standard. Requirements in IRC:112 and Eurocode 2 regarding the mechanical properties are of a similar nature.

4.4 STRESS–STRAIN RELATIONSHIP

Plane mild steel bars are not commonly used nowadays in reinforced concrete because they possess lower yield strength (250 MPa), although the cost is more or less similar to high-strength deformed bars. High-strength deformed bars with significantly higher yield strength than mild steel are now used for new-generation concrete involving higher grades of concrete to avoid the congestion of reinforcements. In the subsequent sections, stress–strain relationships obtained by actual laboratory testing and idealized stress–strain relationships of steel as per different standards of practice are discussed in detail.

The failure of reinforced concrete sections should be ductile in nature as brittle failure will occur suddenly at low values of total strain, causing significant risk of loss of life and property. Reinforced concrete is a combination of two materials, of which concrete is brittle and reinforcements are ductile in nature. Though in a reinforced concrete section the amount of ductile material is very small with respect to the amount of concrete present, it is the duty of the designers to design and detail the sections in such a fashion that the property of the ductile steel material dominates the failure

TABLE 4.2
Mechanical properties of high-strength deformed bars and wires

Property	Fe 415	Fe 415D	Fe 415S	Fe 500	Fe 500D	Fe 500S	Fe 550	Fe 550D	Fe 600	Fe 650	Fe 700
0.2 percent proof stress/yield stress, min. (N/mm^2)	415	415	415	500	500	500	550	550	600	650	700
0.2 percent proof stress/yield stress, max. (N/mm^2)	–	–	540	–	–	650	–	–	–	–	–
TS/YS ratio (N/mm^2)	≥1.10 (but TS not less than 485 N/ mm^2)	≥1.12 (but TS not less than 500 N/ mm^2)	≥1.25	≥1.08 (but TS not less than 545 N/ mm^2)	≥1.10 (but TS not less than 565 N/ mm^2)	≥1.25	≥1.06 (but TS not less than 585 N/ mm^2)	≥1.08 (but TS not less than 600 N/ mm^2)	≥1.06 (but TS not less than 660 N/ mm^2)	≥1.06 (but TS not less than 700 N/ mm^2)	≥1.06 (but TS not less than 770 N/ mm^2)
Elongation percent min., on gauge length, $5.65\sqrt{A}$, where A is the cross-sectional area of the test piece	14.5	18.0	18.0	12.0	16.0	16.0	10.0	14.5	10.0	10.0	10.0
Total elongation at maximum force, percent, minimum on gauge length $5.65\sqrt{A}$, where A is the cross-sectional area of the test piece	–	5	8	–	5	8	–	5	–	–	–

Note: TS/YS ratio refers to the ratio of tensile strength to the actual yield strength or 0.2% of yield stress of the test piece.

condition. Hence the stress–strain relationship of steel as prescribed in the different codes of practice needs to be properly interpreted in the design for enhanced performance of reinforced concrete sections under loads. The subsequent sections attempt to address the stress–strain relationship of steel as per national and international standards.

4.4.1 Engineering stress–strain curves

The basic data for reinforcing steel are obtained by performing a tension test where the sample under test is subjected to gradually increasing axial load until it breaks. The load and elongation are measured at regular intervals during the test. The data that are obtained are plotted in the form of a load–deformation curve and a stress–strain curve. Engineers are usually concerned with the stress–strain curves which form the basis of the design. The tension test is widely used around the world for determining the basic design data of reinforcing steel such as the strength, yield point, ductility, etc. In the tension test, the sample under test is subjected to a uniaxial tensile force which is increased gradually, and the deformations are recorded alongside. The stress–strain relationship is constructed from the recorded load–deformation data. The stress used in determining the stress–strain relationship is the average longitudinal stress in the test specimen. This stress is determined by dividing the load applied on the material by its initial cross-sectional area. The corresponding strain in the stress–strain curve is determined by dividing the increase in the gauge length of the specimen by its original length, which is known as the average linear strain. As both the strain and stress of steel are obtained by dividing the elongation and load by constants, respectively, the nature of the curve remains the same.

In the elastic region of the stress–strain curve, stress is proportional to strain or, in other words, they exhibit a linear relationship. When the yield strength of steel is exceeded, the material undergoes gross plastic deformation. The deformation is permanent in nature in the sense that the deformation remains even after removal of the load. As plastic deformation is continued, stress in the material also increases as the metal undergoes strain hardening. As is evident, there will be no change in the volume of the material and therefore, as the metal elongates, its cross-sectional area keeps reducing. This decrease in area is initially compensated for by strain hardening and, as a result, engineering stress will continue to rise with an increase in strain. However, a point is reached where the rate of decrease in the cross-sectional area will exceed the increase in load-carrying capacity because of strain hardening. This mechanism will be attained by that part of the sample which is weaker compared to the other parts. It is in this part that further plastic deformation will be concentrated, and the sample begins to neck, or there will be a rapid decrease in the cross-sectional area locally.

Because the cross-section area at this point of time decreases far more rapidly than the rate at which the deformation load is increased by strain hardening, the actual load required to deform the specimen decreases and thus the engineering stress also decreases until failure of the sample takes place.

4.4.1.1 Determination of the engineering stress–strain curve

The properties of reinforcing steel are of concern to designers. For determining the stress–strain relationship, engineering stresses and engineering strains are used, i.e., the original cross-sectional area of the sample and the original gauge length of the sample are used. The load–deformation data of the test may be recorded in a Servo-hydraulic Universal Testing Machine. Using this load–deformation data, the engineering stress–strain curve can be plotted for the samples tested. A typical engineering stress–strain relationship obtained by actual testing TMT rebars in our structural engineering laboratory is illustrated in Figure-4.2. The initial distortion in the curve could have occurred due to some initial slippage in gripping the specimen.

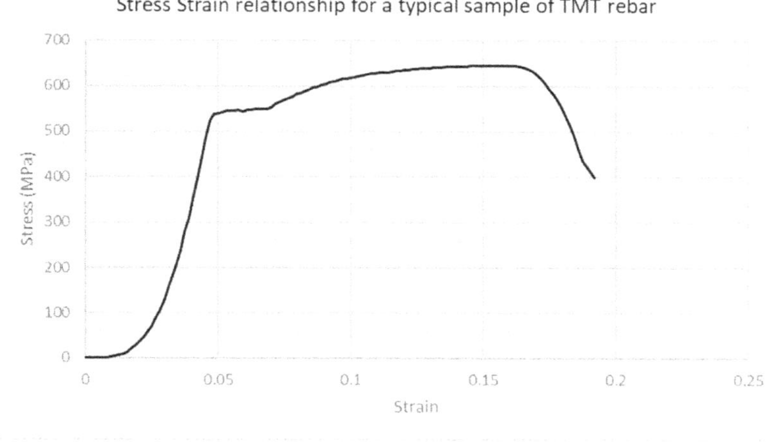

FIGURE 4.2 Actual engineering stress–strain relationship for a typical sample of TMT rebar.

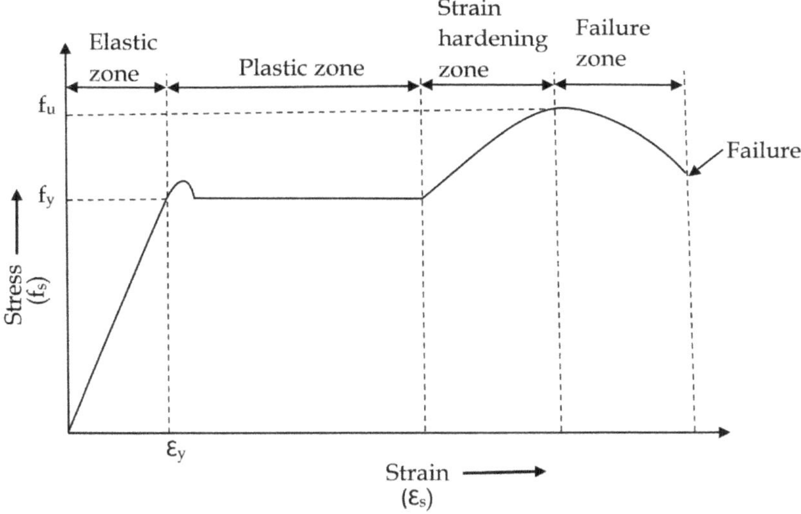

FIGURE 4.3 Stress–strain diagram for steel in tension.

4.4.2 Behaviour of Steel under Stress

In order to investigate the stress–strain relationship of steel in detail, it is segregated into different distinct zones as represented in Figure 4.3.

There are four distinct zones in the stress–strain diagram of steel as shown in Figure 4.3. The first is the elastic zone, which is represented by the straight line. Within this zone strains appear and disappear with the application and removal of stress, and stress in the material is linearly proportional to the strain, i.e., Hooke's law is valid. When steel goes beyond this zone it yields, which signifies an increase in strain with no appreciable change in stress. The onset of yielding is the beginning of the plastic zone. If unloading occurs within this zone, permanent set equal to the inelastic strain will occur and the unloading curve will be parallel to the loading curve. The strain at which steel yields

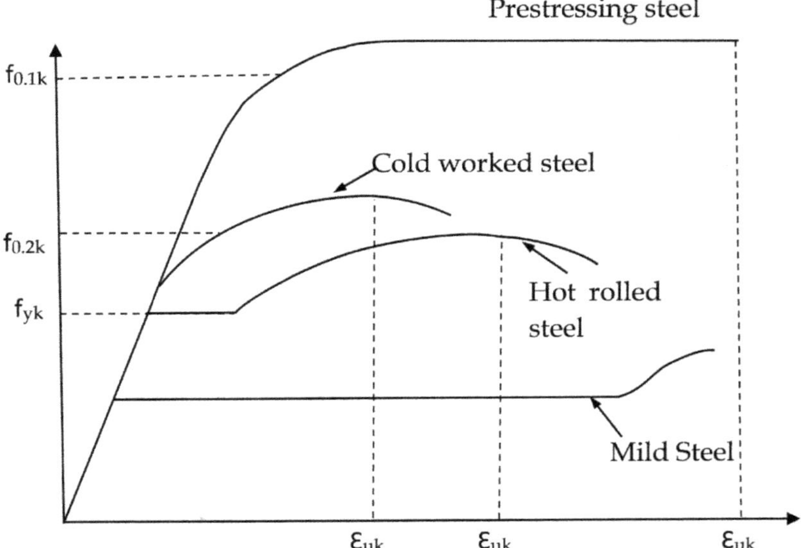

FIGURE 4.4 Typical stress–strain relationship for different types of steel reinforcements.

is termed the yield strain, and the corresponding stress is termed the yield stress. The yield plateau generally ends after 8–10 times the strain at the end of the elastic zone. Thereafter, the curve has an ascending branch indicating an increase in stress with an increase in strain, and finally reaches the value of maximum stress, called the ultimate strength. This curvilinear zone of the stress–strain curve is known as the strain-hardening zone, which is followed by a strain-softening zone exhibited by the descending branch of the stress–strain curve which finally leads to failure. A sharply defined yield point is a characteristic of mild steel. There are other types of steel which do not have a pronounced yield point as shown in Figure 4.4.

For the steels which do not have a definite yield point the stress at which the permanent set or inelastic deformation reaches the value of 0.2% is termed the yield stress or 0.2% proof stress. This has been considered by Indian standards for high-strength deformed bars made using the cold twisted deformed (CTD) process.

4.4.3 Stress–Strain Relationships as per IS:456

IS:456 specifies the stress–strain curve for mild steel bars, as shown in Figure 4.5.

For cold worked bars, stress is proportional to the strain up to a stress of $0.8f_y$, where f_y is the yield stress. Thereafter, the stress–strain curve is defined as per the following relationship (SP:16):

Stress	Inelastic strain
$0.8\ f_y$	Nil
$0.85\ f_y$	0.0001
$0.9\ f_y$	0.0003
$0.95\ f_y$	0.0007
$0.975\ f_y$	0.0010
$1.0\ f_y$	0.0020

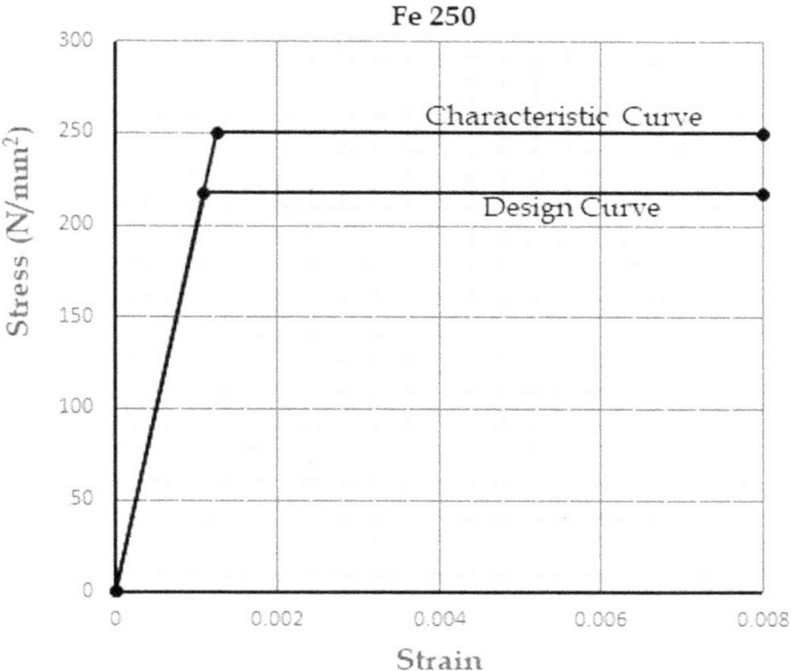

FIGURE 4.5 Stress–strain curve for mild steel.

The stress–strain curve for design purposes is obtained by substituting f_{yd} which is the design stress (obtained by using a material safety factor of 1.15 over the characteristic stress for f_y). This has been illustrated in Figure 4.6.

Nowadays, the process of cold twisting deformation has been replaced by thermo-mechanical treatment, whereby high levels of ductility can be achieved even at higher strength. Thermo-mechanically treated bars (TMT) have definite yield points and a typical relationship, which is bilinear in nature, as shown in Figure 4.7.

4.4.4 Stress–strain relationships as per IRC:112

The stress–strain relationships for different types of steel are shown in Figure 4.8.

For design purposes, any one of the two diagrams, namely, an idealized bilinear or simplified bilinear diagram may be used, as per Figure 4.9.

The idealized bilinear diagram has a sloping top branch joining the yield point with the characteristic strain of reinforcement, ε_{uk}, where the values of stresses and strains refer to the minimum values required by the relevant IS codes and IRC:112. The factored idealized design diagram is obtained by factoring stress values using the partial safety factor and considering the design strain as 0.9 times the characteristic strain $\varepsilon_{ud} = 0.9\ \varepsilon_{uk}$. There is no limit on strain if the horizontal branch of the curve is used (simplified bilinear diagram), as limiting ultimate stress remains as f_{yd}. For grades Fe 415D, Fe 500D and Fe550D the characteristic strain is taken as 5% (maximum) and for grades Fe 415S and Fe500S the value is taken as 8% (maximum). For other grades, it shall be taken as 2.5% (maximum).

4.4.5 Stress–strain relationship as per Eurocode 2

Typical stress–strain relationships for hot rolled and cold worked steel are shown in Figure 4.10.

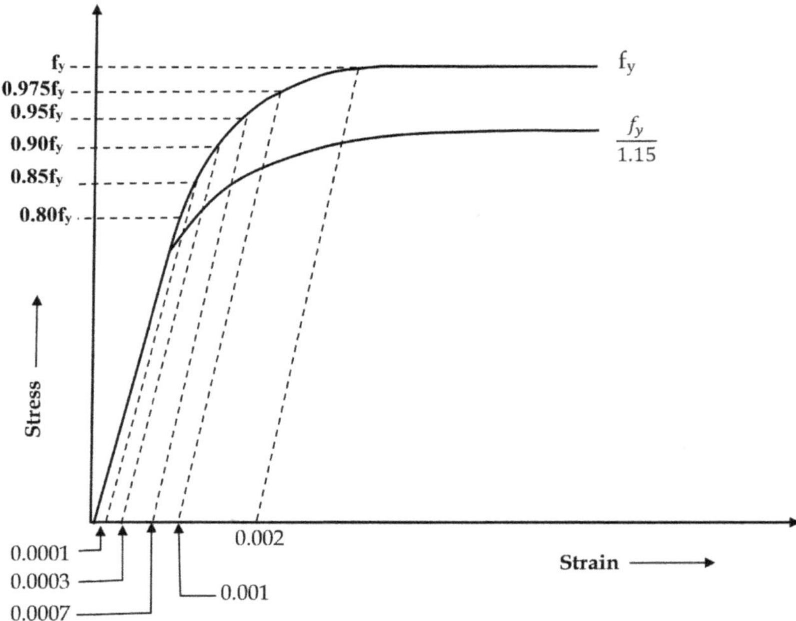

FIGURE 4.6 Stress–strain curve for cold worked steel.

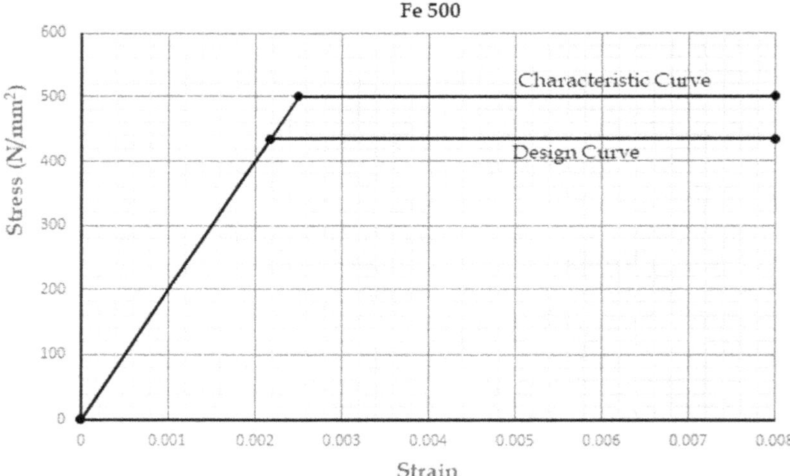

FIGURE 4.7 Stress–strain diagram for TMT bars (Fe 500).

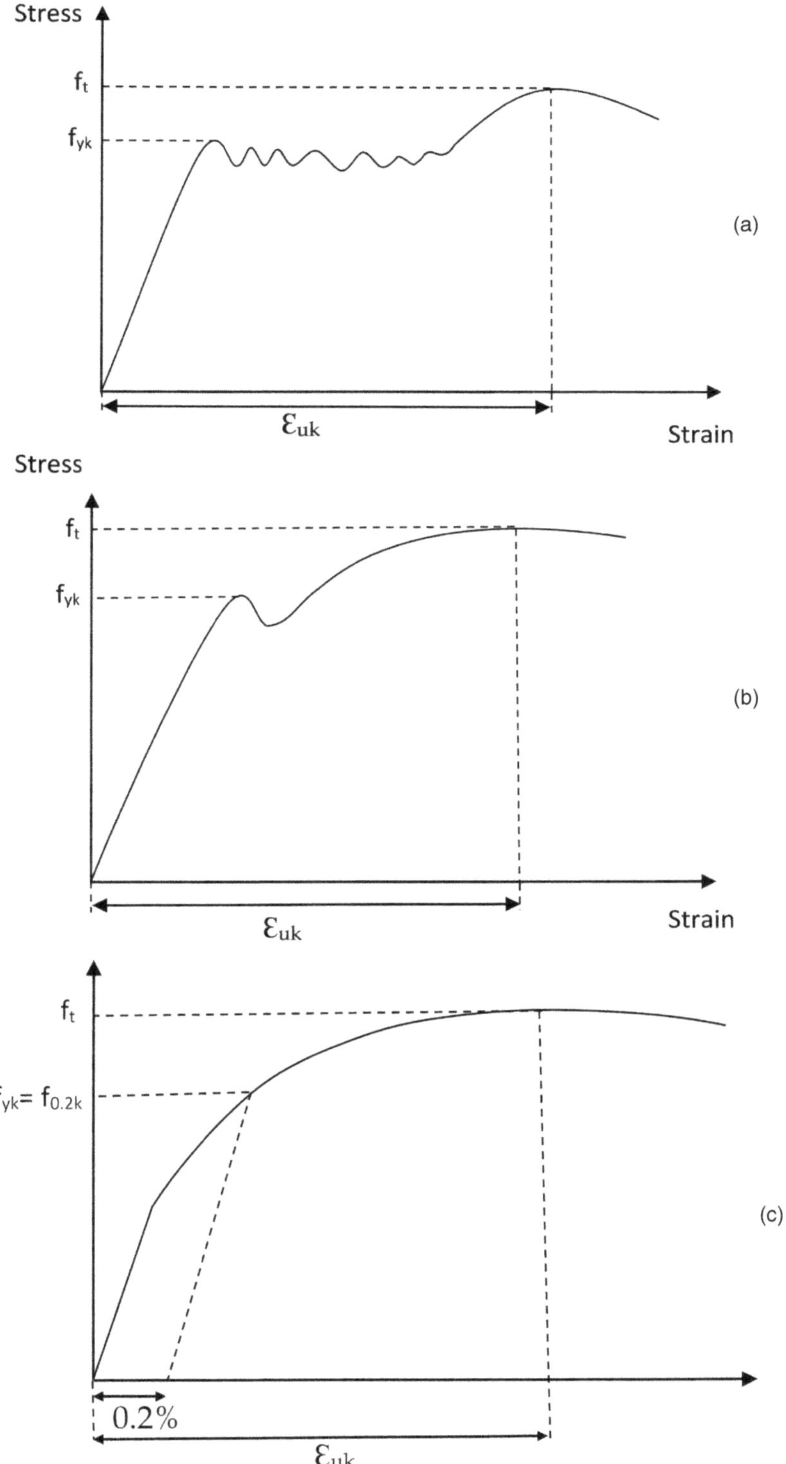

FIGURE 4.8 Stress–strain diagram of untensioned steel: (a) mild steel; (b) hot rolled/heat-treated HSD steel; (c) cold worked HSD steel.

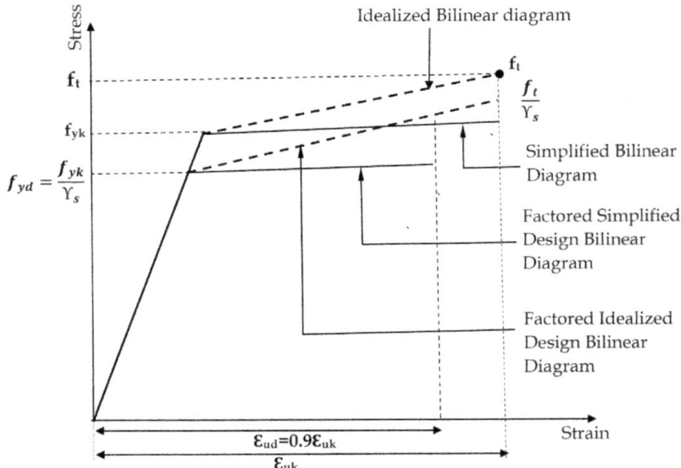

FIGURE 4.9 Bilinear stress–strain diagram of reinforcing steel for design as per IRC:112.

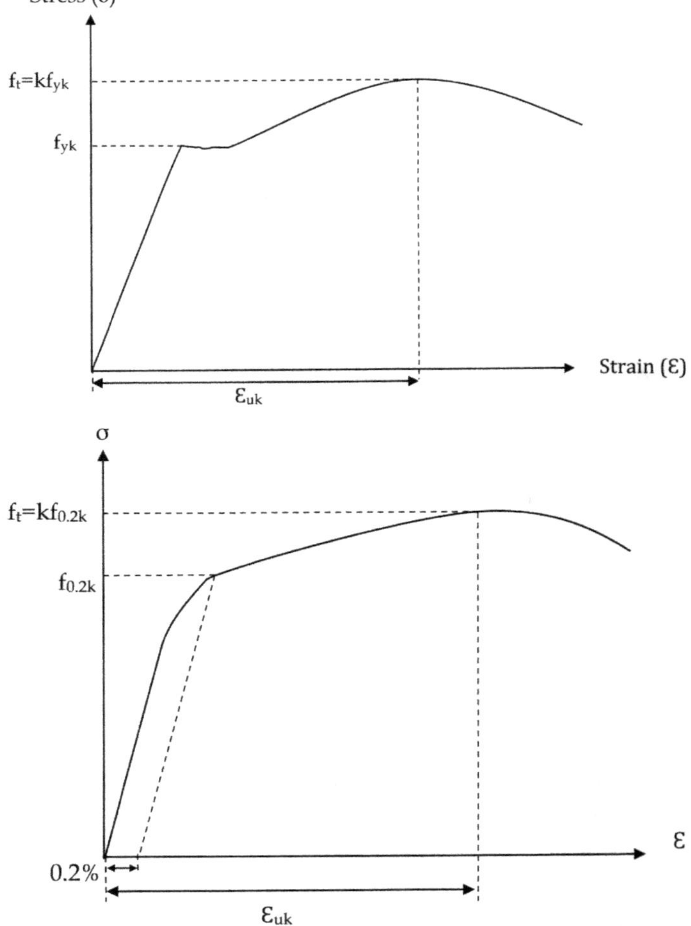

FIGURE 4.10 Stress–strain diagram of typical reinforcing steel: (a) hot rolled steel; (b) cold worked steel.

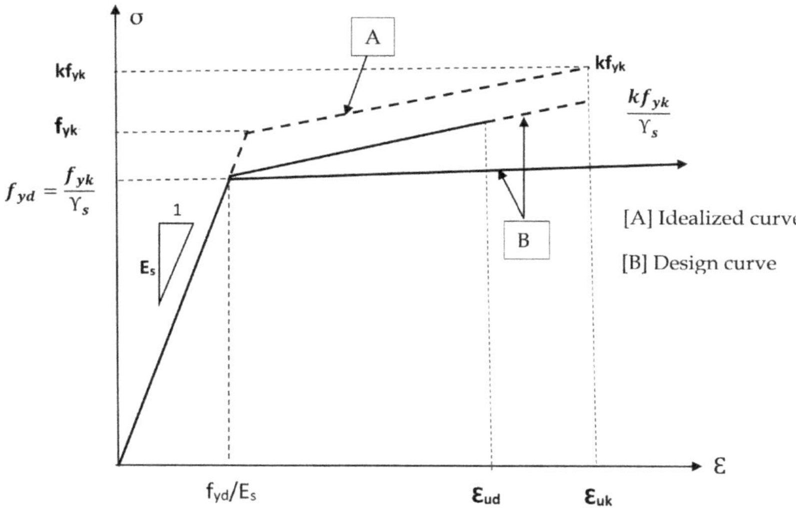

FIGURE 4.11 Idealized and design stress–strain diagrams for reinforcing steel (for tension and compression).

For design, Eurocode 2 prescribes two bilinear stress–strain relationships. The first one, denoted by curve [A] in Figure 4.11, is the idealized curve with an inclined top branch with a strain limit of ε_{ud} and the maximum stress of $\frac{kf_{yk}}{\Upsilon_s}$ at ε_{uk} where $k = \left(\frac{f_t}{f_y}\right)_k$, where:

ε_{uk} = characteristic strain of reinforcement;
ε_{ud} = design strain;
f_{yk} = characteristic yield strength of reinforcement;
f_t = characteristic tensile strength of reinforcement;
f_y = yield strength of reinforcement;
Υ_s = partial factor for reinforcing steel.

The second has a horizontal top branch without the need to check the strain limit. For design, either of the two relationships marked as [B] in Figure 4.11 can be used.

4.5 SUMMARY

In this chapter, the important properties of steel for use in reinforced concrete have been discussed. The current limit state method of design deals with strains and considers strain as the criterion for failure. However, for determining sections' strength, stresses are calculated from the stress–strain diagram of the materials. Engineering stress–strain diagrams for steel have been discussed in detail and the stress–strain relationship provided in national and international standards has been included in this chapter. Based on these relationships, the reinforced concrete design of different types of elements needs to be performed.

BIBLIOGRAPHY

Bhatt, P., MacGinley, T. J. and Choo, B. S. (2005). *Reinforced Concrete Design: Design Theory and Examples*, CRC Press, Boca Raton.

EN 1992-1-1 (2004). *Eurocode 2: Design of Concrete Structures – Part 1-1: General Rules and Rules for Buildings*, European Committee for Standardization, Brussels.

IRC:112-2020 (July 2020). *Code of Practice for Concrete Road Bridges*, First Revision, Indian Road Congress, New Delhi.

IS 456:2000 (2007). *Plain and Reinforced Concrete – Code of Practice*, Fourth Revision, Tenth Reprint April 2007, Bureau of Indian Standards, New Delhi.

IS 1786:2008 (2008). *Indian Standard for High Strength Deformed Steel Bars and Wires for Concrete Reinforcements – Specifications*, Fourth Revision, Bureau of Indian Standards, New Delhi.

Macginley, T. J. and Choo, B. S. (1990). *Reinforced Concrete Design Theory and Examples*, Second Edition. E & FN Spon, London.

Martin, L. and Purkiss, J. (2006). *Concrete Design to EN 1992*, Second Edition, Elsevier, Butterworth-Heinemann, Oxford, UK.

Park, R. and Paulay, T. (1975). *Reinforced Concrete Structures*, Wiley, New York.

Paulay, T. and Priestley, M. J. N. (2018). *Seismic Design of Reinforced Concrete and Masonry Buildings*, Wiley, New York.

SP: 16-1980 (1980). *Design Aids for Reinforced Concrete to IS 456:1978*, Fourth Reprint March 1988, Bureau of Indian Standards, New Delhi.

Timoshenko, Young D. H. (1983). *Elements of Strength of Materials (in MKS Units)*, Fifth Edition, East-West Press, New Delhi.

5 General Considerations for Analysis of RC Structures

5.1 INTRODUCTION

The method of analysis of reinforced concrete structures assumes linear elastic behaviour. Although concrete behaves in a non-linear manner, a linear elastic analysis is generally used as a most important tool for analysis of RC structures. Although wind and earthquake forces are dynamic in nature, they may be represented by equivalent static forces. Hence, the method of analysis is generally static assuming linear elastic behaviour of the structure. However, currently, the standards of different countries are opting for dynamic analysis depending on the height of the structure and the seismic vulnerability of the area in which it is located. With the advent of cheap, efficient and versatile software tools for structural analysis of highly indeterminate structures, the stress resultants for different types of static and dynamic loads and their combinations can easily be determined. Users must have a clear understanding about the load transfer path of different types of loads and accordingly develop a suitable model for computer analysis.

A structure is analysed first to determine the values of stress resultants in the different elements which are then designed on the basis of these results. A structure's response to loading is governed by the elements that take part in the load path. The main structural elements are the slabs, beams, walls, foundations, etc. To simplify the problem of analysis it is assumed that the effects of the non-structural components are small and can be neglected.

5.2 LOADING

A structure must be designed to resist all types of forces, permanent or transient, that can act on it during its life span or design life. The maximum probable values of the different types of loads and their probable combinations must be established before design can be performed. The various loads acting on the structures consist of dead loads, live loads (imposed loads), wind loads, earthquake loads, etc.

With the exception of dead loads, other types of loads cannot be assessed accurately. While maximum gravity live or imposed loads can be anticipated approximately from previous field observations, wind and earthquake loading are random in nature, more difficult to measure, and even more difficult to predict with confidence. The application of probabilistic theory has helped to rationalize the approaches to estimating the wind and earthquake loading.

DOI: 10.1201/9781003415398-5

5.2.1 Dead loads

The dead load in a structure comprises of the weight of structural and non-structural elements. For calculations of weights of different elements, their unit weights are specified in IS:875 (Part 1) 1987.

5.2.2 Imposed loads

The imposed loads are caused by people, furniture, equipment on floors and dust, etc., on roofs. The magnitude of the imposed loads will depend on the type of structure and its intended use.

5.2.3 Wind load

Wind forces and their effects (static and dynamic) should be taken into account when designing structures. Wind speeds vary randomly both in time and space and hence the assessment of wind loads and response predictions are very important in the design of structures. A majority of structures which are generally encountered in practice do not suffer from wind-induced oscillations and generally do not require dynamic analysis. For such structures which are short and heavy, static wind analysis is satisfactory. However, there are various types of structures which require investigation of wind-induced oscillations. For such structures like tall buildings, dynamic analysis is required. Wind loads are covered in IS:875-(Part 3) 2015.

5.2.4 Earthquake force

During an earthquake the ground vibrates. All types of structures have a substructure which rests within the ground and a superstructure which is above the ground. The vibrations of the ground cause the substructure to vibrate. The substructure and superstructure are generally integrally connected, as a result of which the vibrations are transmitted to the superstructure. Structures can be considered as cantilever beams fixed at the base and free at the top. Vibrations in the superstructure result in varying displacements, as a result of which forces are generated or induced in the superstructure as per Newton's laws of motion. While other types of forces are physically present, seismic forces are not physically present but are induced or generated in the form of inertial forces. The structure should transmit these forces to the supporting soil, without affecting their stability, geometry, integrity, safety or serviceability. Though the codes of practice categorically mention that the actual seismic forces can be much higher than the design forces, structures are generally not designed for the actual seismic forces as the cost of the structures would become prohibitive. Earthquake forces are covered in IS:1893 (Part 1-5).

5.2.5 Snow loads

For areas which are subjected to snowfall, structures need to be designed for snow loads on roofs or any other areas above the ground where snow can accumulate.

5.2.6 Special loads

Load or load effects due to temperature changes, soil and hydrostatic pressure, internally generated stresses due to creep, shrinkage, differential settlement, etc., and accidental loads, need to be considered for structures as and when necessary. The nature of the loads to be considered are structure and site specific and should be based on sound engineering judgement.

5.3 CHARACTERISTIC AND DESIGN LOADS

The characteristic or service loads are the actual loads that the structure is generally subjected to. In statistical terms, the characteristic loads have a 95% probability of not being exceeded during the life span of a structure.

The design loads for which structures are to be designed are obtained by multiplying the characteristic loads with the partial safety factor appropriate for that load.

Design load = characteristic load × partial safety factor for load

$$F_d = F_k \times \Upsilon_f \tag{5.1}$$

The partial safety factor, Υ_f, takes into account the possible uncertainties associated with the load for the limit state considered.

5.4 COMBINATIONS OF LOADS

Methods of accounting for the load combinations and their effects on the design of structures vary according to the codes of practice used in the design philosophy adopted. In the limit state method of design, the adequacy of the structure or its members is checked against factored loads corresponding to the different safety and serviceability limit states. The load combination that produces the most unfavourable effect in the structure, foundations, or structural element concerned should be adopted in the design. It should also be recognized that the simultaneous occurrence of maximum values of wind, earthquake, imposed load and snow loads are unlikely.

The following load combinations are generally adopted for design:

- DL;
- DL and IL;
- DL and WL;
- DL and EL;
- DL and TL;
- DL, IL and WL;
- DL, IL and EL;
- DL, IL and TL;
- DL, WL and TL;
- DL, EL and TL;
- DL, IL, WL and TL;
- DL, IL, EL and TL.

(DL = dead load, IL = imposed load, WL = wind load, EL = earthquake load and TL = temperature load.)

5.5 ANALYSIS CONSIDERATIONS

A structure's response to loading is governed by the components that are stressed or which take part in the load path. For the structural analysis, the main structural elements are only considered in the modelling. However, even the non-structural elements can contribute to the structural behaviour, such as unreinforced masonry walls which can act as compression-only bracings under the effects of seismic forces. Nowadays, even the codes of practice prescribe these and accordingly the structures are to be modelled using these considerations.

For performing the analysis of structures, the behaviour of all the structural components and materials are to be accounted for with the help of some simplifying assumptions. The commonly used assumptions can be stated as follows:

a) The material reinforced concrete and the structural components are linearly elastic. Linear methods of analysis are performed for analysing reinforced concrete structures.
b) Only the primary structural elements participate in the overall behaviour. The effect of secondary structural elements and non-structural elements may be significant, such as unreinforced masonry in-fills which can increase the stiffness and thereby change the behaviour of the structure. For accurate analysis these are also necessary to be taken into consideration.
c) Floor slabs are assumed to be rigid in plane. The floor slab is subjected to the horizontal plane rigid body translation and rotation. However, exceptions occur for a very high aspect ratio of floor or discontinuity in the diaphragm.
d) Element stiffnesses of relatively small magnitude are ignored. These include, for example, transverse bending stiffness of slabs, minor axis of stiffness of shear wall, tortional stiffness of beams, columns, and walls, etc.
e) Deformations that are relatively small are ignored. These include axial deformations of beams and columns, in plane bending and shear deformation of floor slabs.
f) The effects of cracking in reinforced concrete elements due to flexural tensile stresses are accounted for by considering a reduced moment of inertia.

5.6 METHODS OF ANALYSIS

For all practical purposes, all structures may be analysed by linear elastic theory to calculate the internal actions produced by design loads. The deflections or deformations are considered to be small. Hence a first-order analysis of a structure may be performed for simultaneously applied gravity and horizontal loading whereby the deflections and forces are obtained by direct superposition of the two types of loading considered separately. Any interaction between the gravity loading and horizontal loading is not accounted for in the analysis. However, if the deformations are large these assumptions are not valid and higher order analysis is to be performed and modified stiffness matrix in the form of geometric stiffness matrix is to be developed for the solution. Nowadays, with the advent of cheap and efficient software, non-linear methods of analysis have become quite popular. The detailed application of material and geometric non-linearity in the analysis are mostly adopted for important structures and for research.

5.7 GUIDELINES FOR ANALYSIS AS PER INDIAN STANDARDS

As per IS:456-2000, all structures may be analysed by the linear elastic analysis. The relative stiffness of the elements may be based on the moment of inertia of the section on the basis of:

I. Gross section;
II. Transformed section;
III. Cracked section.

For deflection calculations, the effective moment of inertia should be used.

As per IS:1893 (Part 1) 2016, to take into consideration the effect of cracking in reinforced concrete, the moment of inertia shall be reduced to 70% and 35% of the gross or uncracked values of columns and beams, respectively.

5.8 STRESS RESULTANTS OF RC STRUCTURAL ELEMENTS

In order to understand the behaviour of a structure under loads, a complete three-dimensional model of the prototype is to be developed using line and area elements. Typical stress resultants of a line element, connected between nodes "j" and "k" of a three-dimensional moment resisting frame is shown in Figure 5.1. For a three-dimensional space frame at any node of a beam element there are six degrees of freedom, three translational and three rotational along the three orthogonal x, y and z axes. The three translational degrees of freedom are indicated as 1, 2 and 3 along the x, y and z axes, respectively. The three rotational degrees of freedom are denoted by 4, 5 and 6 acting along the x, y and z axes, respectively. These are acting at node "j". In a similar manner, at node "k" the six degrees of freedom are indicated. The actions corresponding to these degrees of freedom will give rise to the different stress resultants acting on the beam elements. Considering node "j" force along "1" will result in axial force, force along "2" will produce shear in the vertical plane and force along "3" will result in shear along the horizontal plane. Moment along "4" will result in torsion of the beam, whereas moment along "5" will cause bending of the beam in the horizontal plane and moment along "6" will result in bending in the vertical plane, i.e., normal flexure encountered in transversely loaded beams. Thus, it can be concluded that the six types of actions acting at a node of a beam element of a space frame result in four types of stress resultants: axial, shear, torsion and flexure. In the limit state method of design, the limit state of collapse is of four types: flexure, compression, shear and torsion. Since concrete is very weak in tension but strong in compression, instead of axial, only compression forces are specified.

For an accurate analysis, the structural model should represent all the major structural components of the prototype structure. Not only the structural elements are to be correctly modelled but the load transfer path also needs to be similar to that of the actual prototype. Load paths for gravity and lateral loads need to be correctly incorporated into the model so that the actual behaviour of the prototype can be properly understood. For analysis of reinforced concrete structures, a three-dimensional model with line elements and area elements is generally developed. Line elements are used to represent beams and columns. Area elements are used to model floor slabs and RC walls. Area elements may be represented by membrane, plate and shell elements. Membrane elements can transmit only in-plane forces (not moments) and have no bending stiffness. A membrane is a structural element which has in-plane stiffness but not flexural stiffness. They can resist lateral loads but cannot endure any bending moment. Membrane elements transfer loads directly to supporting

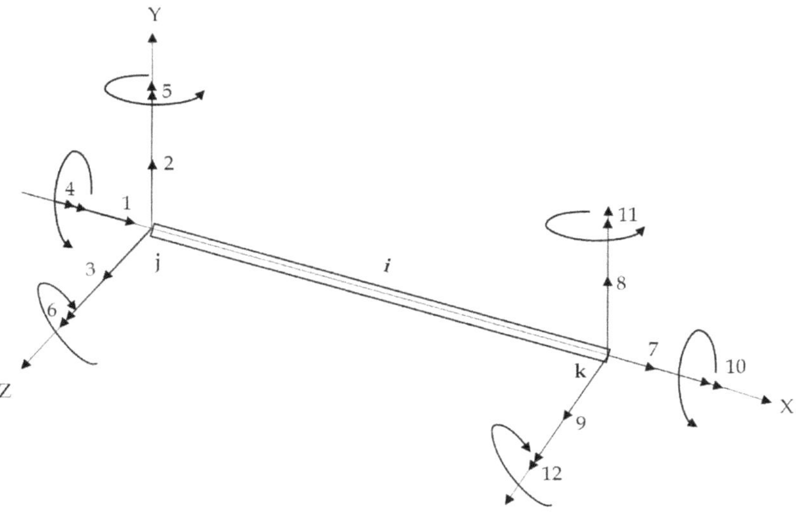

FIGURE 5.1 Stress resultants in a beam element of a three-dimensional moment-resisting frame.

elements. An example of membrane elements is found in the floor slab of a normal RC slab-beam system. Plate elements are used to model thin flat members with significant bending stiffness loaded in the out-of-plane direction. Plates are structural members which add flexural stiffness to the membrane and in addition to transverse load-carrying capacity can withstand lateral loads using their bending stiffness. Nowadays, plate elements are generally replaced with shell elements. Shell elements are combinations of membrane and plate elements, which are used to model thin, flat or curved members with significant bending stiffness. They can be subjected to both in-plane and out-of-plane loads. They have both bending as well as in-plane stiffness. Shells can resist moments and forces from all directions. Shells are further divided into thin and thick shells based on their minimum length to thickness ratio. Thick shells are assumed to have shear deformation, whereas thin shell do not undergo shear deformation. The primary difference between thin shells and plates is in terms of their curvature. Thin shells have curvature, whereas plates are flat. Shear walls, flat slabs, foundation slabs and raft slabs should generally be modelled using shell elements. When a slab is modelled as a membrane element it only receives in-plane forces and does not resist any bending moments, as a result of which all the bending moments are transferred to the beams.

Different types of elements used for modelling reinforced concrete structures are shown in Figure 5.2.

Nowadays, very efficient structural analysis and design software are available for engineers, and the analysis of structures by manual methods is becoming outdated. However, even the best software

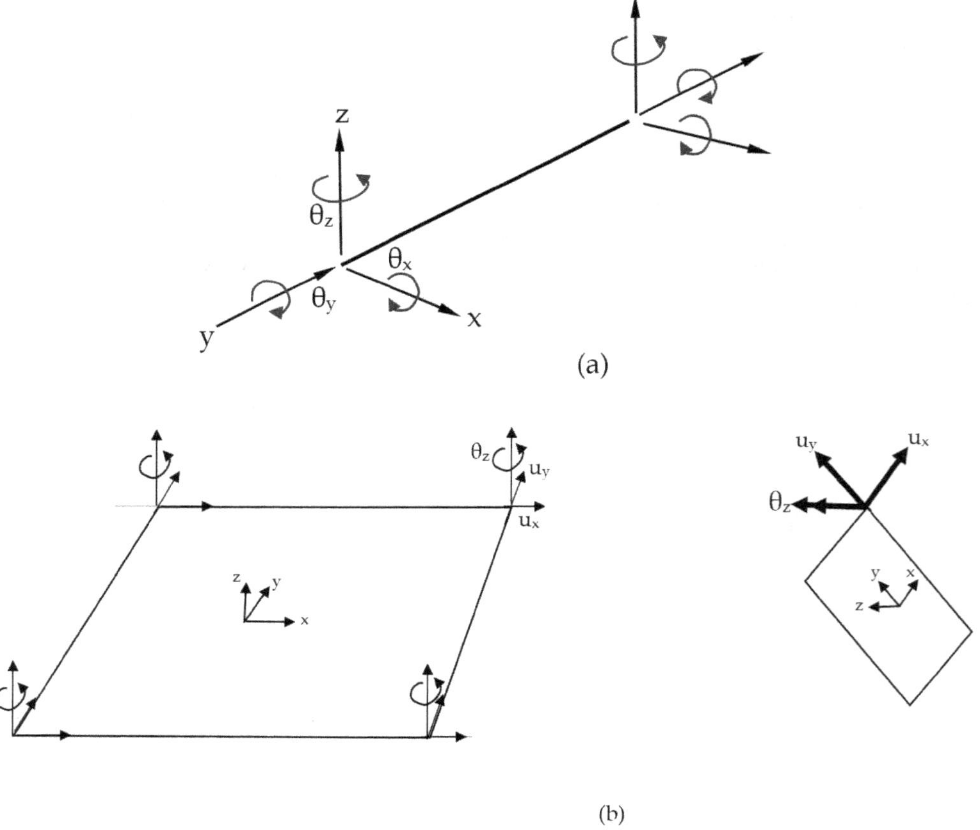

FIGURE 5.2 Stress resultants in elements: (a) beam element; (b) membrane element; (c) plate element; (d) shell element.

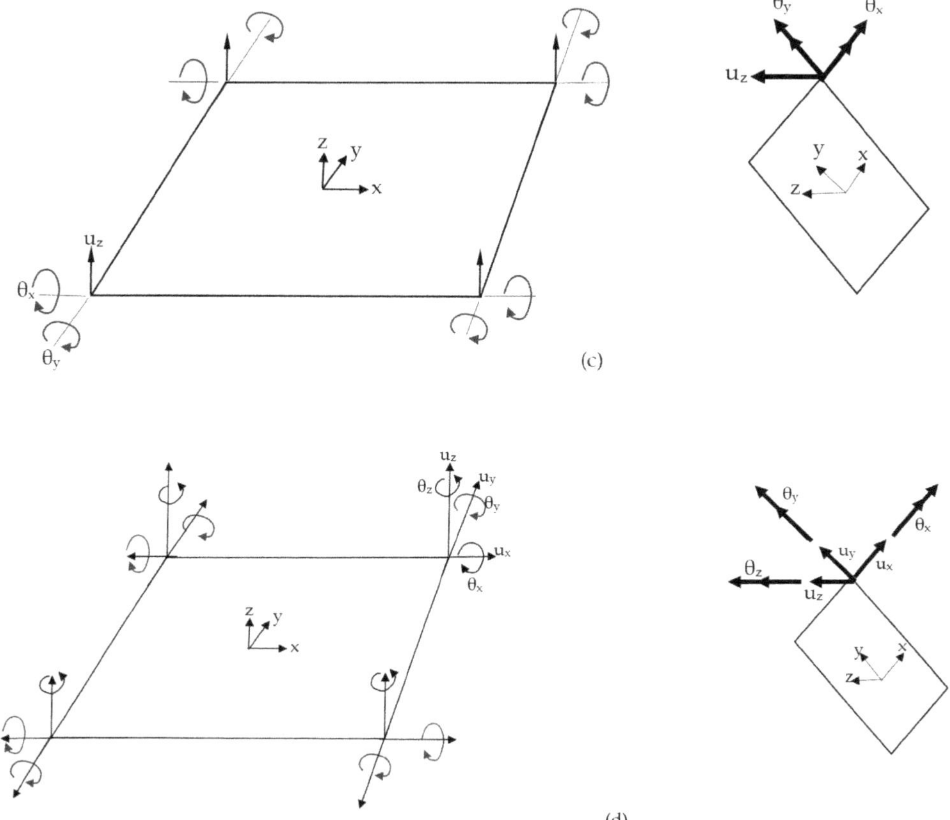

FIGURE 5.2 (Continued)

will not yield satisfactory results if the modelling is not accurate. Designers should make every effort to develop a structural model of the actual prototype in such a manner that there is the maximum level of similitude between the two and the analysis results are able to properly predict the behaviour of the actual structure under loads. The structural system commonly adopted in reinforced concrete buildings has moment resisting frames where the members are connected by rigid joints. In reinforced concrete buildings, slabs are connected between the frames at the floor levels and they play an important role in the load transfer path. The slabs along with their effects also need to be included in the software model. Boundary conditions of the model are enforced by properly assigning supports to the columns by the foundations. Supports may be considered as joints whose degrees of freedoms are defined or known, whereas the beam–column connections are treated as nodes or joints. Hence all supports are nodes, but all nodes are not supports. Beam column joints in moment-resisting frames are termed rigid joints, where the angles between the members before and after deformation are the same. For reinforced concrete sections, the detailing of the reinforcements should be properly performed for fixed or hinged connections.

5.9 SUMMARY

Structural analysis is a prerequisite of structural design wherein the stress resultants on various structural elements are obtained and the overall stability and equilibrium of the entire structure are assessed. Different forms of loadings along with their relevant magnitudes are specified in the relevant codes of practice. The behaviour of the structure under these loads and their combinations are evaluated in detail by performing structural analysis. Nowadays, with the availability of cheap and efficient software, three-dimensional analysis of the entire structure is performed to evaluate the forces and displacements in the structural elements and the structure as a whole. For properly predicting the behaviour of the prototype under loads, a software model is developed which simulates the load-carrying behaviour and load paths in the actual structure. In this chapter, details of loading, analysis methodology and considerations, and modelling features of typical elements have been discussed.

BIBLIOGRAPHY

Bhatt, P., MacGinley, T. J. and Choo, B. S. (2006). *Reinforced Concrete Design Theory and Examples*", Third Edition, Taylor & Francis, Abingdon.

Bryan, S. S. and Allex, C. (1991). *Tall Building Structures, Analysis and Design*, Wiley, USA.

Bungale, T. S. (2010). *Reinforced Concrete Design of Tall Buildings*, CRC Press, Boca Raton/London/New York.

IS 456:2000 (2007). *Plain and Reinforced Concrete – Code of Practice*, Fourth Revision, Tenth Reprint April 2007, Bureau of Indian Standards, New Delhi.

Macginley, T. J. and Choo, B. S. (1990). *Reinforced Concrete Design Theory and Examples*, Second Edition, Spon Press, Boca Raton.

Paulay, T. and Priestley, M. J. N. (2018). *Seismic Design of Reinforced Concrete and Masonry Buildings*, Wiley, New York.

Setareh, M. and Darvas, R. (2007). *Concrete Structures*, Pearson Prentice Hall, Upper Saddle River, New Jersey.

Wilson, E. L. (1996). *Three-Dimensional Static and Dynamic Analysis of Structures*, Computers and Structures, Berkeley, California.

6 General Considerations for Design of RC Structures

6.1 INTRODUCTION

Structural design with reinforced concrete is iterative in nature. The first step is to select the type of structural system and layout of the structural elements with an estimate of their sizes. The design loads which are expected to act on the structure are estimated on the basis of the guidelines available in the relevant codes of practice. The results of structural analysis will provide the stress resultants of the structural elements on the basis of which design is to be performed.

6.2 BASIS OF STRUCTURAL DESIGN – DESIGN PROCESS

Structural design is based on the assessment of structural behaviour under loads. The basis of any structural analysis is centred on the stress–strain relationship of the material with which the structure is composed. Forces applied on structures result in stresses and strains. Strains manifest in the form of deformations or deflections. The criteria for failure are specified in the methodology of design. The basis of any design is to interpret the behaviour of the structure under loads. It will behave in its own natural way and, by adopting any design methodology or process, the designer attempts to predict the actual behaviour of the structure to the best possible extent. A designer should realize that a structural element cannot understand by what methodology it has been designed but it will behave in its own natural manner. However, if the design process is satisfactory the inert structural elements will behave as desired or predicted by the designer.

The first design theory of reinforced concrete, developed near the end of the nineteenth century, was based on the prevailing theory of elasticity. This method assumes that reinforced concrete elements under loads will have stress levels within the elastic zone. At lower values of stresses and strains, the stress–strain relationships of concrete may be considered as linear. Steel also behaves elastically below its yield point. Thus, the concept of working stress design evolved more than a century ago. Though the concept of the working stress method was not an unreasonable methodology, the method has significant drawbacks and limitations and is becoming outdated around the world.

The basis of structural design aims at producing stable, safe, serviceable, economical and functionally sound structures, which should perform efficiently under loads. For more than a century after the invention of reinforced concrete, the design processes have evolved from the traditional working stress method based on an arbitrary allowable stresses approach to the present-day performance-based design which attempt to address the actual failure condition in terms of displacements or deformations based on different performance-based limit states.

DOI: 10.1201/9781003415398-6

6.3 AIMS AND OBJECTIVES OF DESIGN

The aim of design is the achievement of an acceptable probability that the structure will perform satisfactorily during its life. It must carry the loads safely, not deform excessively and have adequate durability and resistance to the effects of misuse and fire. No structure can be made 100% safe, it is only possible to achieve an acceptably low level of failure.

Calculations alone are not sufficient to produce a safe, serviceable and durable structure. The correct selection of materials, quality control, and supervision of construction are equally important.

A civil engineering project involving reinforced concrete structures is a multidisciplinary task where inputs from architecture and other fields of engineering are also involved. The basic functional and aesthetic aspects are covered by the architects, structural engineers deal with the safety, serviceability and performance of the structure under load, whereas other mechanical and electrical aspects such as ventilation and illumination also need to be properly addressed.

The design of a structure must satisfy the following basic requirements:

- Stability to prevent overturning, sliding or buckling of the structure, or parts of it;
- Strength to resist safely the stresses induced by the loads in the various structural members;
- Serviceability to ensure satisfactory performance under service load conditions – adequate stiffness to contain deflections, deformations, vibrations within acceptable limits, providing sufficient durability, fire resistance, etc.;
- Ductility to ensure ductile failure at or beyond ultimate loads;
- Durability – reinforced concrete structures are costly and require a considerable amount of time for construction. Hence to justify the amount of money and effort involved, reinforced concrete structures should be sufficiently durable and should function properly during their life span.

There are other considerations that a sensible designer should bear in mind, such as, economy and aesthetics.

One should keep in mind that the configuration and basic structural system or layout should be chosen properly, otherwise the structure will not behave properly under loads. A poor configuration or layout cannot be improved by good design.

6.4 GENERAL REQUIREMENTS

Design with reinforced concrete is a broad concept that is very difficult to properly define. Architects define in their own perspective, and civil engineers have their own mindset to address this area. In general, for the architects, the requirements of functionality and aesthetics play a key role, whereas for the structural engineers, the requirements of stability, safety and serviceability are paramount, although the considerations of ductility and redundancy are gaining importance currently, especially for seismic-resistant design.

From the conventional concept of strength and serviceability, engineers are nowadays focusing on the actual performance, defined with respect to displacement-based limit state, of the structure under loads. Structures are investigated up to and beyond the design loads. Until recently, the design of civil engineering structures was evaluated on the basis of economy and complying with the design specifications laid out in the relevant standards or codes of practice, whereas nowadays design is judged on how the designers can correctly interpret the performance of the structure under loads, for situations within the design loads and under overloading. A good designer can predict the structural behaviour and design and detail a structure in such a manner that the structure will tread the path, even up to failure, as desired and conceived by the designer.

6.5 PRESCRIPTIVE METHOD OF DESIGN AS PER STANDARDS

A medical practitioner prescribes a prospective mother the good behavioural practices and do's and don'ts which she should follow for giving birth to a healthy baby. Similarly, the codes of practice prescribe the designers with the methodology of design and detailing which should be followed for the birth of a stable, safe, serviceable, functionally sound structure which will perform in the manner desired, up to and beyond the design loads.

The code-based design methodology is based on the assessment of design loads at which level the design is performed. The characteristic loads on the structures are determined first, which when multiplied by the load factors helps in arriving at the design loads. This method is used both for strength and serviceability design. This is the conventional method of design and is followed by different countries around the world. This load-based prescriptive method of design is quite reliable and structures designed using this method have passed the test of the time. This method can predict the behaviour of structures up to the design loads. However, overloading beyond the design loads cannot be addressed. For almost all types of loads other than seismic forces the design values of loads can be predicted quite well by statistical means taking into consideration almost all the uncertainties associated with them. This is valid for dead, imposed, wind, snow loads, etc. Only the seismic design loads on structures can be much less than the actual loads, as designing for actual earthquake forces will result in prohibitive project costs. Hence, under seismic conditions it is taken for granted that overloading can frequently occur and under such conditions the load-based prescriptive method of design will fail. While the loads like dead loads, imposed loads, snow loads, wind loads, etc., are force-based systems, the seismic forces are displacement-based systems. It is a common practice around the world to convert the displacement-based seismic forces into equivalent force-based systems for simplifying the method of analysis. Nowadays, performance-based aseismic design is gaining importance due to its significant advancements over conventional load-based design, and it is slowly infiltrating the design and detailing specifications of the standards or codes of practice.

6.6 STRENGTH AND SERVICEABILITY DESIGN BASED ON LOADS

The limit state method of design requires checking for collapse, i.e., strength design, and limit state of serviceability, i.e., serviceability design. These two designs are based on load. Strength design on ultimate or limiting loads and serviceability design on service or working loads. The design loads are first assessed by the designer and based on this the design is performed. Other than seismic loads, all other types of loads that are commonly encountered are force-based systems. Due to the application of loads, stresses and strains occur and hence load is the cause and the manifestations are the effects. Seismic forces are displacement-based systems whereby vibrations are induced in a structure which results in forces, i.e., displacement is the cause and force is the effect. However, throughout the world for the sake of simplicity, seismic design also deals with force-based systems as the effects due to earthquakes are also expressed in terms of design forces.

However, nowadays, performance-based design is emerging as a very important offshoot of the prescriptive method of design based on loads. Performance-based design deals with deformation or displacement states and assesses the condition of the structure and its damage levels at different performance levels based on deformation or displacement.

6.7 SUMMARY

In this chapter, design considerations have been discussed in detail. Structural analysis is the first step in structural design. Designs performed as per the standards or codes of practice are prescriptive in nature and are based on loads. Both strength and serviceability design are load-based. This has been discussed in detail. The aims and objectives of design have been outlined in this chapter. Nowadays, for aseismic design the methodology is shifting towards performance-based design based on deformations which can predict the structural behaviour in a better manner up to and beyond the design loads.

BIBLIOGRAPHY

Bandyopadhyay, J. N. (2011). *Design of Concrete Structures*, PHI Learning, New Delhi.

Park, R. and Paulay, T. (1975). *Reinforced Concrete Structures*, Wiley, New York.

Pillai, S. U. and Menon, D. (2012). *Reinforced Concrete Design*, Third Edition, Reprint 2012, McGraw Hill, New Delhi.

7 Philosophy of Reinforced Concrete Design

7.1 INTRODUCTION

Structural design with any material involves the use of some assumptions, considerations, laws, axioms, etc. The basic intension is to interpret the behaviour of the structure under loads and to ensure its safety, serviceability and functionality. A design philosophy should aim at assessing the behaviour and performance of the structure under different types of external and internal actions. An efficient designer should design and detail in such a manner that the structural behaviour can be predicted even up to the point of failure, i.e., as if the structure treads the path up to failure as chalked out by the designer. When this requirement is fulfilled along with the basic considerations of stability, safety, serviceability, ductility, aesthetics and economy, the design philosophy can be considered to be a good one. Design aims to assess the behaviour of structures under loads and hence the stress–strain relationship of the construction material forms the basis of any design philosophy.

For designing with any construction material, the material characteristics such as the modulus of elasticity, Poisson's ratio and the relationship between load and deformation need to be known. The different mechanical properties and stress–strain relationships play a key role in the design. Primarily, design needs to be performed satisfying the requirements of strength and serviceability along with stability, aesthetics and economic considerations. Strength is the resistance to rupture and serviceability is governed by different functional aspects such as durability, deflection, vibration, etc. Reinforced concrete design is unique in its approach since the cross-section of the structural element is made of two different materials: concrete and reinforcement. Designing structures with elements of uniform cross-section is comparatively easier. The method of design is conventional, reliable and time tested. Locating the position of the neutral axis in a structural element is one of the initial stages of design as its position governs the strength of the section in bending with or without axial forces. For elements with uniform cross-section, the neutral axis passes through the centroid and hence can easily be located. For elements with non-uniform cross-section, this was a major challenge to engineers for whom design with construction materials of uniform cross-section, i.e., steel, was well known and universally accepted. Thus, initially, when reinforced concrete came into being the already-established theory for the design of steel structures was blindly applied to concrete. However, with time engineers realized that the behaviour of concrete structures was significantly different from that of steel structures, and modifications in the original design concept slowly crept in. The present-day philosophy of design for reinforced concrete structures has evolved significantly over the last 150 years. Any design philosophy should aim towards prediction of the behaviour of a structure under different types of loading and nowadays, due to our ever-increasing demands from concrete, along with the other requirements mentioned above, a structure needs to have sufficient ductility and redundancy, so that it can exhibit enhanced performance under loads at

DOI: 10.1201/9781003415398-7

all stages of loading even up to actual collapse. Along with design, detailing also plays a vital role in a structure's performance. This versatile construction material which has won the confidence of people throughout the world needs to be properly supervised at all stages of production and properly designed and detailed, so that it can continue to serve humanity for years to come.

7.2 BASIC ASSUMPTIONS OF RC DESIGN

When calculating strength and deformations, an essential condition which must be satisfied at any stage of loading is the compatibility of stress and strain in both the concrete and the reinforcements, not only at all points across the sections but also throughout the length of members. It is assumed that steel and concrete in a reinforced concrete element have the same strain. The basic assumptions of RC design can be summarized as follows:

1) Concrete is weak in tension so that all the tension can be assumed to be taken by steel;
2) The transfer of stress between concrete and steel takes place through the bond strength developed between steel and concrete on setting of concrete;
3) The length changes in concrete and in steel, due to the atmospheric change of temperature, are more or less equal as the coefficients of thermal expansion of steel and concrete are approximately the same (i.e., $10–13 \times 10^{-6}$, for concrete and 12×10^{-6}, for steel per degree centigrade).

7.3 UNCONFINED AND CONFINED STRESS–STRAIN DIAGRAMS OF CONCRETE

In general, the prescriptive method of design as per the codes deals with the unconfined stress–strain relationship of concrete as observed in IS:456 and Eurocode 2. The readers will be very familiar with the formulations developed as per the stress–strain relationship for unconfined concrete under uniaxial stress. It should be noted that the stress–strain relationships obtained by actual testing are simplified and idealized for the sake of convenience and ease of computation. The effect of confining the concrete is remarkable in improving its behaviour under loads. In this context, strengths of soil samples tested under unconfined compression test and triaxial test can be compared. The strength of a similar sample extracted from the same soil exhibits significantly higher strength than its unconfined compressive strength when tested in a triaxial test using confining pressure. If concrete can be properly confined the lateral expansion can be significantly restrained, thereby enabling the concrete to reach higher levels of compressive strain before failure. Closely spaced transverse reinforcements in conjunction with longitudinal reinforcements are very effective in confining the core concrete. The areas needing special attention are the beam–column joints where significant inelastic deformation occurs to develop a full hinging mechanism. Buckling of the longitudinal column reinforcements needs to be prevented by using closely spaced transverse reinforcement. Spiral or circular hoops, because of their shape, are subjected to hoop tension by the expanding concrete and thus provide a continuous confining line load around the circumference, which results in much more effective lateral pressure than that provided by rectangular or square links which can apply effective confining pressure near the corners only. Hence, where possible, circular columns are a better alternative to rectangular or square ones.

Tests have shown that the confinement of the concrete by suitable arrangement of transverse reinforcement results in a significant increase in both the strength and ductility of concrete. The strength enhancement from confinement and the slope of the descending branch of the concrete stress–strain curve have a considerable influence on the flexural strength and ductility of the RC sections. The cover concrete will be unconfined and eventually become ineffective, but the core concrete will continue to carry stress at high strains. The compressive stress distributions for the core and cover concrete will be as given by the confined and unconfined concrete stress–strain relations. Good confinement of the core concrete is essential if the element is to have a reasonable plastic rotation capacity to maintain flexural strength at high curvatures. For columns, higher compressive loads will result in lower ductility.

It has been established that the strength and corresponding longitudinal strain of concrete confined by effective lateral confining fluid pressure can be represented as:

$$f'_{cc} = f'_{co} + k_1 f_l \tag{7.1}$$

$$\varepsilon_{cc} = \varepsilon_{co}\left(1 + k_2 \frac{f_l}{f'_{co}}\right) \tag{7.2}$$

where, f'_{cc} and ε_{cc} are the maximum concrete stress and the corresponding strain, respectively, under the lateral fluid pressure f_l, f'_{co} and ε_{co} are the unconfined concrete strength and corresponding strain, respectively, and k_1 and k_2 are the coefficients that are functions of the concrete mix and lateral pressure. The average values of the coefficients have been obtained from actual tests as, $k_1 = 4.1$, and $k_2 = 5k_1$. It has also been observed that the strength of concrete with active confinement by hydrostatic fluid pressure was approximately the same as for concrete with passive confinement pressure from closely spaced circular steel spirals causing an equivalent lateral pressure. Mander et al. have modified the above expressions and proposed a unified stress–strain approach for confined concrete applicable to both circular and rectangular shape transverse reinforcements, as presented in Figure 7.1.

The stress–strain relationship for monotonic loading of confined and unconfined concrete in compression is shown in Figure 7.1, which demonstrates that confinement results in significant improvement in the strains and stresses of concrete, thereby improving its ductility.

For the designer, the significant parameters needed for design are the compressive strength, ultimate compressive strain and the equivalent stress block in compression. The increase in compressive strength is directly related to the confining pressure. Typical values for ultimate compressive

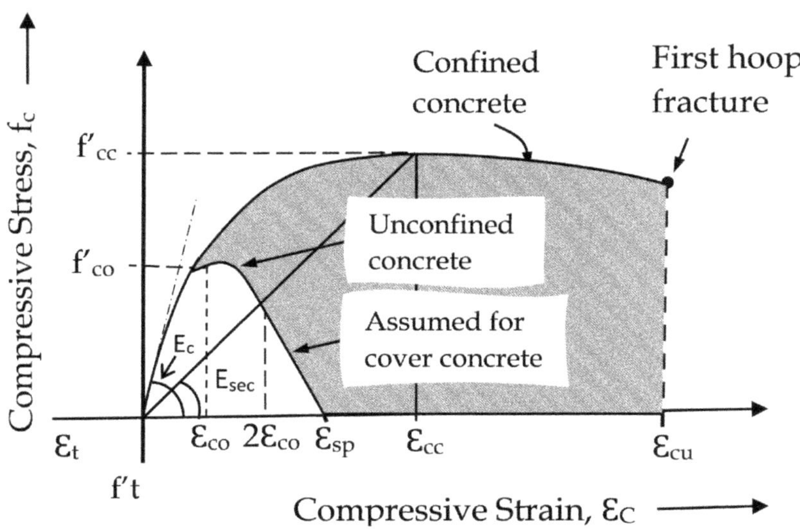

FIGURE 7.1 Stress–strain model proposed for monotonic loading of confined and unconfined concrete.

strain range from 0.012 to 0.05, which is significantly higher than the traditionally assumed value for unconfined concrete. The confined stress–strain relationship of concrete finds application in performance-based design, where the performance-based limit states are assessed on the basis of damage levels at higher levels of deformations.

7.4 STRAIN DISTRIBUTION DIAGRAMS ACROSS THE CROSS-SECTION

The basic assumption for any design methodology with reinforced concrete considers that any section which was a plane before bending will remain a plane after bending. This assumption, Bernoulli's principle, has been proved by careful strain measurements in the laboratory for the case of pure bending. This assumption implies that the longitudinal strain in concrete and steel at the various points across a section is proportional to the distance of the points from the neutral axis of the section. In the case of non-uniform bending, shear forces induced in the section may cause warping. However, elaborate analysis has proved that the normal stresses due to bending are not greatly altered by these shearing stresses. A large number of tests on reinforced concrete beams have proved that this assumption is valid at all loading stages up to flexural failure and both in the compressive and tensile zones. However, for deep beams this assumption is not valid. Figure 7.2 shows the strain distribution diagram measured across the cross-section of an RC beam at different stages of loading which shows linear distribution of strain across the cross-section.

7.5 DEVELOPMENT OF STRESS BLOCKS FROM THE STRESS–STRAIN DIAGRAM

Since the strain in compression concrete is proportional to the distance from the neutral axis, the shape of the stress–strain curve indicates the compressive stress distribution or compressive stress block at various stages of loading. The section will reach its flexural strength or maximum moment of resistance when the extreme concrete compression fibre reaches the ultimate compressive strain, and the stress block fully develops. Research conducted to determine the magnitude and distribution of compressive stress in concrete beams subjected to flexure has proved that the stress distribution in the compression zone of a reinforced concrete beam subjected to flexure is similar to the stress–strain diagram obtained by testing axially loaded concrete cylinders in compression. It is a common question amongst engineers whether, for determining the compressive stress block of members subjected to flexure, it is justified to use the axial stress–strain diagram. The above statement clearly answers the question.

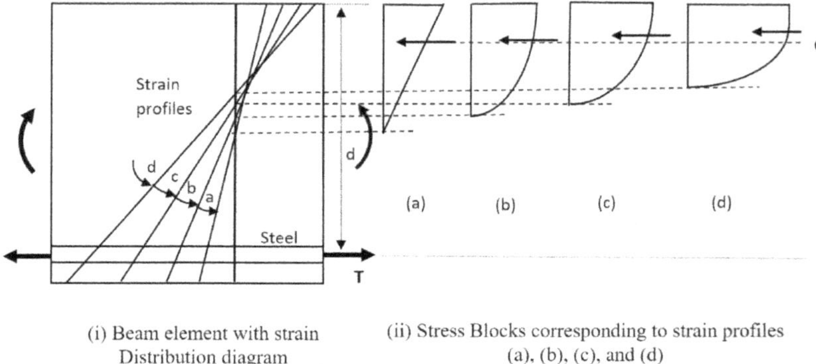

FIGURE 7.2 Strain distributions and corresponding stress blocks of a beam at different stages of loading up to the flexural strength of the section.

The limit state method deals with strains and as per the assumption of plane sections remaining plane after bending, a strain distribution diagram which is linear can be plotted for different types of sections. However, in order to determine the moment capacity of the section, a stress distribution diagram across the cross-section is necessary. Since concrete is effective only in compression, the stress distribution on the compression side is necessary only. Stress distribution on the concrete compression side is obtained by drawing the stress–strain relationship along the depth of the beam with the origin at the neutral axis. Since failure of a section in flexure occurs when the limiting strain in concrete at the outermost fibre is reached, the full stress–strain diagram is developed on the compression side to form the stress block. This is demonstrated in Figure 7.3.

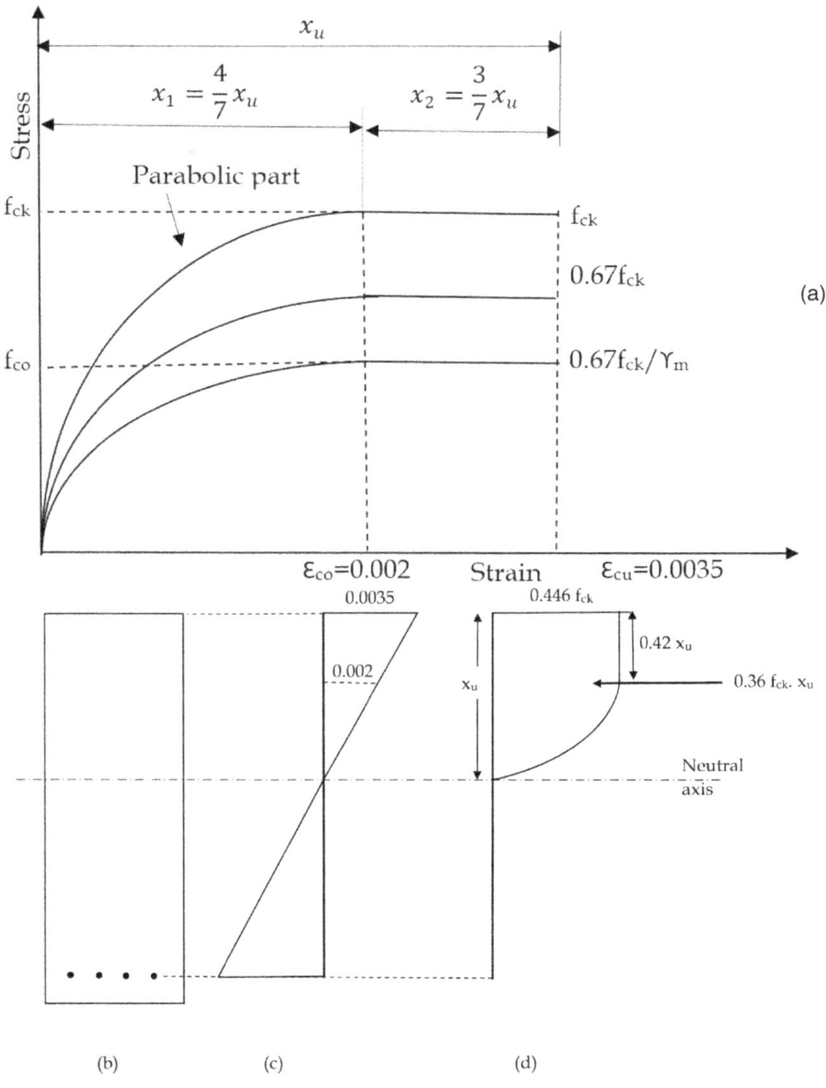

FIGURE 7.3 Stress–strain diagram of concrete and strain and stress distribution in a beam cross-section: (a) stress–strain diagram of concrete; (b) beam cross-section; (c) strain distribution diagram; and (d) stress block.

7.6 HISTORICAL DEVELOPMENT OF DESIGN METHODS

Though there is a notion that concrete has been in use for many centuries, the currently used concrete may be considered to be about 150 years old. The Romans did use some cementitious pozzolanic materials more than 2000 years ago and some Roman concrete structures still exist today. One example is the Pantheon which is located in Rome and was completed in 126 AD. The real breakthrough for concrete was the discovery of Portland cement in 1824 by a British inventor and stonemason called Joseph Aspdin which can be considered as the origin of modern cement concrete. The invention of reinforced concrete by embedding reinforcements in concrete started in the middle of the 19th century. Reinforced concrete as we find it today owes its origin to researchers such as Lambot, Monier, Coignet, Hyatt and numerous scientists who have successfully applied reinforced concrete for making manmade structures. Since the start of the 20th century the use of reinforced concrete structures has been quite rapid.

The Italian architect Ponti remarked the concrete gives us freedom from the rectangle. The ability of concrete to be cast into any shape made it unparalleled amongst all the construction materials. The success of concrete can be easily understood from the fact that this element is the second largest per capita consumed material in the world after water. However, when this composite material was first invented people were very nervous and suspicious about the material. The design philosophy of reinforced concrete members which had cross-sections made of non-uniform materials was not known and was purely imitated from the existing reliable material, steel. Thereafter, with the advent of the modular ratio theory, the reinforced concrete cross-section was transformed into an equivalent single material cross-section, thereby paving the way for proper evaluation of the position of the neutral axis. As concrete proved its worth as an excellent construction material, research and development paved the way for developments in the design philosophy.

7.7 WORKING STRESS METHOD (WSM)

In this method, a structural element is designed in such a manner that the stress resulting from the action of service loads does not exceed a preassigned stress which is kept as some fraction of the ultimate stress to avail a safety margin. The working stress method of design is a simple and somewhat crude method of design. It uses only a small part of the stress–strain diagram of both concrete and steel. This method considers stress as a criterion for failure. However, experimental research has proved that a structural element will not fail if the stress at any point reaches the ultimate value of material strength. The material will fail when the maximum strain reaches the ultimate value, and the limit of safety should be defined by strain and not stress. This method neglects the strain parameters completely and considers permissible stresses only using empirical values of factor of safety. Structures designed by this method are generally very robust and quite safe. Engineers in those times had limited knowledge and hence could not afford to take risks. This method deals with working or permissible stresses only, and when these values are exceeded failure occurs. Hence this method is unable to predict the behaviour of structural elements beyond the permissible stress. The working stress method considers only a small part of the stress–strain diagram of concrete and steel. At this low stress the materials do not yield and hence the concept of ductility cannot be incorporated.

7.8 ULTIMATE LOAD METHOD (ULM)

In this method design is performed to withstand the ultimate load, which is obtained by enhancing the service load by some factor referred to as the load factor for ensuring the desired margin of safety. In this method complete stress–strain diagrams of concrete and steel were considered and hence the prediction of the material behaviour was made more accurately and the theoretical values of failure matched well with the experimental results. This method resulted in sleek, economical structures satisfying the strength requirements but had no provisions for checking serviceability conditions. The structures suffered from excessive deformations and cracking, resulting in severe functionality problems. Hence this method was abandoned.

7.9 LIMIT STATE METHOD (LSM)

The current limit state method of design is a rational combination of both of the two previous methods as it checks the strength of the structures at the ultimate or limiting level and the serviceability at the working level.

Structures designed by this method satisfy both the requirements of strength and serviceability. Strains in the materials are evaluated first and thereafter the stresses are determined from the relevant stress–strain diagrams. The method is slightly more complex than the working stress method and designing from the first principles involves the solution of complex equations and different considerations. Hence design by this method generally requires design aids or tools to simplify the calculations. The method uses the complete stress–strain diagram of the material and considers strain as a criterion for failure which is a more rational approach. Since the limiting strains in the materials are specified at or beyond yield points, ductility can be incorporated in design which is the need of the hour and makes the design philosophy significantly improved over the working stress method.

7.10 SUMMARY

Structural design, which is aimed at ensuring the desirable performance of structures under different types of forces and actions, is based on the stress–strain relationship of the material forming the structure. For reinforced concrete, the stress–strain relationship of concrete and reinforcements plays a vital role in design. The limit state method is based on strains which form the criteria of failure, and stresses are calculated from strains using the stress–strain relationships. In this present chapter, strain and stress distributions of reinforced concrete sections have been discussed in detail while developing the design philosophy. The basics of the different methods of design have also been outlined.

BIBLIOGRAPHY

Baker, L. L. (1956). *The Ultimate-load Theory Applied to the Design of Reinforced and Prestressed Concrete Frames*, Concrete Publications, London.

Bhatt, P., McGinley, T. J. and Choo, B. S. (2006). *Reinforced Concrete, Design Theory and Examples*, Third Edition, E & FN Spon, Boca Raton.

Mander, J. B., Priestley, M. J. N. and Park, R. (1988). "Theoretical stress-strain model for confined concrete", *Journal of Structural Engineering*, Vol. 114, No. 8, 1804–1826.

Park, R. and Paulay, T. (1975). *Reinforced Concrete Structures*, Wiley, New York.

Paulay, T. and Priestley, M. J. N. (2018). *Seismic Design of Reinforced Concrete and Masonry Buildings*, Wiley, New York.

Purushothaman, P. (1984). *Reinforced Concrete Structural Elements, Behaviour, Analysis, and Design*, McGraw Hill, New Delhi.

Ramakrishnan, V. and Arthur, P. D. (1969). *Ultimate Strength Design for Structural Concrete*, Asia Edition, Pitman, London.

8 Working Stress Method

8.1 INTRODUCTION

For a structural element made of a homogeneous cross-section such as timber or steel the neutral axis of the section passes through the centroid. The analysis of a homogeneous beam by the elastic theory of pure bending was proposed by Bernoulli towards the middle of the eighteenth century and can be considered as a stepping-stone for the design of structural elements. When reinforced concrete first came into use, the design philosophy was similar to that used for homogeneous materials like timber or steel. Towards the end of the nineteenth century the concept of modular ratio was introduced for the design of non-homogeneous sections like reinforced concrete. By using this method a non-homogeneous section can be transformed into an equivalent homogeneous or transformed section for which the location of the centroid and hence the neutral axis can be easily determined. The theory of the elastic design of RC sections by the modular ratio method is also called the working stress method as the design is based on safe working stress. This traditional method of design for RC structures, which is more than 100 years old, is still in use in some countries like India, although it has become outdated and deleted from the codes of practice of many countries. The safety of structures is ensured by assigning low values of working stresses. The temples in India, the churches of Europe and the pyramids of Egypt are still standing after several centuries because of the massive constructions resulting from the adoption of very low working stresses. However, the working stress method fails to address the demands of new-generation concrete and is becoming outdated throughout the world.

8.2 DESIGN PHILOSOPHY – STRENGTH AND SERVICEABILITY DESIGN

The code-based design philosophies are termed perspective methods of design and are based on loads. Design is generally performed for strength and serviceability. Strength means resistance to rupture. Strength design means that the sections are strong enough to resist the forces safely. For serviceability design the sections must have sufficient stiffness to restrict deflections, have sufficient durability to restrict crack widths, have sufficient fire resistance, etc. For the working stress method, strength and serviceability design are performed at service or working loads.

8.3 PROVISIONS AS PER IS:456-2000

In the present revision of the code, the working stress method has been denied the status of a section and has been relegated to the Annexure. As per the standard, the working stress method of design should only be adopted when the limit state method cannot be conveniently applied. The domain

DOI: 10.1201/9781003415398-8

of the working stress method of design is becoming increasingly restricted, and even the structures which have been generally designed only by the working stress method until the very recent past are now covered by the limit state design philosophy. Every theory of design is based on a consistent set of axioms, hypotheses, assumptions and approximations. Though the working stress method has many limitations and is becoming outdated throughout the world, a major advantage of this method is that the task of performing design using the basic assumptions or principles of this method is quite simple and straight-forward. In contrary, performing design from the basic principles of the limit state method is quite cumbersome and complex, and design aids are generally required for design. Hence, if design aids or tools are not available, it is considerably difficult for designers to perform design by the limit state method using the fundamental principles. This is a major advantage of the working stress method of design. The working stress concept was introduced in RC design when the theory of elasticity was almost universally accepted at the beginning of the nineteenth century.

8.4 ASSUMPTIONS

In IS:456-2000, the working stress method is based on elastic theory which refers to the method of proportioning the section through the use of permissible stress as in concrete and steel.

The first assumption, i.e., at any cross-section, plane sections before bending remain plane after bending, is based on Bernoulli's principle. This assumption has been experimentally verified both for the compression and tension portions of the section. This is a very important consideration in design because of which the strain distribution diagram across the cross-section is linear and the stress block in concrete can be developed from the stress–strain diagram of concrete. The stress–strain diagram of concrete is rotated by 90° and drawn across the cross-section along the depth of the section with the origin at the neutral axis. This is not valid for deep beams or in regions of high shear.

The second assumption states that all tensile stresses are taken up by reinforcement and none by concrete, except as otherwise specifically permitted. This assumption is primarily for members which predominantly behave in flexure. However, tensile stresses in concrete can be calculated for members in direct tension and for members subjected to combined bending and axial forces. The code allows a small amount of tension in the section for serviceability checks to ensure that the concrete is not unduly cracked. The tensile stresses are not considered for strength design.

The third assumption states that the stress–strain relationship of steel and concrete, under working loads, is a straight line. For low levels of stress (up to one-third of the characteristic strength) the stress–strain relationship of concrete can be considered to be linear. The linearity of the stress–strain relationship of steel is valid for all types of reinforcements, i.e., for mild steel and HYSD steel up to the point of yielding.

The fourth assumption is based on long-term experimentation. It states that the modular ratio of concrete has the value $\frac{280}{3\sigma_{cbc}}$, where σ_{cbc} is the permissible stress in flexural compression in N/mm^2 and is specified in the code. This is valid for normal concrete and not for lightweight concrete.

8.5 STRESS BLOCKS IN CONCRETE

As per Bernoulli's theorem, strain at any point in the cross-section is proportional to the distance of that point from the neutral axis, i.e., the strain distribution diagram is linear. The shape of the stress–strain curve indicates the shape of the compressive stress block of concrete. Thus, the stress block for concrete is its stress–strain diagram rotated by 90° with the depth of the beam as the strain axis with the origin at the neutral axis. The stress block for concrete in the working stress method of design is a triangular one.

The stress–strain diagram of concrete is a non-linear one and is generally idealized as a parabolic one at the initial stages, leading to a horizontal linear part after the characteristic strength is attained.

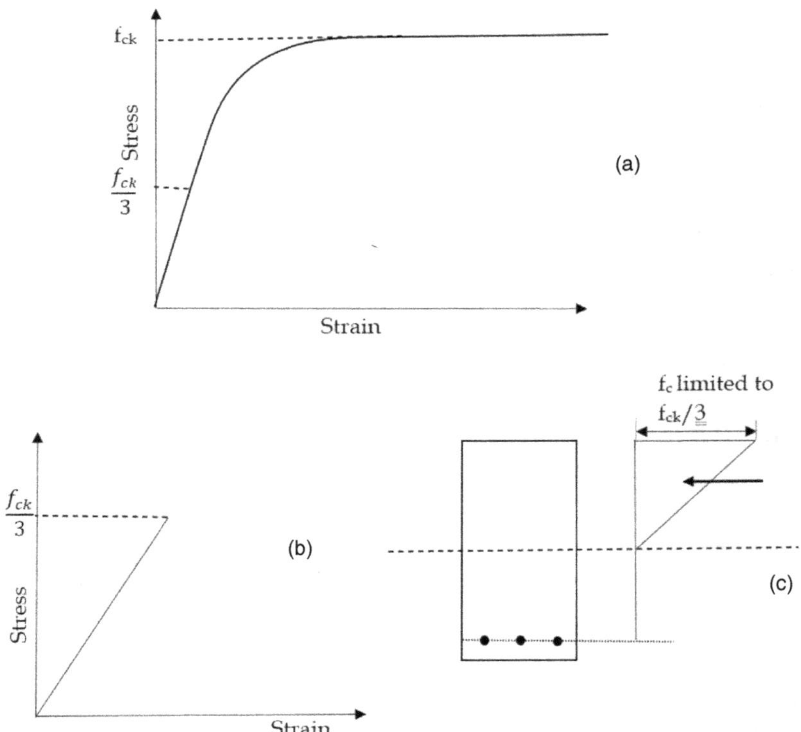

FIGURE 8.1 Stress–strain diagram of concrete and stress block in WSM: (a) idealized stress–strain diagram of concrete as per IS:456-2000; (b) stress–strain diagram of concrete considered in the working stress method; (c) beam cross-section with concrete stress block.

In the working stress method, since the permissible stress is limited to one-third of the characteristic strength, only a very small part of the stress–strain diagram is taken into consideration, which can be considered as a linear one. Thus, the stress–strain diagram of concrete is considered to be linear, which results in a triangular stress block. This is illustrated in Figure 8.1.

8.6 PERMISSIBLE STRESSES

The permissible stresses in concrete in axial tension, axial and flexural compression, bond, and shear for grades of concrete up to M60 are provided in the code. The permissible stresses are kept well below the material strength and the ratio of the strength of the material to the permissible stress is referred to as the factor of safety.

The permissible stresses in steel in axial tension and compression are provided in the code and are obtained by using safety factors. An even lower value of permissible stress in compression has been prescribed in line with the requirements of the limit state method which has a lower value of strain in axial compression than in flexural compression.

Increase in permissible stresses

The code assumes that the coincidence of full horizontal force (i.e., wind or earthquake loads) with dead and live loads is quite unlikely. The horizontal loads are infrequent and temporary in nature. The code specifies that when the effects of wind/earthquake, temperature and shrinkage are combined with those due to dead and imposed loads, an increase in permissible stress is allowed.

8.7 FLEXURAL MEMBERS

For the working stress method of design, the design principles are specified for members subjected to flexure and/or compression. The behaviour of RC sections under flexure can be understood by considering a simply supported beam subjected to pure bending at the central part, as is considered for developing the theory of pure bending. Figure 8.2 shows the cross-section of the beam along with the transformed section, strain and stress distributions.

The first step in design is to convert the non-homogeneous RC section to a transformed one made of concrete only. The equivalent concrete area for A_s area of steel is equal to m A_s area of concrete where m is the modular ratio. Once the section is converted to an equivalent or transformed one, the neutral axis can be determined by calculating the position of the centroid of the section. In the tension zone concrete cannot sustain any stress. The strain distribution diagram has no significance as the working stress method has no relation to strain but only limits the permissible stress, i.e., it deals with stress only. However, as per the first assumption of the working stress method, i.e., Bernoulli's theorem, the strain distribution diagram is linear.

From strain compatibility, strain in concrete = strain in steel

$$\in_c = \in_{st}$$

$$\frac{\sigma_{cbc}}{E_c} = \frac{\sigma_{st}}{E_s}$$

$$\sigma_{st} = \frac{E_s}{E_c} \cdot \sigma_{cbc} = \text{m}.\sigma_{cbc}$$

$$\sigma_{cbc} = \frac{\sigma_{st}}{m}$$

The stress block for concrete will constitute the stress–strain diagram rotated at 90°. Force in compression concrete is the area of the stress block (which is acting on one lamina of the cross-section)

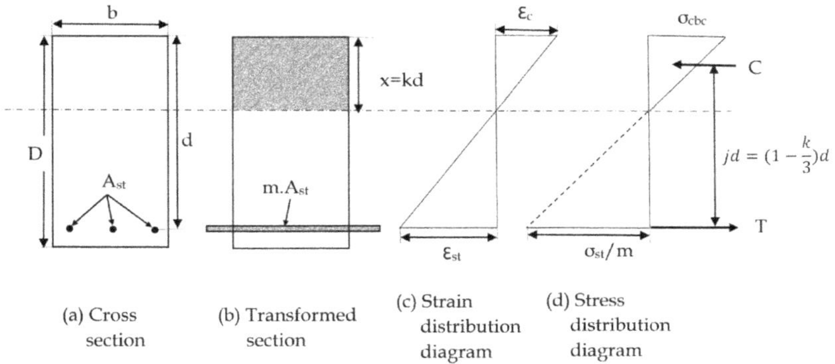

FIGURE 8.2 RC section in flexure.

Note: $x = kd$ = depth of neutral axis; k = depth of neutral axis factor; σ_{cbc} = actual stress in concrete in flexural compression limited to $\sigma_{cbc,p}$; $\sigma_{cbc,p}$ = permissible stress in concrete in flexural compression; σ_{st} = actual stress in steel in axial tension limited to $\sigma_{st,p}$; $\sigma_{st,p}$ = permissible stress in steel in tension; jd = lever arm.

multiplied by the width of the section. Force in tension steel is equal to the equivalent transformed area in concrete multiplied by the hypothetical tensile stress in the equivalent concrete.

From the strain distribution diagram,

$$\frac{\varepsilon_c}{kd} = \frac{\varepsilon_{st}}{d(1-k)}$$

$$or, \frac{\sigma_{cbc}}{k}\frac{E_s}{E_c} = \frac{\sigma_{st}}{(1-k)}$$

$$or, \frac{\sigma_{cbc}}{k}.m = \frac{\sigma_{st}}{(1-k)}$$

$$or, \sigma_{st} = \frac{1-k}{k}.m.\sigma_{cbc}$$

Considering the force equilibrium of the section, force in concrete = force in steel.

From the force distribution diagram,

$$\frac{1}{2}.\sigma_{cbc}.k.bd = \frac{\sigma_{st}}{m}.m.A_{st}$$

$$or, \frac{1}{2}.\sigma_{cbc}.k.bd = \sigma_{st}.p.\ b.\ d$$

A_{st} = pbd/100 where p is expressed as a percentage
= area of steel

$$Hence, \frac{1}{2}.\sigma_{cbc}.k = \left(\frac{1-k}{k}\right).m.\sigma_{cbc}.p.$$

$$or, \frac{1}{2}.k = \left(\frac{1-k}{k}\right).m.p$$

$$or, \frac{1}{2}.\ k^2 = (1-k).m.p$$

$$or, k^2 = 2(1-k).m.p$$

$$or, k^2 - 2.m.p\ (1-k) = 0 \tag{8.1}$$

For balanced conditions, permissible stresses in concrete and steel are reached simultaneously.

$$Hence, k.\ \sigma_{st} = m.\sigma_{cbc} - m.\sigma_{cbc}.k$$

$$or, k.\ \sigma_{st} + m.\sigma_{cbc}.k = m.\sigma_{cbc}$$

$$or,\ k(\sigma_{st} + m.\sigma_{cbc}) = m.\sigma_{cbc}$$

$$or,\ k = \frac{m.\sigma_{cbc,p}}{(\sigma_{st,p} + m.\sigma_{cbc,p})} \tag{8.2}$$

The section conditions can be stated as balanced, under-reinforced or over-reinforced, depending upon the reinforcement provided for the grades of concrete and steel used.

8.8 SINGLY REINFORCED SECTION

A reinforced concrete beam is termed as singly reinforced if longitudinal reinforcements are provided only along one face of the member. However, this consideration is hypothetical and is used for theoretical considerations only, as in real life all reinforced concrete beams have longitudinal reinforcements along both faces.

8.8.1 Balanced section

Refer to Figure 8.3 for an illustration of the balanced section.

8.8.2 Under-reinforced section

Refer to Figure 8.4 for an illustration of the under-reinforced section.

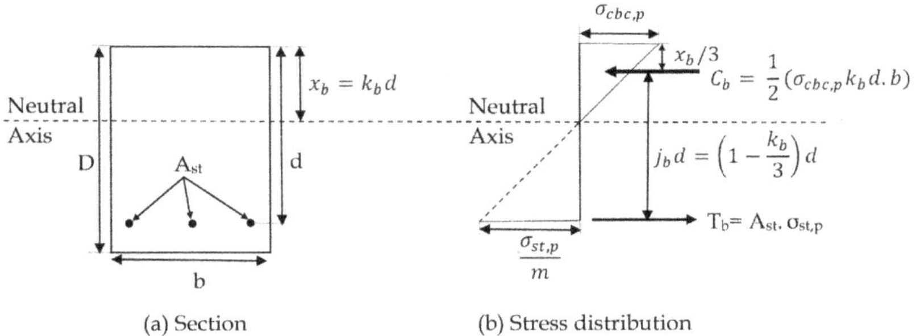

FIGURE 8.3 Singly reinforced balanced section.

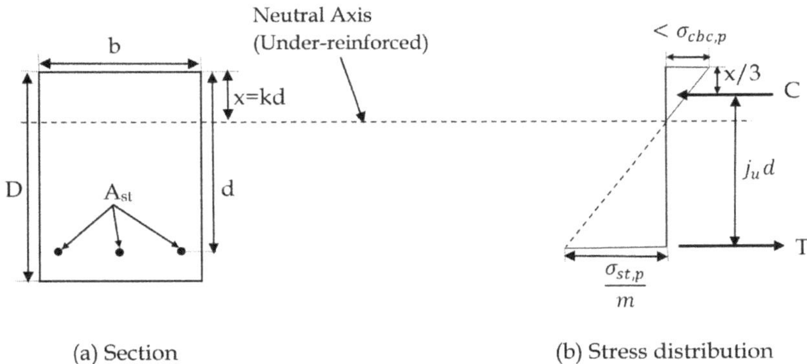

FIGURE 8.4 Singly reinforced under-reinforced section.

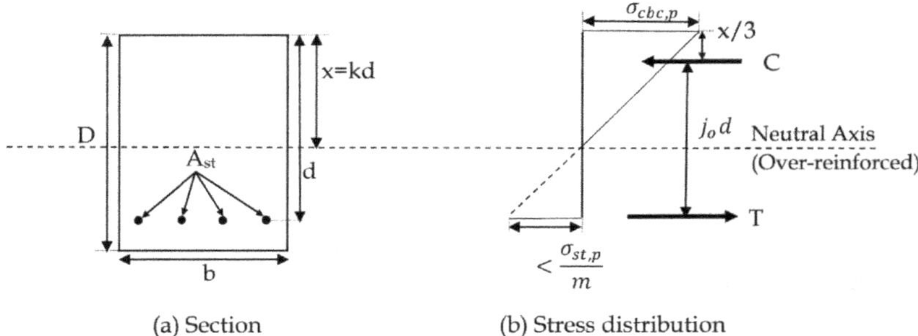

FIGURE 8.5 Singly reinforced over-reinforced section.

8.8.3 Over-reinforced Section

Refer to Figure 8.5 for an illustration of the over-reinforced section.

8.9 TYPES OF SECTIONS – BALANCED, UNDER-REINFORCED AND OVER-REINFORCED

The working stress method of design deals with stresses only and not strains. Hence for defining any type of section the stress distribution diagram needs to be drawn.

For balanced conditions the permissible stresses in concrete and steel are reached simultaneously (refer to Figure 8.3). Hence two points on the stress distribution diagram are defined and known. Strain distribution and stress distribution are both linear for the working stress method. Only one straight line can be drawn between two points and hence the stress distribution diagram and the position of the neutral axis are defined. For an under-reinforced section, permissible stress in steel is reached, with stress in concrete being less than the permissible value. Similarly, for an over-reinforced section, permissible stress in concrete is only reached at failure. These conditions are shown in Figures 8.4 and 8.5. Thus, for conditions other than balanced ones, the stress distribution diagram cannot be obtained from only one stress value on the section. For this condition the neutral axis can be determined by locating the centroid of the transformed section which denotes the location of the neutral axis. From the condition of force equilibrium, the position of the neutral axis can be determined from Equation (8.1) by substituting the value of p as the value of m is constant at $\frac{280}{3\sigma_{cbc}}$ as per IS:456-2000.

8.10 MODE OF FAILURE IN WSM – SINGLE-STAGE FAILURE

Designers should first realize that structural elements are inert and will behave as per their own natural characteristics. It is the duty of the designer to interpret how they will behave under loads. Unfortunately, the working stress method of design does not provide sufficient tools for designers to interpret their behaviour as the entire design ends whenever concrete, steel or both reach the permissible stress. Either one of the materials will reach its permissible stress or both will simultaneously reach their permissible stresses. This is the requirement governing failure in the working stress method which may be termed as single-stage failure. Failure occurs when any of the criteria are reached, and hence it may be termed as single-stage failure. One needs to understand that failure and actual collapse are completely different. Information on how the section will behave beyond the designated failure condition up to the actual collapse state cannot be provided by the working stress method, as the stress–strain diagrams of concrete and steel are defined up to the permissible stress levels only.

8.11 DOUBLY REINFORCED SECTIONS

When the moment of resistance of a section, whose dimensions cannot be increased due to some restrictions, is less than the external moment acting on the section, doubly reinforced sections are resorted to (refer to Figure 8.6). The total moment of resistance of the section is the sum of the balanced moment of resistance of the section plus the moment developed by the force in the compression steel (which is equal to the force in the additional tensile reinforcement over and above the balanced reinforcement). This is demonstrated in Figure 8.6. Hence, in general, all doubly reinforced sections are balanced in nature and permissible stresses in concrete and steel are reached simultaneously. Using the similar triangle consideration for the stress distribution diagram, the stress in compression steel can be calculated. Accordingly, the section is designed by the commonly used steel beam theory and the compression and total tension reinforcement can be calculated. Concrete can undergo an increase in strain at constant stress due to creep, shrinkage and other long-term effects because of which E_c will decrease with time. Hence the value of m in compression is taken as $m_c = 1.5$ m.

8.12 EXAMPLES

8.12.1 Analysis type of problem

In the problem cross-section of the beam, area of tension steel, grades of concrete and steel and working loads are specified.

The first step is to determine the position of the neutral axis.

The RC section is transformed to an equivalent concrete section and the centroid of the equivalent section is calculated to locate the position of the neutral axis.

The neutral axis depth factor, k, is computed (k_{cal}) and compared with k_b, where $k_b = \dfrac{m\sigma_{cbc,p}}{m\sigma_{cbc,p} + \sigma_{st,p}}$ and $m = \dfrac{280}{3\sigma_{cbc,p}}$.

If $k_{cal} = k_b$, the section is balanced.

If $k_{cal} < k_b$, the section is under-reinforced.

If $k_{cal} > k_b$, the section is over-reinforced.

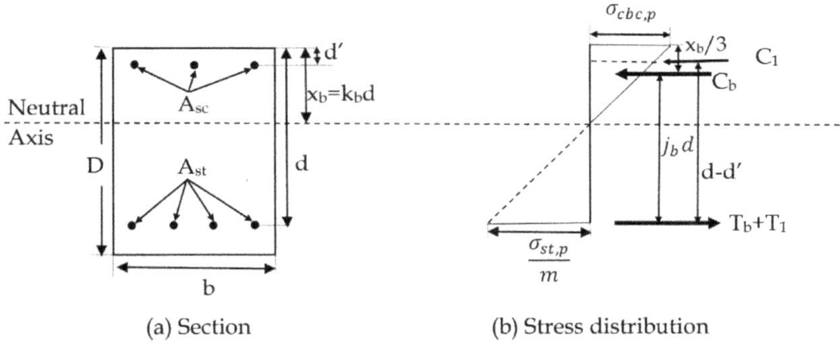

FIGURE 8.6 Stress block for a doubly reinforced section.

For an under-reinforced section, the moment of resistance is calculated from steel stress as the permissible stress in steel is reached.

$$M_{R,ur} = \frac{\sigma_{St,p}}{m}.mA_{St}d\left(1-\frac{k_u}{3}\right) \tag{8.3}$$

k_u = depth of the neutral axis for an under-reinforced section

For over-reinforced condition, the moment of resistance is calculated from the concrete stress, as the permissible stress in concrete is reached.

$$M_{R,ov} = \frac{1}{2}\cdot\left(\sigma_{cbc,p}\right)k_o d\cdot b\left(1-\frac{k_o}{3}\right)d$$

$$= \frac{1}{2}\cdot\left(\frac{f_{ck}}{3}\right)k_o d\cdot b\left(1-\frac{k_o}{3}\right)$$

$$\therefore M_{R,ov} = \frac{1}{6}\cdot f_{ck}\cdot k_o\left(1-\frac{k_o}{3}\right)bd^2 \tag{8.4}$$

k_o = depth of neutral axis for over – reinforced sections

For practical purposes, the differences between k_u and k_o with respect to k_b can be neglected. Otherwise, they can easily be calculated by trial and error based on the calculated values of A_{st} or p.

For balanced condition, the moment of resistance,

$$M_{R,bal} = \frac{1}{2}(\sigma_{cbc,p}).k_b.b.d\left(1-\frac{k_b}{3}\right)d$$

$$= \frac{1}{2}\frac{f_{ck}}{3}.k_b.b.d\left(1-\frac{k_b}{3}\right)d$$

$$= \frac{1}{6}\cdot f_{ck}.k_b\cdot\left(1-\frac{k_b}{3}\right)b.d^2$$

$$\therefore M_{R,bal} = R_b.\,bd^2 \tag{8.5}$$

where, $R_b = \frac{1}{6}\cdot f_{ck}.k_b\cdot\left(1-\frac{k_b}{3}\right)$ = moment of resistance factor. For any combination of grade of concrete and steel, the value of R_b can be determined.

8.12.2 Design type of problem

The width of the beam is assumed from architectural requirements such as being equal to the width of the walls.

Using the moment of resistance factor for the grades of concrete and steel, the depth required for balanced condition is calculated equating the external moment to the moment of resistance of the section

$$M_{ext} = R_b \cdot bd^2$$

$$d_{rqd} = \sqrt{\frac{M_{ext}}{R_b . b}}$$

If, $d_{provided} > d_{required}$, the section is under-reinforced.

If $d_{provided} < d_{required}$, the section is over-reinforced.

If $d_{provided} = d_{required}$, the section is balanced.

The area of steel is calculated considering the permissible stress in steel. If the depth provided is close to the balanced depth calculation can be performed with respect to k_{bal}. If the depth provided is significantly different from the balanced depth, the neutral axis is calculated, and calculations are performed in the same manner as for the analysis problem.

8.12.3 Analysis Type Problem

Example 8.1

Calculate the moment of resistance of a rectangular beam having $b = 250$ mm, $D = 650$ mm, tension reinforcement at bottom $A_{st} = 829$ mm² i.e., (2-20T + 1-16T) using M30 grade concrete and HYSD steel of grade Fe500, as shown in Figure 8.7.

Solution:

For M30 grade concrete and Fe500 steel $\sigma_{cbc} = 10$ N/mm² and $\sigma_{st} = 230$ N/mm² (as per IS:456-2000).

Effective cover for mild exposure and using 8-Tor stirrup and 20Tor main reinforcement = 20 + 8 + 20/2 = 38 ≈ 40 mm

Hence effective depth $d = 610$ mm

$$m = \frac{280}{3\sigma_{cbc}} = \frac{280}{3 \times 10} = 9.333$$

$$k_b = \frac{m\sigma_{cbc}}{m\sigma_{cbc} + \sigma_{st}} = 0.2887$$

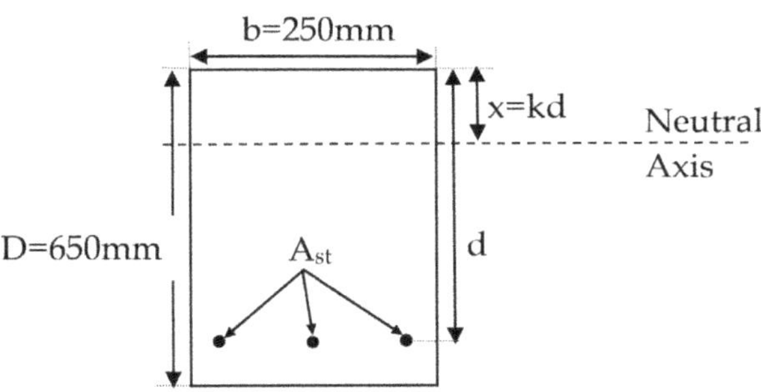

FIGURE 8.7 Beam cross-section.

For the present problem $p = \frac{100A_{st}}{bd} = 0.544\ \%$

Substituting the value of p in Equation (8.1)
$k = 0.272$,
k is less than k_b and hence the section is under-reinforced, and steel reaches the permissible stress,
Lever arm factor=$(1 - \frac{0.272}{3}) = 0.909$. The moment of resistance of the section is

$$M_R = \sigma_{st,p}.A_{st}\left(1 - \frac{k}{3}\right)d$$

$$M_R = 230 \times 829 \times 610 \times 0.909 \text{ N-mm}$$

$$= 105.77\text{kN - m}$$

From Figure 8.4 using the stress distribution diagram, stress in concrete can be determined,

$$\frac{\sigma_c}{0.272d} = \frac{\sigma_{st}/m}{d - 0.272d}$$

$$\sigma_c = 9.21 \text{ MPa} < 10\,\text{MPa}$$

8.12.4 Design type problem

Example 8.2
Design a singly reinforced rectangular beam to carry an external moment of 120 kN-m using M30 grade concrete and Fe 500 steel. Consider the width of the beam as $b = 250$ mm.

Solution:
For M30 grade concrete and Fe 500 grade steel $m = 9.33$ and $k_b = 0.2887$ as obtained from the previous problem.
Lever arm factor for balanced section $j_b = 1 - \frac{k_b}{3} = 0.9038$
Hence the balanced moment of resistance value,

$$M_{R_b} = \frac{1}{6} \times 30 \times 0.2887 \times 0.9038 \times bd^2$$
$$= 1.305bd^2 \text{ N} - \text{mm} = R \times bd^2,$$

where R = moment of resistance factor.
For the present problem depth required,

$$d = \sqrt{\frac{120 \times 10^6}{1.305 \times 250}} = 606.5\ mm$$

Let the overall depth of the section $D = 650$ mm

therefore, $d = 650 - \left(20 + 8 + \frac{20}{2}\right) = 612\text{mm}$

Hence the section is under-reinforced.
The area of tension steel required,

$$A_{st} = \frac{M_{\text{external}}}{\sigma_{\text{st}} \times j_b d}$$

$$= \frac{120 \times 10^6}{230 \times 0.909 \times 612} = 937.9\text{mm}^2$$

The value of j has been considered as that for balanced section j_b, since the difference between the actual and balanced values is not significant. Otherwise, the exact value of j for this section can be calculated for the area of steel provided, i.e., A_{st} using Equation (8.1).

Provide 3-20T bars (area = 942 mm^2).

Note: For the above-mentioned problem checks for minimum and maximum tensile reinforcement, maximum and minimum spacing between bars in beams should be performed as per the requirements of IS:456.

8.13 COMPRESSION MEMBERS

The design of compression members by the working stress method is somewhat empirical. For sections subjected to pure compression or compression with moment due to minimum eccentricity, design is based on permissible stresses in concrete and steel due to pure compression. Hence the design is quite simple and straightforward. As per the code, sections subjected to combined direct load and flexure should be designed by the limit state method. In real life, columns are subjected to axial load along with flexure. Hence, column sections should be designed by the limit state method instead of performing design as uncracked or cracked sections as per the working stress method of design.

8.14 MEMBER SUBJECTED TO DIRECT TENSION

In direct tension, the strength requirement is governed by steel alone. The code provides a limit on the tensile stress in concrete to limit crack width and maintain the bond between steel and concrete.

8.15 SERVICEABILITY REQUIREMENTS

Serviceability checks are performed at working or service stress levels. This is discussed in Chapter 14.

8.16 SHORTCOMINGS OF WSM

In the working stress method, a structural element is so designed that the stress resulting from the action of service loads as computed by linear elastic theory does not exceed a preassigned allowable stress which is kept as some fraction of the material strength to provide a margin of safety. The drawbacks of this method are numerous.

Firstly, the working stress method is logically not applicable to concrete structures since it assumes that both the materials, concrete and steel, are linear and obey Hook's law. The stress–strain relationship of concrete is a non-linear one, but since a very small portion of the curve is considered in design, it is assumed to be linear.

Since the stress–strain diagram of concrete is non-linear, the modulus of elasticity will not be constant and hence a constant value of modular ratio is not acceptable.

Since only a small part of the stress–strain diagram of concrete is utilized, the stress block considered is significantly smaller than the actual one and the design is highly uneconomical.

Experimental research has clearly established that the structure neither fails nor becomes unsafe just by the stress at some point in some member reaching the ultimate value of material strength. It has been proved that the limit of safety of a material is defined by the ultimate strain and not by ultimate stress (i.e., the material fails when the maximum strain reaches the ultimate value). Stress is an abstract mathematical quantity which can neither be measured directly nor be visualized. It is the strain which can be observed and measured.

However, the method is simple and reasonably reliable. The only point in favour of this method is that it is a well-established method having over 100 years of use and it is also simple in application and reasonably reliable.

8.17 WHY WSM IS GETTING OUTDATED THROUGHOUT THE GLOBE

The working stress method deals with permissible stresses which form the criteria for design. The stress–strain curve of concrete is used only up to the permissible stress levels, and it is erased beyond that point. At higher stress levels how the sections will behave under loads or how they will fail actually under loads has not been described. There is a blind alley beyond the permissible stress level, i.e., working loads. The method does not provide any information about strains or deformations and it only deals with stresses. Stress is considered as the criterion for failure.

No information about the failure mechanism is available, whether brittle or ductile.

This is a deterministic method which considers that the material strengths and loads acting on the structure are known accurately and as if there is no chance of deviation. Hence, the chances of overloading or exceeding design loads is not taken into consideration at all.

The present seismic codes categorically mention that the actual loads can be much higher than the design loads, and hence overloading can be routine under earthquake conditions. Thus, in such a situation the WSM of design falls flat and seems to be totally inadequate. Hence, under overloading conditions the behaviour of the sections cannot be predicted or interpreted.

As per the present design philosophy, along with strength and serviceability design, ductility design plays a very vital role, especially in India where more than 60% of the land area is vulnerable to moderate to severe earthquakes.

In simple terms, ductility can be defined as the ratio of ultimate deformation to yield deformation. Hence, for ductility to be incorporated in the section, the material must yield and go beyond the yield point. Reinforced concrete is a combination of two materials, concrete and steel, of which concrete is a brittle and steel a ductile material. In the working stress method of design neither of the materials yields. The permissible stresses are restricted to much lower values than their yield points, almost arbitrarily. Thus, ductility design is not possible in the working stress method of design and thus the failure mode, whether brittle or ductile, cannot be predicted or tailor made to suit the requirements of the designer. In the working stress method of design, strength and serviceability design is possible but ductility design, which is the need of the hour, cannot be performed. As there is no yielding of the materials, failure occurs in a single stage, i.e., on reaching any of the permissible stresses in concrete or steel. This is the major reason why throughout the globe WSM of design is becoming outdated.

There is a common notion amongst civil engineers that for proper understanding of the limit state method of design an introduction to the concept of the working stress method is necessary. The fact that the working stress method is omitted in the codes of practice of different countries proves that this notion is incorrect and the working stress method cannot be considered as a prerequisite of the limit state method of design. The philosophy of the working stress method of design is quite empirical and lacks proper theoretical backing. It is unable to comply with the requirements of present-day new-generation concrete. The concept of ductility design, which is the need of the hour, cannot be addressed at all by this method. The working stress method will find its place in history, whereas the future of concrete will be governed by methods beyond the limit state whereby the strength, serviceability, ductility and performance of structures will play key roles in design.

8.18 SUMMARY

The philosophy and methodology of the working stress method of design have been presented in this chapter. The assumptions of the method as per IS:456 have been discussed. Balanced, under-reinforced and over-reinforced sections have been explained in detail. The mode of failure in the working stress method of design has been critically analysed and it has been demonstrated in detail that the mode of failure in the working stress method of design can be termed as single-stage failure. Stress is considered as the criterion for failure and the method has virtually no connection with strains. Numerical examples have been discussed illustrating both the design and check types of problems. It has been clearly demonstrated that the working stress method of design deals with strength and serviceability designs only and the concept of ductility design is totally missing, which makes it totally unsuitable to meet the requirements of present-day seismic-resistant design.

BIBLIOGRAPHY

Baker, A. L. L. (1956). *The Ultimate-load Theory Applied to the Design of Reinforced and Prestressed Concrete Frames*, Concrete Publications, London.

Bandyopadhyay, J. N. (2011). *Design of Concrete Structures*, PHI Learning, New Delhi.

Gambhir, M. L. (2006). *Fundamentals of Reinforced Concrete Design*, Prentice Hall, New Delhi.

Mallick, S. K. and Gupta, A. P. (1989). *Reinforced Concrete*, Fifth Revision, Oxford & IBH, New Delhi.

Pillai, S. U. and Menon, D. (2012). *Reinforced Concrete Design*, Third Edition, Reprint 2012, McGraw Hill, New Delhi.

Purushothaman, P. (1984). *Reinforced Concrete Structural Elements, Behaviour, analysis, and Design*, McGraw Hill, New Delhi.

Raju, K. N. (2016). *Design of Reinforced Concrete Structures, IS:456-2000*, Fourth Edition, CBS Publishers, New Delhi.

Ramakrishnan, V . and Arthur, P. D. (1969). *Ultimate Strength Design for Structural Concrete*, Asia Edition, Pitman, London.

Shah, V. L. and Karve, S. R. (2016). *Limit State Theory and Design of Reinforced Concrete*', Eighth Edition, Structures Publications, Pune, India.

9 Limit State Method as per IS:456

9.1 INTRODUCTION

Structural design with any construction material requires the interpretation of the behaviour of the structural elements, made of the material, under loads. A structure will respond to external actions, such as loads or displacements, as per their own natural characteristics. A design philosophy aims to predict this response as accurately as possible. With time our knowledge and experience on material behaviour increases and hence the design philosophy gets enriched and continues to change. Over the years, various design philosophies have evolved in different parts of the world. A "design philosophy" is built upon a few fundamental assumptions and is reflective of a way of thinking.

Probability concepts of design have been developed over the years and received a major impetus from the mid-1960s onwards. The philosophy was based on the theory that the various uncertainties in design could be handled more rationally in the mathematical framework of probability theory. Such probabilistic methods came to be known as reliability-based methods.

The past several decades have witnessed an evolution in design philosophy from the traditional working stress method, through the ultimate load method, to the modern limit state method of design.

The salient features and merits of the method are briefly mentioned below.

- ✓ It considers the entire stress–strain diagrams of concrete and steel.
- ✓ It deals with strains and not stresses.
- ✓ It adopts the concept of the fitness of a structure to serve the desired function during the service life span and defines the limiting state of fitness as the "limit state".
- ✓ It attempts to define quantitatively the margins of safety or fitness on some scientific mathematical foundations rather than on an ad hoc basis of experience and judgement.
- ✓ It is based on statistical probabilistic principles.

9.2 OBJECTIVES OF LSM OF DESIGN

The aim of design is the achievement of an acceptable probability that a structure will perform satisfactorily during its life. It must carry all loads safely and should not deform excessively to affect the geometry and integrity of the structure and should have adequate durability and resistance to the effects of misuse and fire. The primary objective is to reduce the probability of failure to an acceptably low level. The criterion for a safe design is that the structure should not reach the limit state during its design life, i.e., it should not become unfit for use.

 DOI: 10.1201/9781003415398-9

9.3 LIMIT STATES

Limit state design philosophy is an advancement over the working stress concept. The behaviour of RC sections under loads can be better predicted using this method of design. As per IS:456-2000, the limit state is an impending state of failure and the aim of design is to ensure that the structure should not reach the limit state. The method is based on some predefined criteria for failure, i.e., a condition or state is defined based on certain considerations. If the structure reaches that state or condition, it becomes unfit for use, i.e., it has failed. However, failure and actual collapse conditions are totally different, and the designer must be well aware of this.

The working stress method of design is deterministic in nature. It considers that the design loads calculated and the permissible stress of the materials derived from the characteristic strength by using the relevant factor of safety are correct and there is no chance of any deviation. The working stress method checks the safety and serviceability of the structure at working or service loads.

The limit state method of design is probabilistic in nature, whereby uncertainties in design loads and material strengths are addressed by resorting to statistical principles.

The theoretical approach to design was given a realistic concept by adopting a statistical basis which includes the assessment of characteristic values of strengths and loads acting on the structure. The term limit state connotes, in a statistical sense, a limiting condition of a structure which may be considered to be appropriate as a limiting condition in design. When calculating the strength and deformation in RC sections, an essential condition which must be satisfied at any limit state is the compatibility of strains in both the concrete and reinforcement, not only at all points across the section but also throughout the length of the members. When steel and concrete are bonded together it is assumed that they have the same strain. The standards or codes of practices generally deal with prescriptive methods of design based on loads.

9.4 TYPES AND CLASSIFICATION OF LIMIT STATES

The limit state method can be classed into the following two states:

(a) Limit states of collapse deal with the strength and stability of the structure and need to ensure that the whole structure and its elements will be sufficiently strong and stable when subjected to the design loads. Failure is to be checked regarding the overall stability of a structure with respect to buckling, sliding and overturning. The strength of a member at the limit state of collapse is known as the ultimate strength. Regarding strength, failure can occur either in flexure, shear, bond, torsion, axial compression or tension, or by a combination of these actions.

(b) Limit states of serviceability deal with deflection, cracking, durability and fire resistance, and should ensure that the structure does not become unfit for use due to excessive deflection, cracking, vibration or durability problems.

For RC structures, normal practice is to design for the limit state of collapse and check the limit state of serviceability and take all necessary precautions to ensure durability. IS:456-2000 specifies the factor of safety values for over-turning and sliding. The structure should be able to transmit all loads acting on it during its lifetime safely to the supporting soil. Though the structure must be designed for all relevant limit states, designers mostly deal with limit states of collapse only because if the relevant codal clauses are properly adhered to the requirements of the limit state of serviceability are generally satisfied. For example, if the span to effective depth ratio is kept within the limiting values, the deflection requirements are generally complied with, and if the maximum distance between bars and proper detailing are provided as per the codal requirements, cracking is expected to be within permissible limits. Hence, the designer must keep in mind that both the requirements of the limit states of collapse and limit states of serviceability must be satisfied for safe and serviceable design.

9.5 CHARACTERISTIC AND DESIGN VALUES OF LOADS AND MATERIAL STRENGTHS

The characteristic or service loads are actual loads that the structure is expected to carry and can be considered as the maximum loads which will generally not be exceeded during the life of the structure. It is that value of load which has a 95 percent of probability of not being exceeded during the life of the structure. For dead loads, the requirements are provided in IS:875 part 1, imposed loads in IS:875 part 2, wind loads in IS:875 part 3, snow loads in IS:875 part 4 and seismic loads in IS:1893. Design load is calculated as characteristic load × partial safety factor for loads. The partial safety factors Υ_f for loads are given in Table 18 of IS:456, which is shown in Table 9.1

As per IS:456-2000 the term characteristic strength means that the value of strength of the material below which not more than 5 percent of test results are expected to fall. The characteristic strength for concrete is specified by f_{ck}, which is the specified compressive strength of standard 150 mm cubes at 28 days expressed in N/mm^2 . The grades of concrete permitted by IS:456 are from M10 to M100, where M refers to the mix and the number refers to the specified compressive strength of 150 mm cubes at 28 days expressed in N/mm^2. For structural concrete, the minimum grade permitted is M20 and the design parameters given in IS:456 are applicable up to concrete grade M60. However, for IRC112:2020, design parameters are provided up to M90 grade concrete and as per Eurocode 2 design parameters are provided up to a cube compressive strength of 105 N/mm^2.

For reinforcement, IS:456-2000 allows mild steel and medium tensile steel bars conforming to "IS 432 – Specifications for mild steel and medium steel bars and hard drawn steel wire for concrete reinforcement: Part 1" and high-strength deformed bars conforming to "IS 1786 – Specifications for high strength deformed bars and wires for concrete reinforcement-specification". As per IS:456-2000, the characteristic strength for reinforcing steel shall be assumed as the minimum yield stress/ 0.2 percent proof stress specified in the relevant Indian standards.

IRC 112-2020 permits mild steel (MS), Grade 1 and high-strength deformed steel (HSD) as per the following grades: Fe415, Fe415D, Fe415S, Fe500, Fe500D, Fe500S, Fe550, Fe550D and Fe600.

Eurocode 2 [EN:1992-1-1 (2004) – Design of concrete structures, part 1-1: General rules and rules for buildings] permits bars and de-coiled rods and wire fabrics of class A, B, C having characteristic yield strength f_{yk} or $f_{0.2k}$ ranging from 400 to 600 MPa.

Unfortunately, IS:456-2000 only refers to IS:1786, whereby designers can use any grade of steel within the range of Fe 415 to Fe 700 as no specific capping on strength is provided. As per "IS:13920-2016 – Ductile design and detailing of reinforced concrete structures subjected to

TABLE 9.1
Values of partial safety factors Υ_f for loads

	Limit state of collapse			Limit state of serviceability (for short-term effects only)		
Load combination	**DL**	**IL**	**WL**	**DL**	**IL**	**WL**
DL+IL	1.5	1.5	–	1.0	1.0	–
DL+WL	1.5 or 0.9*	–	1.5	1.0	–	1.0
DL+IL+WL	1.2	1.2	1.2	1.0	0.8	0.8

Note: DL = dead load; IL = imposed load; WL = wind load. While considering earthquake effects, WL is to be replaced by EL, EL = earthquake load.

* This value is to be considered when stability against overturning or stress reversal is critical.

seismic forces – codes of practice", maximum grades of HYSD bars to be used in earthquake zones III, IV and V should be limited to Fe550 having elongation of at least 14.5 percent, the ratio of ultimate stress to 0.2 percent of proof stress shall not exceed 1.25 and the ratio of ultimate stress to 0.2 percent proof stress shall be at least 1.15. However, for earthquake zone II no such capping for maximum grade of steel is provided. In this context it may be mentioned that India is divided into four earthquake zones (II, III, IV and V), with increasing risk of seismic vulnerability, as per "IS 1893 (Part 1): 2016, Criteria for Earthquake Resistant Design of Structures".

The design strength of material is obtained by dividing the characteristic strength by Υ_m, where Υ_m is the partial safety factor appropriate to the material and limit state being considered. When assessing the strength of a structure or structural member for the limit state of collapse, the values of partial safety factor Υ_m should be taken as 1.5 for concrete and 1.15 for steel.

9.6 PROVISION OF IS:456 FOR LSM OF DESIGN

As per IS:456-2000, all structures may be analysed by linear elastic theory to calculate internal actions produced by design loads. The aim of analysis is to determine the values of stress resultants and deformations. The effective spans to be considered for analysis of different types of structural elements or systems are specified in the standards. The member stiffness to be used may be based on the moment of inertia of the section determined on the basis of: (a) gross sections ignoring reinforcement; (b) transformed section – concrete cross-section + the area of reinforcement transformed on the basis of modular ratio; and (c) cracked sections. For all practical purposes, the gross concrete section is generally used for calculating concrete stiffness. However, for analysis with seismic loads, the cracked moment of inertia should be considered in RC structures whereby the cracked moment of inertia should be taken as 70 percent of the gross moment of inertia for columns and 35 percent of the gross moment of inertia for beams. This has been stipulated in IS:1893 (Part 1) 2016 since in seismic conditions the induced forces can be very high and under those conditions reinforced concrete sections are expected to crack and the entire cross-sections are not expected to be effective. With the availability of present-day cheap and efficient software, the simplified analysis methodology prescribed in IS:456 for frames and beams are not of much use. Reinforced concrete structures are generally monolithic in nature; critical sections for moments may be computed at the face of the supports. Regarding the critical section for shear design, the codes stipulate that when the reaction in the direction of the applied shear introduces compression into the end region of the member, sections located at a distance less than *d* from the face of the support may be designed for the same shear as that computed at distance *d*. Regarding shear design, IS:456 allows for enhanced shear strength of sections close to supports. However, designers should keep in mind that reinforced concrete sections can be made ductile only in flexure. All other modes of failure are brittle in nature. Hence the objective of the designer should be that failure should occur in flexure first, so that inhabitants get sufficient warning before actual collapse and can vacate the structure to save their lives. Therefore, the designer should prioritize the modes of failure and ensure that the sections fail in flexural mode. The current concept of shear design ensures that the sections should be very strong in shear and capacity design is to be performed for shear whereby plastic hinges are to be considered at the ends of the members based on which shear design is performed. The designer should ensure that failure will never occur in shear before flexure. IS:456-2000 permits redistribution of the calculated moments in continuous beams and frames. For this the section should be sufficiently ductile, which is ensured by imposing restrictions on the depth of neutral axis. For frames which provide the lateral stability, redistribution of moments is not very desirable and is restricted by the code, especially for multistorey buildings. For predominantly two-dimensional elements, that is floor slabs, the yield line theory may be used for analysis, alternatively, the provision given in IS:456 may also be followed.

Reinforced concrete is a combination of two materials, concrete and steel, of which concrete is highly brittle, and steel is highly ductile. Ductile failures are always preferable to brittle failures since they are gradual in nature and will provide sufficient warning before failure, which will result in a reduction in the loss of life and property. In a reinforced concrete element, the percentage of steel is significantly low with respect to the volume of the concrete. The designer should ensure that the mode of failure is governed by steel rather than concrete. The duty of the designer is not restricted to merely calculation of the area of concrete and steel required in the sections but to ensure that the failure mode will be ductile. The structural elements are inert and will act under loads as per their own natural characteristics. It is the responsibility of the designer to design and detail the sections in such a manner that the failure modes are ductile. The codes of practice deal with prescriptive methods of design based on loads. However, the designer should use his/her knowledge and foresightedness, and perform proper design and detailing so that the behaviour of the elements under the design loads and beyond can be properly interpreted. A properly designed and detailed element will move along with the loading up to failure as per the path chalked out by the designer.

IS456:2000 specifies four limit states of collapse: flexure, compression, shear and torsion. In the limit state method, design in flexure and compression are based on the assumptions provided in the code and the effect of torsion is segregated into flexure and shear. For shear design, the necessary requirements are specified in the standards. Nowadays for shear design, instead of load design, capacity design is resorted to as one must ensure that shear failure will not precede flexural failure.

As per IS:456, the limit state of serviceability is subdivided into two main states: deflection and cracking. These requirements are generally satisfied if the general provisions as mentioned in the standard are properly complied with. In special cases, calculations for deflection and crack width for RC sections need to be actually performed as per the guidelines provided in the standard.

9.7 MODE OF FAILURE – MULTISTAGE FAILURE

In the limit state method of design, failure occurs in the post-yield stage. Hence it may be stated that failure is occurring in multiple stages – the first stage being the yield stage or Stage I and the final stage is the ultimate stage or Stage II. Stage I corresponds to the stage when the material (steel or concrete) reaches the yield stage and Stage II corresponds to the condition when the failure strain in concrete is reached. For steel, the yield point corresponds to the yield stress or proof stress and for concrete, it may be considered to yield beyond the point after which there is an increase in strain without increase in stress. This corresponds to the concrete reaching a strain value of 0.002 as observed from the stress–strain diagram of concrete. Reinforced concrete sections generally fail by crushing of concrete as the strain corresponding to the snapping of steel is almost two orders of magnitude higher than the yield strain. The flexural strength of concrete is reached when the compressive strain at the outermost compressive fibre of the concrete reaches a strain value of 0.0035 as per IS:456-2000 and concrete may be considered to yield at a strain level of 0.002.

However, for pure axial compression, failure occurs at a strain value of 0.002 as per the code, which means that the yield and the ultimate stages are coincident, and failure is occurring in a single stage. However, whenever flexure is introduced into a section with compression, failure occurs in two stages. For columns where the neutral axis lies beyond the section, the failure strain in concrete in flexural compression ranges from 0.002 to 0.0035. However, for neutral axis positions within the section of the column, the failure strain in concrete is 0.0035.

Thus, it may be stated that apart from failure in pure compression only, failure of reinforced concrete sections in flexure or compression with flexure occurs in two stages: the yield stage and the ultimate stage. By carefully designing the section, a designer can increase the gap between Stage I and Stage II, and hence increase the ductility of the section. These are explained in detail in Chapters 10 and 17.

Strain distributions corresponding to a typical under-reinforced section are shown in Figure 9.1.

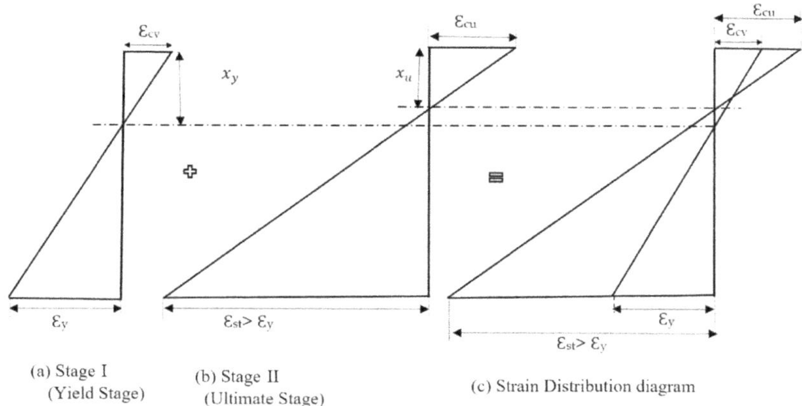

FIGURE 9.1 Strain distribution diagrams of a typical under-reinforced section showing multistage failure in limit state method: (a) stage I – yield stage; (b) stage II – ultimate stage; (c) combination of stages I and II.

9.8 SUMMARY

In this chapter an overview of the limit state method of design as per IS:456-2000 has been presented. The salient features of the method have been discussed along with the types and classifications of different limit states. The relationships between characteristic and design values of loads and material strengths have been highlighted. The significant design provisions as laid down in the code have also been included. In order to differentiate between working stress and limit state methods, the modes of failure have been explicitly dealt with highlighting the fact that multistage failure of the limit state method of design allows incorporation of ductility into the section. This is not possible in the working stress method of design where failure occurs before the materials reach their yield points and incorporation of ductility is not possible in this type of single-stage failure.

BIBLIOGRAPHY

Baker, A. L. L. (1956). *The Ultimate-load Theory Applied to the Design of Reinforced and Prestressed Concrete Frames*, Concrete Publications, London.

Bandyopadhyay, J. N. (2011). *Design of Concrete Structures*, PHI Learning, New Delhi.

Gambhir, M. L. (2006). *Fundamentals of Reinforced Concrete Design*, Prentice Hall, New Delhi.

Mallick, S. K. and Gupta, A. P. (1989). *Reinforced Concrete*, Fifth Revision, Oxford & IBH, New Delhi.

Park, R. and Paulay, T. (2018). *Reinforced Concrete Structures*, Wiley, New York.

Paulay, T. and Priestley, M. J. N. (2018). *Seismic Design of Reinforced Concrete and Masonry Buildings*, Wiley, New York.

Pillai, S. U. and Menon, D. (2012). *Reinforced Concrete Design*, Third Edition, Reprint 2012, McGraw Hill, New Delhi.

Priestley, M. J. N, Calvi, G. M. and Kowalsky. (2007). *Displacement Based Seismic Design of Structures*, IUSS Press, Italy.

Purushothaman, P. (1984). *Reinforced Concrete Structural Elements*, TMH, Torsteel Research Foundation in India, New Delhi.

Raju, N. K. (2016). *Design of Reinforced Concrete Structures, IS:456-2000*, Fourth Edition, CBS Publishers, New Delhi.

Shah, V. L. and Karve, S. R. (2016). *Limit State Theory and Design of Reinforced Concrete*, Eighth Edition, Structures Publications, Pune, India.

10 Limit State of Collapse in Flexure as per IS:456

10.1 INTRODUCTION

Any design methodology is based on some assumptions, principles and limitations. The design methodologies prescribed in the codes of practice follow the prescriptive method of design based on loads. By performing analysis, loads are evaluated which are converted into design loads for performing design as per the relevant considerations prescribed in the codes of practice. Thus, the prescriptive method of design basically aims at understanding the behaviour of the structural elements under design loads. The behaviour of a material under the action of loads is best predicted by the stress–strain relationship and the designer must appreciate the fact that the stress–strain relationship of the material forms the basis of design.

In reinforced concrete, the flexural mode of failure only can be made ductile, since all other modes of failure such as shear, compression, bond, etc., are brittle in nature. Hence flexure plays a very important role in reinforced concrete design. The limit state method of design deals with strains. The basic methodology and assumptions need to be properly understood for performing design. By flexure design one should not only aim to perform calculations for singly and doubly reinforced sections but also try to explore the philosophy and principles involved so that the design becomes an optimum one.

10.2 STRESS–STRAIN RELATIONSHIPS OF CONCRETE AND STEEL

10.2.1 Stress–strain relationships of concrete

In reinforced concrete design, the stress–strain relationships of concrete and steel play a very vital role. For design, the actual stress–strain relationships of materials obtained by laboratory testing are simplified and converted to idealized stress–strain relationships.

The stress–strain relationship of concrete is non-linear and depends on many variables such as the strength of concrete, rate of loading, types of testing machine used, moisture condition of the specimen, etc. IS:456 prescribes a parabolic-rectangular stress–strain relationship for design, although it is specified in the standards that other relationships such as rectangular, trapezoidal, parabolic or any other shape which results in prediction of the strength in substantial agreement with the test results may be used. The idealized curve is parabolic up to a strain of 0.002 and thereafter it becomes horizontal with the maximum strain value of 0.0035. Though the stress–strain curve of concrete varies significantly at different strength levels, the code prescribes a uniform curve for all grades of structural concrete up to M60. Since the strength of concrete existing in an actual structure and the strength of concrete as tested in small, companion samples in the laboratory are significantly

DOI: 10.1201/9781003415398-10

different due to a large number of reasons such as structure size effect, variability of strength within the section, etc., a factor of 0.67 is multiplied with the characteristic strength to arrive at a curve having a peck strength of $0.67f_{ck}$. Over this, the factor of safety of 1.5 is applied to arrive at the design curve (refer to Figure 10.1). This stress–strain relationship is for uniaxial compression; however it has been proved by detailed experimental studies that the stress–strain relationship of a concrete section in flexure is almost to that obtained by uniaxial compression.

If the strain at the end of the parabolic curve is taken as, $\varepsilon_{c2} = 0.002$ and the maximum strain as $\varepsilon_{cu2} = 0.0035$, the equation of the idealized stress–strain curve of concrete can be written in the form,

$$\sigma_c = f_{cd}\left[1-\left(1-\frac{\varepsilon_c}{\varepsilon_{c2}}\right)^2\right]$$

$$\text{or,}\quad \sigma_c = 0.446\times f_{ck}\times\left[2\times\left(\frac{\varepsilon_c}{0.002}\right)-\left(\frac{\varepsilon_c}{0.002}\right)^2\right]\quad \text{for } 0\le\varepsilon_c\le\varepsilon_{c2} \tag{10.1}$$

$$\text{and,}\quad \sigma_c = f_{cd}$$

$$\text{or,}\quad \sigma_c = 0.446\times f_{ck}\ \text{for } \varepsilon_{c2}\le\varepsilon_c\le\varepsilon_{cu2} \tag{10.2}$$

where, ε_c is strain in concrete,

σ_c = compressive stress in the concrete

$f_{cd} = \dfrac{\alpha f_{ck}}{\gamma_m}$,

where,

$\alpha = 0.67$,

$\gamma_m = 1.5$

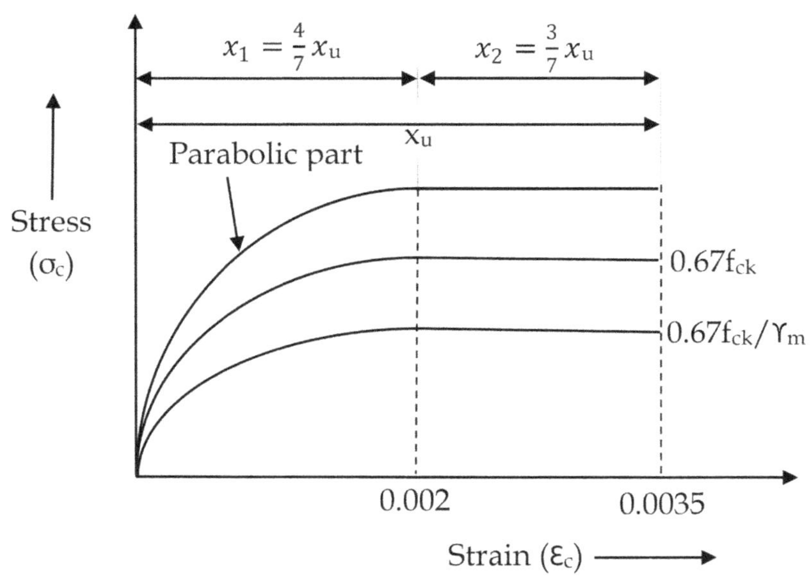

FIGURE 10.1 Design stress–strain curve for concrete as per IS:456-2000.

The shape of the stress–strain curve of concrete can be approximated to a second-degree parabola up to the maximum stress beyond which there is a falling branch in the curve which is idealized as a horizontal part. The small variation in the actual and idealized curves results in significant simplification in design calculations.

As per the limit state method of IS:456-2000, the maximum compressive strain has been considered as the failure criterion for concrete in flexure and compression. In real-life structures concrete may be subject to axial forces (compression or tension), biaxial stress and also triaxial stress states and the failure criteria will depend on the stress states. Restraint on the surrounding concrete of the loaded area, that is when the lateral expansion of the concrete is restrained, result in significant improvement in compressive strength and ductility. This restraint can be provided by the surrounding concrete by friction developed at the end of the loaded area or by confining the concrete with binding reinforcement. Though generally this effect is not considered in design, it results in significant improvement of the reserve strength of the elements.

10.2.2 Stress–strain relationships of reinforcing steel

For steel lacking a well-defined yield point, the yield stress is taken as the stress corresponding to 0.2 percent inelastic strain. This stress is termed proof stress. The stress–strain curves for steel in tension and compression are assumed to be identical, which has been proved by actual experimentation. For design it is necessary to idealize the shape of the stress–strain curve. As per IS:456-2000, the stress–strain curve for mild steel is a bilinear one and for HYSD bars it is linear up to $0.8f_y$ and horizontal beyond f_y with the intermediate portion joined by a smooth curve. This relationship is generally obtained for HYSD bars made by the CTD process. Nowadays, HYSD bars are produced by the TMT process and these steels have a bilinear stress–strain relationship (refer to Figures 10.2–10.3).

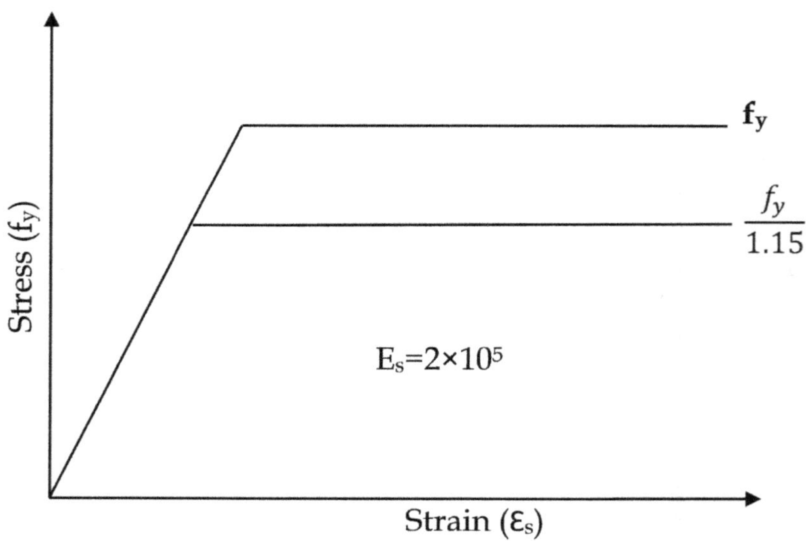

FIGURE 10.2 Stress–strain curve for steel with definite yield point (Ref. IS:456-2000).

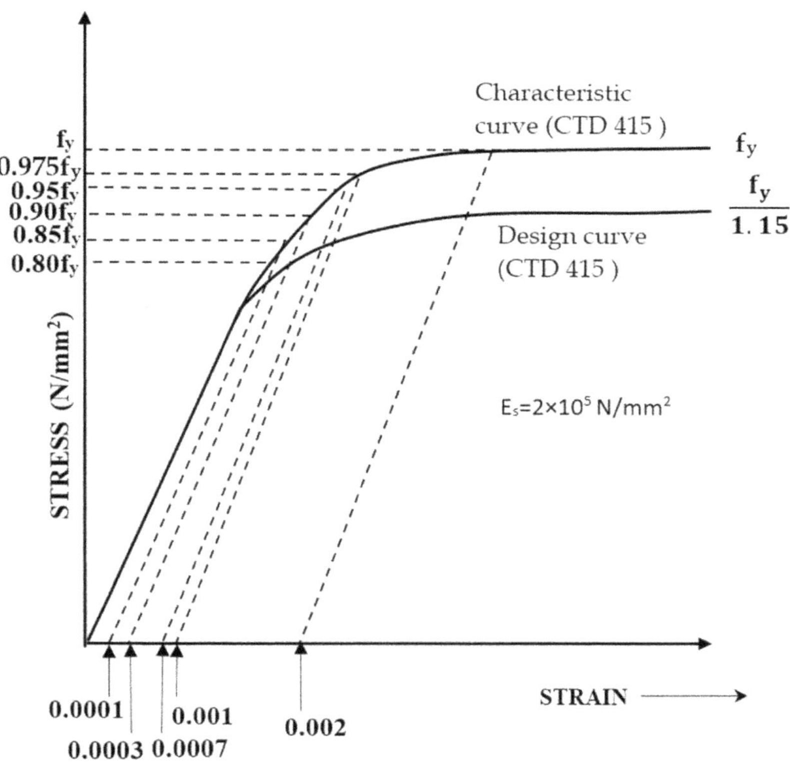

FIGURE 10.3 Stress–strain curve for steel for CTD 415.

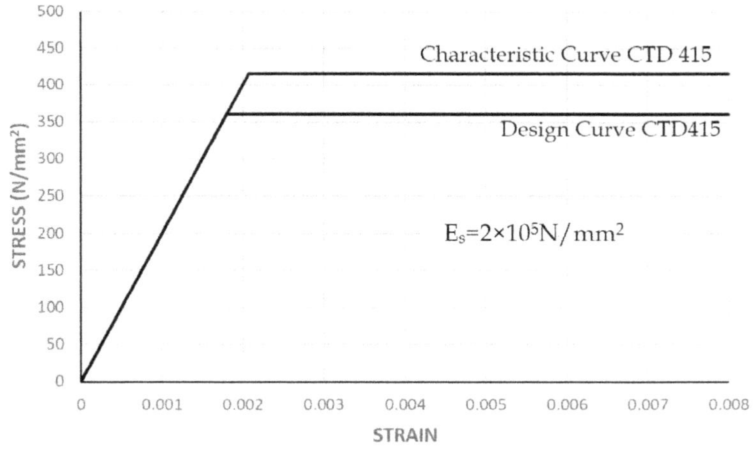

FIGURE 10.4 Stress–strain curve for steel for TMT 415.

10.3 STRAIN DISTRIBUTION DIAGRAM AND STRESS BLOCKS

In all methods of design for members in flexure and compression, the first assumption states that the plane section remains plane after bending. This assumption is the well-known Bernoulli's hypothesis which implies that the strain at any point in a section is proportional to the distance of that point from the neutral axis where the strain is zero. This will ensure that the strain distribution diagram will be a linear one, having zero value at the neutral axis and maximum compressive strain at the outermost concrete compression fibre. This will result in a triangular strain distribution across the section. Bernoulli's principle has been experimentally validated in RC members both in the compression and tension zone, provided good bonds exist between concrete and steel. Though cracks will develop in the tension zone, which may result in slip between concrete and steel, it has been found that Bernoulli's principle is applicable even in the tension zone. For deep beams, the effect of shear, being significantly higher, results in warping of the section and this principle is not valid and thus the flexure theory cannot be applied. Since the strain distribution diagram is linear, that is strains are proportional to the distance from the neutral axis, the shape of the stress–strain curve will indicate the shape of the compressive stress block across the beam cross-section. The area of the stress block is the force acting on a lamina of the cross-section. The area of the stress block multiplied by the width of the section results in the force on the compression concrete. Though strains form the failure criteria, it is the stress block which gives us force in compression concrete and moment of resistance of the section and this is considered to be a very important step in design. Knowing the stress values and areas of reinforcements, tensile forces at those section can be computed and hence the moment of resistance can be easily obtained. Thus, the stress–strain diagram of concrete forms the concrete stress block. Since as per IS:456-2000 a parabolic rectangular stress–strain diagram is specified, the stress block is a parabolic-rectangular one, however, as per IRC:112-2020 and international standards like Eurocode 2, different types of stress–strain diagrams of concrete are specified which will result in different types of concrete stress blocks. In 1930, a simplified method was proposed by Whitney whereby a rectangular compressive stress distribution paved the way for replacing the parabolic rectangular stress blocks. This is also used in many countries throughout the globe. Figure 10.5 shows the strain distribution and stress block across a beam section. Designers must understand that the flexural strength of a section is reached when extreme concrete compression fibre reaches the ultimate compressive strain $\varepsilon_{cu} = 0.0035$ and the stress block in concrete fully develops. This is valid for all types of beam sections, including under-reinforced, balanced and over-reinforced sections (however, over-reinforced sections are not permitted in flexure as per IS:456).

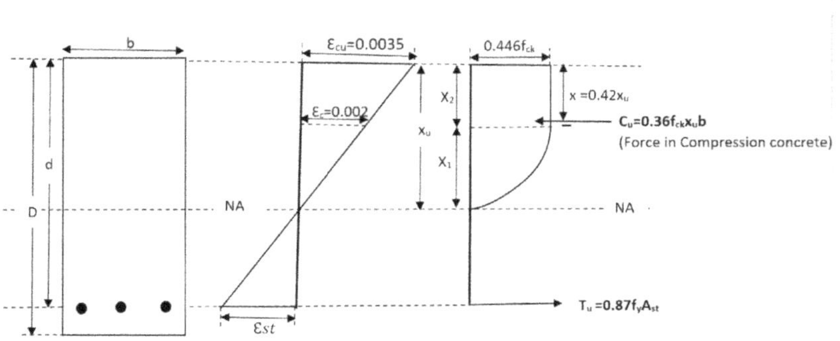

FIGURE 10.5 Strain distribution diagram and stress block of a rectangular section.

10.4 ASSUMPTIONS

As per IS:456, the design for limit state of collapse in flexure is based on the following assumptions:

a) Plane sections normal to the axis remain plane after bending. This assumption is based on the well-known Bernoulli's hypothesis. This assumption ensures that the cross-section of the member does not warp due to application of the external load. Thus, the strain at any point on the cross-section is directly proportional to its distance from the neutral axis, i.e., the strain distribution diagram is linear.
b) The maximum strain in concrete at the outermost compression fibre is taken as 0.0035 in bending. This assumption clearly defines the limiting strain of concrete in bending compression, by reaching which the concrete will be taken as having reached its limit state of collapse. It also clearly states that the failure criteria or the criteria for collapse in concrete depend on strain, not stress, which is a much more realistic and rational approach than the working stress method. IS code prescribes the maximum compressive strain in concrete as 0.0035 even though it varies with the grade of concrete and ranges between 0.003 and 0.005.
c) The relationship between the compressive stress distribution and strain in concrete may be assumed to be a rectangle, trapezoid, parabola, or any other shape which results in prediction of strength in substantial agreement with the test results. An acceptable stress–strain curve is given in Figure 10.1. For design purposes, the compressive strength of concrete in the structure shall be 0.67 times the characteristic strength. The partial safety factor $\Upsilon_m = 1.5$ shall be applied in addition to this. IS:456 provides an acceptable stress–strain curve of concrete which may be used in the design of reinforced concrete structures by the limit state method. Initially, the curve is parabolic and then linear. The stress–strain curve for concrete defines the magnitude and distribution of compressive stress at various stages of loading. As the bending moment in the section increases, the stresses and strains in concrete increase and the stress blocks develop from an initial triangular shape slowly to the complete parabolic rectangular one with limiting strain of 0.0035.
d) The tensile strength of concrete is ignored. Concrete has very little tensile strength. Under tension, steel reinforcements are assumed to resist the tensile stresses fully. The stresses in the reinforcement are derived from the representative stress–strain curve for the type of steel used. Typical curves are given in Figures 10.2–10.4. For design purposes the partial safety factor Υ_m equal to 1.15 shall be applied.

   Based on the metallurgy of steel, even a lower grade of steel has the same initial modulus of elasticity as a higher grade of steel. However, a lower grade concrete has a lower modulus. Hence, the factor of safety on the stress–strain curve is used only on the inelastic part of the curve for steel, whereas for concrete it is applied for the entire curve right from the origin.
e) The design stress in steel reinforcement is obtained from the strain at the reinforcement level using an idealized stress–strain curve for the type of reinforcement used. Normally, strain hardening is observed in all types of steel but the point at which the strain hardening begins is not stipulated in specifications for steel properly. Hence, the code ignores the strain hardening part. However, current standards provide different relationships with and without strain hardening, both of which may be used for design.
f) According to the IS code, the maximum strain in steel in tension shall not be less than $\frac{f_y}{1.15E_s} + 0.002$ at collapse.

IS code prescribes the condition that the maximum strain in steel in tension at the limit state of flexural collapse shall not be less than $\frac{f_y}{1.15E_s} + 0.002$, which is the yield stress for CTD bars. This condition ensures that the flexural failure must be initiated by yielding of steel in tension. The strain in steel, therefore, reaches at least the yield value at failure.

In the above-mentioned assumption, the term maximum is used with strain. Since the strain distribution diagram is linear, this term signifies that if more than one layer of tension reinforcement is provided in the section the outermost line should yield whereby the strain in the inner layer will be less than the yield value. The strain value mentioned in the assumption corresponds to the strain at 0.2 percent proof stress for HYSD bars made using the CTD process, which do not have a definite yield point. However, this assumption is valid for all types of steel and even for mild steel which has a definite yield point. Thus, the code is conservative in nature and imposes upon the designer that even for mild steel this expression is to be used.

This assumption ensures ductile failure, whereby the reinforcement must yield at failure. This condition is valid for balanced and under-reinforced sections only and invalid for over-reinforced sections. Hence this assumption ensures that designers can go for only balanced and under-reinforced sections. Over-reinforced sections for which failure is brittle and failure is initiated in concrete are not permitted in flexure design of the limit state method. Although over-reinforced sections are prohibited in flexure, the Indian code of practice does not emphasize that the under-reinforced sections are more desirable than balanced ones. This type of restriction directly prohibits the use of over-reinforced section, which are nowadays not available in the international standards. Instead, balanced section design is not always permitted, and designers generally opt for under-reinforced sections. Adequate ductility must be built into the section and a hierarchy must be formed among the different failure modes by virtue of which ductile failure should always occur before brittle failure. Especially for earthquake-resistant design, the maximum area of tension steel is generally restricted and incorporation of compression steel is made mandatory in flexural elements so that the neutral axis depth of the section is limited. In some standards this depth of the neutral axis is limited to 75 percent of the depth under balanced condition. Thus, designers should be encouraged to go for under-reinforced section design only. The level of under-reinforcement will govern the ductility of the section and its behaviour up to and beyond design loads. However, in limit state of collapse in compression, IS:456 allows all types of sections: balanced, under-reinforced and over-reinforced. Otherwise, the domain of design will become highly restricted as is observed from the P–M interaction charts used for the design of sections subjected to flexure with compression. This has been demonstrated in Chapter 11.

10.5 FAILURE MODES IN BALANCED, UNDER-REINFORCED AND OVER-REINFORCED SECTIONS – YIELD AND ULTIMATE STAGES (TWO-STAGE FAILURE)

As per IS:456, failure in the limit state of collapse occurs in two stages: Stage I, i.e., the yielding stage, and Stage II, the ultimate or failure stage. This is generally true for limit state of collapse in flexure and compression except under pure compression when failure occurs in a single stage.

10.5.1 Modes of Failure

The behaviour of the flexural members is governed by the amount of reinforcement provided in the section. Based on the percentage of steel, three failure modes occur, namely balanced failure, under-reinforced failure and over-reinforced failure. The corresponding sections are defined as balanced, under-reinforced and over-reinforced sections (refer to Figure 10.6).

Failure in the limit state method occurs in two stages as described below:

Stage I – Initiation of failure, yielding of the section – yielding of steel or concrete.
Stage II – Actual failure – reaching the limiting or maximum strain in concrete, i.e., 0.0035.

Curvature ductility is defined as the ratio of ultimate curvature to yield curvature.

The greater is the gap between these two stages the greater is the ductility of the section and the better is the performance of the section, although plastic rotation is also higher and the section has to rotate by this large amount to mobilize this ductility. This has been discussed in detail in Chapter 17.

This two-stage failure is a significant advancement over the failure mode in the working stress method whereby failure occurs in one stage, i.e., when the permissible stress in concrete (over-reinforced), or steel (under-reinforced) or both (balanced) is reached. While performing design, designers should not merely calculate the concrete and steel areas but should ensure that sufficient ductility has been incorporated into the section so that the behaviour of the section will help in the preservation of life and property. A design is rated as good or bad mostly based on economy, but the efficacy of the design should be investigated with respect to these two failure stages. This is discussed in detail in the subsequent section.

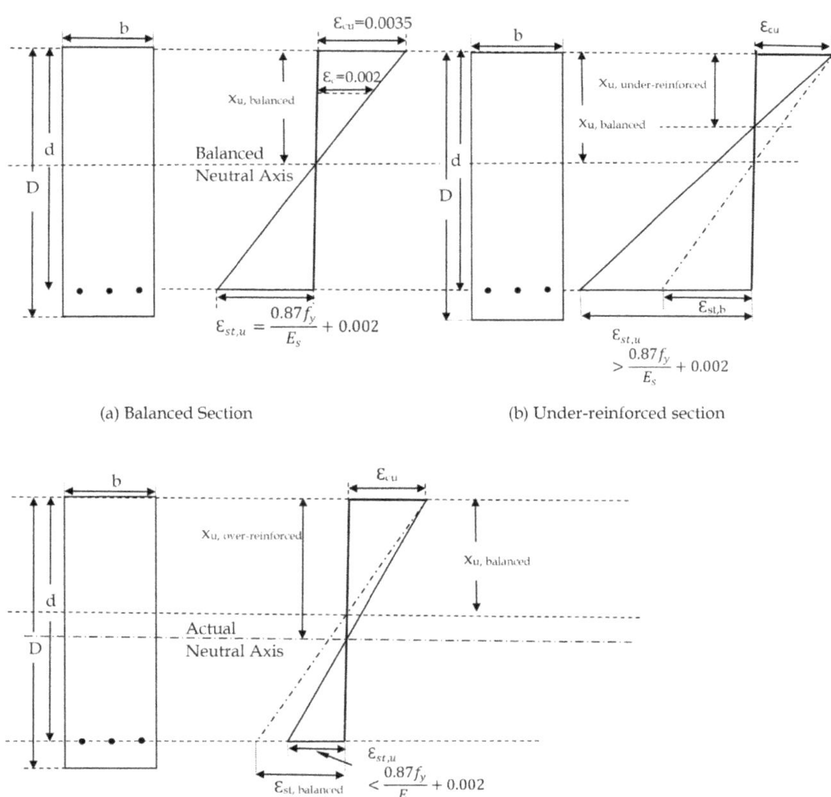

FIGURE 10.6 Failure of different types of sections in flexure: (a) balanced; (b) under-reinforced; and (c) over-reinforced.

10.5.1.1 Balanced section or critical section

When the tensile reinforcement is such that the strains in steel and concrete reach their limiting values simultaneously, the section is defined as a balanced or critical section. Thus, in a balanced section, strain in steel, $\varepsilon_{su} = \frac{f_y}{1.15E_s} + 0.002$ and strain in concrete, $\varepsilon_c = \varepsilon_{cu} = 0.0035$ are reached simultaneously.

The two stages of failure in balanced section are described below.

Stage I – At yield stage – concrete reaches a strain of 0.002 beyond which there is no increase in stress with an increase in strain, as concrete yields beyond a strain of 0.002. For this stage one can consider that concrete is yielding, and steel has not reached the yield point.

Stage II – At failure stage – concrete reaches a strain of 0.0035 and steel reaches the limiting strain i.e., yield strain. At this stage the flexural strength of the section is reached, failure occurs by crushing of concrete and full stress block of concrete develops. The interval between Stages I and II, which occurs due to an increase in strain of concrete from 0.002 to 0.0035 (as if due to yielding of concrete), is quite small. This yielding is quite insignificant, hence balanced sections are not desirable. Balanced sections also possess ductility, although the values are very small. It is observed from Figure 10.7 that the depth of the neutral axis is reduced from Stage I to Stage II. At any stage the force equilibrium condition needs to be satisfied. As the strain increases from Stage I to Stage II, the stress block in concrete slowly and finally takes the final parabolic rectangular shape. The balanced section stress in steel in Stage I is less than the yield stress and the stress block in concrete is only a parabolic one (up to a strain value of 0.002). Even though forces in steel and concrete increase in Stage II, the depth of the neutral axis gets reduced. This has been demonstrated in Chapter 17. For an under-reinforced section, force in steel reaches the yield value in Stage I and hence the tension force reaches the maximum value and remains constant even at Stage II. As the force in concrete is also constant at Stage I and Stage II, the shape changes from a small parabolic portion of the curve to a complete parabolic rectangular one, and the depth of the neutral axis get reduced from Stage I to Stage II.

10.5.1.2 For under-reinforced sections

Stage I – Initiation of failure, steel yields at this condition and strain in steel at the outermost layer is equal to $f_y/1.15E_s + 0.002$.

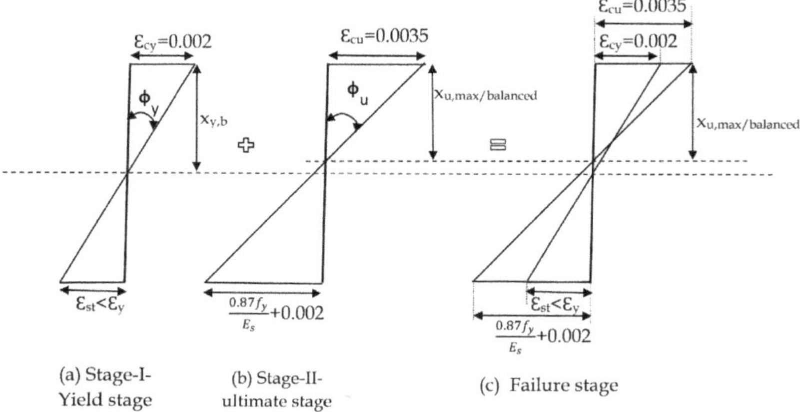

FIGURE 10.7 Strain distribution diagram for different failure stages of a balanced section: (a) Stage I – yield (b) Stage II – ultimate; and (c) – Combination of Stage-I and Stage-II.

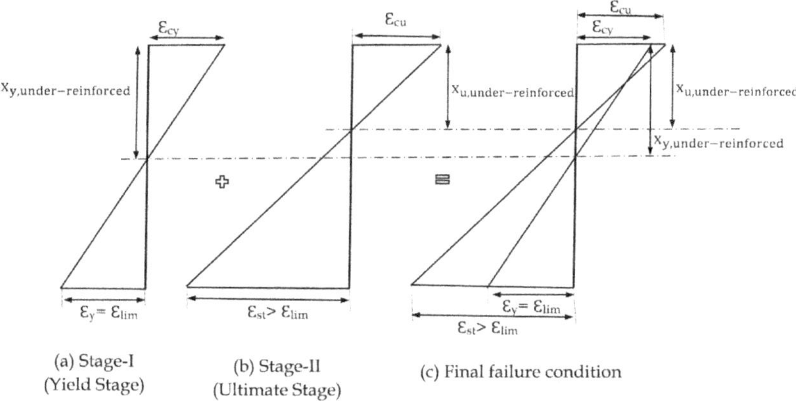

FIGURE 10.8 Strain distribution diagrams for different failure stages of an under-reinforced section.

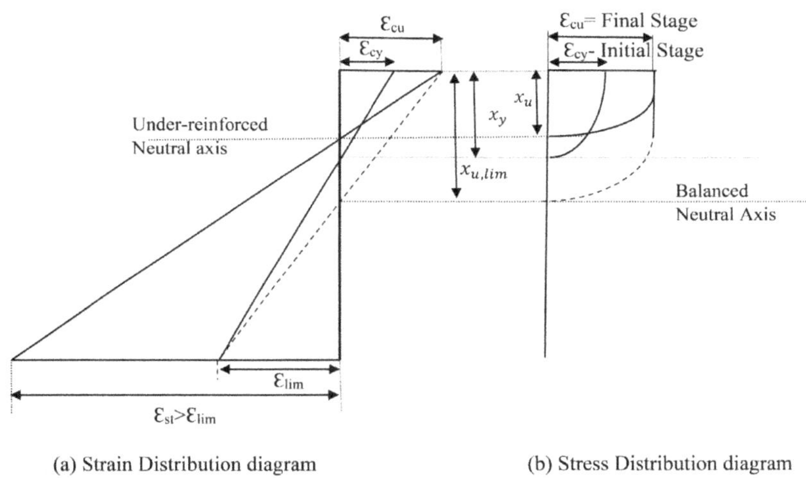

FIGURE 10.9 Development of stress blocks in concrete in an under-reinforced section.

The neutral axis moves up for an under-reinforced section with respect to a balanced section. The line joining the limiting steel strain with the neutral axis position leads to a concrete strain of less than 0.0035.

Stage II – This is the ultimate stage where the so-called failure occurs. To reach this stage steel goes on yielding until the strain in concrete reaches 0.0035 and the outermost fibre of concrete fails in compression, hence final failure occurs by crushing of concrete.

The full stress block in concrete develops. Hence, for under-reinforced sections the failure mode is ductile in nature as a significant increase in inelastic strain of steel occurs.

With an increase in the level of under-reinforcement, the yielding of steel will increase. Figures 10.8 and 10.9 show the strain distribution diagrams and stress blocks for different stages of failure of an under-reinforced section.

Since over-reinforced sections are not permitted as per IS:456-2000, failure modes under this condition have not been demonstrated.

10.6 SHORTCOMINGS/LIMITATIONS OF THE ASSUMPTIONS AS PER INDIAN STANDARDS

As per IS:456-2000, assumptions for performing design in the limit state of collapse in flexure and compression are provided. Assumptions and design methodology of the limit state of collapse in compression are presented in Chapter 11. The assumptions for flexure, although presented in a detailed manner, have some limitations. They do not properly highlight the failure condition, that the limiting strain in concrete in flexural compression, $\varepsilon_{cu} = 0.0035$, has to be reached at failure otherwise the full stress block in concrete cannot develop. The code prohibits the use of over-reinforced sections by compulsorily initiating failure in steel. However, this assumption is a bit confusing as it reads that the maximum strain in tension reinforcement in the section at failure shall not be less than $\dfrac{f_y}{1.15E_s} + 0.002$. If the reinforcement on the tension side of flexural members is placed in multiple layers it only signifies that the outermost layer in steel will yield and reach the maximum stress, whereas the inner layers will not reach the yield stage. The minimum limiting strain in steel is specified, however the maximum permissible strain in steel has not been restricted, keeping the strain values open ended. Snapping of steel occurs at a strain value about two orders of magnitude higher than the yield strain, as a result of which reinforced concrete sections fail due to crushing of concrete but not due to rupture of steel. Minimum tension reinforcement requirements in beams as per IS:456 call for very high elongations of steel. To accommodate these deformations in steel, the plastic rotations in the robust RC sections may result in exceeding the limit states and can cause actual collapse, which has been discussed in detail in Chapter 17.

The code prohibits over-reinforced sections but as per present international standards even balanced sections are not considered desirable and under-reinforced sections are always preferred from the enhanced ductility point of view.

The stress–strain diagram of steel provided in IS:456-2000 for HYSD bars conforms to the requirements of the CTD process which does not have a definite yield point, because of which the concept of proof stress has been introduced. However, nowadays, HYSD reinforcements are made by the TMT process whereby the steels have definite yield points with a bilinear stress–strain diagram. As per IS:456-2000, incorporating the inelastic strain of 0.002 in the design for both mild steel and HYSD steel results in a significantly conservative design and does not predict the true behaviour of the material. As per the assumptions of the limit state of collapse in compression, all types of sections including under-reinforced, balanced, and over-reinforced, are permitted. This seems to be quite logical as permitting only under-reinforced and balanced sections would have restricted the domain of design significantly as is observed in the interaction charts (provided in Chapter 11). However, with an increase in axial compression the failure modes of the columns will move towards a more brittle nature, significantly reducing the ductility values. Some restrictions on maximum axial stress that can act on a section could have been imposed in the code.

IS:456-2000 deals with strength and serviceability design only. However, for a country like India where seismic forces mostly dominate the design, some guidelines to incorporate ductility in design would have been helpful. In an add-on code, IS:13920-2016 entitled "Ductile Design and Detailing of Reinforced Concrete Structures Subjected to Seismic Forces", some guidelines for incorporating ductility in RC sections have been provided.

10.7 ASSUMPTIONS OF LSM OF DESIGN AS PER INTERNATIONAL GUIDELINES

In Eurocode 2, the limit states philosophy is covered in two basic groups: (a) ultimate limit states (ULS) and (b) serviceability limit states (SLS). IRC:112-2020 – "Code of Practice for Concrete Road Bridges", also prescribes the same two limit states for design. The design of sections as per

ULS of Eurocode 2 deals with bending with or without axial forces. IRC:112-2020 covers the design principles under ultimate limit state of linear elements for bending and axial forces.

As per international standards and specialist literature, in the design of sections, the assumptions made in assessing the flexural strength of concrete elements such as Bernoulli's principle, perfect bond between concrete and steel, ignoring tensile strength of concrete, stress–strain relationship of concrete, etc., are in line with the assumptions provided in IS:456-2000. Hence these are not repeated here. IS:456 deals with flexure and compression separately but the current standards deal with the design of sections simultaneously subjected to flexure and/or axial forces. Stress–strain relationships form the basis of design. Although the stress–strain relationship of concrete as per IS:456-2000 is in line with the present national and international standards, the relationship for HYSD steel for CTD bars given in the code seems to be outdated. Two types of relationships for steel have been prescribed in the Eurocode. If the idealized bilinear stress–strain relationship is considered, i.e., bilinear with inclined plastic branch, then for the design of a section subjected to bending with or without axial forces, the possible range of strain distributions is shown in Figure 10.10.

From Figure 10.10 it is observed that strain is considered as the criterion for failure, and three failure conditions are specified in the form of A, B and C. Condition C denotes the concrete pure compression strain limit. Point B denotes the maximum compressive strain in concrete, i.e., when the section is failing in flexural compression. Condition A is the reinforcing steel tension strain limit which is specified for the idealized bilinear stress–strain diagram of steel, and this condition refers to failure of the section due to steel attaining the limiting strain. However, this condition is occurring along with point B, i.e., steel strain limit and concrete strain limit are being attained simultaneously. No further possible distribution of strain beyond this line on the tension side has been provided in the standard. This may be attributed to the fact that with concrete, being a material strong in compression but very weak in tension, it is not desirable to have increased tensile stresses on the section. When the horizontal plastic branch of the stress–strain diagram of steel is considered for design, there is no maximum strain limit, and the Figure 10.10 will not have the failure condition specified by point A. The failure conditions will be specified by points C and B, ensuring that failure occurs only because of crushing of concrete. From the above-mentioned discussions, it is clear that when the section is subjected to flexure with or without axial forces, failure will occur when the limiting strain in concrete reaches the maximum code-specified value and the concrete stress block will develop fully as per the entire stress–strain relationship of concrete provided in the standard. IRC:112-2020 provides a strain distribution diagram quite similar to that of Eurocode 2, along with some additional distribution of strains dealing with higher tensile stresses in the section.

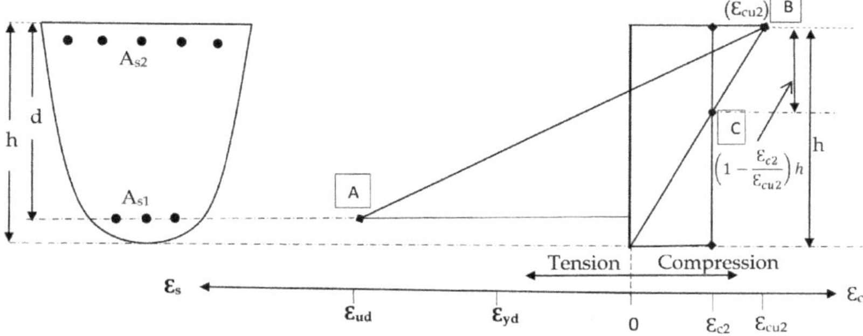

FIGURE 10.10 Possible strain distributions in the ultimate limit state: (A) reinforcing steel tension strain limit; (B) concrete compression strain limit; and (C) concrete pure compression strain limit.

10.8 SINGLY REINFORCED SECTIONS

A flexural member is termed a singly reinforced one, when reinforcements are placed only on the tensile side of the cross-section. Designers generally prefer singly reinforced sections unless there are restrictions on the depth of the beams. While designing by the limit state method, development of the strain distribution diagram is of paramount importance. Once the strain distribution diagram is obtained, the concrete stress block in compression can be drawn and the stresses in reinforcement can be calculated. The forces in concrete and steel and their point of application can be determined. The moment of resistance of the section can be calculated, thereby completing the process of flexural design.

10.9 COMPUTATION OF DESIGN PARAMETERS FOR FLEXURAL MEMBERS

The limit state method deals with strains and considers strain as the criterion for failure. Hence, the development of a strain distribution diagram across a cross-section can be considered as a first step in design. If the strains at different points on the cross-section can be determined, the corresponding stresses can be calculated for both concrete and steel from the stress–strain diagram of the respective materials. From Bernoulli's principle it can be specified that the strain distribution diagram will be linear. Hence at least strain values at two points across a cross-section must be known to draw the strain distribution diagram. Through two points only one straight line can pass. Hence the line joining these two points will represent the strain distribution diagram across the cross-section from which the position of the neutral axis that is the point where the strain is zero can easily be determined. For all types of sections failure occurs by crushing of concrete, due to concrete reaching the limiting strain 0.0035 for flexural compression. Therefore, for all types of section this point is constant in the strain distribution diagram and another point on the cross-section is to be determined where the strain value is known. For a balanced section the limiting strains in concrete and steel are reached simultaneously, hence the second point on the strain distribution diagram is the strain corresponding to the limiting/yield strain in steel. Thus, the strain distribution diagram can be drawn and the depth of the neutral axis denoting the point where the strain in cross-section is zero, can be determined. For under-reinforced sections, from force equilibrium condition and considering the amount of reinforcement provided in the section, the depth of the neutral axis can be determined. Joining the point corresponding to the neutral axis depth with the limiting strain in concrete, the strain distribution diagram can be plotted. For an over-reinforced section, the neutral axis depth can be calculated in the same manner and the strain distribution diagram can be determined. However, since the strain in steel does not reach the yield value, the steel stress is unknown and needs to be calculated by successive iteration.

As per IS:456-2000, flexure design is limited to two terminal conditions: (a) balanced condition and (b) maximum under-reinforced condition with minimum percentage of tension reinforcement permitted by the standard. Any section designed in flexure as per IS:456 will lie within these two terminal conditions. The different design parameters corresponding to these two limiting conditions are specified in the subsequent sections.

10.9.1 Balanced section

Maximum stress in concrete $= \dfrac{0.67 f_{ck}}{1.5} = 0.446 f_{ck}$

Let x_u = the depth of the neutral axis

As shown in Figure 10.11, $x_u = x_1 + x_2$

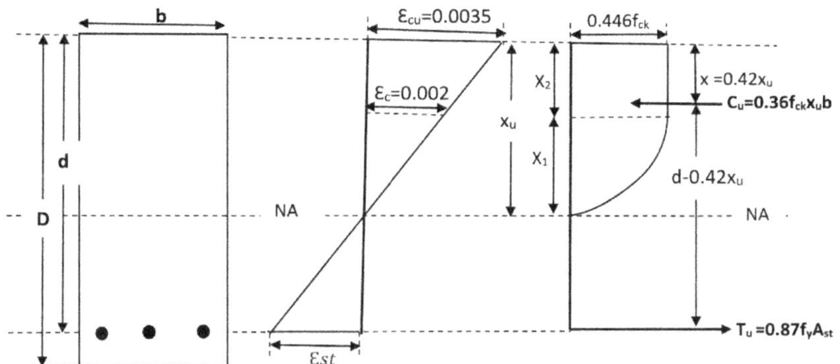

FIGURE 10.11 Strain distribution diagram and stress blocks for a balanced section.

From similar triangles,

$$\frac{x_1}{0.002} = \frac{x_u}{0.0035} \tag{10.3}$$

$$or,\ x_1 = \frac{0.002}{0.0035} x_u = \frac{4}{7} x_u$$

$$\therefore x_2 = \frac{3}{7} x_u$$

Let C_u = compressive force. Thus, for a single lamina of the beam cross-section the compressive force is given by:

$$C_u = 0.446 f_{ck} \times \frac{3}{7} x_u + \frac{2}{3} \times 0.446 f_{ck} \times \frac{4}{7} x_u = 0.36 f_{ck} x_u \tag{10.4}$$

Thus, for a section having cross-sectional width b, the compressive force will be:

$$C_u = 0.36 f_{ck} x_u b$$

Let, A_{st} = area of tensile reinforcement provided. Stress in steel at yield = $0.87 f_y$. Thus, the tensile force, T_u is given by:

$$T_u = 0.87 f_y A_{st}$$

The line of action of C_u is at the centroid of the stress block located at a distance $\bar{x}$ from the outermost concrete fibre subjected to maximum compressive strain.

Considering moments of the compressive forces about the outermost concrete compressive fibre, where C_1, C_2 are the forces acting on the parabolic and rectangular part of the compressive stress block, respectively, it can be stated that,

$$0.36 f_{ck} x_u \times \bar{x} = 0.446 f_{ck} \times \frac{3}{7} x_u \times \frac{3}{14} x_u + \frac{2}{3} \times 0.446 f_{ck} \times \frac{4}{7} x_u \times \left(x_u - \frac{5}{8} \times \frac{4}{7} x_u \right)$$

$$\therefore \bar{x} = 0.416 x_u$$

Let z denote the lever arm of the couple formed by the two equal and opposite forces C_u and T_u. Thus, the lever arm is given by:

$$z = d - 0.416 x_u$$

For any section, force equilibrium across the section is to be satisfied. Hence,

$$C_u = T_u$$

$$C_u = 0.36 f_{ck} x_u b$$

$$T_u = 0.87 f_y A_{st}$$

Therefore, the actual depth of the neutral axis,

$$x_u = \frac{0.87 f_y A_{st}}{0.36 f_{ck} b} \tag{10.5}$$

where, x_u is the depth of the neutral axis. This equation is valid for under-reinforced and balanced conditions.

10.9.1.1 Limiting values of neutral axis depth

From Figure 10.11,

$$\frac{x_u}{0.0035} = \frac{d}{0.0035 + \varepsilon_{st}} \tag{10.6}$$

$$\text{or, } x_u = \frac{0.0035}{0.0035 + \varepsilon_{st}} \times \text{d}$$

For balanced condition, the depth of neutral axis can be denoted as $x_{u,\max}$

$$x_{u,max} = \frac{0.0035}{0.0035 + \frac{0.87 f_y}{E_s} + 0.002} \times \text{d}$$

$$\text{or, } x_{u,max} = \frac{0.0035}{0.0055 + \frac{0.87 f_y}{E_s}} \times \text{d}$$

TABLE 10.1
Limiting strain and limiting depth of neutral axis

Grade of steel	E_s (N/mm²) (modulus of elasticity)	f_y (N/mm²) (yield stress)	$\varepsilon_{st,min} = \frac{0.87f_y}{E_s} + 0.002$	$\frac{x_{umax}}{d}$
Fe 415	2×10^5	415	0.0038	0.4791
Fe 500	2×10^5	500	0.004175	0.4560
Fe 550	2×10^5	550	0.004392	0.4435
Fe 600	2×10^5	600	0.00461	0.4316

TABLE 10.2
p_{tlim} for different grades of steel and concrete

	p_{tlim} for different grades of steel and concrete			
Grade of steel	**Fe 415**	**Fe 500**	**Fe 550**	**Fe 600**
M20	0.95543	0.7548	0.66735	0.59532
M25	1.19429	0.9434	0.83418	0.74415
M30	1.43314	1.1322	1.00102	0.89298
M35	1.672	1.3208	1.16786	1.04181
M40	1.91086	1.5095	1.33469	1.19064
M45	2.14972	1.6982	1.50153	1.33947
M50	2.38857	1.8869	1.66837	1.4883
M55	2.62743	2.0756	1.8352	1.63713
M60	2.86629	2.2643	2.00204	1.78596

or, $\frac{x_{u,max}}{d} = \frac{0.0035}{0.0055 + \frac{0.87f_y}{E_s}}$ = limiting or balanced depth of neutral axis (10.7)

The values of limiting strain and limiting depth of neutral axis for different grades of steel are provided in Table 10.1. These values are independent of the grade of concrete.

10.9.1.2 Balanced percentage of tension reinforcement (Table 10.2)

$$x_u = \frac{0.87 f_y A_{st}}{0.36 f_{ck} b}$$

$$\frac{x_{u,max}}{d} = \frac{0.87 f_y}{0.36 f_{ck}} \times \frac{p_{t,lim}}{100}$$

$$\text{or, } p_{t,lim} = 41.38 \times \left(\frac{f_{ck}}{f_y}\right)\left(\frac{x_{u,max}}{d}\right) \tag{10.8}$$

TABLE 10.3
Limiting moment of resistance for different grades of steel

Grade of steel	$\frac{x_{u,max}}{d}$	$R=\frac{M_{u,lim}}{f_{ck}bd^2}$
Fe 415	0.4791	0.13777
Fe 500	0.4560	0.13272
Fe 550	0.4435	0.12992
Fe 600	0.4316	0.12721

10.9.1.3 Computation of limiting moment of resistance

The limiting moment of resistance of the section is shown in Table 10.3.

$$M_{u,lim} = 0.36 f_{ck} x_{u,max} b \times (d - 0.42 x_{u,max}) \tag{10. 9}$$

$$\frac{M_{u,lim}}{f_{ck} bd^2} = 0.36 \times \frac{x_{u,max}}{d} \times \left(1 - 0.42 \frac{x_{u,max}}{d}\right) = \text{R} \tag{10.10}$$

where, R = limiting moment of resistance factor

The limiting moment of resistance can be calculated from the steel side also, whereby the equation will be in the following form:

$$M_{u,lim} = 0.87 f_y A_{st} \times (d - 0.42 x_{u,max}) \tag{10.11}$$

10.9.2 Maximum under-reinforced section

The minimum area of tension reinforcement is provided in a maximum under-reinforced section. Figure 10.12 shows the strain distribution across a maximum under-reinforced section.

As per IS:456, the minimum area of tension reinforcement is given by $A_{st,min} = \frac{0.85}{f_y} bd$

$$p_{t,min} = \frac{100}{bd}\left(\frac{0.85}{f_y} bd\right) = \frac{85}{f_y}$$

The minimum reinforcement percentages for different grades of steel as per IS:456-2000 are given in Table 10.4.

From force equation

$$0.87\text{f}_y\left(\frac{0.85}{f_y} bd\right) = 0.36 f_{ck} x_u b$$

$$\frac{x_u}{d} = \frac{(0.87 \times 0.85)}{0.36 f_{ck}}$$

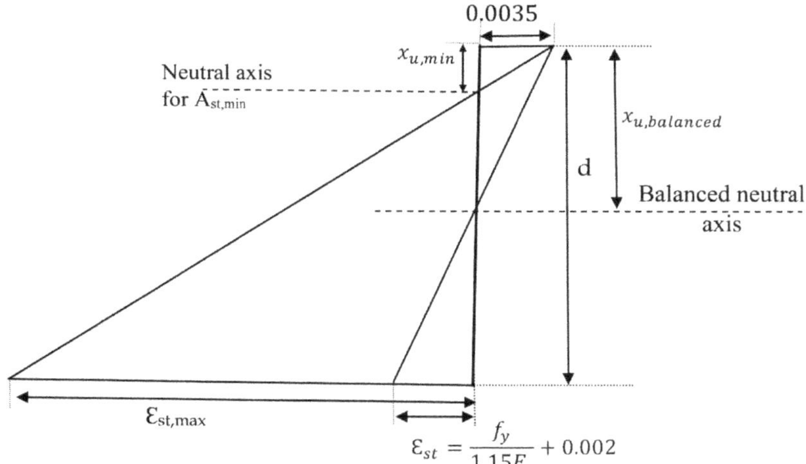

FIGURE 10.12 Strain distribution diagram for a maximum under-reinforced section showing minimum depth of neutral axis and maximum strain in steel.

TABLE 10.4
Minimum reinforcement (%) as per IS:456-2000

	$p_{t,min} = 100A_{st}/bd$			
	Grade of steel			
Grade of concrete	**Fe 415**	**Fe 500**	**Fe 550**	**Fe 600**
M20 to M60	0.2048	0.1700	0.1545	0.1416

$$\text{or, } \frac{x_u}{d} = \frac{2.054}{f_{ck}} = \left(\frac{x_u}{d}\right)_{min}$$

From the strain distribution diagram in Figure 10.12

$$\frac{0.0035}{x_{u,min}} = \frac{\varepsilon_{st,max}}{d - x_{u,min}}$$

$$\frac{d}{x_{u,min}} - 1 = \frac{\varepsilon_{st,max}}{0.0035}$$

$$\frac{f_{ck}}{2.054} = \frac{\varepsilon_{st,max}}{0.0035} + 1$$

$$\varepsilon_{st,max} = \left(\frac{f_{ck}}{2.054} - 1\right) \times 0.0035 \tag{10.12}$$

TABLE 10.5
Minimum depth of the neutral axis and maximum strain in steel

	Minimum tension reinforcement provided as per IS:456-2000	
Grade of concrete	Minimum neutral axis depth $(x_{u,min})\frac{2.054}{f_{ck}}$	Maximum strain in steel $(\epsilon_{st,max})$
M20	0.1027d	0.0305
M25	0.08216d	0.0391
M30	0.06846d	0.0476
M35	0.05868d	0.0561
M40	0.05135d	0.0647
M45	0.0456d	0.0732
M50	0.04108d	0.0817
M55	0.0373d	0.0902
M60	0.0342d	0.0987

The minimum depth of the neutral axis and maximum strain in steel are given in Table 10.5.

10.10 SINGLY REINFORCED SECTIONS – AN IDEALIZED OR PURELY THEORETICAL CONDITION

Flexural members or beams are primarily subjected to bending moments and in real-life situations these moments are never constant along the length of the beam. Hence shear forces come into play. Longitudinal reinforcements are provided to resist the bending moment and transverse reinforcements resist shear forces. Thus, to carry the transverse reinforcement or stirrups, minimum reinforcements are required on the compression side as hanger bars. Moreover, the codes of practice of different countries prescribe a minimum amount of reinforcement on the compression side of the section due to the reversibility of moments which are regular features for lateral loads such as wind and earthquakes. Hence, all beams are doubly reinforced, and singly reinforced beams are purely theoretical in nature. It is common practice to ignore the reinforcements provided on the compression side which is mandatorily to be provided as per codal requirements. Nowadays, to comply with the requirements of seismic-resistant design, a certain percentage of the maximum support reinforcements are provided on both the faces of beams all along their lengths. Hence, the concept of singly reinforced beams seems to be a hypothetical or idealized one.

10.11 DOUBLY REINFORCED SECTIONS

Doubly reinforced sections are those in which, in addition to tension reinforcement, compression reinforcements are also provided. Since a certain percentage of supports or span reinforcements are to be extended all through the section length, as per requirements of the standards, all sections in flexure have both tension and compression reinforcements. However, conventionally designers opt for a doubly reinforced section only when the external moment on the section exceeds the balanced moment of resistance capacity of the section.

10.12 INTRODUCTION AND NEED FOR DOUBLY REINFORCED SECTIONS

The design of doubly reinforced sections is similar to those of singly reinforced ones, the only difference being the presence of reinforcement on the compression side. Forces in compression steel are calculated considering the strain compatibility, that is strain in steel is equal to strain in the surrounding concrete. Thus, in the force equilibrium equation, the force in the tension steel should be equal to the sum of the forces in concrete and compression steel, and the moment of resistance of the section needs to be calculated accordingly. A designer generally resorts to a doubly reinforced section when the limiting/critical/balanced moment of resistance of the section as a singly reinforced one with the specified grades of concrete and steel are exceeded. Under such conditions, since over-reinforced sections are not permitted by the code, the designer has to opt for a doubly reinforced section.

10.13 ASSUMPTIONS

The stress–strain relationships for steel in tension and compression are similar. Hence, compression steel also yields at $0.87f_y$. Due to strain compatibility, strain in compression steel will be similar to the strain in the surrounding concrete. From the strain distribution diagram across the cross-section the strain in compression steel can be determined and then using the stress–strain relation the corresponding stress in the compression reinforcement can be calculated.

For a doubly reinforced section, $M_{external} = (M_{u,lim} + M_{u2}) = (R_{balanced} \times bd^2) + [(C_c \times \text{corresponding lever arm})$ or $(f_{st} \times A_{st2} \times \text{corresponding lever arm})]$.

where,

$M_{external}$ = External moment acting at the section
$M_{u,lim}$ = Limiting/balanced moment of resistance of the section
$M_{u,2}$ = Additional moment of resistance required beyond the balanced value
$= M_{external} - M_{u,lim}$
$R_{balanced}$ = Limiting moment of resistance factor
C_c = Force in compression steel = $A_{sc}\,(f_{sc} - f_{cc})$,
f_{sc} = Stress in compression steel
f_{cc} = Stress in concrete at the level of compression steel
A_{st2} = Additional tension reinforcement required to resist C_c
f_{st} = Stress in tension steel which is equal to yield stress
$A_{st,balanced}$ = Steel required for balanced section = $A_{st,lim}$
A_{sc} = Area of compression steel

Since the section is balanced, the strain distribution diagram and depth of the neutral axis are known. Hence, the strain in concrete at the level of compression steel can easily be calculated using strain compatibility. Using the stress–strain diagram of steel, stress in compression steel can be calculated. As the tension steel yields, stress will be equal to $0.87f_y$. Hence, the amount of compression steel and additional tension steel required can be determined based on $M_{u,2}$, which is the additional moment of resistance of the section beyond its balanced singly reinforced value.

10.14 ARE ALL DOUBLY REINFORCED SECTIONS BALANCED IN NATURE?

As discussed in the previous section, reinforcements on the compression side are to be routinely provided in beams. These compression reinforcements are generally not considered in strength design and are totally ignored in the calculations, as they contribute to the additional reserve strength of the section only. However, as a beam is inert, it cannot understand that some of its resources have not been considered by the designer. Under loads it will behave as per its own natural characteristics, whereby the reinforcements available in the section are expected to act with the concrete to resist external actions. As per EN 1998-1, Eurocode 8: Design of structures for earthquake resistance, Part 1, it is specified that at the compression zone reinforcement of not less than half of the reinforcement provided in the tension zone is to be placed in addition to any compression reinforcement needed. Similar requirements are found in IS:13920-2016. As per the standard, longitudinal steel on bottom face of a beam framing into a column (at the face of the column) shall be at least half

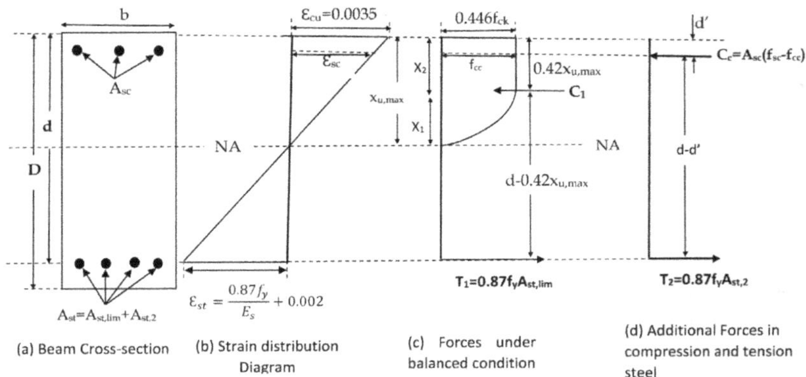

FIGURE 10.13 Balanced doubly reinforced beam: (a) beam cross-section; (b) strain distribution diagram; (c) forces under balanced condition; and (d) additional forces in compression and tension steel.

the steel on its top face at the same section. Longitudinal steel in beams at any section on the top or bottom face shall be at least one-quarter of longitudinal steel provided at the top face of the beam at the face of the column; when the top longitudinal steel in the beam at the two supporting column faces is different, the larger of the two shall be considered. Hence, with the provision of additional compression reinforcement, a balanced doubly reinforced section will be converted into an under-reinforced one. Therefore, the notion that all doubly reinforced sections are balanced does not seem to be justified.

10.15 BASIC PRINCIPLES OF UNDER-REINFORCED DOUBLY REINFORCED SECTIONS

The moment of resistance, M_u, of a doubly reinforced section will consist of (a) moment of resistance of singly reinforced beams, $M_{u,sr}$, and (b) M_{u2} due to forces in compression steel and additional tension steel. If the section is under-reinforced, the position of neutral axis is to be calculated with respect to the tension and compression forces in the section. The total moment of resistance will be due to the force contributed by the compression concrete and the force in the compression steel. The lever arm will be the distances of the point of application of these forces from the tension force at the level of tension reinforcement. The method of computation of moment of resistance is somewhat cumbersome and is explained with the help of a numerical example in the subsequent sections.

10.16 DETERMINATION OF STRAIN DISTRIBUTION AND STRESS BLOCKS

The strain distribution diagram, stress block, and forces are presented in Figure 10.14. This depicts the actual section condition in real-life situations but for simplicity and easier calculation the compression reinforcements are generally ignored in load design. Detailed discussions will be presented with the help of numerical examples.

10.17 REAL-LIFE BEAMS – ALL ARE DOUBLY REINFORCED

As per the discussions presented in the previous sections, designers should realize that all beams are doubly reinforced in real-life situations. In order to simplify the design calculations, the compression reinforcements which are to be mandatorily provided as per the codal requirements are ignored in the load design. When a beam is designed at a particular section, say at support or at span, the maximum hogging and sagging moments are obtained from analysis using the design loads. The

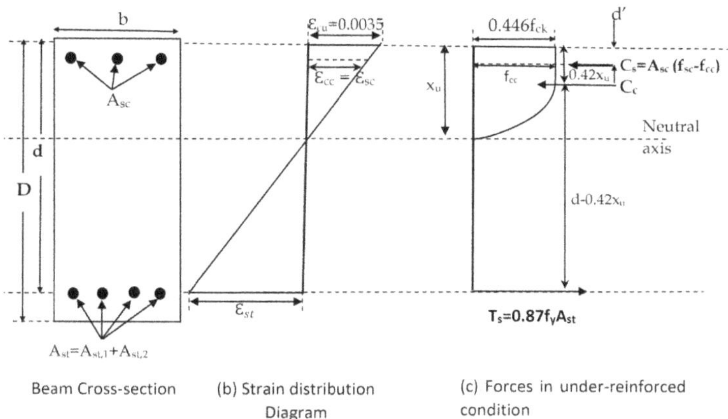

FIGURE 10.14 Strain distribution diagram, stress blocks and forces in under-reinforced doubly reinforced section.

sections are designed for reinforcement required at the tension side. If the external moment is less than the balanced or critical moment of resistance of the section only tension reinforcements are provided, otherwise both tension and compression reinforcements are required.

10.18 LOAD DESIGN – IGNORE REINFORCEMENTS ON COMPRESSION SIDE

The codes of practice deal with a prescriptive method of design based on loads, and the design loads form the basis of design. For flexural members design against external maximum moments is the primary requirement and thereafter the beam is checked for shear. The maximum design bending moment at a section is evaluated from analysis, and reinforcement required for that moment is calculated. However, beams always have reinforcement along either faces as per mandatory requirements of the standards. The reinforcements on the compression side due to the codal requirements are totally ignored in design, however nowadays all the top and bottom reinforcements of a framed beam are to be provided with necessary development lengths. The aim of design should be to properly interpret the behaviour of a structural element under loads as per its own natural characteristics. The structural elements being inert cannot understand how they have been designed. In a country like India, where a major part is vulnerable to medium to strong earthquakes and wind forces are also quite significant, the reinforcements to be continued as per codal requirements are quite high and should never be considered as insignificant.

10.19 CAPACITY DESIGN APPROACH – CONSIDERING REINFORCEMENT BOTH ON THE TENSION AND COMPRESSION SIDES

For capacity design, all the reinforcements provided at a particular section are taken into consideration. The requirements for satisfying the condition that shear failure should never occur before flexural failure and checking for the flexural strength ratio between columns and beams need to be checked during design. These checks are performed considering the actual reinforcements provided at a section. These conditions are especially important for earthquake-resistant design and have become a routine practice of engineers nowadays. Hence, it appears that dual standards are used in design as a complete demarcation between load and capacity design cannot be enforced under the present situation. By ignoring the compression reinforcement in load design, the design becomes significantly conservative. Though one may try to maintain that this will result in higher reserve strength of the sections, it needs to be kept in mind that the aim of the design should be to produce weaker flexural members so that the failure will always be ductile in nature.

10.20 IN RC SECTIONS ONLY FLEXURE FAILURE CAN BE MADE DUCTILE

For ductility to be incorporated in a reinforced concrete section either concrete or steel must yield. This has been discussed in detail while proposing the two-stage method of failure in the limit state method in earlier sections. If concrete is to yield and reach from Stage I to Stage II, i.e., from a strain of 0.002 to 0.0035, as happens for a balanced and over-reinforced section, the gap between these two stages is very small, leading to lower values of ductility. However, if the steel yields, very high values of ductility can be achieved, especially at higher levels of under-reinforcement, i.e., with lower tension reinforcements and higher compression reinforcements. With incorporation of axial force the ductility always decreases, as is discussed in Chapter 17. However, designers should always try to design sections in such a fashion that flexure failure precedes all other types of failures.

10.21 ILLUSTRATIVE EXAMPLES

Types of problems
Two types of problems are possible:

(i) Design type
In the design type of problems, the designer has to determine the beam dimensions and area of steel. These need to be determined from architectural considerations and the necessary provisions prescribed in the standards. The width and depth are to be checked as per specifications provided in the standards to ensure lateral stability and to control deflection. Reinforcements provided in beams should be in the range between the minimum and maximum values permitted by the code.

Tips to the designer for efficient design
While adopting the beam dimensions the designer should keep in mind that the current trend of the design needs weak beams and strong columns. Hence, large beam dimensions are not desirable. For proper load transfer between beams and columns, the width of the beams should preferably be less than the width of the columns into which they are being framed. The minimum distance between bars to ensure proper flow of concrete and the maximum distance between bars in tension for crack control needs to be properly checked. Though a minimum amount of transverse reinforcements is to be mandatorily provided in beams, designers should keep in mind that closer transverse reinforcements will help in confining the core concrete and will significantly improve the strength and ductility of the sections. It needs to be clearly understood that a high percentage of tension reinforcement is detrimental to the beam regarding its ductility, whereas compression reinforcements are highly beneficial in this regard. Designers should always aim at providing under-reinforced sections to improve section ductility. Thus, lower tension reinforcements need to be provided but, instead of increasing the depth of the section, compression reinforcements should be provided to increase the moment of resistance of the section. This will result in weaker beams with lower dimensions, which will have higher plastic rotation capacity than deeper sections and higher section ductility due to the incorporation of compression reinforcements.

In the prescriptive method of design as per the standards, designers do not have significant freedom and are shackled by the requirements of do's and don'ts of the standards. Even then, a good designer can yield an efficient design which will comply with the requirements of strength, serviceability, economy and performance of the sections up to and beyond the design loads. A good design doesn't only mean that it is economical and has sufficient strength and stiffness, but also that the section should have sufficient ductility, so that it performs satisfactorily up to and beyond the limit states. Thus, it can be inferred that design not only requires calculations of reinforcements but also a significant amount of foresightedness on the part of the designer. Last but not least, the grades of concrete and steel should be selected on the basis of proper exposure conditions of the site. Though higher grades of concrete are beneficial for the requirements of strength and stiffness, high grades of steel have low ductility and can result in inferior performance of the sections. As per IS:13920-2016, the maximum grade of steel permitted in vulnerable earthquake zones is Fe 550.

In a nutshell, it can be concluded that before venturing into design all of the above-mentioned features are to be considered by the designer so that the design meets the desired requirements.

(ii) Analysis/check type problems

It may be required to determine the moment of resistance of a beam whose dimension, area of steel, and grades of concrete and steel are all provided. In such a situation the designer has very little scope to use his or her expertise in design and just needs to perform some calculations. The first step is to obtain the strain distribution diagram of the section. As per IS:456-2000, flexural design can be performed between the limiting conditions of maximum under-reinforced and balanced situations as over-reinforced sections are not permitted. The flexural strength of the section is reached when the limiting strain in concrete at the outermost fibre reaches a value of 0.0035. This point on the strain distribution diagram is known and another point needs to be determined so that the linear strain distribution diagram can be plotted. Only for the balanced condition the strain in steel is defined, and the strain distribution diagram is unique for any combination of the grades of concrete and steel. For all other conditions less than balanced, strains in steel will be higher than the yield values and are unknown. Hence, the position of the neutral axis needs to be determined so that the line joining the maximum compressive strain in concrete with the neutral axis depth can result in the strain distribution diagram. When the section is defined the first stage requires calculation of the neutral axis depth. In the limit state method of design, the depth of the neutral axis is calculated by considering force equilibrium across the section, whereby the compressive force is equated to the tensile force. For an under-reinforced condition, stress in steel is constant and equal to the design stress of $0.87f_y$ and hence if the area of tension steel is known the total tensile force in the section can be easily calculated. Regarding the compression side, for failure to occur the full stress block in concrete should develop, whose parameters being known, the depth of the neutral axis can be determined.

$$0.36 f_{ck}.x_u.b = 0.87 f_y.\, A_{st}$$

$$or,\ \frac{x_u}{d} = \frac{0.87 f_y.\, A_{st}}{0.36 f_{ck}.b.d}$$

If the under-reinforced section has compression reinforcement, the problem becomes a bit complicated as using the force equilibrium equation depth of neutral axis cannot be directly calculated. With the force in compression steel and the depth of the neutral axis being unknown, the strain at the level of compression steel cannot be determined. Using the force equilibrium condition and the strain distribution diagram, the depth of the neutral axis can be determined by adopting a trial-and-error method using the process of iteration. This process is outlined below.

Iteration process

Step 1. Assume a value of stress in compression steel (f_{sc1}) (however strain value can also be assumed).

Step 2. Using the stress–strain diagram of steel (which is similar in tension and compression) the strain corresponding to the assumed value of stress can be determined (ε_{sc1}).

Step 3. This is also the value of strain in the surrounding concrete (ε_{cc1}).

Step 4. The stress in concrete corresponding to this value of strain is determined from the stress–strain diagram of concrete (f_{cc1}).

Step 5. Using these values of stresses, the depth of the neutral axis can determined (x_{u1}) from the force equilibrium equation.

Step 6. From the strain distribution diagram using this value of the depth of the neutral axis the strain at the level of compression steel is determined as $\varepsilon_{sc2} = \varepsilon_{cc2}$.

Step 7. Corresponding to ε_{sc2}, f_{sc2} can be calculated from the stress–strain diagram.

Step 8. In general, f_{sc1} and f_{sc2} will not match, and a new value of stress in steel will be considered as $f_{sc3} = \frac{f_{sc1} + f_{sc2}}{2}$. Repeat the process in the same manner as mentioned above until the stresses in compression steel considered and calculated, converge.

This task can be performed by using software, however for the present problem the trial-and-error method has been adopted for arriving at the solution.

10.22 NUMERICAL EXAMPLES

10.22.1 Problem 10.1

A real-life reinforced concrete multistorey building consisting of moment-resisting frames has been analysed using software. Design bending moment values (factored) of a typical end span of a beam are shown in Figure 10.15.

The beam cross-section adopted in analysis is depth = 650 mm, width = 250 mm, grades of concrete and steel of M30 and Fe 500, respectively, and type of exposure condition is mild. Negative means hogging moment and positive means sagging moment.

Solution:

Size of beam: 250 × 650 mm
Concrete grade: M30
Steel grade: Fe 500
Exposure condition = mild

Step 1 – Determination of effective depth

As per IS:456-2000, the nominal cover for mild exposure is 20 mm or diameter of the bar, whichever is higher. Nominal cover refers to cover to any reinforcement.

Assume diameter of the main bar to be equal to 25 mm and that for stirrup as 8 mm.

The effective cover = 20 (or bar diameter) + diameter of stirrup + ½ the diameter of the largest longitudinal bar

Effective cover $d' = 25 + 8 + ½ \times 25 = 45.5$ mm ≈ 45 mm
Effective depth $d = 650 - 45 = 605$ mm

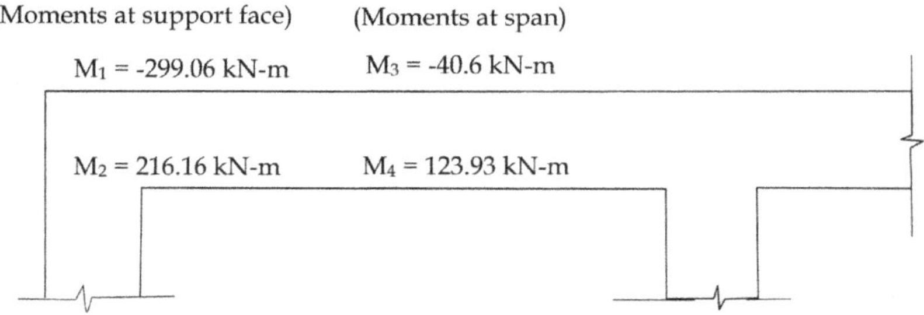

FIGURE 10.15 Beam section.

Step 2 – Determination of limiting moment of resistance of the section
From Equation (10.10) and Table 10.3,

$$M_{u,lim} = R \times f_{ck} \times bd^2$$

$$= 0.1327 \times 30 \times 250 \times 605^2$$

$$= 364.28 \times 10^6 \text{ N-mm} = 364.28 \text{ kN-m}$$

Design for support section

1) Maximum negative support moment = –299.06 kN-m

$$M_{udesign} < M_{u,lim}$$

Hence the section is an under-reinforced one.

For under-reinforced condition,

$$M_u = 0.87 f_y A_{st} \mathrm{d}\left(1 - \frac{f_y A_{st}}{bdf_{ck}}\right) \quad (10.13)$$

or, $299.06 \times 10^6 = 0.87 \times 500 \times A_{st} \text{ x } 605 \times (1 - \frac{500 \times A_{st}}{250 \times 605 \times 30})$

$A_{st} = 1331 \text{ mm}^2$

Check

Now provide 3-25Φ $A_{st} = 1472.622 \text{ mm}^2$ ($A_{st,required} < A_{st,provided}$)

$A_{st,max} = 0.04bD = 0.04 \times 250 \times 650 = 6500 \text{ mm}^2$

$A_{st,min} = \frac{0.85}{f_y} bd = 0.0017 \times 250 \times 605 = 257.13 \text{ mm}^2$

$A_{st,max} > A_{st,provided} > A_{st,min}$, hence OK

Using three number 25 Tor bars, in a width of 250 mm, the minimum and maximum spacing are both satisfied.

2) Maximum positive support moment (factored) = +216.16 kN-m

$M_{udesign} < M_{u,lim}$

Hence the section is an under-reinforced one.

Calculating in a similar manner,

$A_{st} = 913.26 \text{mm}^2$

Provide 2-25Φ, $A_{st} = 981.75 \text{mm}^2$

The reinforcement provided lies within the minimum and the maximum permitted values as per the code.

Design of span section

3) For factored moment = +123.03 kN-m

$$M_{udesign} < M_{u,lim}$$

Hence the section is an under-reinforced one.

$$M_u = 0.87 f_y A_{st} d \left(1 - \frac{f_y A_{st}}{bdf_{ck}}\right)$$

$$123.03 \times 10^6 = 0.87 \times 500 \times A_{st} \times 605 \times \left(1 - \frac{500 \times A_{st}}{250 \times 605 \times 30}\right)$$

$$A_{st} = 498.26 \text{ mm}^2$$

Provide 2-25Φ, A_{st}=981.74mm^2
The reinforcement provided lies within the minimum and maximum permitted values as per the code.

4) For factored moment = –40.06 kN-m
Though the reinforcement required will be much less, at least two 25Tor bars are to be provided for maintaining continuity of bars from the support end (refer to Figure 10.16).

As per IS:456, at least one-third of the total reinforcement provided for negative moment at the support shall extend beyond the point of inflection for a distance not less than the effetcive depth of the member or 12φ or 1/16th of the clear span, whichever is greater. Regarding positive moment reinforcement, at least one-fourth of that provided in continuous members shall extend along the same face of the member into the support to a length equal to one-third of the development length. For aseismic design as per IS:13920-2016, longitudinal steel on the bottom face of a beam (framing into a column) at the face of a column shall be at least half the steel on its top face and at the exterior joint the anchorage length calculation for this bottom steel should be based on the development length in tension. Longitudinal steel in beams at any section on the top or bottom face shall be at least one-fourth of the maximum longitudinal steel provided at the top of the beam at the column face.

Complying with these requirements the details of the beam cross-sections are shown in Figure 10.17.

The above-mentioned example has been performed on the basis of load design whereby for calculation of reinforcement at a section, only the sagging or hogging moment has been considered at a time, and reinforcements have been calculated along one face ignoring the reinforcements on the other face which are required for resisting the opposite moment at that section or to comply with the requirements of the standards.

However, the structural elements being inert cannot understand that some of their resources are not to be used but will behave as per their own natural characteristics. To take this effect into consideration, reinforcements on both sides of the section need to be used for determining the moment capacity of the section. The sections and reinforcements being known, the capacity of the sections can be determined in a manner similar to the analysis/check type of problem.

When reinforcements are placed along both faces of the beam cross-section, the beam is considered as a doubly reinforced one. Since as per IS:456-2000 over-reinforced sections are not permitted, when the external moment exceeds the balanced or singly reinforced moment capacity of the section (for the grades of concrete and steel and the beam cross-section considered) designers need to opt for doubly reinforced sections, hence it is almost a rule that all doubly reinforced sections are balanced and the additional moment requirement is made by providing compression reinforcement and additional tension reinforcement over and above the tension reinforcement required for a balanced section. Thus, the amount of compression reinforcement is always generally less than the

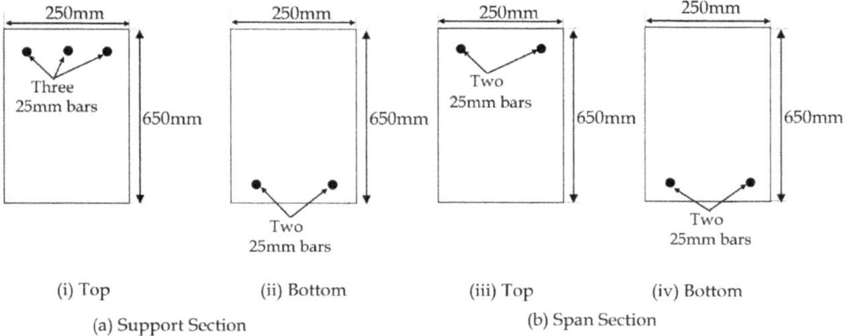

FIGURE 10.16 Cross-section of the beam at support and span.

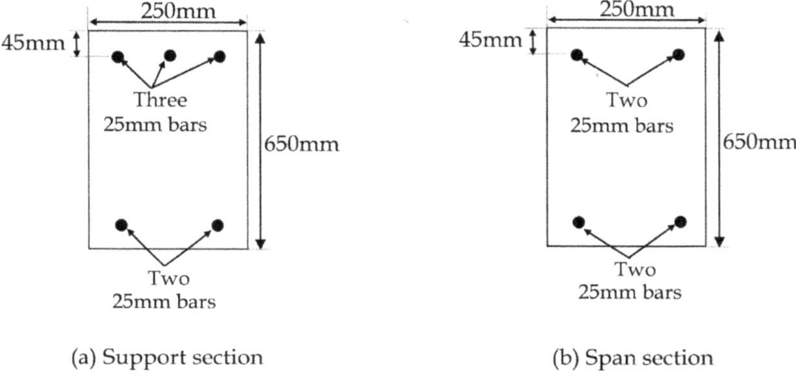

FIGURE 10.17 Beam cross-section with reinforcements.

tension reinforcement provided in the section. This is also observed in the design aids available for doubly reinforced sections, for all grades of concrete and steel.

It also needs to be understood that at beam sections where compression is occurring at the top, the slab can act as a flange of the beam. Generally, the external moments at the span and sagging moments at supports do not render the beam as truly flanged one, i.e., neutral axis passing through the beam rib, and hence the sections are generally designed as rectangular ones, ignoring the contribution of the flange.

The compression reinforcements are generally ignored in load design but if they are considered the sections will become even more under-reinforced. This is the general practice of designers, but for aseismic design since brittle shear failure should never precede ductile flexural failure, shear reinforcements are designed considering the development of plastic hinges at the ends of the beams of moment-resisting frames. Thus, at the beam ends the plastic or capacity moments are calculated whereby all the reinforcements provided are taken into consideration. To ensure the strong-column weak-beam condition, the capacities of columns and beams at joints are compared to comply to a specific ratio as per the standards. For all these conditions moment capacities are calculated on the basis of all reinforcements provided in the sections without ignoring the reinforcement actually available in the compression side. Ignoring the compression reinforcement actually provided for load design and considering it for capacity design leads to actually stronger beams resulting in even stronger columns. As per IS:13920-2016, at each beam–column joint of a moment-resisting frame the sum of nominal design strength of columns meeting at that joint along each principal plane shall

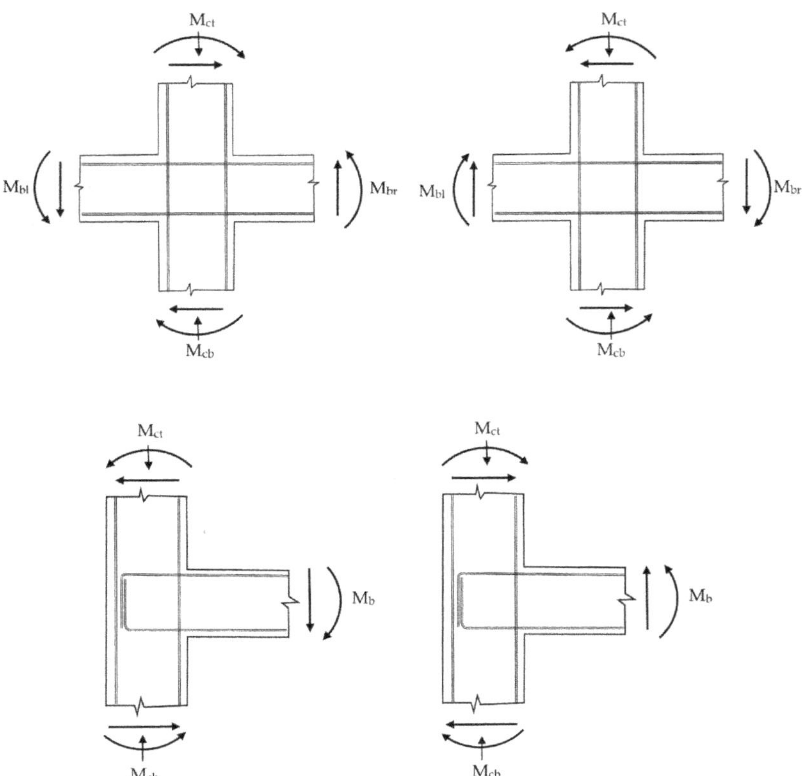

FIGURE 10.18 Strong-column weak-beam requirement.

be at least 1.4 times the sum of the nominal design strength of beams meeting at that joint in the same plane (refer to Figure 10.18).

$$\left(\Sigma_{M_c} \geq 1.4M_b\right) where,\ M_C = M_{Ct} + M_{Cb} \text{ and } M_b = M_{bl} + M_{br}$$

10.22.2 Problem 10.2

For the beam whose details of support and span sections are shown in Figure 10.17 calculate the moment capacities at the support and span sections and compare the values with those obtained from analysis.

Solution

From the load design as illustrated in problem no. 10.1, it is observed that at the support section three 25Tor bars are provided at top and two 25Tor bars at the bottom of the section. At the span both top and bottom layers constitute of two 25Tor bars (refer to Figure 10.16).

Calculations of moment capacities of the support section
For hogging moment

For the hogging moment required from the analysis the section was designed as an under-reinforced one with three 25Tor bars at the top. For the section if additional compression reinforcement is provided, the section will become more under-reinforced. For the reinforcement provided, the position of the neutral axis needs to be determined using the force equilibrium condition:

Force in tension steel = force in compression concrete + force in compression steel

As the section is under-reinforced, strain in the tension steel will be equal to its yield stress. From the consideration of failure of the section in flexure, strain at the outermost fibre of concrete in flexural compression will be equal to 0.0035. If the depth of the neutral axis is considered as x_u the force equilibrium equation can be written as:

$$0.87f_y.A_{st} = 0.36f_{ck}.x_u.b + (f_{sc} - f_{cc}).A_{sc}$$

where the symbols have their standard meanings.

From the strain distribution diagram similar to Figure 10.13(b) it can be stated that

$$\frac{0.0035}{x_u} = \frac{\varepsilon_{cc}}{x_u - d'}$$

Area of steel, cross-section of the beam and stress–strain relationship of concrete and steel are known. Unknowns are the depth of the neutral axis (x_u) and stresses in the concrete and steel at the level of compression steel. Due to the strain compatibility, the corresponding stresses in concrete and steel at the level of compression steel will be equal. A trial-and-error method as discussed above may be adopted for the solution. A typical set of sample calculations for determining the hogging moment capacity of the section is presented below.

10.22.2.1 Capacity design approach

For hogging moment at support (tension at top)

For the present section (refer to Figure 10.19)

A_{sc} provided two 25Tor = 981.74 mm^2

A_{st} provided three 25Tor = 1472 mm^2

Effective cover = 45 mm

Effective depth = 650 – 45 = 605 mm

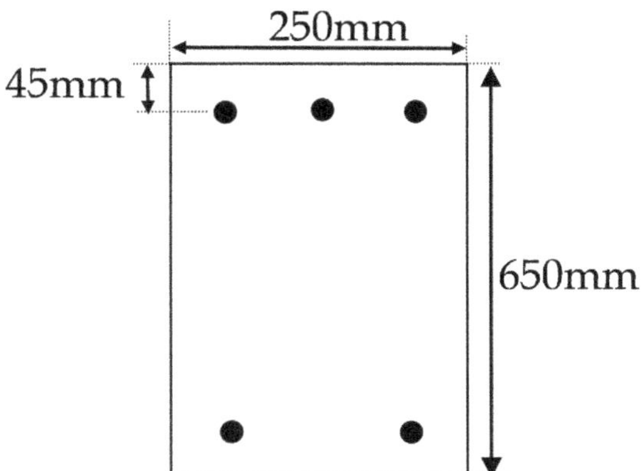

FIGURE 10.19 Cross-section of beam with reinforcement.

Step 1. Stress in the compression steel is assumed as $f_{sc1} = 400$ MPa.

Step 2. From the stress–strain diagram of HYSD steel, the inelastic strain corresponding to the stress of $0.8f_y$ is nil. Hence, strain at this stress level is elastic, having a value of $\frac{400}{2\times10^5} = 0.002$:

$$\varepsilon_{cs1} = 0.002.$$

Step 3. From strain compatibility, $\varepsilon_{cs1} = \varepsilon_{cc1} = 0.002$.

Step 4. The stress in concrete corresponding to this value of strain is determined from the stress–strain diagram of concrete = $0.446f_{ck} = 0.446 \times 30 = 13.5$ N/mm^2.

Step 5. Substituting these values of stresses in the force equilibrium equation, the depth of neutral axis can be computed as follows:

$$0.36f_{ck}x_{u1}\text{b} + A_{sc}(f_{sc1} - f_{cc1}) = 0.87f_yA_{st}$$

$$0.36 \times 30 \times x_{u1} \times 250 + 981.74\,(400 - 13.5) = 0.87 \times 500 \times 1472.62$$

$$\text{Hence, } (x_u)_1 = 96.72 \text{ mm}$$

Step 6. Using this value of neutral axis depth, strain at the level of compression steel can be determined (ε_{sc2}):

$$\frac{0.0035}{x_u} = \frac{\varepsilon_{sc}}{x_u - d'}$$

Hence, $\varepsilon_{sc2} = 0.001872$.
From strain compatibility $\varepsilon_{sc2} = \varepsilon_{cc2}$.

Step 7. Corresponding to ε_{sc2} using the stress–strain diagram of steel, stress in the compression steel (f_{sc2}) is calculated.

In this context it may be mentioned that the stress–strain diagram of HYSD steel from the elastic to the yield point has been defined by SP16 – Design aids for reinforced concrete to IS:456-1978, with the help of inelastic strain values of 0.0001, 0.0003, 0.0007, 0.0010 and 0.0020 at stress levels of $0.8f_y$, $0.85f_y$, $0.9f_y$, $0.95f_y$, $0.975f_y$ and $1f_y$ respectively, allowing linear interpolation between the intermediate values. This has been discussed in detail in Chapter 4. Hence, for calculating the stress corresponding to any strain level necessary interpolation is to be performed between the two adjacent points within which the strain value lies. However, the present strain value lies within the elastic part of the stress–strain diagram. Hence, for the present case where the total strain in compression steel has been considered as 0.001872, the corresponding stress is obtained as 374.3187 MPa $= f_{sc2}$.

Step 8. Since f_{sc2} is not equal to the assumed value of f_{sc1}, a new value of stress in compression steel is to be considered as $f_{sc3} = \frac{f_{sc1} + f_{sc2}}{2} = 387.1595\,\text{MPa}$.

With this new value of f_{sc3} the process of iteration will be continued in a similar manner as mentioned above until the stress values converge, i.e., $f_{scn} = f_{sc(n-1)}$.

For the present problem, after four sets of iterations the values have converged and the final values are as follows:

$$f_{sc} = 387.75 \text{ MPa};$$

$$f_{cc} = 12.69 \text{ MPa};$$

$$x_u = 100.88 \text{ mm}.$$

The final strain distribution, stress block and forces in concrete and steel are shown in Figure 10.20.

Calculation of moment of resistance of the section based on the capacity design approach:

Moment of resistance capacity of this section is

$0.36\, f_{ck} x_u b \times \text{Lever arm 1} + A_{sc}(f_{sc} - f_{cc}) \times \text{Lever arm 2}$

$= 0.36\, f_{ck} x_u b \times (d - 0.42x_u) + A_{sc}(f_{sc} - f_{cc}) \times (d - d')$

$= 0.36 \times 30 \times 100.88 \times 250 \times (605 - 0.42 \times 100.88) + 981.74 \times (387.7467 - 12.6855) \times (605 - 45)$

= 359.45 kN-m.

Hence, hogging moment of resistance capacity of the section is obtained as 359.45 kN-m. This is the capacity moment of resistance of the section considering both the top and bottom reinforcements simultaneously.

As per analysis, the maximum or design value of the hogging moment of this section is 299.06 kN-m.

As per the load design, three 25Tor bars were provided at the support top. Area of tension reinforcement provided = 1472.62 mm^2.

$$p_{t,provided} = \frac{100A_{st}}{bd} = 0.973\%$$

M_u of the section for A_{st} = 1472.62 mm^2 can be calculated from Equation (10.13) and is obtained as 323.72 kN-m.

For this condition, the reinforcement on the compression side was totally ignored in the calculations.

Considering both tension and compression reinforcement, which is the actual situation, moment of resistance of the section calculated from capacity design approach = 359.45 kN-m.

Thus, the actual moment of resistance provided in the section is about 11% higher than that calculated for load design and 20 % higher than that required.

Moment required from analysis = design moment = 299.06 kN-m

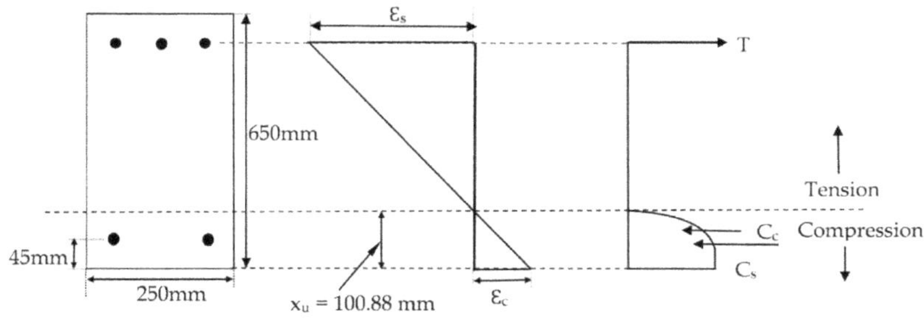

FIGURE 10.20 Strain distribution and stress distribution diagram of the section.

Moment of resistance as per load design = 323.72 kN-m
(As singly reinforced under-reinforced section)
Actual moment of resistance provided = 359.45 kN-m
(As doubly reinforced under-reinforced section)

In an attempt to reduce the amount of reinforcement it has been observed that the moment of resistance of the section using four 20Tor bars at the top and three 20Tor bars at the bottom, the depth of the neutral axis is obtained as 81.62 mm (under-reinforced section) with a moment of resistance calculated from the capacity design approach 308.48 kN-m (greater than the external or design moment) and hence the section is safe. Hence, instead of five 25Tor bars (area = 2455 mm^2), seven 20Tor bars (area = 2199 mm^2) are required, with a saving of about 10% in the area of reinforcement.

The amount of calculations involved is extremely cumbersome, if the capacity based design approach is adopted instead of load design. This methodology is adopted for design of sections subjected to flexure with axial forces viz. columns. The benefit derived is quite significant for the present problem. Hence instead of performing capacity design at each and every step, load design seems to be justified for normal reinforcement calculations. However, a detailed comparison of the load- and capacity-based design approaches needs to be performed for different types of flexural members to select the optimum method of design. But for calculating the shear reinforcement or checking the strong-column weak-beam condition, the method of capacity design as described needs to be followed to obtain the correct results.

The aim of design should be to reduce the assumptions and to simulate the actual behaviour of sections as accurately as possible. The present problem is a humble effort in this direction. However, designers should keep in mind that since reinforcement at both the faces is simultaneously taken into consideration all the reinforcement should be provided with the requisite development length in tension. The above problem exhibits the optimum design of beams in flexure and can be designated as under-reinforced doubly reinforced sections. This will put an end to the existing notion that all doubly reinforced sections are balanced.

Sagging moment of resistance of the support section based on capacity design approach

When the support section is subjected to a sagging moment the top portion will be in compression and the bottom in tension. Hence, for the present problem compression reinforcement, A_{sc}, constitutes three 25Tor bars and tension reinforcement, $A_{st,}$ of two 25Tor bars. A_{sc} is higher than A_{st}. For this condition the force equilibrium equation leads to the depth of neutral axis less than the effective cover of compression reinforcement, which is untenable as the force equilibrium equation will become invalid. Hence the section cannot be designed as a doubly reinforced one and the moment capacity may be determined considering a flanged beam or a rectangular beam as a singly reinforced one with tension reinforcement only ignoring the compression reinforcement.

10.23 SUMMARY

Throughout the world, the limit state method of design is generally adopted for the design of reinforced concrete sections. All design philosophies are based on some laws, axioms, hypothesis, assumptions, limitations, etc. Limit state philosophy, which is a definite advancement over the working stress concept, is based on a number of assumptions specified in our standards. The entire methodology rests on these assumptions and in this chapter an attempt has been made to explain the underlying considerations behind these assumptions. Numerical examples have also been included to explain the concept more clearly. The limit state method defines failure in two stages: Stage I, the yield stage, and Stage II, the ultimate or failure stage. This multistage failure occurs as materials have to yield before failure occurs and since the materials are going to the post-yield stage, ductility can be incorporated in the design. In this chapter, the limit state method of design as per IS:456-2000 has been discussed in detail, with an introduction to other national and international standards. The stress–strain relationships of concrete and steel which form the basis of design have been explored in detail. The strain and stress distributions of different types of sections have been discussed in this chapter in detail. As per theoretical considerations, all doubly reinforced sections are balanced. However, in real-life situations, doubly reinforced sections are under-reinforced in nature. This has been explained with the help of numerical examples.

BIBLIOGRAPHY

Bhatt, P., MacGinley, T. J. and Choo, B. S. (2005). *Reinforced Concrete Design: Design Theory and Examples*, CRC Press, Oxon.

IRC:112-2020 (July, 2020). *Code of Practice for Concrete Road Bridges*, First Revision, Indian Road Congress, New Delhi.

IS 456:2000 (2007). *Plain and Reinforced Concrete – Code of Practice*, Fourth Revision, Tenth Reprint April 2007, Bureau of Indian Standards, New Delhi.

IS 13920:2016 (July 2016). *Ductile Design and Detailing of Reinforced Concrete Structures Subjected to Seismic Forces – Code of Practice*, First Revision, Bureau of Indian Standards, New Delhi.

Jain, Ashok K. (2002). *Reinforced Concrete Limit State Design*. Nem Chand, Roorkee, India.

Paulay, T. and Priestley, M. N. (1992). *Seismic Design of Reinforced Concrete and Masonry Buildings* (Vol. 768), Wiley, New York.

Pillai, S. U. and Menon, D. (2012). *Reinforced Concrete Design*, Third Edition, Reprint 2012, McGraw Hill, New Delhi.

Purushothaman, P. (1984). *Reinforced Concrete Structural Elements, Behaviour, Analysis, and Design*, McGraw Hill.

Shah, V. L. and Karve, S. R. (2016) *Limit State Theory and Design of Reinforced Concrete*, Eighth Edition, Structures Publications, Pune, India.

11 Limit State of Collapse in Compression as per IS:456

11.1 INTRODUCTION

Columns are structural elements which behave predominantly in compression. Usually columns are subjected to flexure along with compression, and columns subjected to pure axial compression are almost hypothetical. Since they are subjected to flexure and compression, the design methodology of columns has considerable similarity with flexural members. However, the introduction of compressive force results in normal stresses similar to flexure and changes the behaviour of the columns and their failure modes.

11.2 CLASSIFICATION AND DEFINITIONS

Depending upon the cross-section, columns may be classed as circular, square, rectangular, hexagonal, T-shaped, L-shaped, cross-shaped, etc. Columns also may be classified as per transverse reinforcement. When the longitudinal reinforcements are tied by separate closed loops it is termed as a tied column. In a circular column the longitudinal bars may be held in position by a continuous spiral line of reinforcements when it is termed as a spiral column (refer to Figure 11.1). Columns may also be classified according to their length. As per the Indian standard, when the ratio of the effective length to the least lateral dimension is less than three, the column is termed as pedestal. A column is described as a short column when the ratio of its effective length to its least lateral dimension is less than 12 ($L_{eff}/D \leq 12$). On the other hand, if the ratio of the effective length to its least lateral dimension is greater than 12 the column is described as a long column ($L_{eff}/D > 12$).

The vertical distance between the point of inflection of a column under loading is termed as effective length (L_{eff}), which is a function of the unsupported length, column dimension and end restraints. $L_{eff} = k.L_{un}$, where k = effective length factor, L_{un} = unsupported length of the column and is a very important parameter for column design. Methodologies for determination of k and its probable values are provided in the different codes of practice.

Short columns fail by crushing of the material, whereas long columns fail due to buckling. An intermediate mode of failure can occur due to a combination of crushing and buckling.

11.3 BRACED AND UNBRACED COLUMNS

A structural element subjected to compression and bending is defined as a wall when the ratio of its length to its thickness is more than or equal to four. Otherwise, it will be termed as a column. A column may be considered braced in a given plane if the lateral loads are resisted by walls or some other form of bracing. That is, the lateral stability of a structure as a whole is provided by walls or

DOI: 10.1201/9781003415398-11

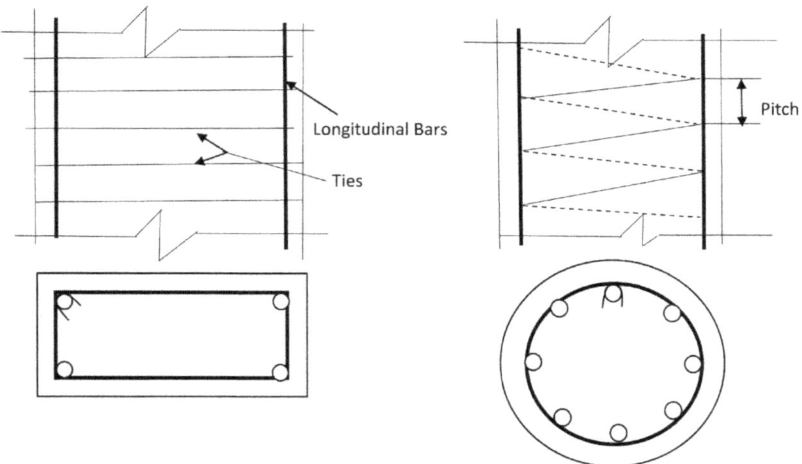

FIGURE 11.1 Rectangular columns with ties and circular columns with spirals.

bracing. Braced columns are not designed to resist lateral forces. For a braced column, the side sway is not very significant. Examples of braced columns are the columns of a water tank or shear wall buildings where the shear walls carry all the lateral loads.

A column is defined as an unbraced one where the lateral loads are resisted by the bending action of the columns. A column will be considered unbraced in a given plane if the lateral stability of the structure as a whole is provided by columns only. They are designed to resist lateral loads. Unbraced columns are subjected to a significant amount of side sway. Columns of multi-storeyed buildings without shear walls or bracings are examples of unbraced columns.

11.4 LONGITUDINAL REINFORCEMENT

The longitudinal reinforcing bars carry the axial loads acting on the section and have to carry both compressive and tensile forces. The requirements regarding the minimum and maximum amounts of reinforcement, number of bars, spacing, etc. are covered in the codes of practice which are to be followed by the designer. In order to satisfy strain compatibility, strains in reinforcement at any location should be the same as the strain in the adjacent concrete.

11.5 TRANSVERSE REINFORCEMENT

The role of the transverse reinforcement is to hold the longitudinal reinforcements in their position and prevent them from buckling or bulging out of concrete. The requirements of bar diameter, spacing, configuration, etc. are provided in the codes of practice. The transverse reinforcements provide confinement of the core concrete and if properly detailed can significantly improve the strength and ductility of the sections.

11.6 DESIRABLE CHOICES OF SECTION

For columns, rectangular or square cross-sections are very common and designers generally prefer these as the connectivity with the beams is very simple. However, designers need to keep in mind that the basic stress–strain relationship of concrete governs the design as it directly controls the stress blocks developed in concrete. The stress–strain diagrams of unconfined and confined concrete are

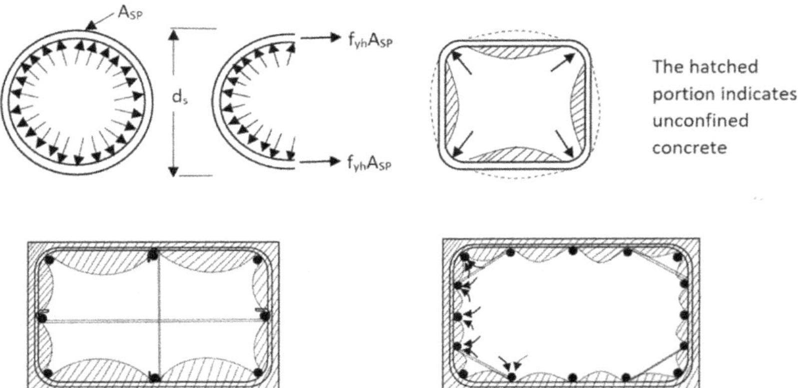

FIGURE 11.2 Confinement of concrete by circular and square hoops.

significantly different as both the stress and strain levels of confined concrete are much higher than their unconfined counterparts. The confinement provided by the spirals in circular columns is much more effective than those provided by the ties in rectangular or square sections. The spirals being subjected to hoop tension provide much better all-round confinement to the core concrete than the ties which are more effective at the four corners of a square or rectangle (refer to Figure 11.2). Thus, if the designer has a choice it is always better to opt for circular sections which will have higher ductility and reserve strength than square and rectangular columns. The confinement provided by the rectangular or square hoops can be significantly improved by the use of overlapping hoops or hoops with cross ties. Spacing of transverse reinforcement should be relatively close to ensure better confinement. Nature has made tree trunks of circular cross-sections only.

The effect of confinement is to increase the compressive strength and ultimate strain of concrete. This has been illustrated in Chapter 7 (refer to Figure 7.1).

11.7 DIFFERENCE BETWEEN FLEXURE DESIGN AND COMPRESSION DESIGN

Columns are subjected to compression and flexure, both of which result in normal stresses and the section can fail in flexure only, compression only, or a combination of compression and flexure. Beams are subjected to flexure only and need to be checked for the bending moments only. Both columns and beams are subjected to shear as bending moments are generally not constant over their length, but shear stresses act orthogonally to the normal stresses and need to be checked separately. The effect of buckling in columns may also govern the design. Designers should keep in mind that reinforced concrete can be made ductile only in the flexural mode of failure and hence failure of beams should always precede failure of columns. Incorporation of axial forces along with flexure reduces the ductility of the sections.

11.8 WHY DESIGN CHARTS ARE NEEDED FOR COLUMN DESIGN

Columns are to be checked for two stress resultants simultaneously, compression and flexure, unlike beams which are checked for flexure only. Columns are subjected to compression and flexure, both of which result in normal stresses. For beams, checking for moment or flexure is only performed for design.

For columns, compressive force (P) and moment (M) occur simultaneously and hence checking should be done for P and M. The section can fail due to only P, only M, or a combination thereof. Since two parameters are to be simultaneously checked, interaction diagrams or charts are required for design, unlike beams which are to be checked for only one parameter M and need a table only, relating M to the area of reinforcement (A_{st}) required. But for columns, A_{st} needs to be calculated for P and M simultaneously, and thus interaction charts relating P and M with A_{st} are required for design.

11.9 DESIGN OF AXIALLY LOADED COLUMNS

An axially loaded short column will fail when the concrete crushes or reinforcement yields, or both occur simultaneously. As per basic assumptions of the limit state of collapse in compression of IS:456-2000, the maximum compressive strain in concrete in axial compression is 0.002. However, for practical purposes the minimum eccentricity is always considered in design. To take care of this effect the stresses in concrete and steel are both reduced, and a simplified expression has been proposed in the code in the following form:

$$P_u = 0.4 f_{ck} A_c + 0.67 f_y A_{sc} \tag{11.1}$$

where A_c = area of concrete and A_{sc} = area of steel and other symbols have their standard meaning.

11.10 COLUMNS UNDER AXIAL LOAD AND UNIAXIAL BENDING

In real-life situations, almost all columns are subjected to combined axial loads and bending moments. Under this condition, RC sections are subjected to compression and flexure and hence should be designed as per the requirements of the limit state of collapse in compression, whereby almost all the assumptions in the limit state of collapse in flexure are considered along with additional assumptions for compression. Assumptions for the limit state of collapse in flexure have already been discussed in Chapter 10, and hence they are not repeated here. Only the last assumption in flexure which ensured that failure should only be initiated in steel because of which over-reinforced sections were prohibited have been deleted, allowing design of all types of sections in compression. The first assumption in compression limits the axial compressive strain in concrete to 0.002.

In beams, the neutral axis lies within the section, whereas in columns, the neutral axis of the section can lie anywhere within the section or beyond the section. The compressive strain in concrete ranges from 0.002 in axial compression to a value of 0.0035 in flexural compression. The second assumption in compression deals with the intermediate strain values between 0.002 and 0.0035 at the maximum compressed fibre of the section considering linear strain distribution as per Bernoulli's theorem. This is illustrated in Figure 11.3.

Columns may be subjected to pure compression or pure tension. However, since concrete is very weak in tension, the limit state method of design as per IS:456-2000 does not cover the limit state of collapse in tension. Hence, column design is restricted between the limiting conditions of pure compression to steel beam condition, where the concrete is totally ineffective and the section acts as a steel beam.

11.11 MODES OF FAILURE IN COLUMNS

Beams behave predominantly in flexure, as a result of which the neutral axis will lie inside the section. Due to the simultaneous action of compression and flexure, the neutral axis of a column section can vary from an infinite distance from the section, that is when the section is under pure compression, to the steel beam condition. Under the steel beam condition the concrete is considered to be fully ineffective. The reinforcement cage results in the moment capacity of the section with equal tensile and compressive forces, resulting in zero axial load capacity. Hence the failure mode considered can vary from pure axial compression when there is no flexure in the section to pure

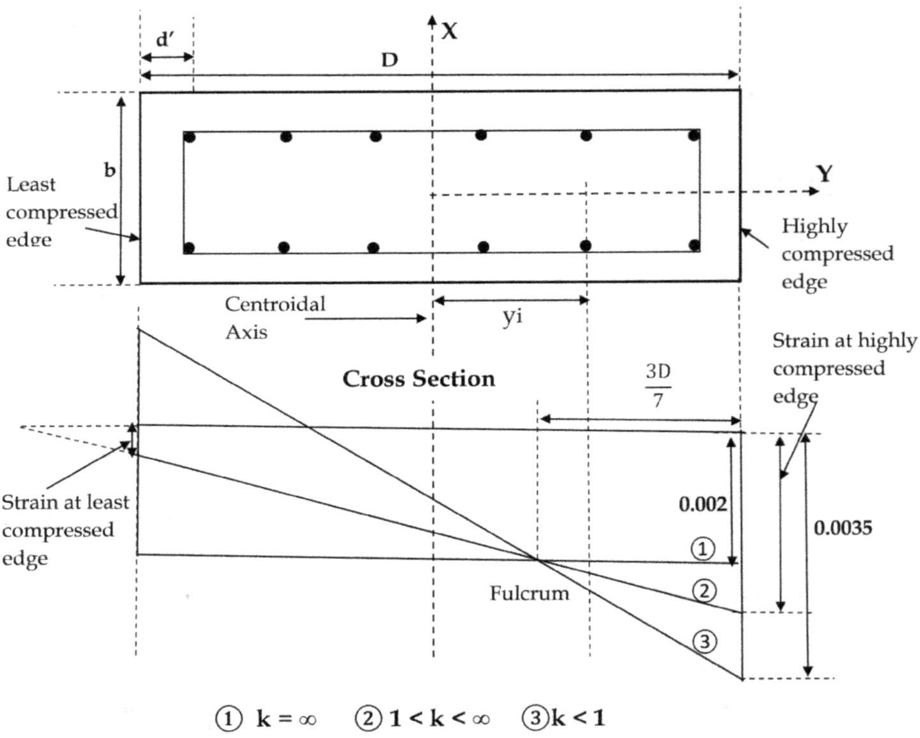

FIGURE 11.3 Strain profile for different positions of neutral axis.

Note: k denotes the neutral axis depth factor, i.e., kD = depth of neutral axis.

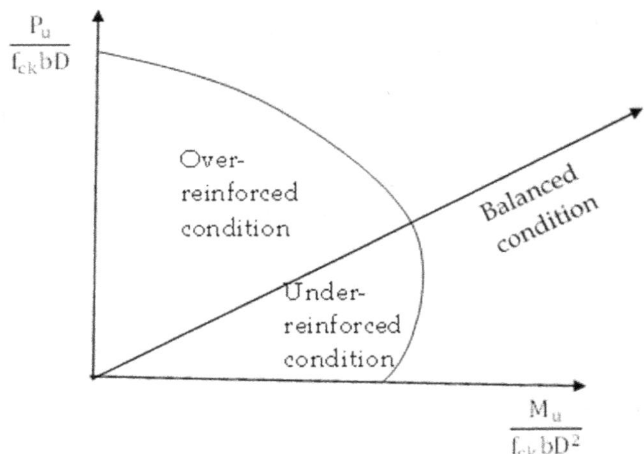

FIGURE 11.4 Failure in under-reinforced, over-reinforced and balanced conditions.

flexure when there is no axial force in the section. Beam sections are classed as under-reinforced, balanced and over-reinforced depending upon the reinforcement provided in the section based on which the position of the neutral axis varies within the section. For columns, sections cannot be classed as under-reinforced, balanced and over-reinforced on the basis of the reinforcement provided. A section with a certain percentage of reinforcement can behave as an under-reinforced, balanced, or over-reinforced section depending upon the loading that is axial load, P, and bending moment, M. Hence, column sections cannot be classified as beams, rather the failure of the column sections is considered to have occurred in under-reinforced, balanced or over-reinforced condition. Only the balanced failure condition of a column section is uniquely defined. The balanced depth of the neutral axis is constant for a particular grade of concrete and steel. Any depth of neutral axis higher than this up to infinity will result in failure under over-reinforced condition, that is compression failure, and any depth lower than this will result in failure under under-reinforced condition that is tension failure (refer to Figure 11.4).

Under limit states of collapse, how the section will behave under tension with or without bending, no guidance is available in the standard, and hence these conditions will not be considered for design.

11.12 FAILURE UNDER BALANCED CONDITION

Under this mode of failure, the outermost compression fibre of the column will reach the maximum compressive strain of 0.0035 and on the tension side, the outermost row of longitudinal reinforcement will yield. As per IS:456-2000, since the HYSD bars do not have a definite yield point, the concept of proof stress is considered as the yield stress which corresponds to the stress at an inelastic strain of 0.002.

11.13 FAILURE IN OVER-REINFORCED CONDITION OR COMPRESSION FAILURE

When the depth of the neutral axis of the column is more than the balanced depth, the failure occurs in over-reinforced condition, i.e., compression failure.

11.14 FAILURE IN UNDER-REINFORCED CONDITION OR TENSION FAILURE

This is the condition when the neutral axis depth of the section is less than the balanced neutral axis depth and the failure under this condition is termed as tension failure.

11.15 EQUILIBRIUM EQUATIONS FOR COLUMN DESIGN

In reinforced concrete column sections, axial loads (P) and bending moments (M) act either in isolation or simultaneously. The two equations of equilibrium can be written as:

$$\Sigma V = 0 \text{ and}$$

$$\Sigma M = 0$$

where, V and M refer to axial forces and moments acting on the section, respectively.

External force acting on the column section = algebraic sum of forces acting on concrete and steel:

$$P_{external} = C_c + C_{sc} + (-C_{st}) \tag{11.2}$$

where, C_c = force in concrete, C_{sc} and C_{st} refer to force in compression steel and force in tension steel, respectively.

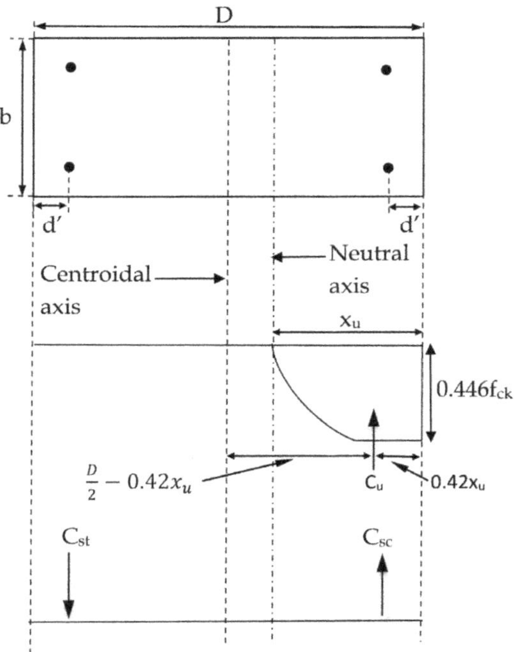

FIGURE 11.5 Column cross-section, stress distribution in concrete and forces in steel reinforcement.

External moment acting on the column section = algebraic sum of moments of all forces acting about the centroidal axis of the section. From Figure 11.5.

$$M_{external} = C_u \cdot \left(\frac{D}{2} - 0.42x_u\right) + C_{sc} \cdot \left(\frac{D}{2} - d'\right) + (-C_{st}) \cdot \left\{-\left(\frac{D}{2} - d'\right)\right\} \tag{11.3}$$

A negative sign has been used for force in tension and for distance measured on the tensile side of the centroidal axis, and other symbols have their standard meanings.

11.16 NEED FOR P–M INTERACTION CHARTS FOR COLUMN DESIGN

A couple is formed by multiplying two equal and opposite forces with the distance between their points of application. For beams, since the axial force acting on the section is zero or negligible, the compression and tension forces acting along any section must be equal in order to satisfy the force equilibrium. Thus, the moment of resistance of a beam is obtained by multiplying the equal and opposite tension and compression forces with the distance between them. For columns which are subjected to a significant amount of axial forces along with bending, the tensile and compressive forces acting at a column cross-section are not equal, rather their algebraic sum should be equal and opposite to the external force for maintaining force equilibrium. Hence, unlike beams, for columns the moment capacity of a section is obtained as the algebraic sum of moment of all forces about the centroidal axis of the section, assigning a positive sign to the compressive forces and a negative sign to the tensile forces.

In the case of beams, members are subjected to flexural load only, and they can fail due to flexure only. Therefore, for this reason, to perform a beam design checking for moment or flexure only is performed, i.e., checking for only one parameter.

Columns are different from beams. For column sections, both flexure and compression occur simultaneously. Therefore, column sections can fail due to compression only, flexure only, or a combination of both compression and flexure. Hence for design, checking should be done for both axial load (P) and bending moment (M) acting either in isolation or simultaneously.

Since beams behave predominantly in flexure, the strains and stresses in a beam cross-section occur because of the action, i.e., flexure. As no axial forces are acting, in order to maintain the force equilibrium, the tensile and compressive forces (developed due to normal stresses acting on the section) should be equal.

Force in tension = Force in compression

Area of steel provided × stress developed in steel = Area of stress block of concrete × width of the section

For beam design, based on the external moments acting on the section, the strain and stress distribution diagrams are developed and accordingly the moment of resistance of the section is calculated.

Failure in reinforced concrete sections occurs by crushing of concrete, and the failure strain in concrete in axial and flexural compression is defined by the codes of practice.

For columns, the section capacities are governed by the percentage and distribution of longitudinal reinforcements. Reinforcements may be provided in two or four faces, and their strains and stresses cannot be calculated unless the position of the neutral axis is known.

For safe design, external force acting on the section ≤ algebraic sum of forces acting on concrete and steel.

External moment acting on the section ≤ algebraic sum of moment of all the forces acting on the section about the centroidal axis.

Thus, the unknowns involved in the equations are too many and solving them even by successive iterations for each and every load case does not seem to be feasible. Hence, instead of working with the external forces and moments, the capacities of the sections in axial force and moment are determined assuming arbitrary positions of the neutral axis with an assumed percentage of longitudinal reinforcements. Thus a safety profile of the section corresponding to that reinforcement percentage can be generated. Using different percentages of reinforcements, a numbers of safety profiles can be drawn which will constitute the P–M interaction chart for the column, that can be used for design.

11.17 DEVELOPMENT OF STRAIN DISTRIBUTION DIAGRAM AND STRESS BLOCKS FOR DIFFERENT ARBITRARY POSITIONS OF THE NEUTRAL AXIS

The design of a section subjected to the combined action of axial load and flexure should be in accordance with the fundamental principles of the limit state method and are to be performed by using P–M interaction charts.

11.17.1 When neutral axis is outside the section

Figure 11.6 shows a typical strain and stress distribution diagram of a column section. These strain and stress distributions are similar to those for flexural members or beams. The basic differences are the presence of axial force and distribution of reinforcements across the section. Even for columns, the strain distribution diagram will be drawn in a similar manner as beams by using a straight line distribution. For balanced condition, the strain distribution diagram is obtained by joining the points corresponding to the limiting strain in concrete in flexural compression with the tensile strain at the outermost layer in steel equal to the yield strain, i.e., $\frac{f_y}{1.15E_s}+0.002$.

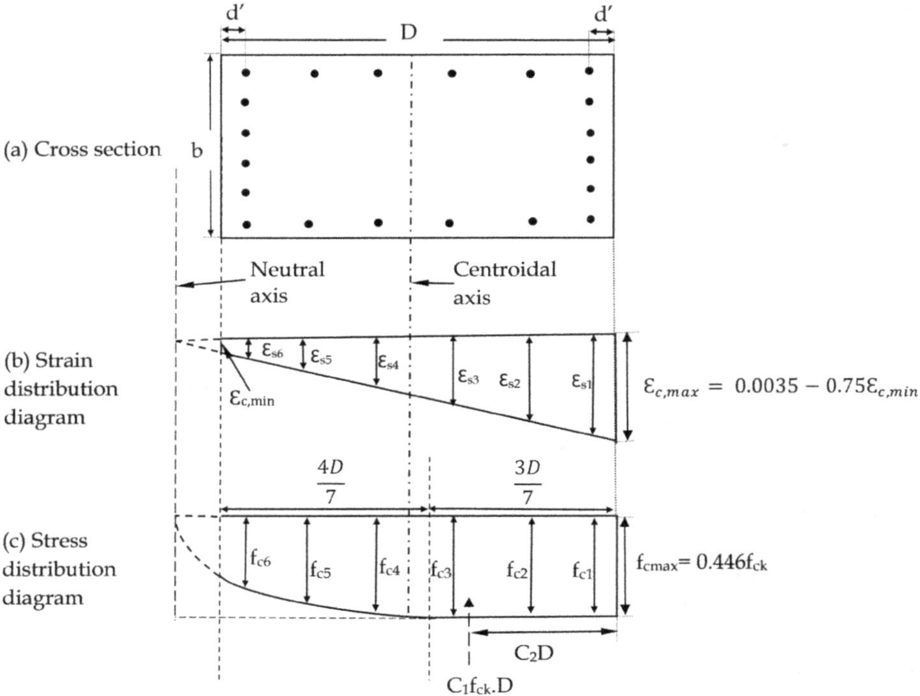

FIGURE 11.6 Stress block and strain profile when neutral axis is outside the section $\left(\frac{x_u}{D} \geq 1\right)$.

Thus, the balanced depth of the neutral axis will be a function of the grade of steel only, as is observed in beams. However, the axial compression and the moment capacities under balanced condition will depend on the percentage and distribution of longitudinal reinforcement. If the depth of the neutral axis is more than the balanced depth for that reinforcement percentage the failure will occur under over-reinforced condition. And if the neutral axis depth is less than the balanced value, failure will occur in the under-reinforced condition.

Figure 11.6 represents the strain profile and corresponding stress block for neutral axis outside the sections, $kD > D$. When the neutral axis is outside the section the limiting condition ranges from infinity ($k = \infty$) to the position when it is located at the face of the section ($k = 1$). Accordingly, the stress blocks can be determined. Under these conditions the entire cross-section is in compression.

When the neutral axis is at infinity, the stress block consists of a rectangular one with a stress value of $0.446\,f_{ck}$ corresponding to a strain distribution with a constant strain of 0.002. When the neutral axis is at the face of the section, the strain distribution diagram is a triangular one with a maximum compressive strain of 0.0035 and zero value at the least compressed edge. The stress block consists of the entire parabolic rectangular stress–strain diagram of concrete having an area of $0.36\,f_{ck}.x_u$ and distance of the centroid as $0.42\,x_u$ from the maximum compressed fibre. For these positions of the neutral axis outside the section, if the area of the stress block and distance of its centroid from the maximum compressed fibre are expressed as $C_1.f_{ck.}D$ and C_2D, respectively, the values of the co-efficients C_1 and C_2 can be easily obtained by using simple mathematics and are presented in Table 11.1. The values are shown in Figure 11.6.

When the neutral axis is at infinity, the values of C_1 and C_2 will be 0.446 and 0.5, respectively. When the neutral axis is at the face of the section, the corresponding value will be 0.361 and 0.416, respectively. When the neutral axis is at a distance of $4D$ from the maximum compression face, the area of the stress block within the section becomes almost rectangular. Hence, from this position of the neutral axis to the position at infinity the area of the stress block and the location of its centroid

TABLE 11.1
Stress block parameters when neutral axis lies outside the section

k	C_1	C_2
1	0.361	0.416
1.05	0.374	0.432
1.1	0.384	0.443
1.2	0.399	0.458
1.3	0.409	0.468
1.4	0.417	0.475
1.5	0.422	0.480
2	0.435	0.491
2.5	0.440	0.495
3	0.442	0.497
4	0.444	0.499

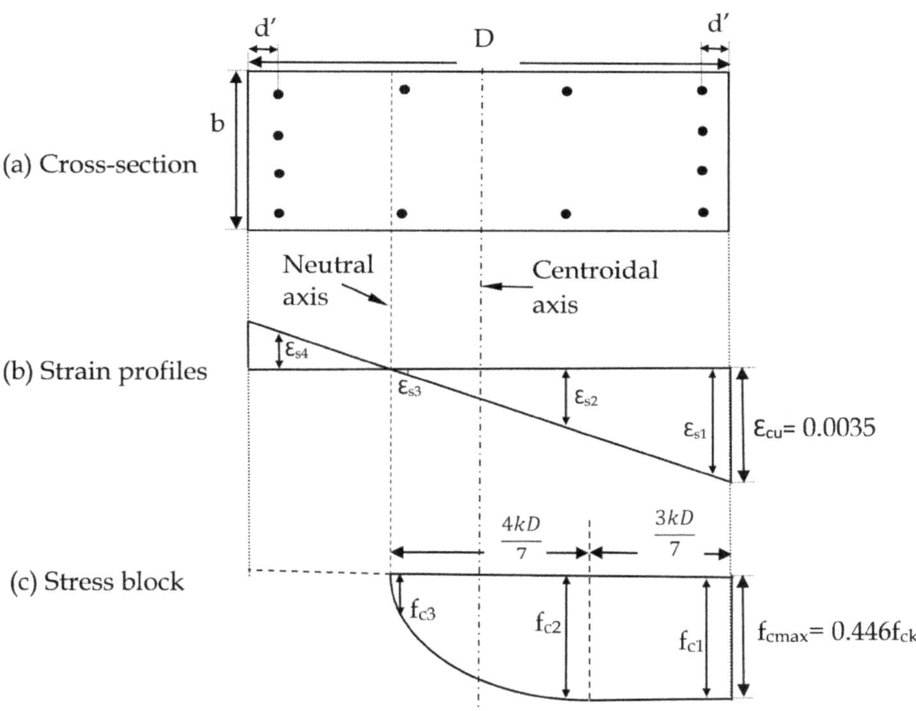

FIGURE 11.7 Stress block and strain profile when neutral axis is within the section $\left(\frac{x_u}{D} < 1\right)$.

becomes almost constant. These calculations being quite simple and commonly available are not elaborated further here.

11.17.2 When the Neutral Axis Is within the Section

When the neutral axis is within the section, the strain profile of the section is shown in Figure 11.7. The area of the stress block and the position of its centroid can be easily evaluated using the equation of the stress–strain relationship of concrete. These are explained with the help of numerical examples.

11.18 DEVELOPMENT OF SAFETY PROFILE OF A COLUMN SECTION FROM PURE AXIAL LOAD TO STEEL BEAM CONDITION WITH A CERTAIN PERCENTAGE OF LONGITUDINAL REINFORCEMENT

For developing the strain distribution diagrams, stress block for concrete, and stresses along the reinforcements, the same principles followed in flexure have been used in compression. Hence they are not repeated in this section.

11.18.1 Determination of Stress and Strain Values in Reinforcements from Stress–Strain Relationships

The stress–strain relationships of steel are provided in IS:456-2000 and SP16-1980. The stress–strain diagram for mild steel is a bilinear one, whereas for HYSD bars the relationship is linear up to $0.8f_y$ and horizontal beyond f_y with an inelastic range in between, whereby the stresses and strains are shown in Table 11.2.

Four types of HYSD steel, Fe415, Fe500, Fe550 and Fe600, have been considered in this chapter, and the corresponding values are shown in Table 11.3.

11.18.2 Development of the Safety Profile of a Column Section with Typical Percentage of Longitudinal Reinforcements

A column is generally subjected to compression along with flexure. The axial load and moment capacity of a column section will depend upon a number of parameters such as the grade of concrete and steel, percentage of longitudinal reinforcement and its distribution across the section and cover of the reinforcing bars. Knowing these parameters, values of axial and moment capacities are obtained at different arbitrary points ranging from pure compression to pure flexure, and these points are plotted to obtain the axial force–moment relationship for the section. This corresponds to the safety profile of the section with that percentage of the reinforcement. In order to plot the safety profiles, non-dimensional parameters, namely $\frac{P_u}{f_{ck}bD}$ and $\frac{M_u}{f_{ck}bD^2}$ have been used as they are available in the design aids with which designers are familiar.

TABLE 11.2
Total strain in HYSD steel corresponding to different values of stress between elastic and yield points

Stress level	Elastic strain	Inelastic strain as per IS:456-2000	Total strain (elastic strain + inelastic strain)
$0.8f_{yd} = 0.8 \times 0.87f_y$	$0.8f_{yd}/E_s$	0	$0.8f_{yd}/E_s + 0$
$0.85f_{yd} = 0.85 \times 0.87f_y$	$0.85f_{yd}/E_s$	0.0001	$0.85f_{yd}/E_s + 0.0001$
$0.9f_{yd} = 0.9 \times 0.87f_y$	$0.9f_{yd}/E_s$	0.0003	$0.9f_{yd}/E_s + 0.0003$
$0.95f_{yd} = 0.95 \times 0.87f_y$	$0.95f_{yd}/E_s$	0.0007	$0.95f_{yd}/E_s + 0.0007$
$0.975f_{yd} = 0.975 \times 0.87f_y$	$0.975f_{yd}/E_s$	0.001	$0.975f_{yd}/E_s + 0.001$
$1.0f_{yd} = 1.0 \times 0.87f_y$	$1.0f_{yd}/E_s$	0.002	$1.0f_{yd}/E_s + 0.002$

TABLE 11.3
Total strains at different stress levels for different grades of HYSD reinforcement

	Fe415		Fe500		Fe550		Fe600	
Different stress level	**Stress (f_s) (N/mm²)**	**Strain (ε_s)**	**Stress (f_s) (N/mm²)**	**Strain (ε_s)**	**Stress (f_s) (N/mm²)**	**Strain (ε_s)**	**Stress (f_s) (N/mm²)**	**Strain (ε_s)**
$<0.8f_{yd}$	$\varepsilon_s \times E_s$	<0.00144	$\varepsilon_s \times E_s$	<0.00174	$\varepsilon_s \times E_s$	<0.00191	$\varepsilon_s \times E_s$	<0.00209
$0.8f_{yd}$	288.84	0.00144	348	0.00174	382.8	0.00191	417.6	0.00209
$0.85f_{yd}$	306.8925	0.00163	369.75	0.00195	406.725	0.00213	443.7	0.00232
$0.9f_{yd}$	324.945	0.00192	391.5	0.00226	430.65	0.00245	469.8	0.00265
$0.95f_{yd}$	342.9975	0.00241	413.25	0.00277	454.575	0.00297	495.9	0.00318
$0.975f_{yd}$	352.0237	0.00276	424.125	0.00312	466.537	0.00333	508.95	0.00354
f_{yd}	361.05	0.0038	435	0.00417	478.5	0.00439	522	0.00461

11.18.3 Different Arbitrary Positions of the Neutral Axis

In order to develop the safety profile, a set of arbitrary positions of the neutral axis are considered, which are as follows:

1) When the neutral axis is at infinity i.e., $k = \infty$, in that case the section has only axial capacity;
2) When the neutral axis is outside the sections, i.e., $\infty > kD > D$;
3) When the neutral axis is inside the sections. i.e., $kD < D$;
4) When a column behaves like a steel beam.

For each case the capacities are to be developed and then plotted on a force–moment interaction diagram.

Case 1: When the neutral axis is at infinity i.e., $k = \infty$

$$\frac{x_u}{D} = \infty \text{ i.e., } \frac{kD}{D} = \infty$$

In this case the whole column section is subjected to pure axial load.

The compressive strain in the section is 0.002, corresponding to which the section is under a constant stress of $0.446f_{ck}$. Total compressive force in concrete is $C_c = 0.446f_{ck} \times b \times D$.

Now, the steel bars of area $\left(\frac{pbD}{100}\right)$ are subjected to a constant strain of 0.002, corresponding to which the stress level will depend upon the grades of steel as obtained from the stress–strain relationship of steel. Hence force in steel, $C_s = \left(\frac{pbD}{100}\right) \times \left(f_{sc} - 0.446f_{ck}\right)$

Force in compression concrete is obtained by considering the gross cross-section including reinforcements. Since the area of reinforcement has not been deducted for the force calculation in concrete, the effect is nullified by deducting the stress in surrounding concrete, while calculating the force in reinforcement (refer to Figure 11.8).

Axial force capacity of the section,

$$P_u = C_c + C_s$$

$$\text{or, } P_u = \left[0.446f_{ck}bD\right] + \left(\frac{pbD}{100}\right) \times \left(f_{sc} - 0.446f_{ck}\right)$$

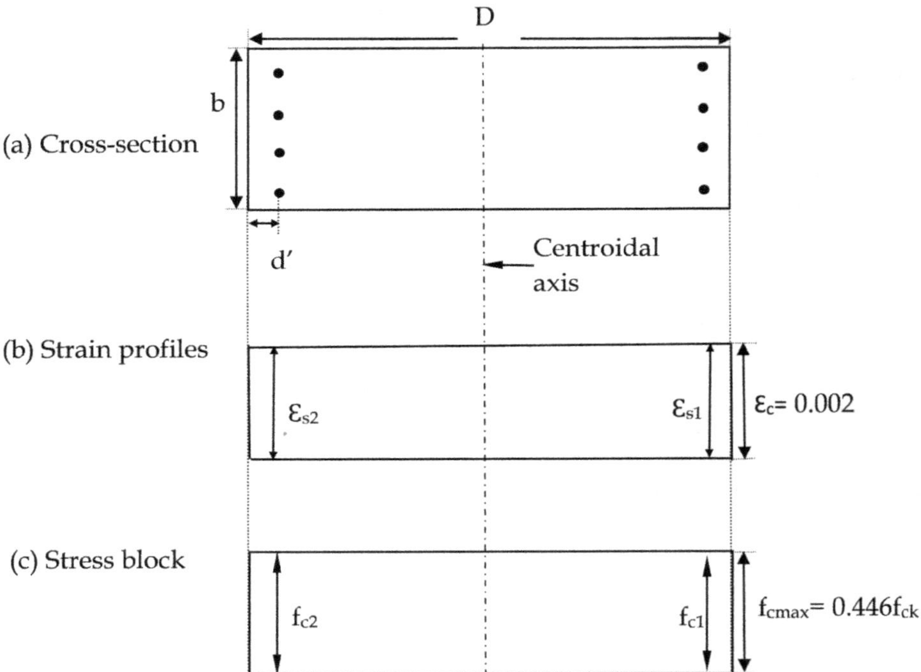

FIGURE 11.8 Strain diagram and stress-block when neutral axis is at infinity.

$$\text{or,}\quad \frac{P_u}{f_{ck}bD} = 0.446 + \left(\frac{p}{100f_{ck}}\right) \times \left(f_{sc} - 0.446f_{ck}\right) \tag{11.4}$$

The compressive force acts along the line of the centroid and hence the moment capacity of the section is zero.

$$\text{So,}\quad \frac{M_u}{f_{ck}bD^2} = 0 \tag{11.5}$$

Case 2: When the neutral axis is outside the sections ($kD > D$)

The strain and stress distribution diagram is shown in Figure 11.6.

Axial force capacity of section,

$$P_u = C_c + C_s$$
$$C_c = C_1 f_{ck} bD$$

where C_1 is the coefficient of area of the stress block as indicated in Table 11.1.

C_s is the force in steel. In this case the whole section is in compression. Hence the axial load capacity of the column in compression may be written as:

$$P_u = C_1 \times f_{ck} bD + \sum_{i=1}^{n} \frac{p_i bD}{100} \times \left(f_{si} - f_{ci}\right)$$

where f_{si} is the strain at the i-th line of reinforcement and f_{ci} is the strain in concrete along that line.

Hence,
$$\frac{P_u}{f_{ck}bD} = C_1 + \sum_{i=1}^{n} \frac{p_i}{100 f_{ck}} \times (f_{si} - f_{ci}) \tag{11.6}$$

and
$$M_u = C_1 f_{ck} bD\,(0.5 - C_2)D + \sum_{i=1}^{n} \frac{p_i bD}{100} \times (f_{si} - f_{ci}) \times y_i$$

where C_2 is the coefficient relating the distance of the compressive force of concrete from the extreme compression fibre and is shown in Table 11.1.

y_i is the distance between the centroidal axis and the i-th line of reinforcement.

Hence, $$\frac{M_u}{f_{ck}bD^2} = C_1(0.5 - C_2) + \sum_{i=1}^{n} \frac{p_i}{100 f_{ck}} \times (f_{si} - f_{ci}) \times \left(\frac{y_i}{D}\right) \tag{11.7}$$

Case 3: When the neutral axis is within the section ($kD < D$)

The strain distribution diagram and the stress block in concrete are indicated in Figure 11.7.

$$C_c = 0.36 f_{ck} \times bD \times k$$
$$C_s = \sum_{i=1}^{n} \frac{p_i bD}{100} \times (f_{si} - f_{ci})$$

Hence,
$$P_u = (0.36 f_{ck} \times bD \times k) + \sum_{i=1}^{n} \frac{p_i bD}{100} \times (f_{si} - f_{ci})$$

or,
$$\frac{P_u}{f_{ck}bD} = 0.36k + \sum_{i=1}^{n} \frac{p_i}{100 f_{ck}} \times (f_{si} - f_{ci}) \tag{11.8}$$

and,
$$M_u = 0.36(k) f_{ck} bD\{0.5 - (0.42 \times k)\}D + \sum_{i=1}^{n} \frac{p_i bD}{100} \times (f_{si} - f_{ci}) \times y_i$$

Hence
$$\frac{M_u}{f_{ck}bD^2} = 0.36(k)\{0.5 - (0.42 \times k)\} + \sum_{i=1}^{n} \frac{p_i}{100 f_{ck}} \times (f_{si} - f_{ci}) \times \frac{y_i}{D} \tag{11.9}$$

A negative sign has been considered for force in tension and for distance measured on the tension side of the centroidal axis and other symbols have their standard meanings.

Case 4: When the column section behaves like a steel beam

As the depth of the neutral axis is reduced, the net force acting on the section slowly turns tensile from compressive. Tensile concrete will be ineffective, but the steel cage will be effective in carrying loads and finally the reinforcements will yield in the tension and compression side resulting in the formation of a hypothetical steel beam. It is a special case when the column has moment capacity only, and the axial load resistance becomes zero. Longitudinal steel in the section is generally symmetrical on the two sides of the centroidal axis. The reinforcement cage is only effective. Hence, the moment capacity may be stated as:

$$M_u = \sum_{i=1}^{n} \left(\frac{p_i bD}{100}\right)(0.87 f_y)(y_i)$$

$$\frac{M_u}{f_{ck} bD^2} = \sum_{i=1}^{n} \left(\frac{p_i}{100 f_{ck}}\right)(0.87 f_y)\left(\frac{y_i}{D}\right) \tag{11.10}$$

$$\frac{P_u}{f_{ck} bD} = 0 \tag{11.11}$$

These formulae are used to prepare design charts for different grades of concrete and steel and different d'/D ratios.

11.19 P–M INTERACTION DIAGRAMS

A number of safety profiles with various percentages of longitudinal reinforcements ranging from zero to maximum percentage permitted by the code are plotted to develop the P–M interaction diagrams corresponding to the grade of concrete and steel and having section dimensions b and D, where b is the width and D is the depth of the rectangular section.

In the present section, P–M interaction charts have been developed for reinforcements equally distributed on four sides of the column. Cases of reinforcements equally distributed on two sides of the column, being quite simple and commonly available, have not been included.

11.19.1 Different steps for the preparation of charts

In the present section, six reinforcement bars have been considered on each face, resulting in 20 bars distributed on four sides of the column.

Considering different arbitrary positions of the neutral axis, a total of nine cases have been considered, which are discussed below.

Case 1: When neutral axis is at infinity.

Case 2: When neutral axis is outside the sections $\left(\frac{x_u}{D}\right) \geq 1$, i.e., k≥1

Assumed value of $\frac{x_u}{D} = 1.1$.

Case 3: When tensile stress at the outermost layer of the longitudinal steel is zero, $f_{st} = 0$.

Case 4: When the tensile stress at the outermost layer of the longitudinal steel is $-0.4 f_{yd} = -0.4 \times 0.87 \times f_y$.

Case 5: When the tensile stress at the outermost layer of the longitudinal steel is $-0.8 f_{yd} = -0.8 \times 0.87 \times f_y$.

Case 6: When the tensile stress at the outermost layer of the longitudinal steel is $-f_{yd} = -0.87 \times f_y$ and strain $= -0.87 f_y / E_s$ (i.e., the initial yield point).

Case 7: When the tensile stress at the outermost layer of the longitudinal steel is $-f_{yd} = -0.87 \times f_y$ and strain $= -\{(0.87 f_y / E_s) + 0.002\}$ (i.e., the final yield point).

Case 8: When $\frac{x_u}{D} = 0.25$.

Case 9: When the column behaves like a steel beam.

These nine points are plotted in a graphical format, whereby $\frac{P_u}{f_{ck}bD}$ is plotted along the Y-axis and $\frac{M_u}{f_{ck}bD^2}$ along the X-axis for a constant percentage of reinforcement to develop the safety profile of the section corresponding to that reinforcement percentage. With different reinforcement percentages, a number of safety profiles for the section can be drawn which will form the P–M interaction chart. A typical numerical problem demonstrating the entire process of development of a safety profile is illustrated in the following example.

11.19.2 Problem Example

Prepare a P–M interaction diagram for a rectangular column section of dimensions b and D with 2.5 percent of longitudinal steel using the following data.

Grade of concrete M40

Grade of steel Fe600 and $\frac{d'}{D} = 0.10.$

Solution: The solution to the problem requires calculations which are lengthy and cumbersome. Hence software has been used for the solution and the results have been directly adopted, because of which the numerical values are provided up to a number of digits beyond the decimal point.

Given data: Grade of concrete (f_{ck}) = 40 N/mm²

Grade of reinforcement (f_y) = 600 N/mm²

$d'/D = 0.10$

Percentage of steel (p_i) = 2.5

Consider six bars along each face, as shown in Figure 11.9.

Total number of reinforcements = 20

Calculations for percentage of steel (p_i) in each row of reinforcement and corresponding distance from the centroidal axis (y_i/D) are shown in Table 11.4.

$$\text{Spacing between bars} = \frac{1 - d' - d'}{6-1}D = \frac{1 - 0.1 - 0.1}{6-1}D = 0.16D \text{ (assuming } d' = 0.1D)$$

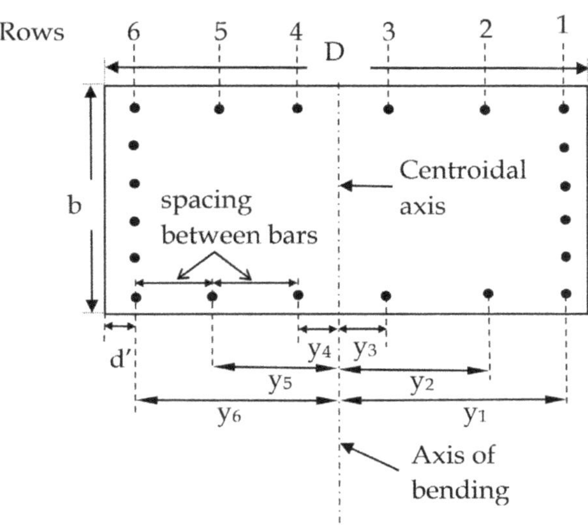

FIGURE 11.9 Reinforcement distribution for rectangular section with reinforcements equally distributed on four sides.

TABLE 11.4
Detailing of reinforcement distribution within the rectangular section

	Number of bars	Percentage of steel in each row (p_i)	Distance from centroidal axis $\left(\frac{y_i}{D}\right)$
Row 1	6	$p_1=\left(\frac{2.5}{20}\right)\times 6 = 0.75\%$	$\frac{y_1}{D}=\frac{1-0.1-0.1}{2}=0.40$
Row 2	2	$p_2=\left(\frac{2.5}{20}\right)\times 2 = 0.25\%$	$\frac{y_2}{D}=0.4-0.16=0.24$
Row 3	2	$p_3=\left(\frac{2.5}{20}\right)\times 2 = 0.25\%$	$\frac{y_3}{D}=0.24-0.16=0.08$
Row 4	2	$p_4=\left(\frac{2.5}{20}\right)\times 2 = 0.25\%$	$\frac{y_4}{D}=0.08-0.16=-0.08$
Row 5	2	$p_5=\left(\frac{2.5}{20}\right)\times 2 = 0.25\%$	$\frac{y_5}{D}=-0.08-0.16=-0.24$
Row 6	6	$p_6=\left(\frac{2.5}{20}\right)\times 6 = 0.75\%$	$\frac{y_6}{D}=-0.24-0.16=-0.40$

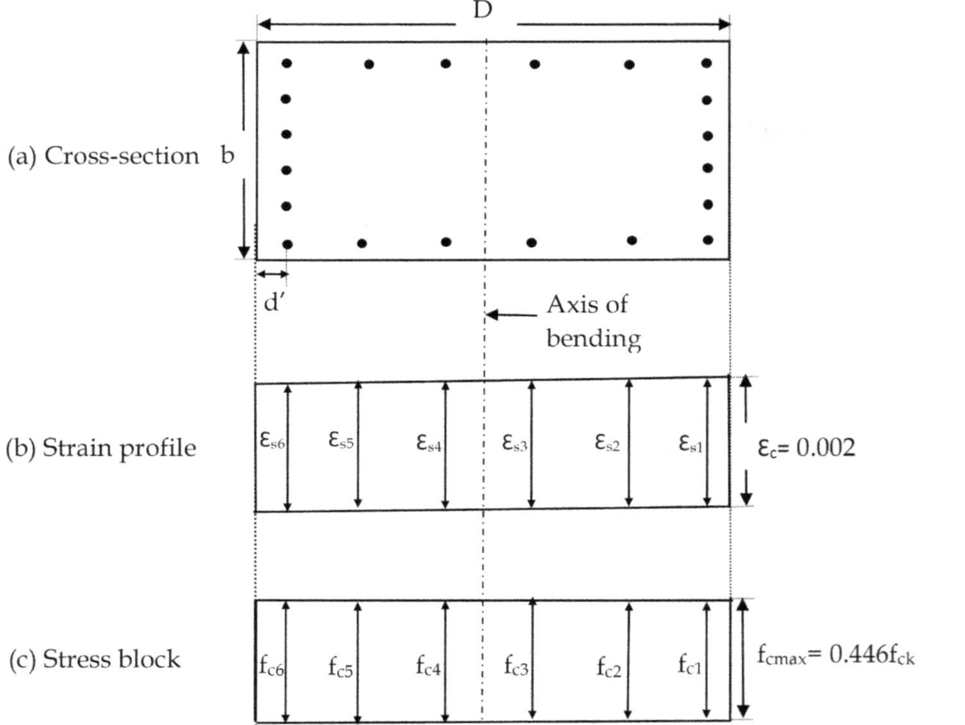

FIGURE 11.10 Strain distribution diagram and stress block for neutral axis at infinity.

Case 1: When neutral axis is at infinity,

$x_u/D = \infty$

In this case, the whole column section is subjected to pure axial load.

Strain in concrete = $\varepsilon_c = 0.002$, stress in concrete is constant at $0.446f_{ck}$.

Strain in steel is also constant for all the bars. From the stress–strain relationship as per Table 11.3, the corresponding stresses in steel (f_{sc}) can be calculated.

$$f_{sc} = 0.002 \times 200000 = 400 \text{ N/mm}^2$$

$$\frac{P_u}{f_{ck}bD} = 0.446 + \left(\frac{p}{100f_{ck}}\right) \times \left(f_{sc} - 0.446 f_{ck}\right)$$

$$\frac{P_u}{f_{ck}bD} = 0.446 + [\left(\frac{2.5}{100 \times 40}\right) \times \left(400 - 0.446 \times 40\right)]$$

$$= 0.446 + 0.23885$$

$$= 0.68485$$

$$\text{And, } \frac{M_u}{f_{ck}bD^2} = 0$$

Case 2: When neutral axis is outside the section ($\frac{x_u}{D} > 1$)

For, $\frac{x_u}{D}$ = 1.1 (refer to Figure 11.6)

For this case, strain at maximum compressed concrete fibre is $\varepsilon_{c,max} = 0.0035 - 0.75\ \varepsilon_{c,min}$, where $\varepsilon_{c,min}$ is the strain at the least compressed fibre.

Hence,

$$\frac{\varepsilon_{c,min}}{(1.1-1)D} = \frac{\varepsilon_{c,max}}{1.1D}$$

or,

$$\frac{\varepsilon_{c,min}}{0.1} = \frac{0.0035 - 0.75\varepsilon_{c,min}}{1.1}$$

$$or, 1.1 \times \varepsilon_{c,min} = 0.1 \times \left(0.0035 - 0.75\ \varepsilon_{c,min}\right)$$

$$or,\ \varepsilon_{c,min} = \frac{(0.0035 \times 0.1)}{1.1 + 0.075}$$

or,

$$\varepsilon_{c,min} = 0.000297$$

Hence,

$$\varepsilon_{c,max} = 0.0035 - \left(0.75 \times \varepsilon_{c,min}\right) = 0.0035 - (0.75 \times 0.000297)$$

$$= 0.003276$$

Again, from similar triangles, strain in concrete in each row of reinforcement may be denoted as ε_{c1}, ε_{c2}, ε_{c3}, ε_{c4}, ε_{c5}, ε_{c6}. (The suffixes 1 to 6 indicate the number of rows of reinforcement.)

$$\varepsilon_{c1} = [0.003276 \times (1.1 - 0.10)]/1.1 = 0.002978$$

$$\varepsilon_{c2} = [0.003276 \times (1.1–0.1–0.16)]/1.1 = 0.002502$$

$$\varepsilon_{c3} = [0.003276 \times (1.1–0.1–0.16–0.16)]/1.1 = 0.002025$$

$$\varepsilon_{c4} = [0.003276 \times (1.1–0.1–0.16–0.16–0.16)]/1.1 = 0.001548$$

$$\varepsilon_{c5} = [0.003276 \times (1.1–0.1–0.16–0.16–0.16–0.16)]/1.1 = 0.001072$$

$$\varepsilon_{c6} = [0.003276 \times (1.1–0.1–0.16–0.16–0.16–0.16–0.16)]/1.1 = 0.000595$$

From strain compatibility, strain in steel (ε_s) is equal to the strain in the surrounding concrete (ε_c). Hence $\varepsilon_{c1} = \varepsilon_{s1}$, $\varepsilon_{c2} = \varepsilon_{s2}$, $\varepsilon_{c3} = \varepsilon_{s3}$, $\varepsilon_{c4} = \varepsilon_{s4}$, $\varepsilon_{c5} = \varepsilon_{s5}$, $\varepsilon_{c6} = \varepsilon_{s6}$.

Using the stress–strain relationship of corresponding grades of steel, the stresses can be calculated, as shown below.

$$f_{s1} = 485.98808 \text{ N/mm}^2$$

$$f_{s2} = 458.10464 \text{ N/mm}^2$$

$$f_{s3} = 405.10638 \text{ N/mm}^2$$

$$f_{s4} = 309.78 \text{ N/mm}^2$$

$$f_{s5} = 214.468 \text{ N/mm}^2$$

$$f_{s6} = 119.149 \text{ N/mm}^2$$

Similarly, stresses in concrete can be determined from the stress–strain relationship of concrete as per IS:456-2000.

$f_{c1} = 0.446 f_{ck} = 0.446 \times 40 = 17.84 \text{ N/mm}^2$

$f_{c2} = 0.446 f_{ck} = 0.446 \times 40 = 17.84 \text{ N/mm}^2$

$f_{c3} = 0.446 \times f_{ck} = 0.446 \times 40 = 17.84 \text{ N/mm}^2$

$$f_{c4} = 0.446 \times f_{ck} \times \left[2 \times \left(\frac{\varepsilon_{c4}}{0.002}\right) - \left(\frac{\varepsilon_{c4}}{0.002}\right)^2\right] = 0.446 \times 40 \times 0.949325 = 16.93596 \text{ N/mm}^2$$

$$f_{c5} = 0.446 \times f_{ck} \times \left[2 \times \left(\frac{\varepsilon_{c5}}{0.002}\right) - \left(\frac{\varepsilon_{c5}}{0.002}\right)^2\right] = 0.446 \times 40 \times 0.784428 = 14.001936 \text{ N/mm}^2$$

$$f_{c6} = 0.446 \times f_{ck} \times \left[2 \times \left(\frac{\varepsilon_{c6}}{0.002}\right) - \left(\frac{\varepsilon_{c6}}{0.002}\right)^2\right] = 0.446 \times 40 \times 0.5070 = 9.044957 \text{ N/mm}^2$$

From Table 11.1,

for $\frac{x_u}{D} = 1.1$, $C_1 = 0.384$ and $C_2 = 0.443$

Hence, $\dfrac{P_u}{f_{ck}bD} = C_1 + \sum_{i=1}^{n} \dfrac{p_i}{100 f_{ck}} \times (f_{si} - f_{ci})$

$$= 0.384 + \frac{1}{100} \times [\{\frac{0.75}{40} \times (485.988 - 17.84)\} + \{\frac{0.25}{40} \times (458.104 - 17.84)\}$$
$$+ \{\frac{0.25}{40} \times (405.106 - 17.84)\} + \{\frac{0.25}{40} \times (309.78 - 16.935)\}$$
$$+ \{\frac{0.25}{40} \times (214.468 - 14.002)\} + \{\frac{0.75}{40} \times (119.149 - 9.044957)\}]$$

$$= 0.384 + 0.190975$$

$$= 0.574975$$

$$\frac{M_u}{f_{ck}bD^2} = C_1(0.5 - C_2) + \sum_{i=1}^{n} \frac{p_i}{100 f_{ck}} \times (f_{si} - f_{ci}) \times y_i$$
$$= [0.384 \times (0.5 - 0.443)] + \frac{1}{100} \times [\{\frac{0.75}{40} \times (485.98808 - 17.84) \times 0.4\}$$
$$+ \{\frac{0.25}{40} \times (458.10464 - 17.84) \times 0.24\} + \{\frac{0.25}{40} \times (405.106 - 17.84) \times 0.08\}$$
$$+ \{\frac{0.25}{40} \times (309.78 - 16.93596) \times (-0.08)\} + \{\frac{0.25}{40} \times (214.468 - 14.002) \times (-0.24)\}$$
$$+ \{\frac{0.75}{40} \times (119.149 - 9.0449) \times (-0.4)\}]$$
$$= 0.021888 + 0.030922$$
$$= 0.052810$$

Hence the coordinates of the point representing this condition are

$$\frac{P_u}{f_{ck}bD} = 0.574975$$

And $\dfrac{M_u}{f_{ck}bD^2} = 0.05281036$

Case 3: when tensile stress at outermost layer of longitudinal steel is zero, $f_{st} = 0$

$$x_u = D - d'$$

Since $d'/D = 0.1$, $\dfrac{x_u}{D} = 1 - 0.1 = 0.9$

Therefore, $\varepsilon_{c6} = \varepsilon_{s6} = 0$

$\varepsilon_{c,max} = 0.0035$

From similar triangles,

$$\frac{\varepsilon_{c1}}{(1 - 0.1 - 0.1)D} = \frac{\varepsilon_{c,max}}{(1 - 0.1)D}$$

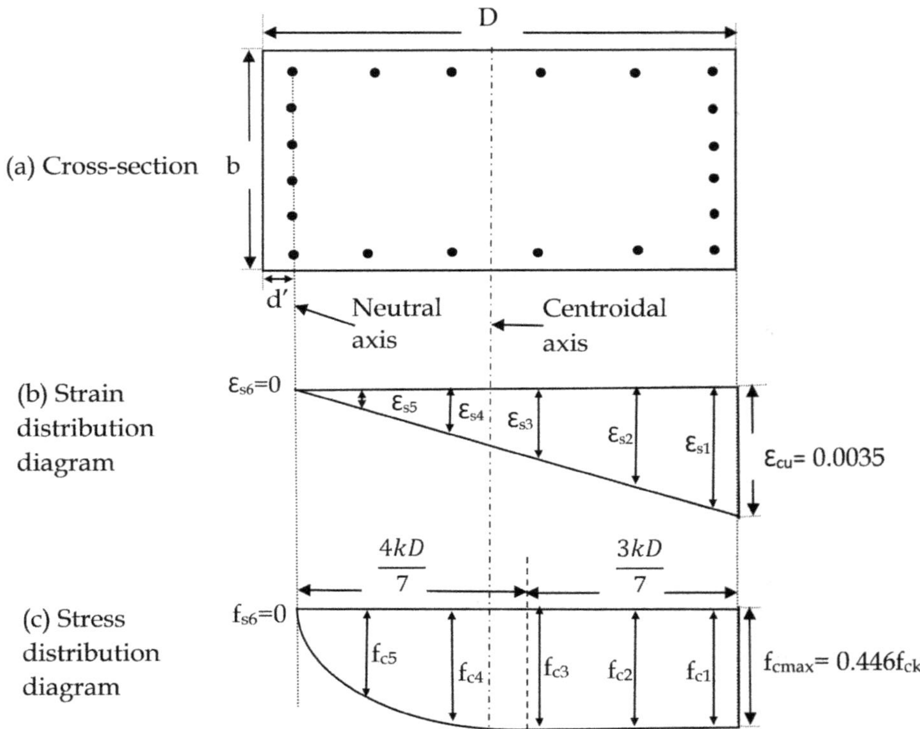

FIGURE 11.11 Strain distribution and stress block of concrete at $x_u/D = (D - d')/D$.

$$\text{Therefore, } \varepsilon_{c1} = \frac{\{0.0035 \times (0.90 - 0.10)\}}{0.90} = 0.00311$$

$$\varepsilon_{c2} = [0.0035 \times (0.9 - 0.1 - 0.16)]/0.9 = 0.002488$$

$$\varepsilon_{c3} = [0.0035 \times (0.9 - 0.1 - 0.16 - 0.16)]/0.9 = 0.001866$$

$$\varepsilon_{c4} = [0.0035 \times (0.9 - 0.1 - 0.16 - 0.16 - 0.16)]/0.9 = 0.001244$$

$$\varepsilon_{c5} = [0.0035 \times (0.9 - 0.1 - 0.16 - 0.16 - 0.16 - 0.16)]/0.9 = 0.000622$$

$$\varepsilon_{c6} = [0.0035 \times (0.9 - 0.1 - 0.16 - 0.16 - 0.16 - 0.16 - 0.16)]/0.9 = 0$$

From strain compatibility, $\varepsilon_{c1} = \varepsilon_{s1}$, $\varepsilon_{c2} = \varepsilon_{s2}$, $\varepsilon_{c3} = \varepsilon_{s3}$, $\varepsilon_{c4} = \varepsilon_{s4}$, $\varepsilon_{c5} = \varepsilon_{s5}$, $\varepsilon_{c6} = \varepsilon_{s6}$.

Using the stress–strain relationship corresponding stresses are calculated which are as follows:

$$f_{s1} = 492.50755 \text{ N/mm}^2$$

$$f_{s2} = 457.05758 \text{ N/mm}^2$$

$$f_{s3} = 373.3333 \text{ N/mm}^2$$

$$f_{s4} = 248.88889 \text{ N/mm}^2$$

$$f_{s5} = 124.44444 \text{ N/mm}^2$$

$$f_{s6} = 0 \text{ N/mm}^2$$

Stresses in concrete calculated from strains are as follows:

$$f_{c1} = 0.446 f_{ck} = 0.446 \times 40 = 17.84 \text{ N/mm}^2$$

$$f_{c2} = 0.446 f_{ck} = 0.446 \times 40 = 17.84 \text{ N/mm}^2$$

$$f_{c3} = 0.446 \times f_{ck} \times \left[2\times\left(\frac{\varepsilon_{c3}}{0.002}\right)-\left(\frac{\varepsilon_{c3}}{0.002}\right)^2\right]$$

$$= 0.446 \times 40 \times \left[2\times\left(\frac{0.001866}{0.002}\right)-\left(\frac{0.001866}{0.002}\right)^2\right] = 0.446 \times 40 \times 0.995555$$

$$= 17.7607 \text{ N/mm}^2$$

$$f_{c4} = 0.446 \times f_{ck} \times \left[2\times\left(\frac{\varepsilon_{c4}}{0.002}\right)-\left(\frac{\varepsilon_{c4}}{0.002}\right)^2\right]$$

$$= 0.446 \times 40 \times \left[2\times\left(\frac{0.001244}{0.002}\right)-\left(\frac{0.001244}{0.002}\right)^2\right] = 0.446 \times 40 \times 0.857286$$

$$= 15.29398 \text{ N/mm}^2$$

$$f_{c5} = 0.446 \times f_{ck} \times \left[2\times\left(\frac{\varepsilon_{c5}}{0.002}\right)-\left(\frac{\varepsilon_{c5}}{0.002}\right)^2\right]$$

$$= 0.446 \times 40 \times \left[2\times\left(\frac{0.000622}{0.002}\right)-\left(\frac{0.000622}{0.002}\right)^2\right] = 0.446 \times 40 \times 0.525432$$

$$= 9.373717 \text{ N/mm}^2$$

$$f_{c6} = 0 \text{ N/mm}^2$$

$$\text{Now } \frac{P_u}{f_{ck}bD} = 0.36 \times\left(\frac{x_u}{D}\right) + \sum_{i=1}^{n} \frac{p_i}{100 f_{ck}} \times \left(f_{si} - f_{ci}\right)$$

$$= (0.36\times0.9) + \left(\frac{1}{100}\right)\times[\{\frac{0.75}{40}\times(492.50755-17.84)\} + \{\frac{0.25}{40}\times(457.05758-17.84)\}$$

$$+ \{\frac{0.25}{40}\times(373.3333 - 17.7607)\} + \{\frac{0.25}{40}\times(248.88889 - 15.29398)\}$$

$$+ \{\frac{0.25}{40}\times(124.44444\text{-}9.373717)\} + \{\frac{0.75}{40}\times(0 - 0)\}]$$

$$= 0.324 + 0.160466$$

$$= 0.484466$$

$$\frac{M_u}{f_{ck}bD^2} = 0.36 \times \left(\frac{x_u}{D}\right) \times (0.5 - 0.42 \times \left(\frac{x_u}{D}\right)) + \sum_{i=1}^{n} \frac{p_i}{100 f_{ck}} \times \left(f_{si} - f_{ci}\right) \times \frac{y_i}{D}$$

$$= 0.36 \times 0.9 \times (0.5 - 0.42 \times 0.9) + \left(\frac{1}{100}\right) \times [\{\frac{0.75}{40} \times (492.50755 - 17.84) \times 0.4\}$$

$$+ \{\frac{0.25}{40} \times (457.05758 - 17.84) \times 0.24\} + \{\frac{0.25}{40} \times (373.3333 - 17.7607) \times 0.08\}$$

$$+ \{\frac{0.25}{40} \times (248.88889 - 15.29398) \times (-0.08)\}$$

$$+ \{\frac{0.25}{40} \times (124.44444 - 9.373717) \times (-0.24)\}$$

$$+ \{\frac{0.75}{40} \times (0 - 0) \times (-0.4)\}]$$

$$= 0.039528 + 0.041072$$

$$= 0.080600$$

Hence the coordinates of the point representing this condition are

$$\frac{P_u}{f_{ck}bD} = 0.484466 \text{ and } \frac{M_u}{f_{ck}bD^2} = 0.0806001$$

Case 4: When the tensile stress at the outermost layer of longitudinal steel is $-0.4f_{yd} = -0.4 \times 0.87 \times f_y$

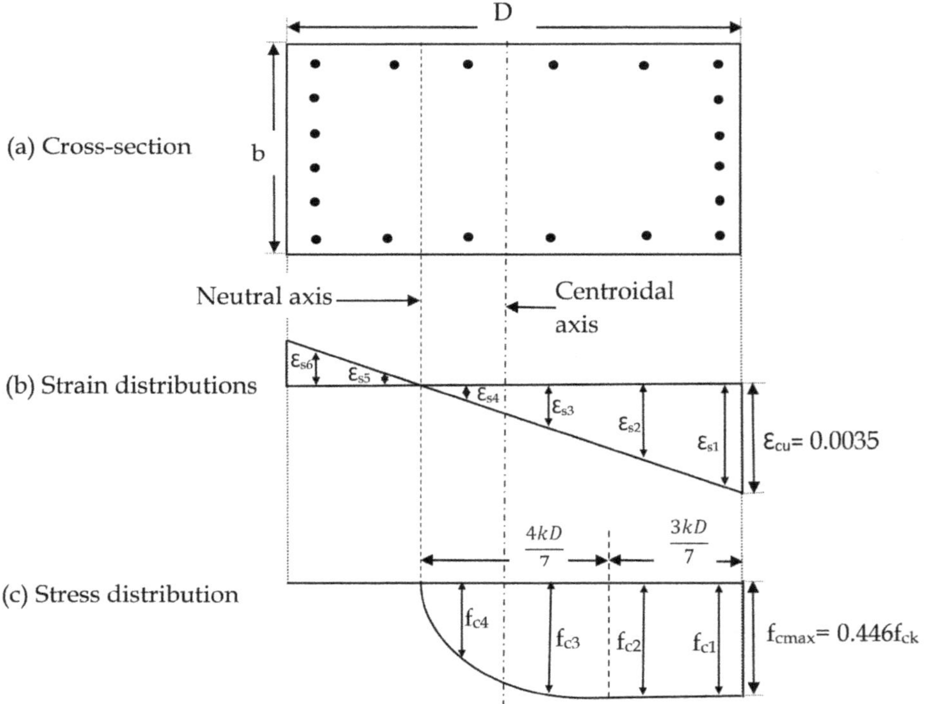

FIGURE 11.12 Strain distribution diagram and stress block when the tensile stress at the outermost layer of longitudinal steel is $-0.4f_{yd} = -0.4 \times 0.87 \times f_y$.

Here, f_{st} at the sixth row of reinforcement = (–0.4 × 0.87 × 600) = (–208.8) N/mm^2 (tensile).

Now from the stress–strain relation strain can be determined at position of longitudinal steel at outermost edge.

$$\varepsilon_{s6} = \left(\frac{0.87 \times 600}{200000}\right) \times 0.4 = 0.001044 \text{ (tensile)}$$

From similar triangles,

$$\frac{0.0035}{\frac{x_u}{D}} = \frac{0.001044}{\left\{1 - 0.10 - \left(\frac{x_u}{D}\right)\right\}}$$

$$\text{or, } \frac{x_u}{D} \times 0.001044 = 0.0035 \times (1 - 0.10) - \left\{0.0035 \times \left(\frac{x_u}{D}\right)\right\}$$

$$\text{or, } \frac{x_u}{D} = \frac{0.0035 \times 0.90}{0.001044 + 0.0035} \quad \text{or, } \frac{x_u}{D} = 0.693221$$

Similarly,

$$\frac{\varepsilon_{c1}}{0.69322 - 0.10} = \frac{\varepsilon_{c,max}}{0.69322}$$

$$\text{Therefore, } \varepsilon_{c1} = \frac{\{0.0035 \times (0.69322 - 0.10)\}}{0.69322} = 0.002995$$

$$\varepsilon_{c2} = [0.0035 \times (0.69322 - 0.1 - 0.16)]/0.69322 = 0.002187$$

$$\varepsilon_{c3} = [0.0035 \times (0.69322 - 0.1 - 0.16 - 0.16)]/0.69322 = 0.001379$$

$$\varepsilon_{c4} = [0.0035 \times (0.69322 - 0.1 - 0.16 - 0.16 - 0.16)]/0.69322 = 0.000571$$

$$\varepsilon_{c5} = [0.0035 \times (0.69322 - 0.1 - 0.16 - 0.16 - 0.16 - 0.16)]/0.69322 = -0.000236$$

$$\varepsilon_{c6} = [0.0035 \times (0.69322 - 0.1 - 0.16 - 0.16 - 0.16 - 0.16 - 0.16)]/0.69322 = -0.001044$$

From strain compatibility, $\varepsilon_{c1} = \varepsilon_{s1}$, $\varepsilon_{c2} = \varepsilon_{s2}$, $\varepsilon_{c3} = \varepsilon_{s3}$, $\varepsilon_{c4} = \varepsilon_{s4}$, $\varepsilon_{c5} = \varepsilon_{s5}$, $\varepsilon_{c6} = \varepsilon_{s6}$.

From the stress–strain diagram of steel, the corresponding stresses are calculated and mentioned below.

$$f_{s1} = 486.79509 \text{ N/mm}^2$$

$$f_{s2} = 428.64017 \text{ N/mm}^2$$

$$f_{s3} = 275.8933 \text{ N/mm}^2$$

$$f_{s4} = 114.32888 \text{ N/mm}^2$$

$$f_{s5} = -47.23555 \text{ N/mm}^2$$

$$f_{s6} = -208.60019 \text{ N/mm}^2$$

Similarly, stresses in concrete at the different reinforcement lines have been calculated and are shown below,

$$f_{c1} = 0.446 f_{ck} = 0.446 \times 40 = 17.84 \text{ N/mm}^2$$

$$f_{c2} = 0.446 f_{ck} = 0.446 \times 40 = 17.84 \text{ N/mm}^2$$

$$f_{c3} = 0.446 \times f_{ck} \times \left[2 \times \left(\frac{\varepsilon_{c3}}{0.002} \right) - \left(\frac{\varepsilon_{c3}}{0.002} \right)^2 \right] = 0.446 \times 40 \times 0.903732 = 16.122587 \text{ N/mm}^2$$

$$f_{c4} = 0.446 \times f_{ck} \times \left[2 \times \left(\frac{\varepsilon_{c4}}{0.002} \right) - \left(\frac{\varepsilon_{c4}}{0.002} \right)^2 \right] = 0.446 \times 40 \times 0.489944 = 8.7407 \text{ N/mm}^2$$

$$f_{c5} = f_{c6} = 0$$

Using the similar expressions of $\frac{P_u}{f_{ck} bD}$ and $\frac{M_u}{f_{ck} bD^2}$ as mentioned in the previous case, coordinates of the point representing this condition are,

$$\frac{P_u}{f_{ck} bD} = 0.343913 \text{ and } \frac{M_u}{f_{ck} bD^2} = 0.110576$$

Case 5: When the tensile stress in longitudinal steel at the outermost layer is $-0.8 f_{yd} = -0.8 \times 0.87 \times f_y$

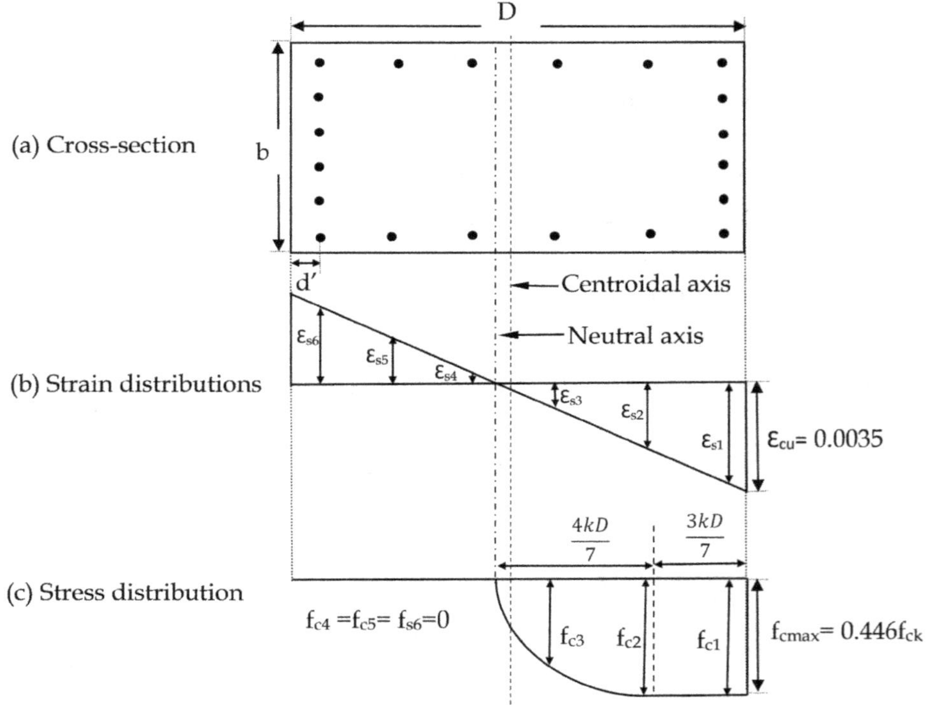

FIGURE 11.13 Stress block and strain profile when the tensile stress at the outermost layer of longitudinal steel is $-0.8 f_{yd} = -0.8 \times 0.87 \times f_y$.

Performing calculations in a similar manner as shown in the previous case, the strains and stresses in concrete and steel have been calculated and are mentioned below.

For the stress at the outermost layer of steel = $-0.8f_{yd}$,

$$\frac{x_u}{D} = 0.563707$$

Strains at the different line of reinforcement are as follows:

$$\varepsilon_{c1} = \frac{\{0.0035 \times (0.5637 - 0.10)\}}{0.5637} \quad \text{or, } \varepsilon_{c1} = 0.002879$$

$$\varepsilon_{c2} = [0.0035 \times (0.5637 - 0.1 - 0.16)]/0.5637 = 0.001885$$

$$\varepsilon_{c3} = [0.0035 \times (0.5637 - 0.1 - 0.16 - 0.16)]/0.5637 = 0.000892$$

$$\varepsilon_{c4} = [0.0035 \times (0.5637 - 0.1 - 0.16 - 0.16 - 0.16)]/0.5637 = -0.000101$$

$$\varepsilon_{c5} = [0.0035 \times (0.5637 - 0.1 - 0.16 - 0.16 - 0.16 - 0.16)]/0.5637 = -0.001094$$

$$\varepsilon_{c6} = [0.0035 \times (0.5637 - 0.1 - 0.16 - 0.16 - 0.16 - 0.16 - 0.16)]/0.5637 = -0.002088$$

The corresponding stresses in steel and concrete are as follows:

$$f_{s1} = 481.08264 \text{ N/mm}^2$$

$$f_{s2} = 376.77688 \text{ N/mm}^2$$

$$f_{s3} = 178.28256 \text{ N/mm}^2$$

$$f_{s4} = -20.21175 \text{ N/mm}^2$$

$$f_{s5} = -218.70607 \text{ N/mm}^2$$

$$f_{s6} = -417.20038 \text{ N/mm}^2$$

Hence,

$$f_{c1} = 17.84 \text{ N/mm}^2$$

$$f_{c2} = 17.781697 \text{ N/mm}^2$$

$$f_{c3} = 12.36689 \text{ N/mm}^2$$

$$f_{c4} = f_{c5} = f_{c6} = 0$$

Hence, performing calculations in a similar manner, coordinates of the point representing this condition are:

$$\frac{P_u}{f_{ck}bD} = 0.2294$$

$$\frac{M_u}{f_{ck}bD^2} = 0.12905$$

Case 6: When the tensile stress in the outermost layer of longitudinal steel is $-f_{yd} = -0.87 \times f_y$ and strain = $-0.87f_y/E_s$ (i.e., the initial yield point)

Therefore, $f_{s6} = -(0.87 \times 600) = (-522)$ N/mm² (tensile)

For this case, $\frac{x_u}{D} = 0.515548$

Strains at the different lines of reinforcement are as follows:

$$\varepsilon_{c1} = \varepsilon_{s1} = 0.002821$$

$$\varepsilon_{c2} = \varepsilon_{s2} = 0.001734$$

$$\varepsilon_{c3} = \varepsilon_{s3} = 0.000648$$

$$\varepsilon_{c4} = \varepsilon_{s4} = -0.000437$$

$$\varepsilon_{c5} = \varepsilon_{s5} = -0.001524$$

$$\varepsilon_{c6} = \varepsilon_{s6} = -0.002610$$

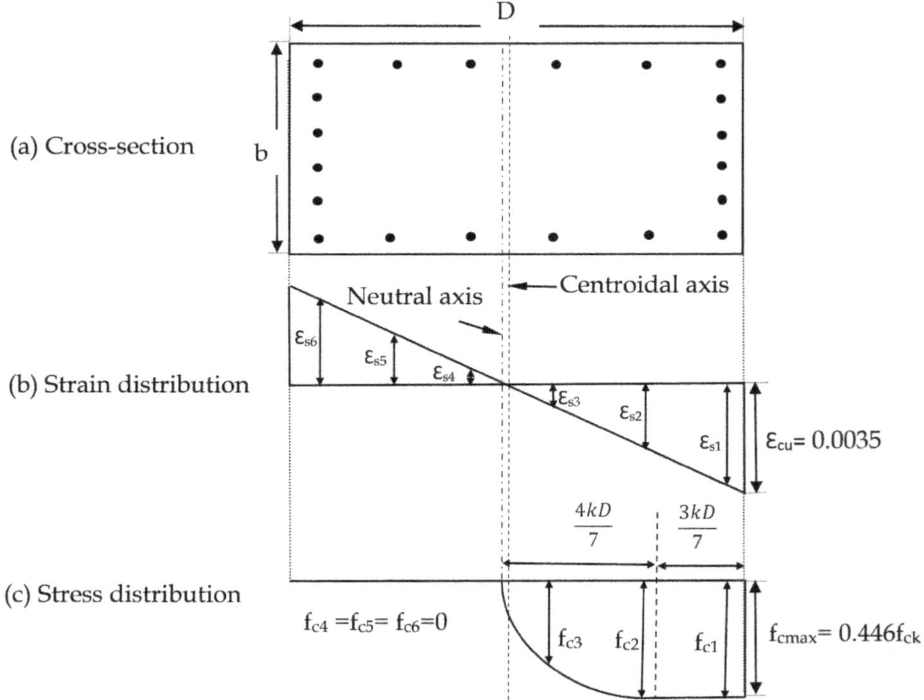

FIGURE 11.14 Strain distribution and stress block when the tensile stress at the outermost layer of longitudinal steel is $-f_{yd} = -0.87 \times f_y$ and strain = $-0.87f_y/E_s$.

The corresponding stresses in steel and concrete are shown below:

$$f_{s1} = 478.226415 \text{ N/mm}^2$$

$$f_{s2} = 346.9777 \text{ N/mm}^2$$

$$f_{s3} = 129.73333 \text{ N/mm}^2$$

$$f_{s4} = -87.427368 \text{ N/mm}^2$$

$$\text{f}_{s5} = -304.463923 \text{ N/mm}^2$$

$$f_{s6} = -522 \text{ N/mm}^2$$

Stresses in concrete are as follows:

$$f_{c1} = 17.84 \text{ N/mm}^2$$

$$f_{c2} = 17.52614 \text{ N/mm}^2$$

$$f_{c3} = 9.692366 \text{ N/mm}^2$$

$$f_{c4} = f_{c5} = f_{c6} = 0$$

Hence, coordinates of the point representing this condition are

$$\frac{P_u}{f_{ck}bD} = 0.177621$$

$$\frac{M_u}{f_{ck}bD^2} = 0.136841$$

Case 7: When the tensile stress at the outermost layer of longitudinal steel is $-f_{yd} = -0.87 \times f_y$ and strain $= -\left\{\left(\frac{0.87f_y}{E_s}\right) + 0.002\right\}$ (i.e., the final yield point) (balanced condition) (Refer Figure 11.15)

$$f_{s6} = -(0.87 \times 600) = (-522) \text{ N/mm}^2 \text{ (tensile)}$$

$$\frac{x_u}{D} = 0.388409$$

Strains at the different lines of reinforcement are as follows:

$$\varepsilon_{c1} = \varepsilon_{s1} = 0.002598$$

$$\varepsilon_{c2} = \varepsilon_{s2} = 0.001157$$

$$\varepsilon_{c3} = \varepsilon_{s3} = -0.000284$$

$$\varepsilon_{c4} = \varepsilon_{s4} = -0.001726$$

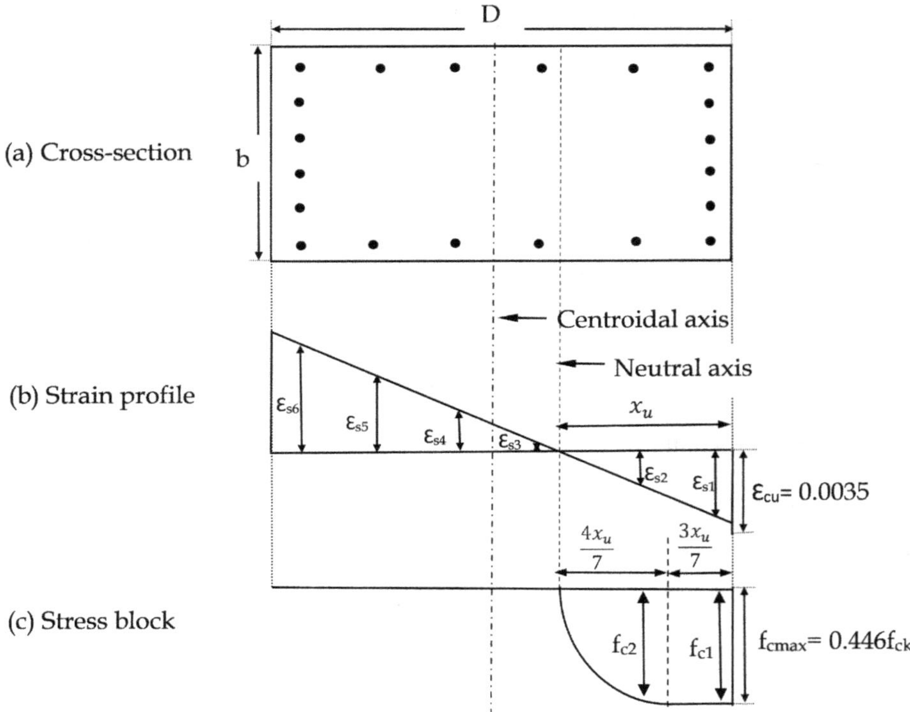

FIGURE 11.15 Strain distribution and stress block when the tensile stress at the outermost layer of longitudinal steel is $-f_{yd} = -0.87 \times f_y$ and strain = $-((0.87f_y)/E_s) + 0.002$.

$$\varepsilon_{c5} = \varepsilon_{s5} = -0.003168$$

$$\varepsilon_{c6} = \varepsilon_{s6} = -0.004610$$

Stresses in steels are:

$$f_{s1} = 465.757575 \text{ N/mm}^2$$

$$f_{s2} = 231.200765 \text{ N/mm}^2$$

$$f_{s3} = -56.878851 \text{ N/mm}^2$$

$$f_{s4} = -344.958468 \text{ N/mm}^2$$

$$f_{s5} = -495.32 \text{ N/mm}^2$$

$$f_{s6} = -522 \text{ N/mm}^2$$

Stresses in concrete are:

$$f_{c1} = 17.84 \text{ N/mm}^2$$

$$f_{c2} = 14.670915 \text{ N/mm}^2$$

$$f_{c3} = f_{c4} = f_{c5} = f_{c6} = 0$$

Hence coordinates of the point representing this condition are

$$\frac{P_u}{f_{ck} bD} = 0.083397$$

$$\frac{M_u}{f_{ck} bD^2} = 0.131965$$

Case 8: When the depth of the neutral axis, $\frac{x_u}{D}$ = 0.25 (under-reinforced condition)

Failure in concrete will occur at a flexural compressive strain of 0.0035, joining this strain value with the neutral axis depth by a straight line, the strain distribution diagram of the section can be obtained. Accordingly, the strain and stresses have been calculated and are shown below (Refer Figure 11.16).

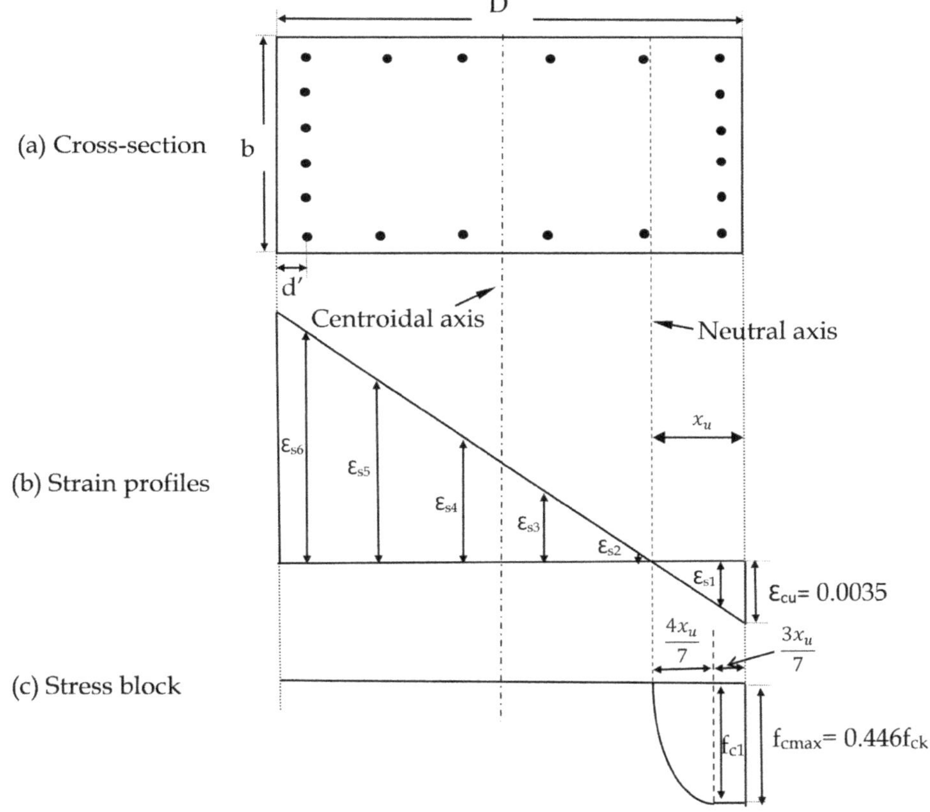

FIGURE 11.16 Strain distribution and stress block and when the depth of neutral axis: $\frac{x_u}{D} = 0.25$.

$$\varepsilon_{c1} = \varepsilon_{s1} = 0.0021$$

$$\varepsilon_{c2} = \varepsilon_{s2} = -0.00014$$

$$\varepsilon_{c3} = \varepsilon_{s3} = -0.00238$$

$$\varepsilon_{c4} = \varepsilon_{s4} = -0.00462$$

$$\varepsilon_{c5} = \varepsilon_{s5} = -0.00686$$

$$\varepsilon_{c6} = \varepsilon_{s6} = -0.0091$$

Stresses in steel:

$$f_{s1} = 418.734782 \text{ N/mm}^2$$

$$f_{s2} = -27.973205 \text{ N/mm}^2$$

$$f_{s3} = -448.445454 \text{ N/mm}^2$$

$$f_{s4} = -522 \text{ N/mm}^2$$

$$f_{s5} = -522 \text{ N/mm}^2$$

$$f_{s6} = -522 \text{ N/mm}^2$$

Stresses in concrete:

$$f_{c1} = 17.84 \text{ N/mm}^2$$

$$f_{c2} = f_{c3} = f_{c4} = f_{c5} = f_{c6} = 0$$

Hence, coordinates of the point representing this condition are

$$\frac{P_u}{f_{ck}bD} = -0.027733$$

$$\frac{M_u}{f_{ck}bD^2} = 0.112545$$

Case 9: When a column behaves like a steel beam

This is the specific condition when the section behaves as a hypothetical steel beam and the concrete becomes completely ineffective. Steel on either side of the centroidal axis yields and reaches a constant stress of $0.87f_y$. Hence, force in steel reinforcement on either side of the centroidal axis is equal in magnitude but opposite in sign. Therefore, the axial load capacity of the columns is zero but the moment capacity is developed by compressive and tensile forces in the reinforcement on either side of centroidal axis.

Therefore, $M_u = \sum_{i=1}^{n} \left(\frac{p_i bD}{100}\right)(0.87 f_y).y_i$

Hence, $\frac{M_u}{f_{ck} bD^2} = \sum_{i=1}^{n} \left(\frac{p_i}{100 f_{ck}}\right)(0.87 f_y)\left(\frac{y_i}{D}\right)$

$$= \frac{1}{100} \times (0.87 \times 600) \times [\{\frac{0.75}{40} \times 0.4\} + \{\frac{0.25}{40} \times 0.24\} + \{\frac{0.25}{40} \times 0.08\}$$

$$+ \{\frac{0.25}{40} \times 0.08\} + \{\frac{0.25}{40} \times 0.24\} + \{\frac{0.75}{40} \times 0.4\}]$$

$= 0.09918$

Hence, the coordinates of the point representing this condition are:

$$\frac{P_u}{f_{ck} bD} = 0$$

$$\frac{M_u}{f_{ck} bD^2} = 0.09918$$

Table 11.5 shows coordinates of the nine points which will constitute the safety profile of the section corresponding to the reinforcement percentage considered, i.e., 2.5.

These nine pairs are now plotted in the graph in Figure 11.17 to prepare the safety profile of the section.

TABLE 11.5
Values of non-dimensional parameters for the nine different section conditions for plotting the axial force – moment safety profile

Cases	Illustrations of case	$\frac{P_u}{f_{ck} bD}$	$\frac{M_u}{f_{ck} bD^2}$
Case 1	When neutral axis is at infinity	0.6848	0
Case 2	When neutral axis is outside the section $\left(\frac{x_u}{D} \geq 1\right)$, $\frac{x_u}{D} = 1.1$	0.5749	0.0528
Case 3	When tensile stress at outermost layer of longitudinal steel is zero, $f_{st} = 0$	0.4844	0.0806
Case 4	When the tensile stress at outermost layer of longitudinal steel is $-0.4f_{yd} = -0.4 \times 0.87 \times f_y$	0.3439	0.1106
Case 5	When the tensile stress at outermost layer of longitudinal steel is $-0.8f_{yd} = -0.8 \times 0.87 \times f_y$	0.2294	0.1291
Case 6	When the tensile stress at outermost layer of longitudinal steel is $-f_{yd} = -0.87 \times f_y$ and strain $= -0.87f_y/E_s$	0.1776	0.1368
Case 7	When the tensile stress at outermost layer of longitudinal steel is $-f_{yd} = -0.87 \times f_y$ and strain $= -\{((0.87f_y)/E_s) + 0.002\}$	0.0834	0.1319
Case 8	When the depth of neutral axis $\frac{x_u}{D} = 0.25$	-0.0277	0.1125
Case 9	When a column behaves like a steel beam	0	0.0992

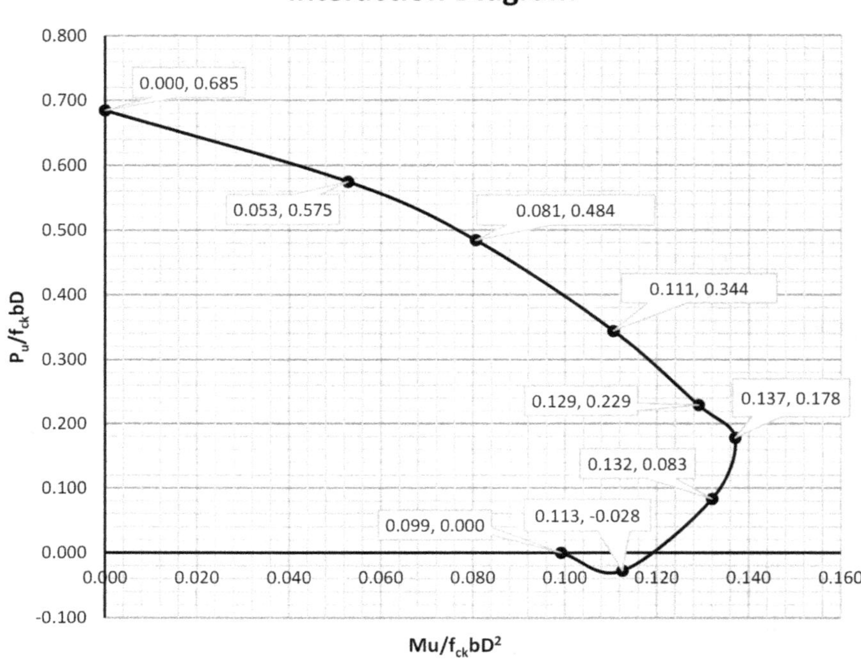

FIGURE 11.17 Safety profile of the section with 2.5 percent of longitudinal reinforcement prepared from values of Table 11.5.

From Figure 11.17 it is observed that a portion of the curve goes below the moment axis indicating tension in the cross-section along with bending. This portion can be ignored as the basic intention is to develop charts for compression with bending. It is to be noted that the interaction charts are never prepared for any external load cases, rather they are derived corresponding to the axial load and moment capacities of the section.

11.20 DESIGN AIDS FOR COMPRESSION DESIGN AND SHORTCOMINGS OF SP16

As discussed in the previous sections, different percentages of longitudinal reinforcements may be assumed ranging from zero to the maximum and the corresponding safety profiles can be drawn. IS:456:2000 allows nine grades of structural concrete ranging from M20 to M60 and grades of HYSD steel generally used are of Fe 415, Fe 500, Fe 550 and Fe600. These steel grades are also permitted in IRC:112-2020. Values of *d'/D* can range from 0.05 to 0.2 with 0.05 as the common interval, hence the total number of interaction charts for reinforcement distributed on two sides will be equal to $4 \times 9 \times 4$, which is 144.

For any specific number of bars equally distributed on four sides, the number of charts required will be 144. In order to reduce the number of charts, the interaction diagrams have been prepared with only M40 grade of concrete (intermediate grade between M20 and M60) with p/f_{ck} values. The variations in values obtained from charts for M20 and M60 grades of concrete from these charts have been observed to be significantly small.

11.20.1 Shortcomings of SP16 with Respect to Design Charts in Compression

SP16 was developed for the third revision of IS:456, i.e., IS:456-1978, which allowed structural concrete grades up to M40. The interaction charts for column design presented in SP16 have been prepared for M20 grade concrete and HYSD steel grades of Fe415 and Fe500 only. Nowadays, higher grades of concrete and steel are being commercially used, and hence the use of these charts has been significantly reduced. It may be mentioned that the SP16 has been withdrawn by the Bureau of Indian Standards.

11.21 INTERACTION CHARTS FOR RECTANGULAR COLUMN SECTIONS VALID FOR M20–M60 GRADES OF CONCRETE AND FE 415, FE 500, FE 550 AND FE 600 STEEL WITH *d′/D* VALUES OF 0.05–0.20 AS PER THE FUNDAMENTAL PRINCIPLES OF THE LIMIT STATE METHOD

Since the variation in the results is not significant with a change in grades of concrete, in order to reduce the number of charts, interaction charts have been developed only for the intermediate grade of concrete, that is M40, and reinforcement percentages have been furnished as multiples of f_{ck} so that the desired result for any grade of concrete can easily be obtained. In the subsequent section, a total of 32 charts have been provided for reinforcements equally distributed on two sides (Charts 11.1–11.16) and four sides (Charts 11.17–11.32). For distribution on four sides, six bars have been considered on each face. Four sets of d'/D values, namely 0.05, 0.1, 0.15 and 0.2, have been considered in the development of these charts.

For each of the two and four sides' distribution of reinforcement, four d'/D values and four grades of steel have been considered. Hence, these charts can comply to the requirements of all grades of structural concrete and commonly used HYSD bars and can act as very useful tools for the design of sections subjected to compression with bending.

A close observation of these interaction charts reveals that the failure of the sections in under-reinforced condition occupies only a small lower portion of the diagram and the majority of failures will occur in over-reinforced condition. Whenever axial load is introduced into the section along with flexure there is a reduction in the ductility of the section, as indicated in Chapter 17. Wherever possible designers should try to select the column dimension in such a manner that the effect of axial load is reduced, i.e., the value of $P_u/f_{ck}bD$ is less so that even if the section is not under-reinforced the effect of the axial force can be reduced so that the reduction in ductility is not very significant.

11.22 INTERACTIONS CHARTS FOR A RECTANGULAR SECTION WITH REINFORCEMENT DISTRIBUTED ON TWO SIDES (CHARTS 11.1–11.16)

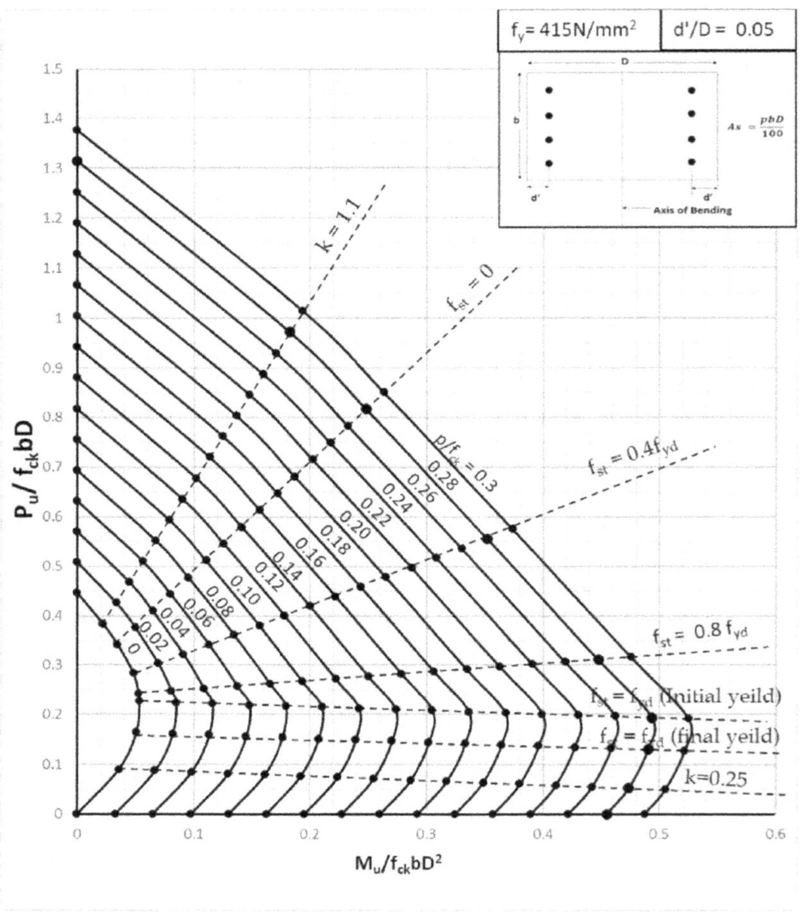

CHART 11.1 Compression with bending – rectangular section - reinforcement distributed equally on two sides.

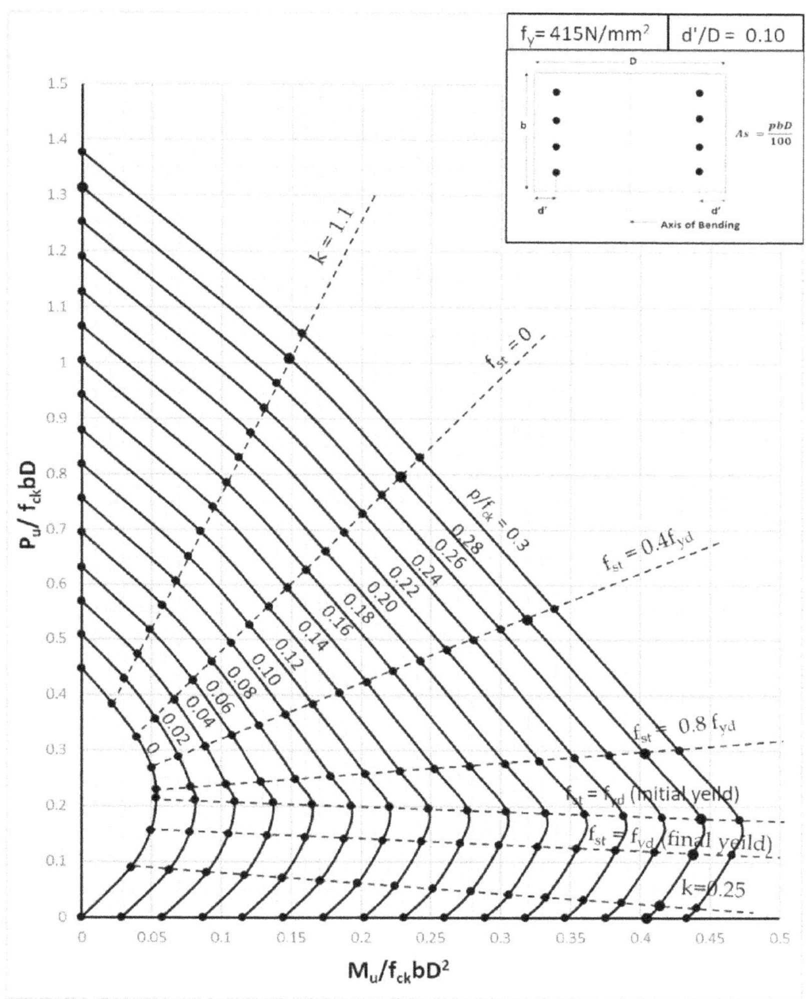

CHART 11.2 Compression with bending – rectangular section – reinforcement distributed equally on two sides.

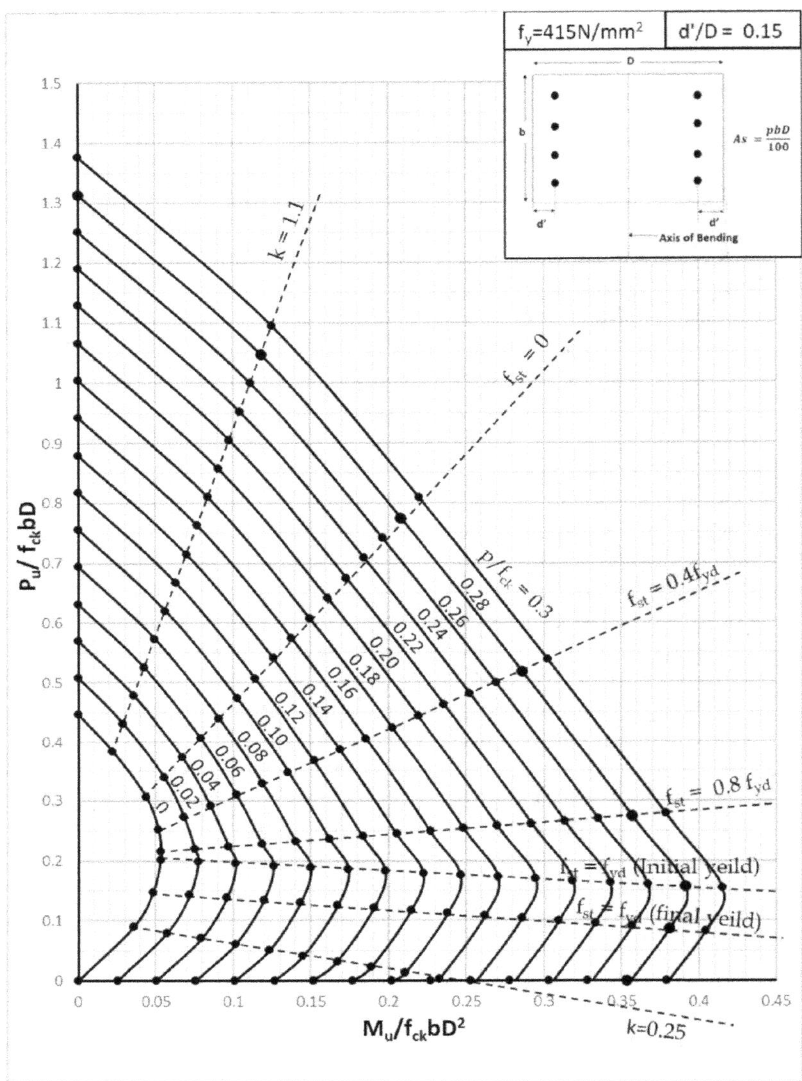

CHART 11.3 Compression with bending – rectangular section – reinforcement distributed equally on two sides.

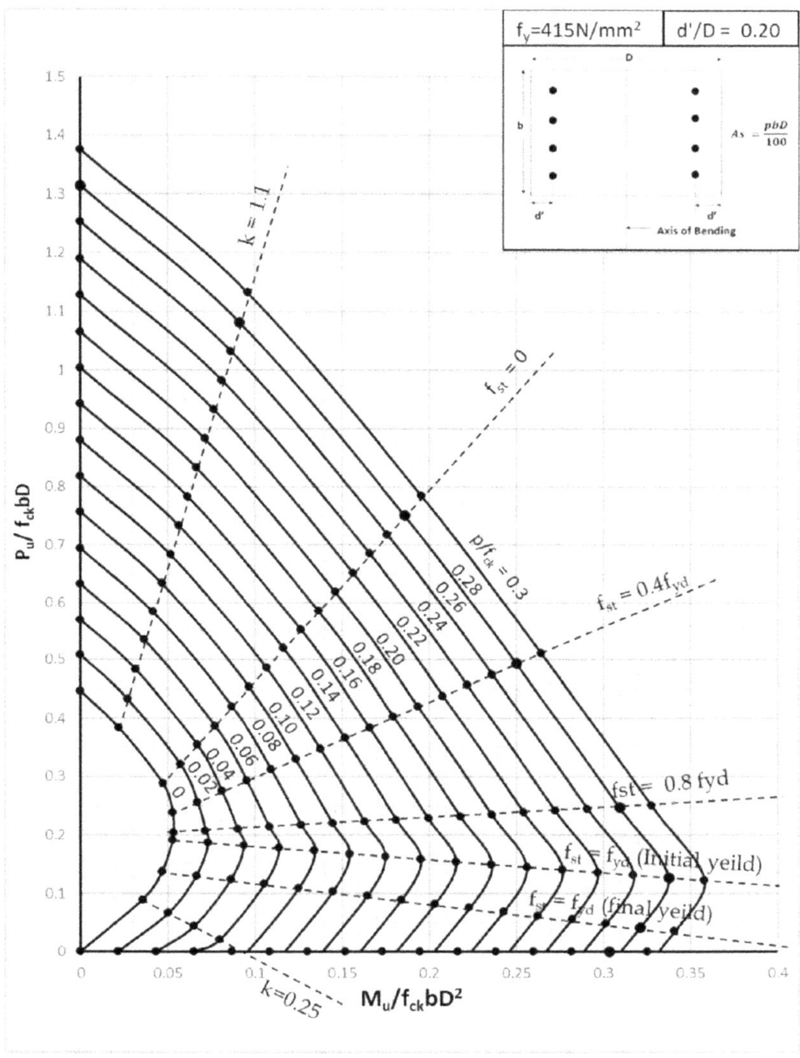

CHART 11.4 Compression with bending – rectangular section – reinforcement distributed equally on two sides.

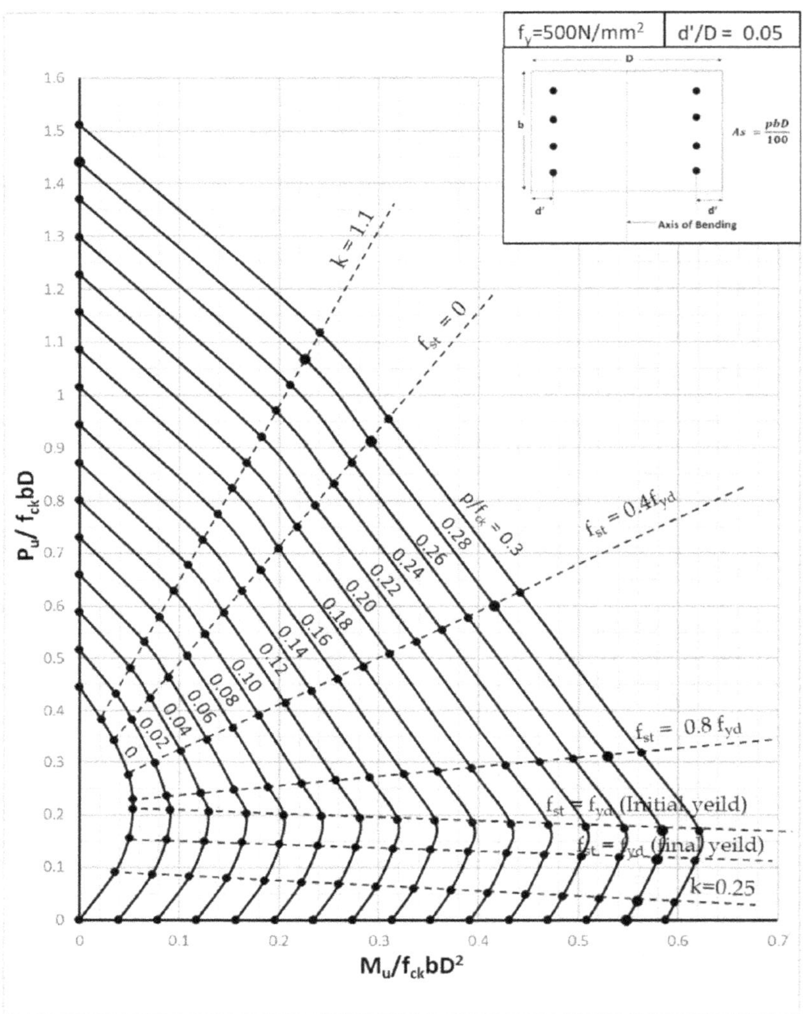

CHART 11.5 Compression with bending – rectangular section – reinforcement distributed equally on two sides.

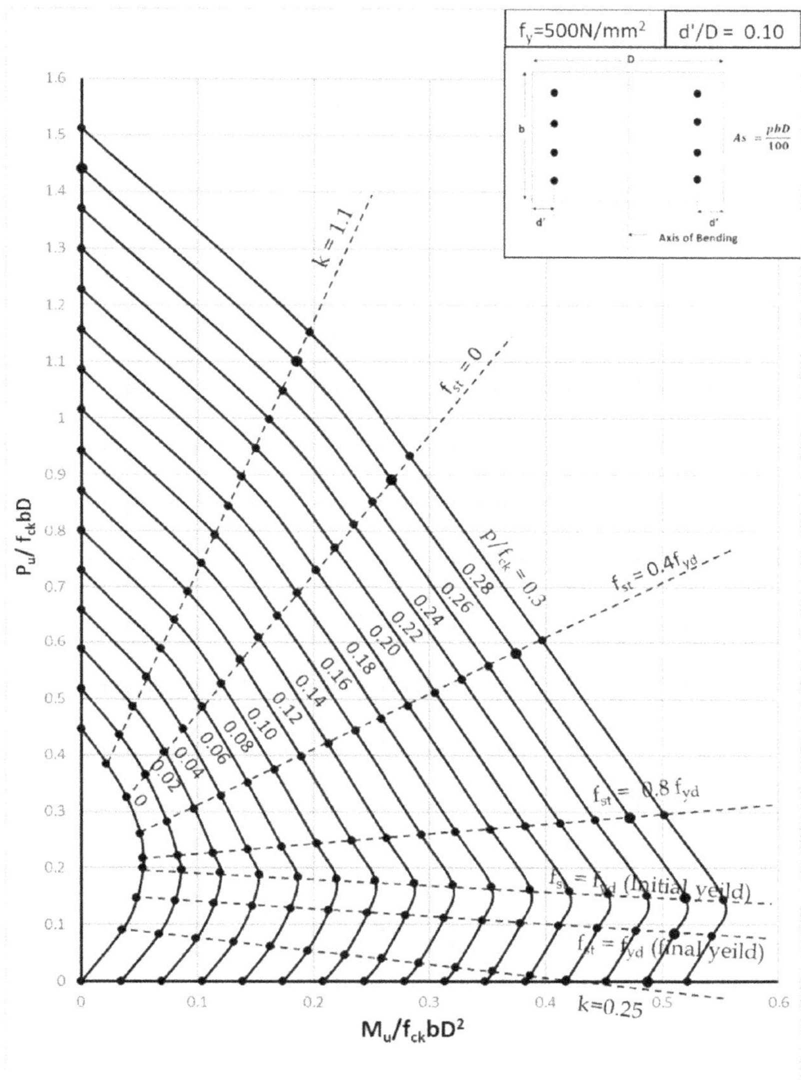

CHART 11.6 Compression with bending – rectangular section – reinforcement distributed equally on two sides.

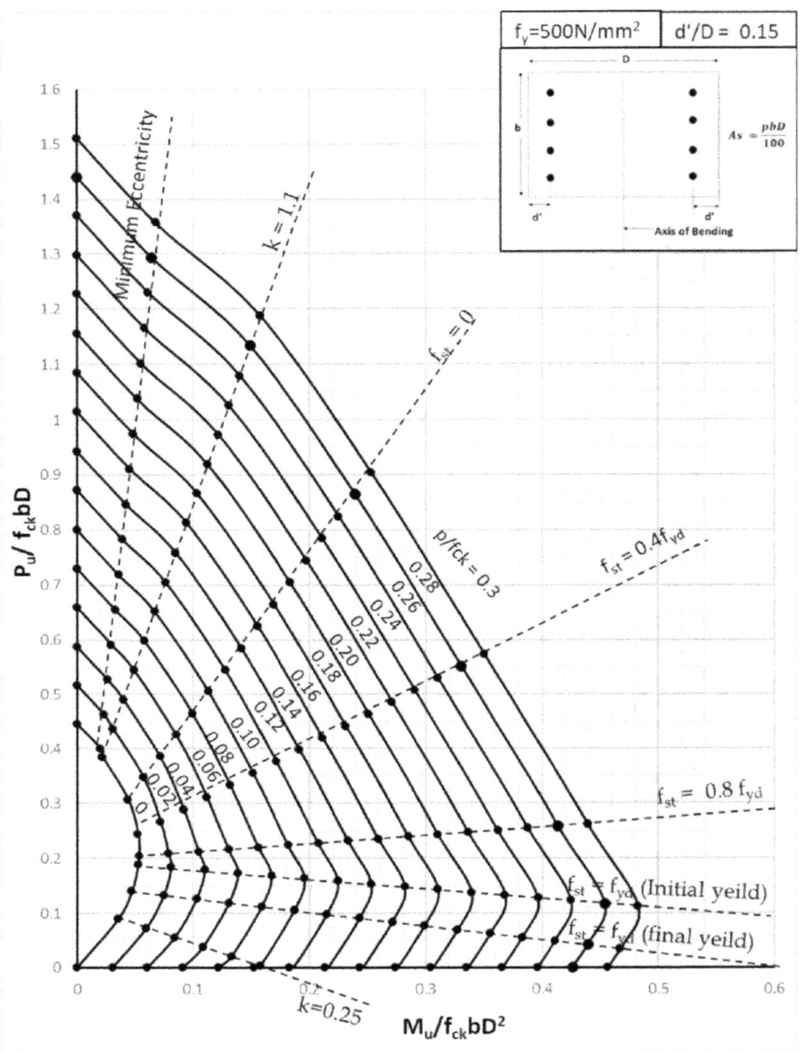

CHART 11.7 Compression with bending – rectangular section – reinforcement distributed equally on two sides.

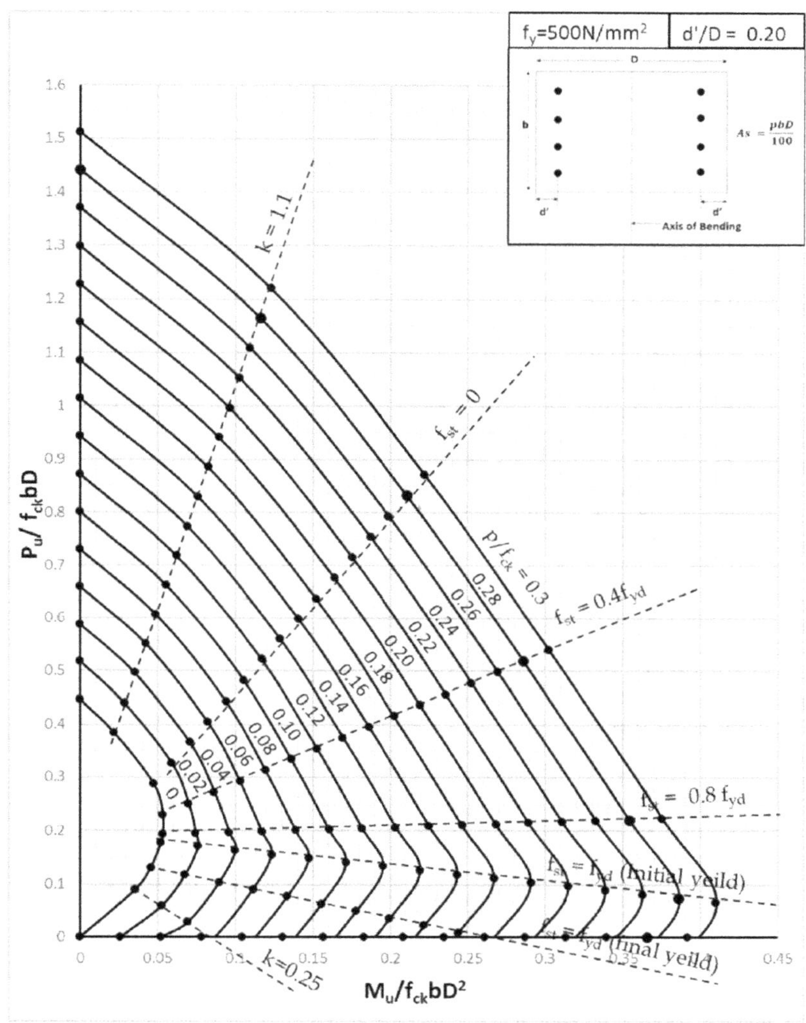

CHART 11.8 Compression with bending – rectangular section – reinforcement distributed equally on two sides.

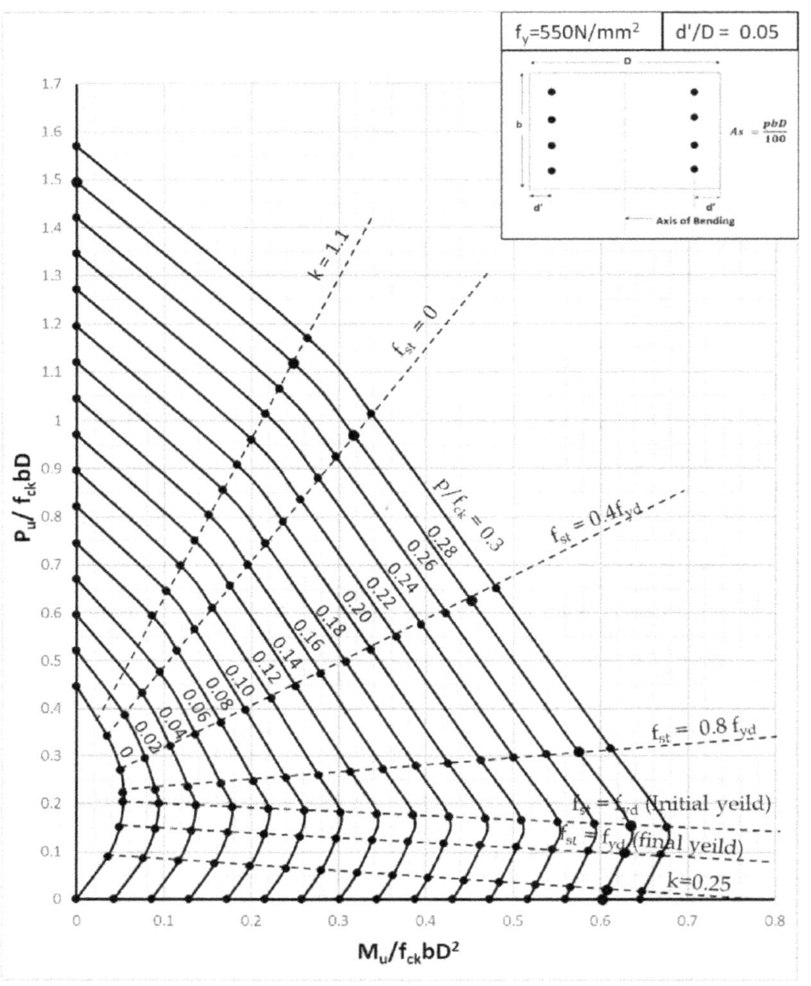

CHART 11.9 Compression with bending – rectangular section – reinforcement distributed equally on two sides.

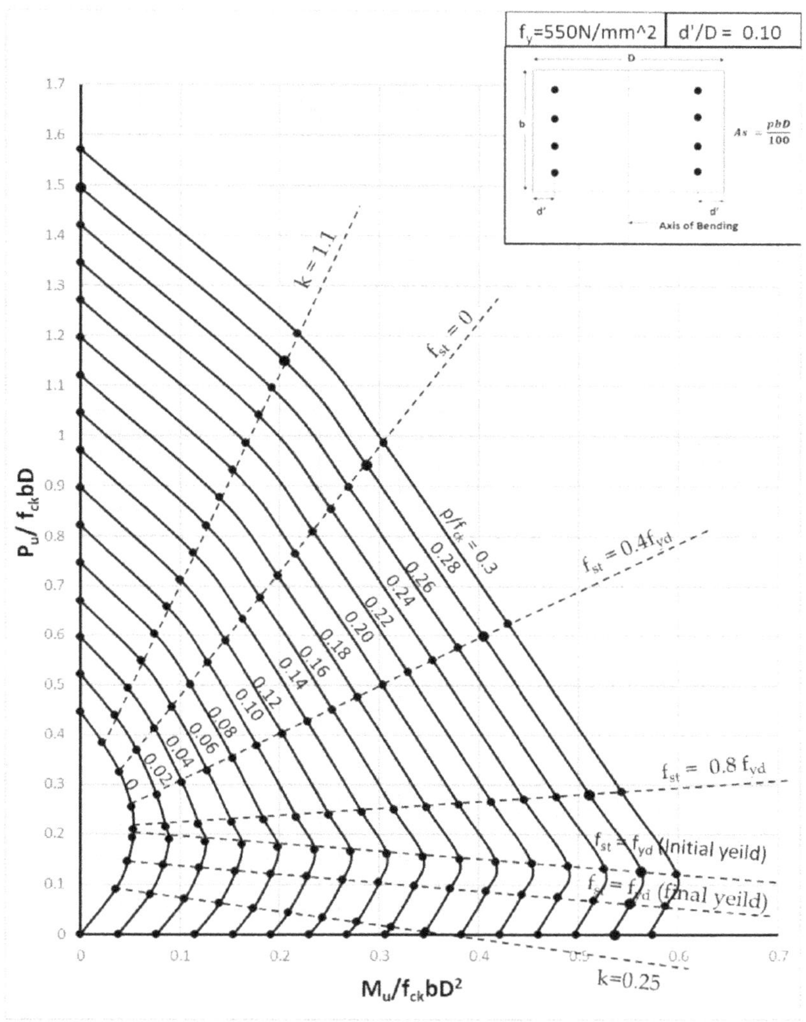

CHART 11.10 Compression with bending – rectangular section – reinforcement distributed equally on two sides.

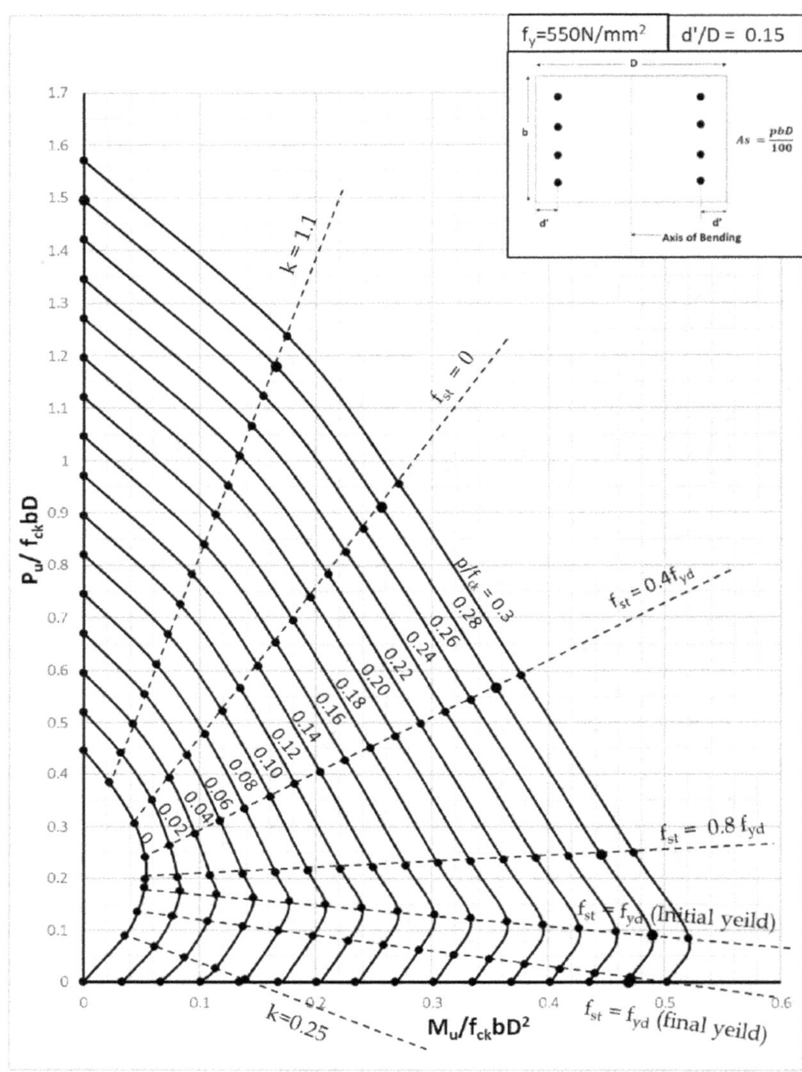

CHART 11.11 Compression with bending – rectangular section – reinforcement distributed equally on two sides.

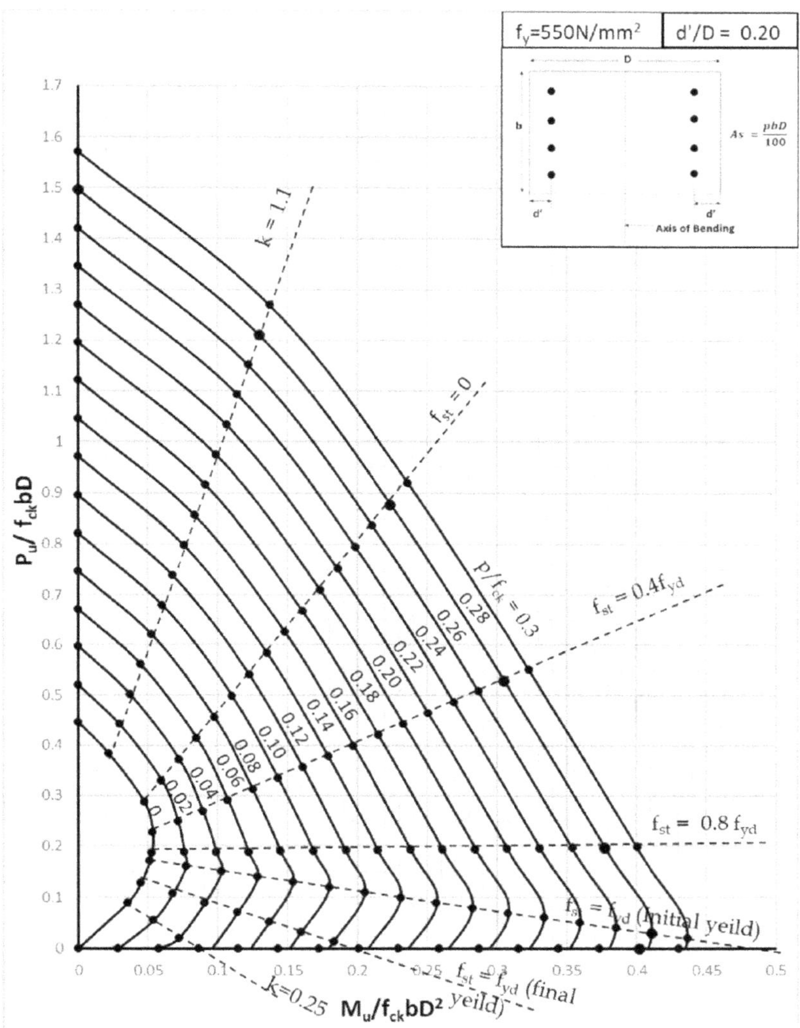

CHART 11.12 Compression with bending – rectangular section – reinforcement distributed equally on two sides.

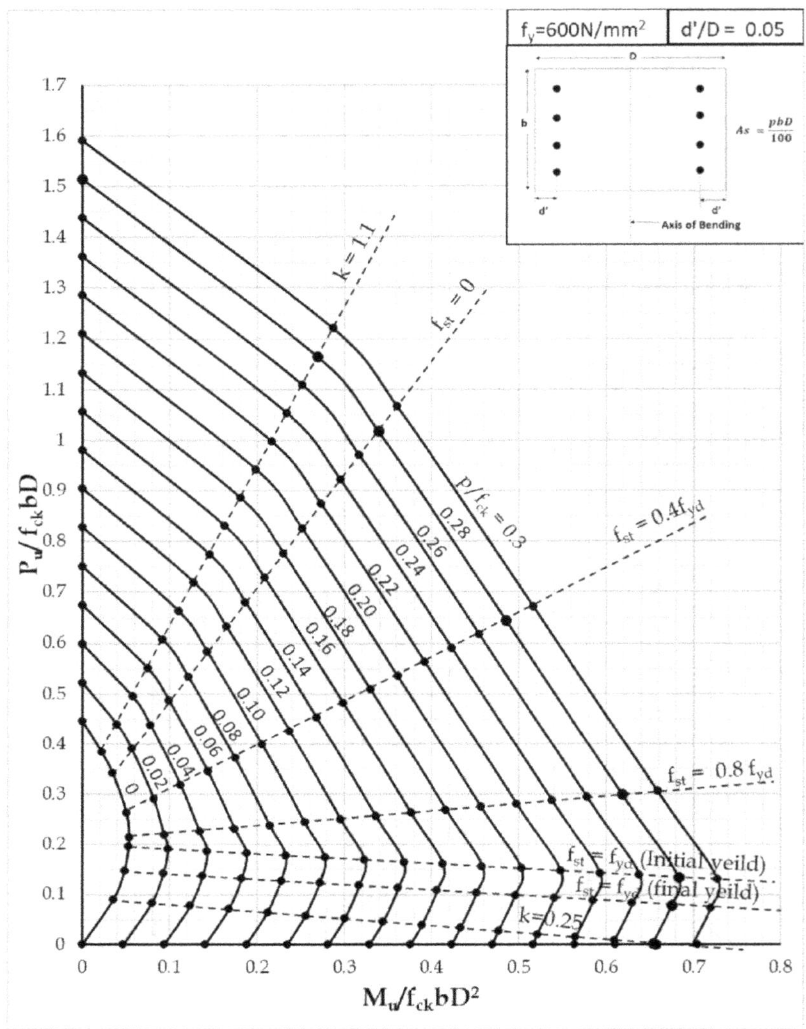

CHART 11.13 Compression with bending – rectangular section – reinforcement distributed equally on two sides.

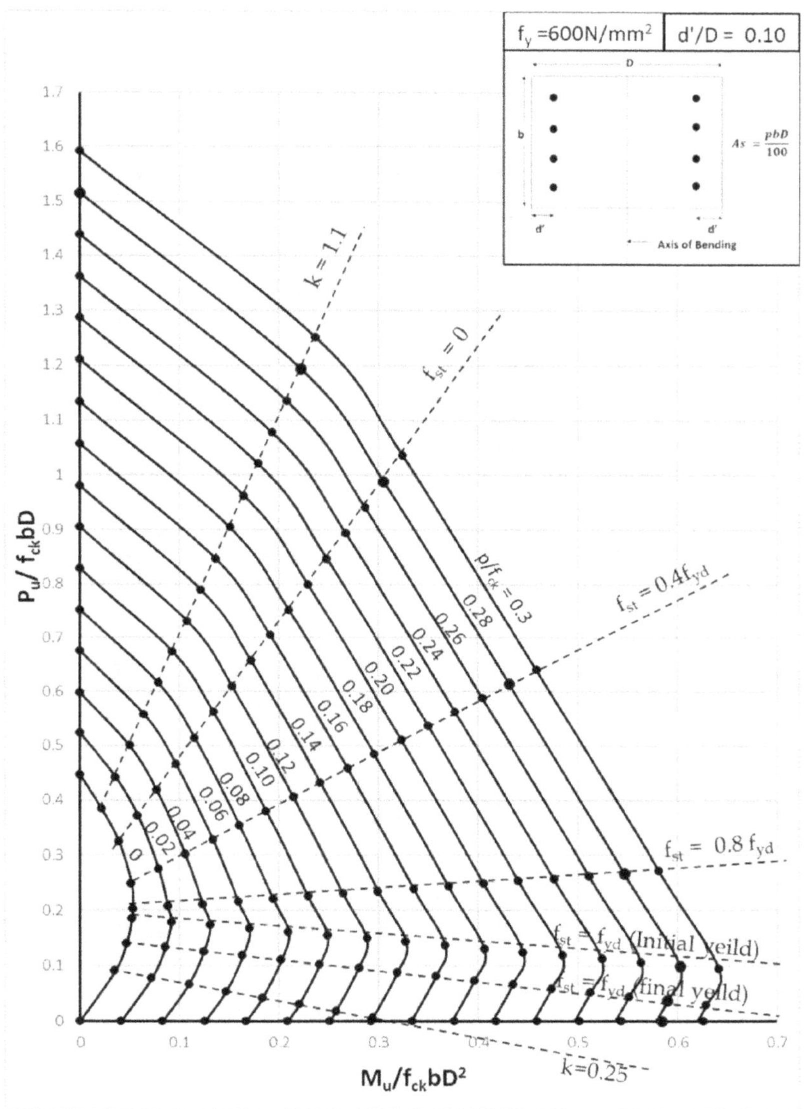

CHART 11.14 Compression with bending – rectangular section – reinforcement distributed equally on two sides.

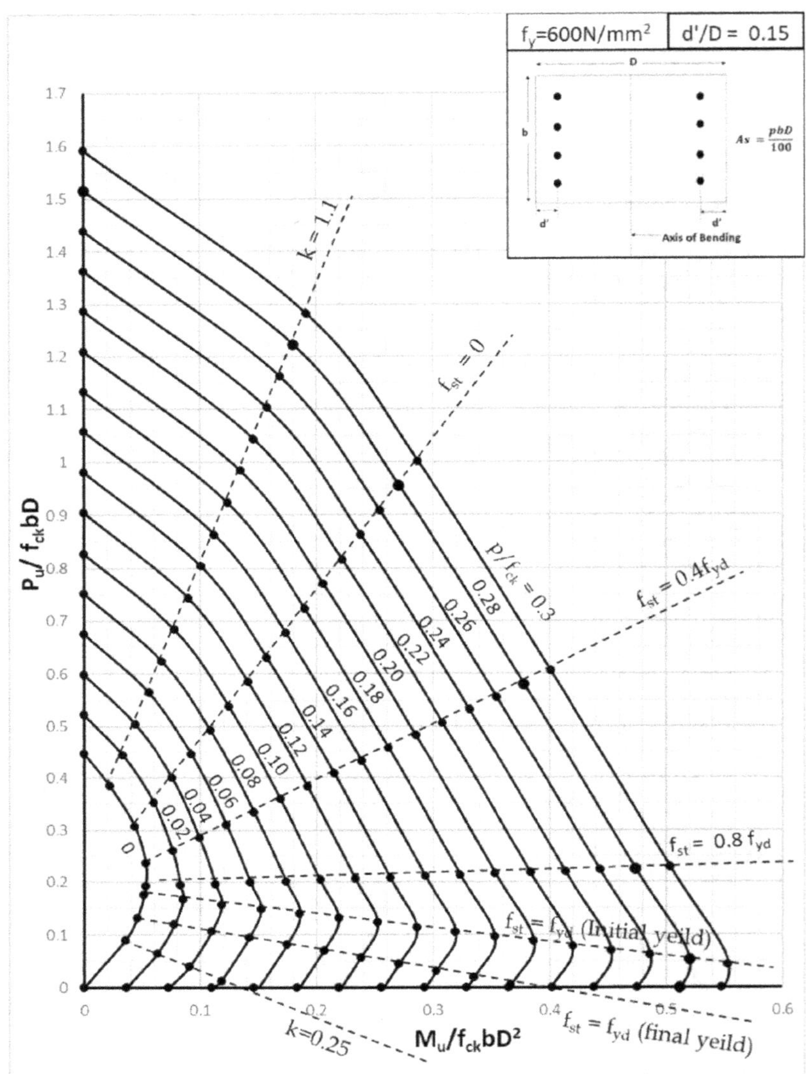

CHART 11.15 Compression with bending – rectangular section – reinforcement distributed equally on two sides.

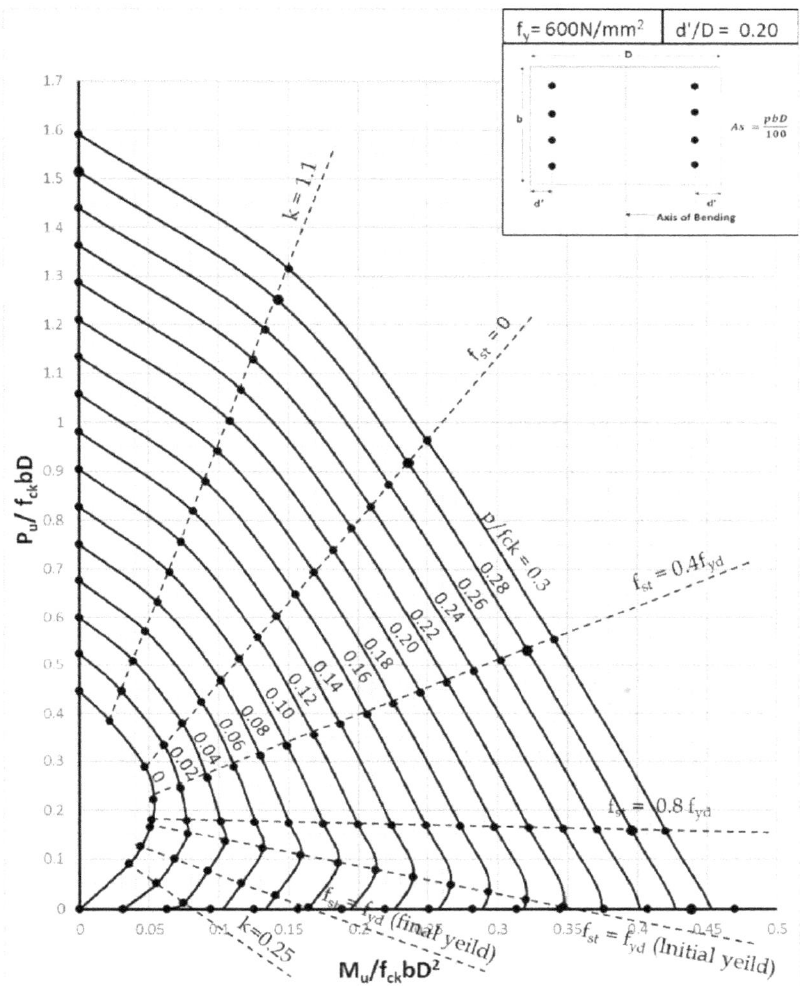

CHART 11.16 Compression with bending – rectangular section – reinforcement distributed equally on two sides.

11.23 INTERACTION CHARTS FOR A RECTANGULAR COLUMN SECTION WITH REINFORCEMENTS EQUALLY DISTRIBUTED ON FOUR SIDES (CHARTS 11.17–11.32)

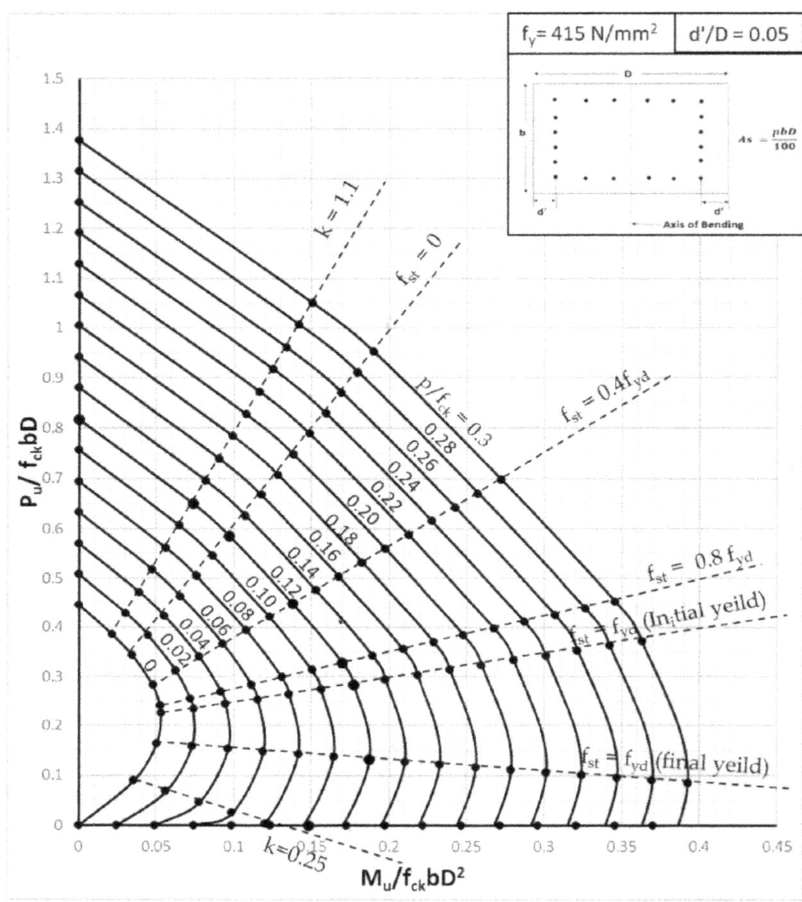

CHART 11.17 Compression with bending – rectangular section – reinforcement distributed equally on four sides.

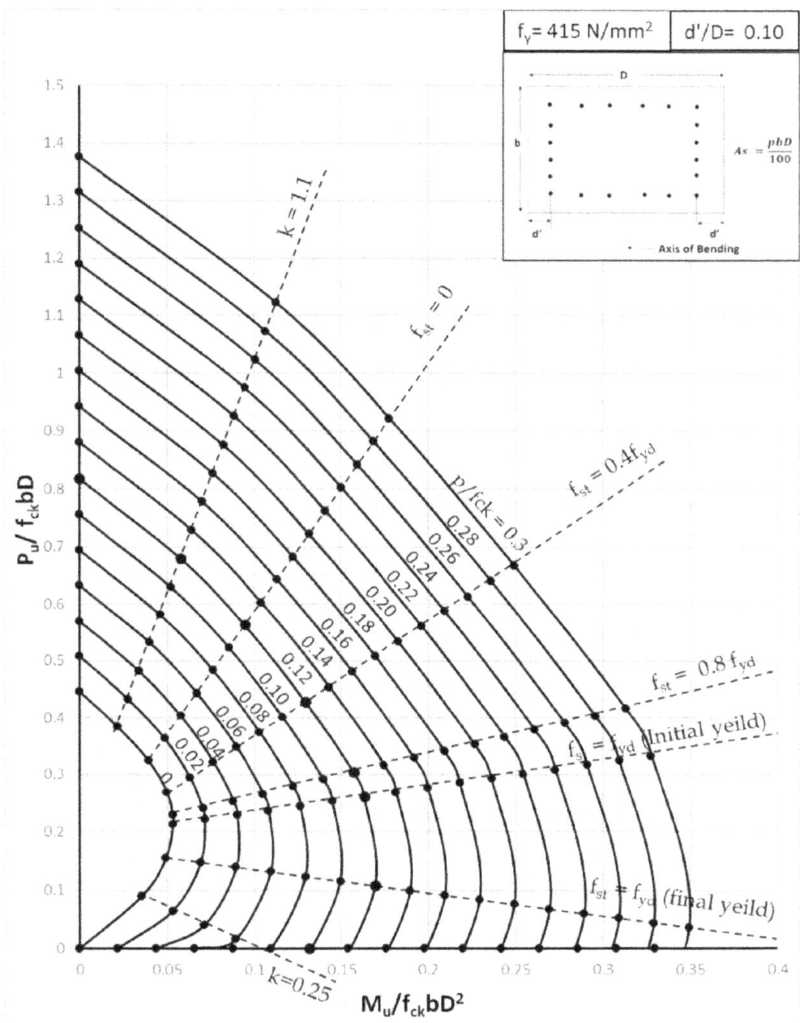

CHART 11.18 Compression with bending – rectangular section – reinforcement distributed equally on four sides.

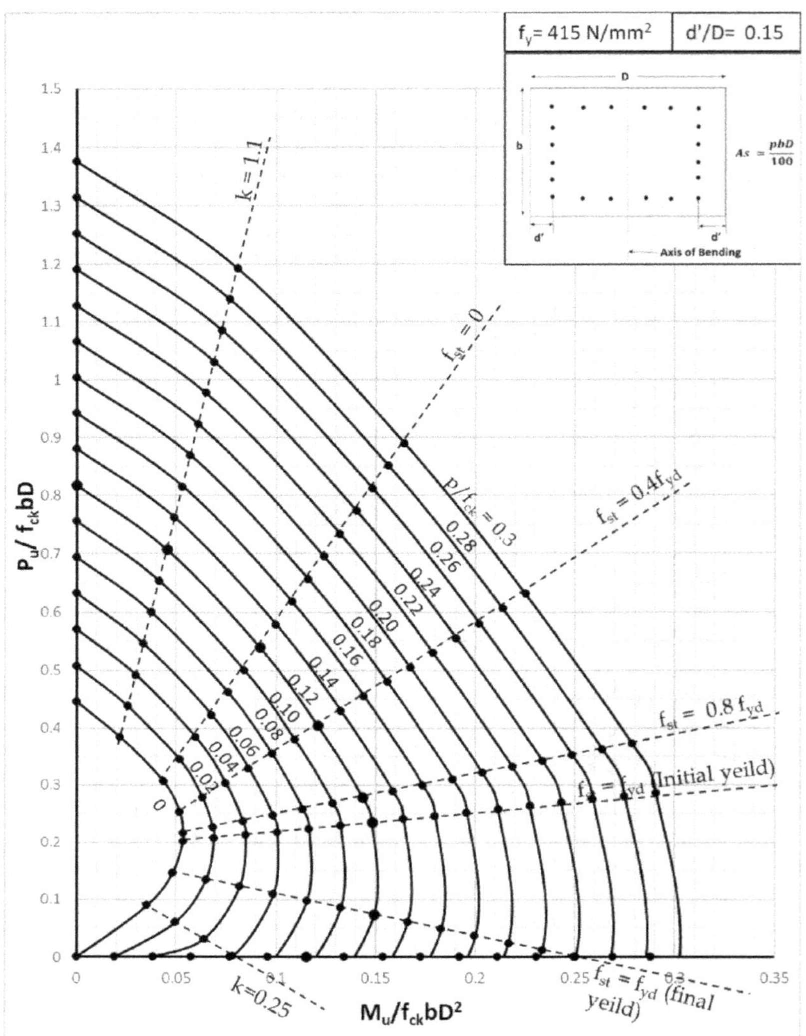

CHART 11.19 Compression with bending – rectangular section – reinforcement distributed equally on four sides.

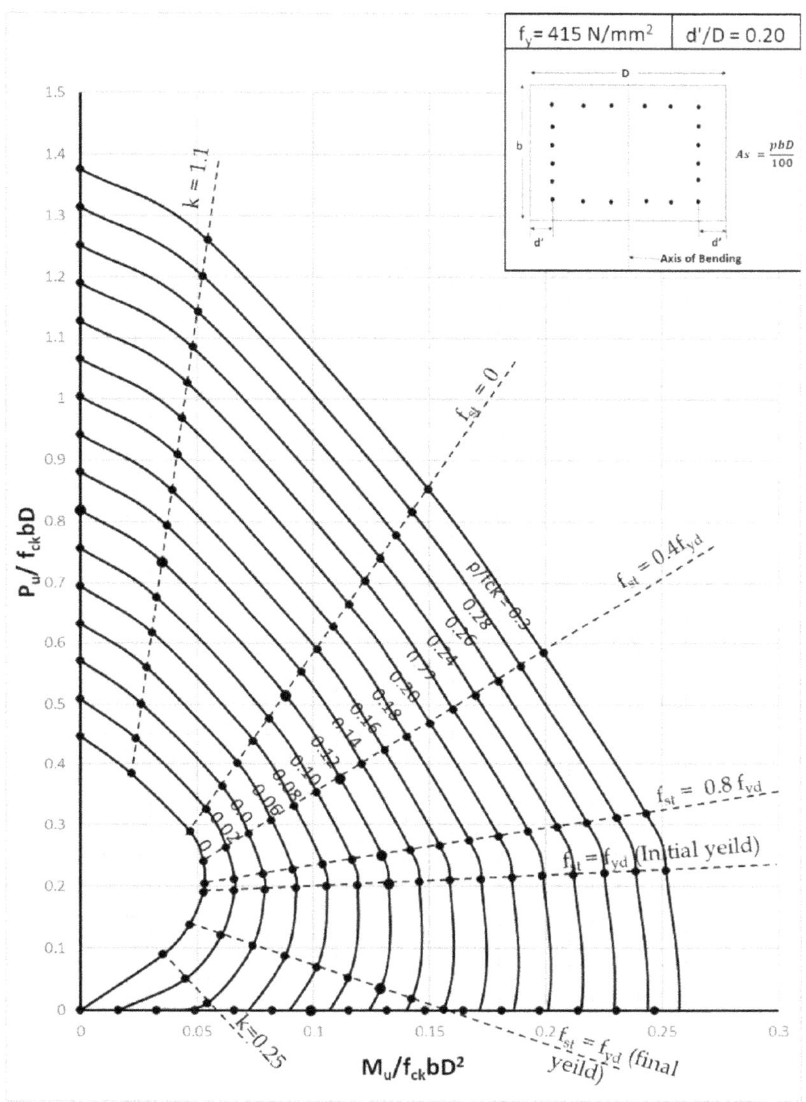

CHART 11.20 Compression with bending – rectangular section – reinforcement distributed equally on four sides.

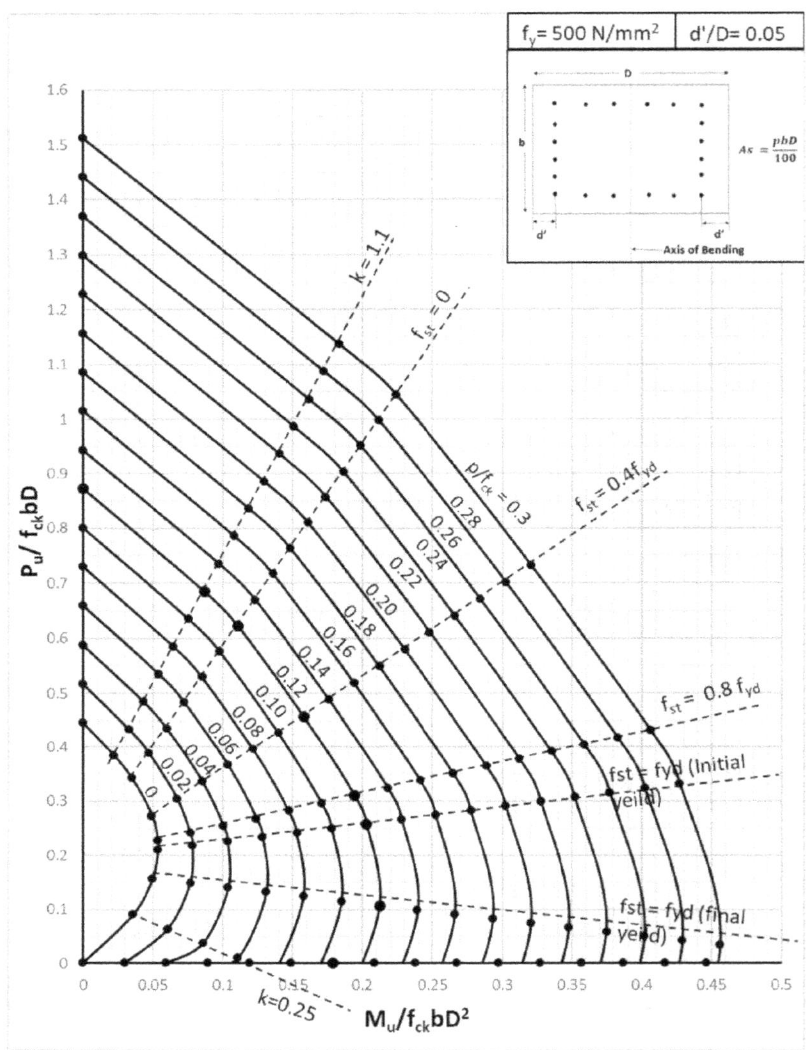

CHART 11.21 Compression with bending – rectangular section – reinforcement distributed equally on four sides.

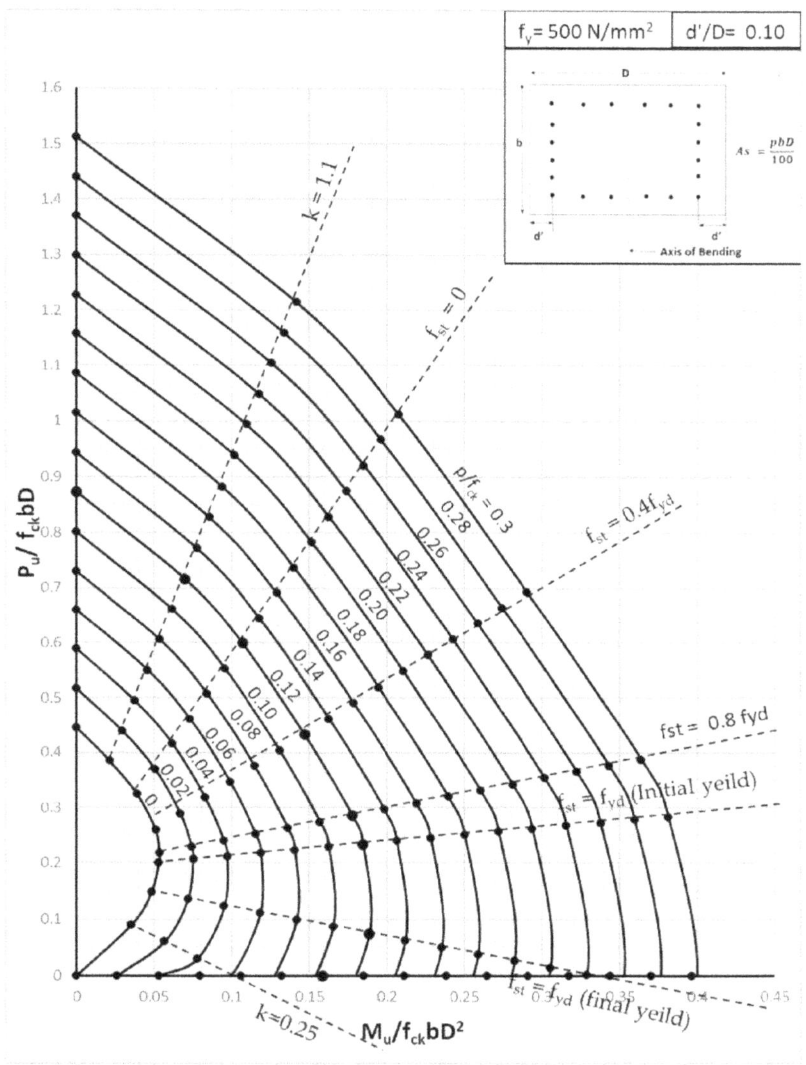

CHART 11.22 Compression with bending – rectangular section – reinforcement distributed equally on four sides.

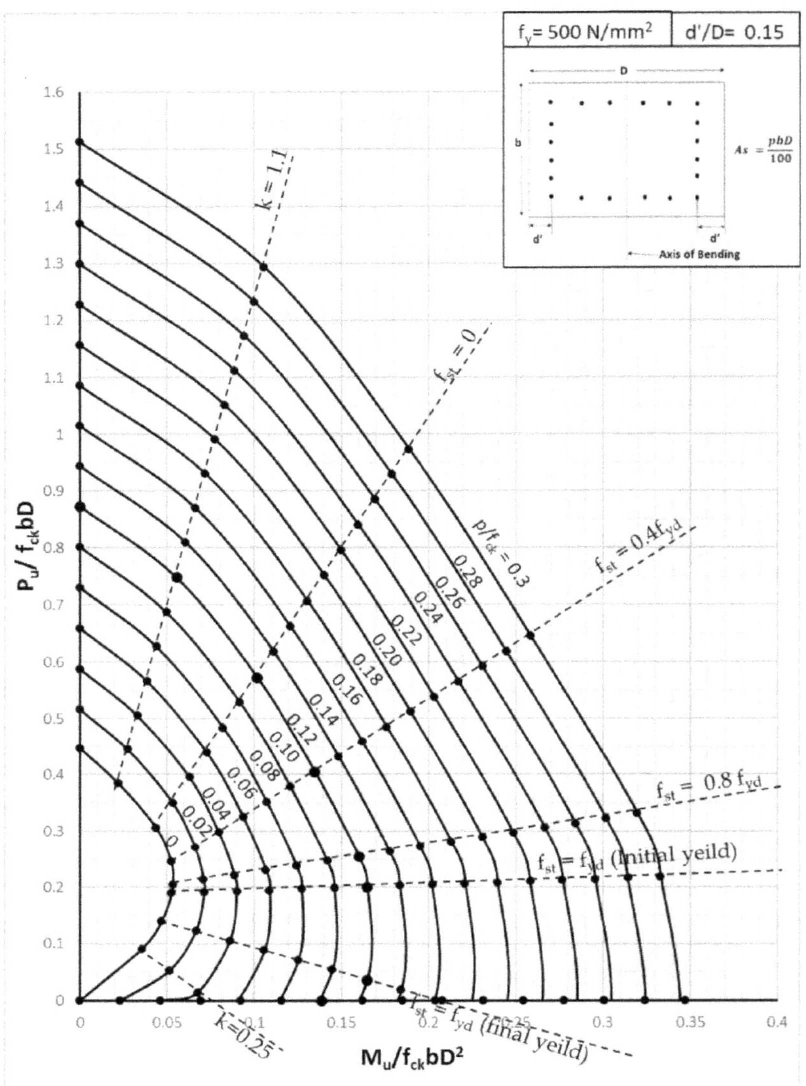

CHART 11.23 Compression with bending – rectangular section – reinforcement distributed equally on four sides.

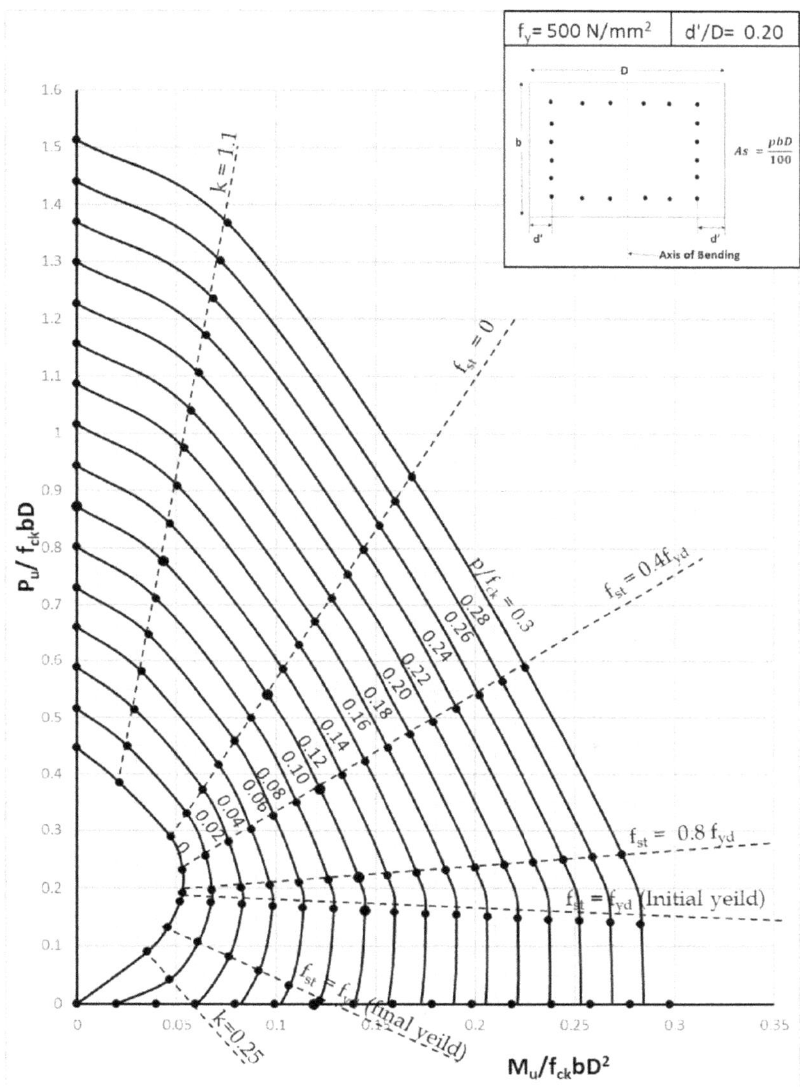

CHART 11.24 Compression with bending – rectangular section – reinforcement distributed equally on four sides.

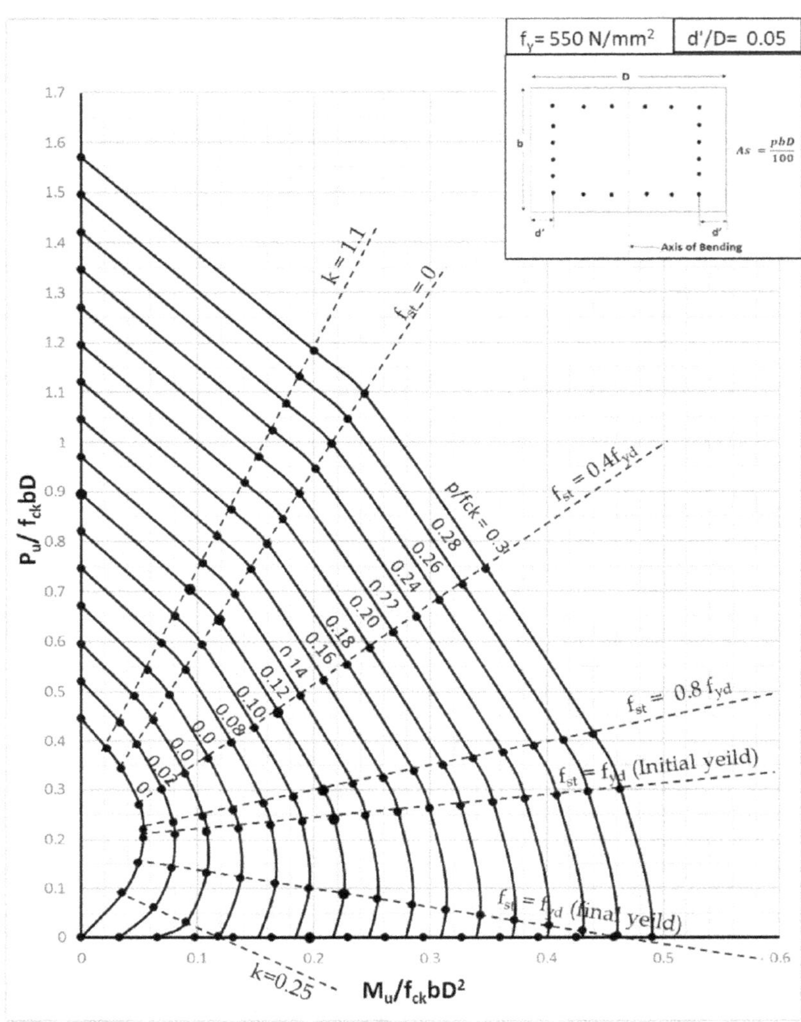

CHART 11.25 Compression with bending – rectangular section – reinforcement distributed equally on four sides.

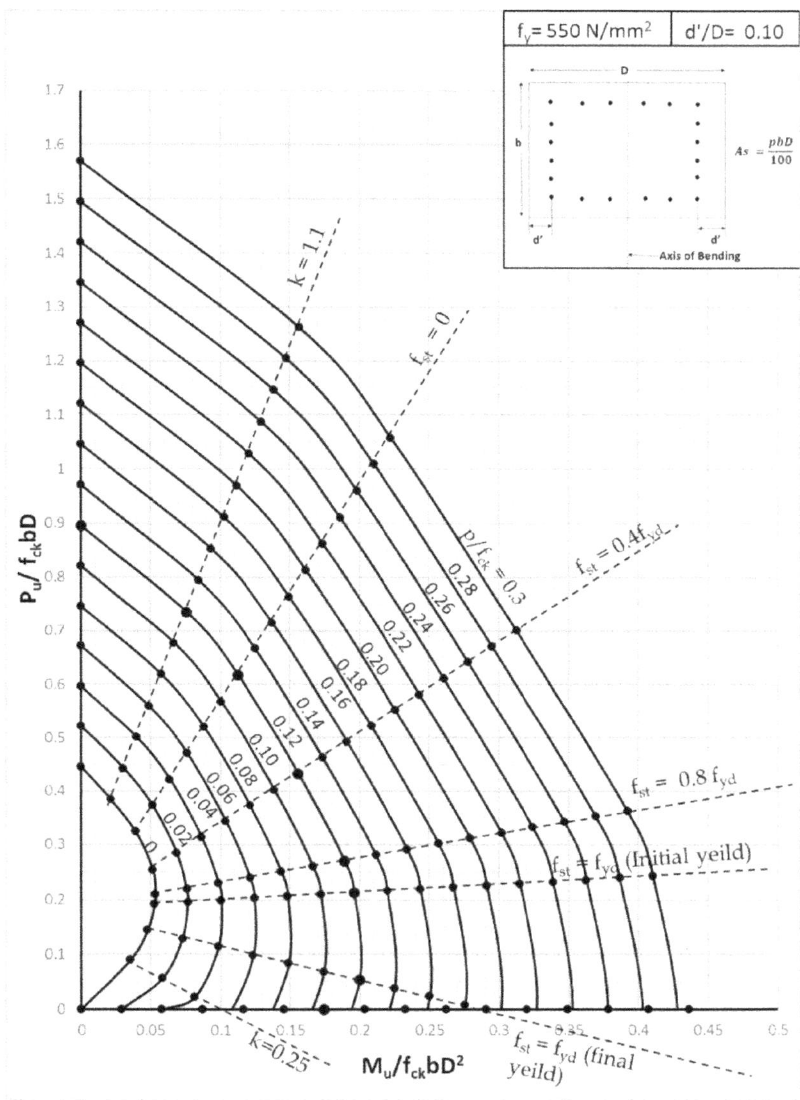

CHART 11.26 Compression with bending – rectangular section – reinforcement distributed equally on four sides.

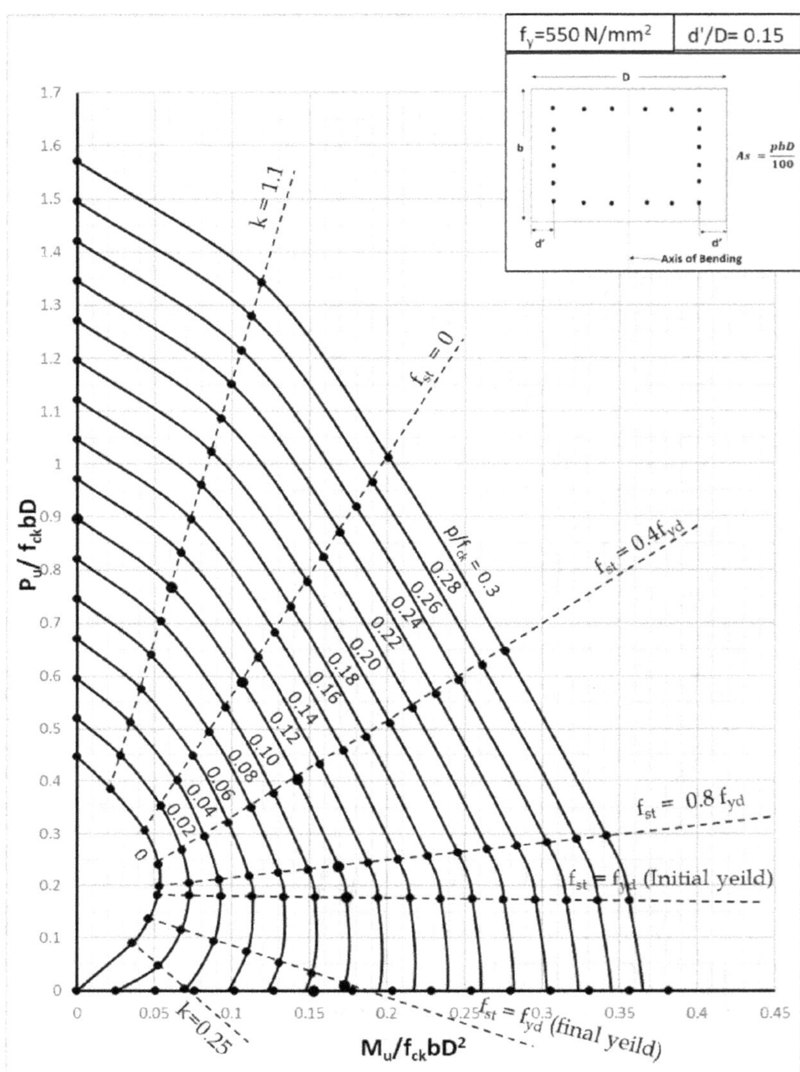

CHART 11.27 Compression with bending – rectangular section – reinforcement distributed equally on four sides.

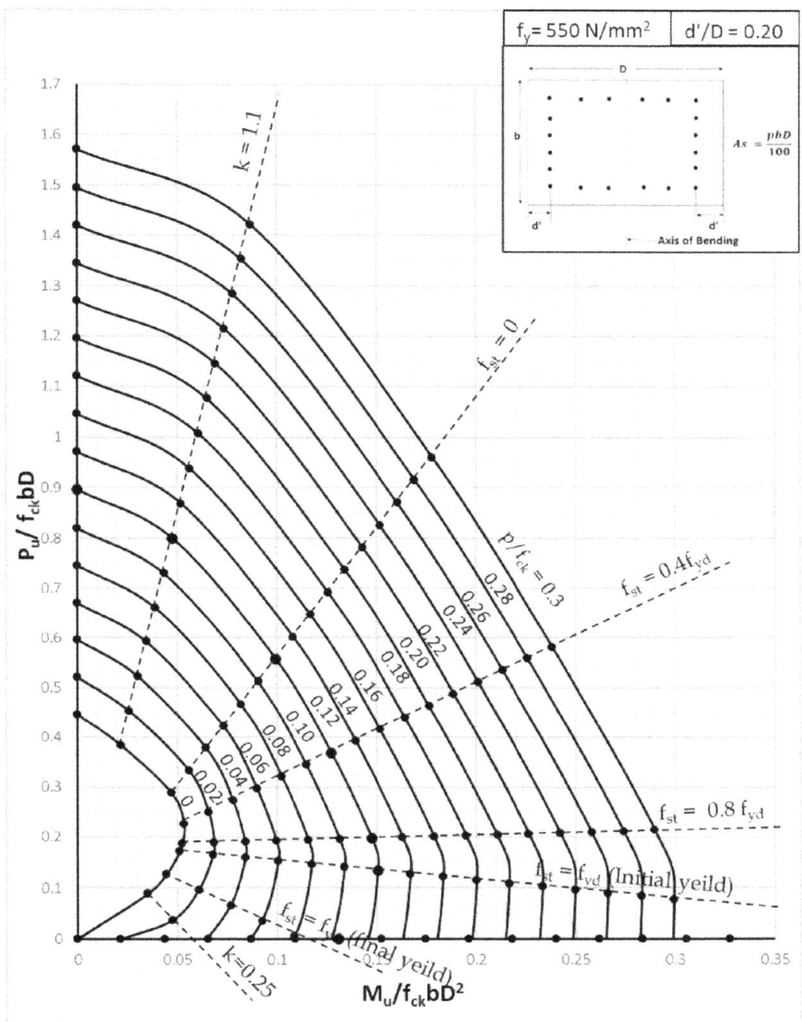

CHART 11.28 Compression with bending – rectangular section – reinforcement distributed equally on four sides.

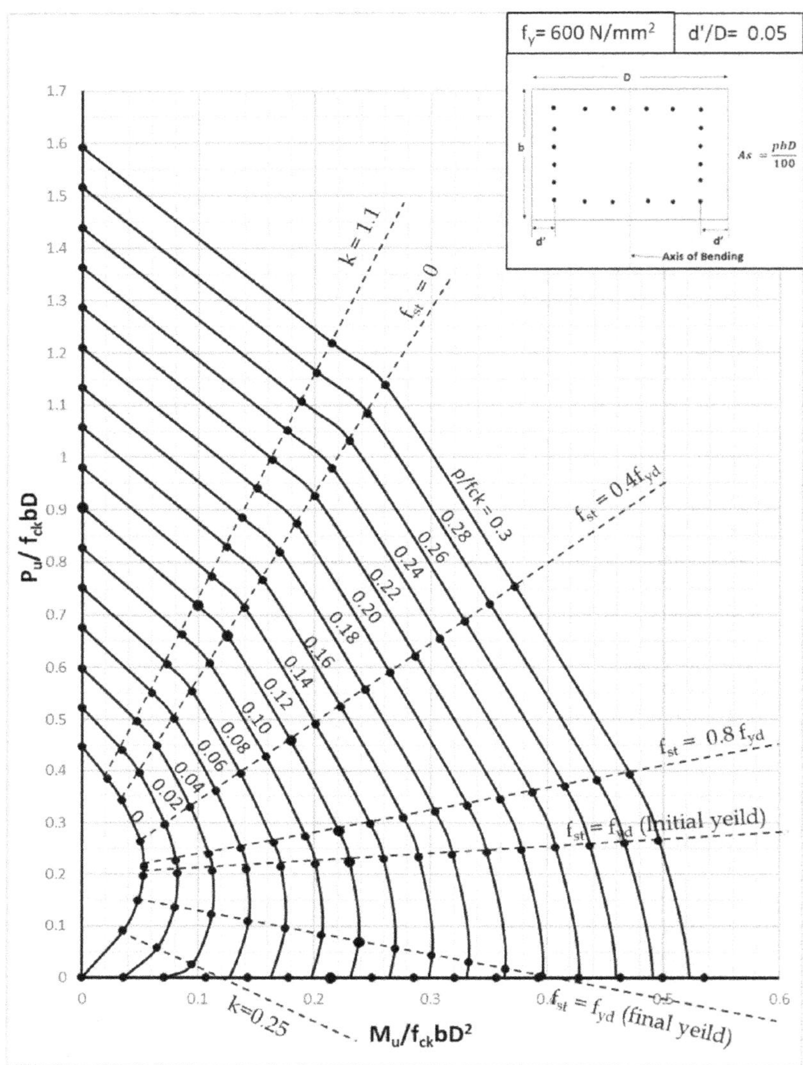

CHART 11.29 Compression with bending – rectangular section – reinforcement distributed equally on four sides.

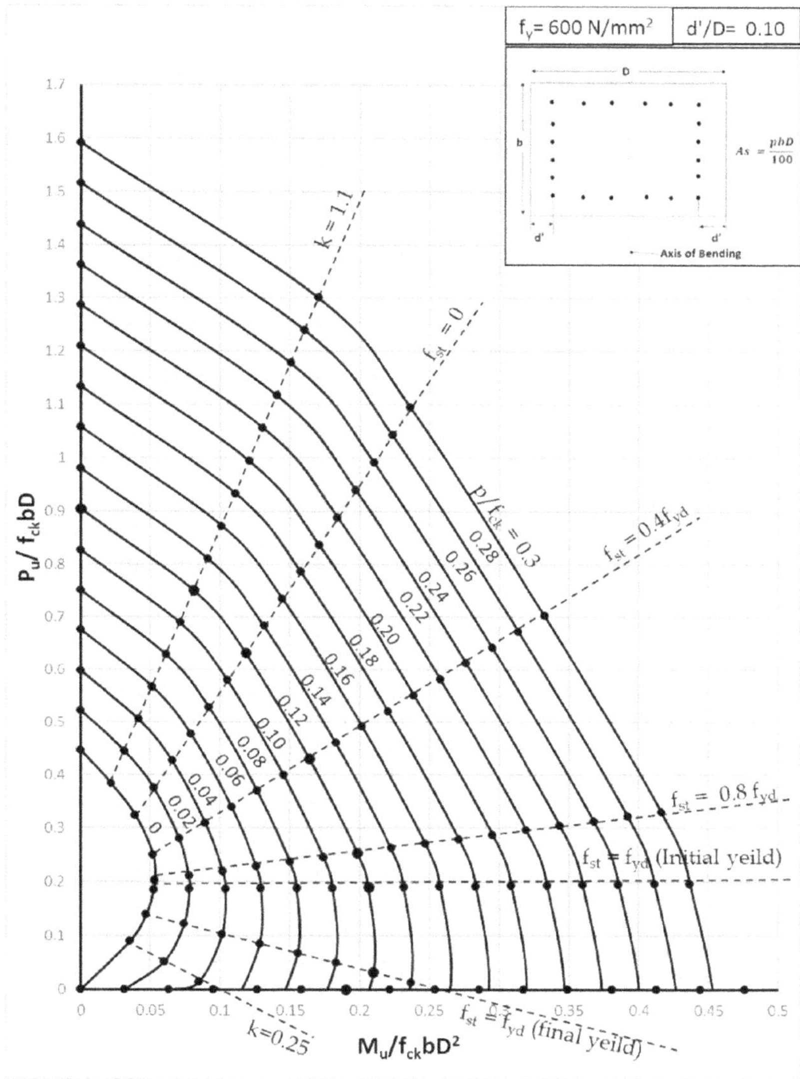

CHART 11.30 Compression with bending – rectangular section – reinforcement distributed equally on four sides.

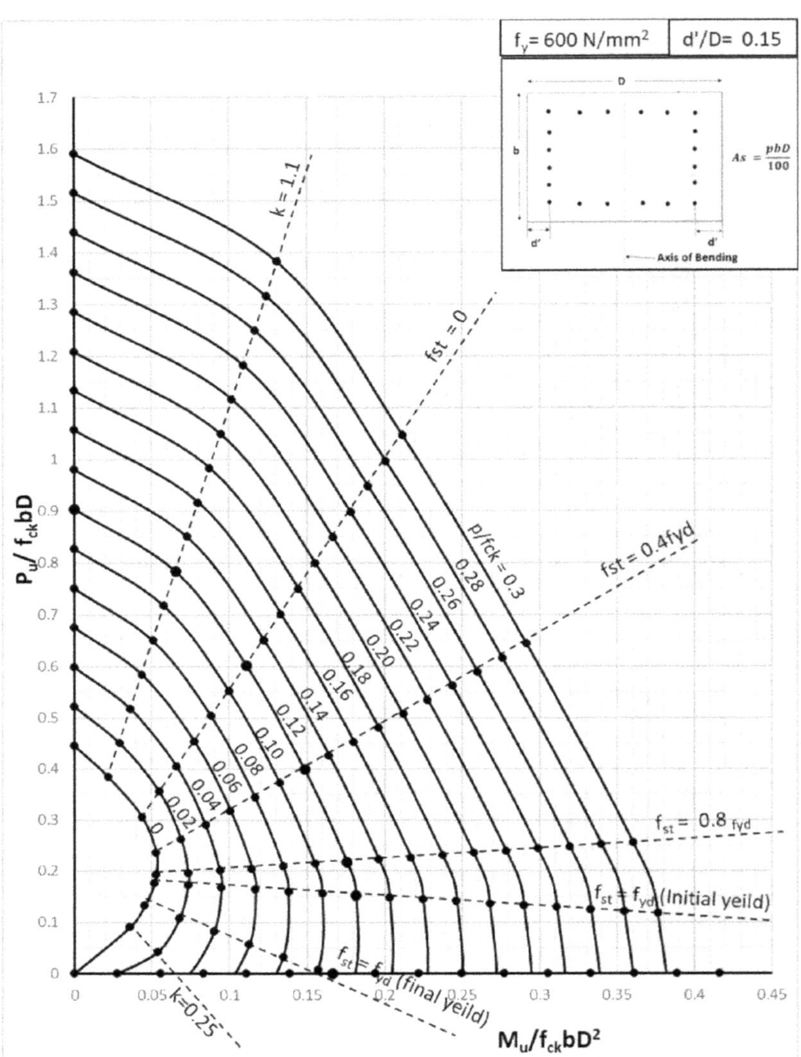

CHART 11.31 Compression with bending – rectangular section – reinforcement distributed equally on four sides.

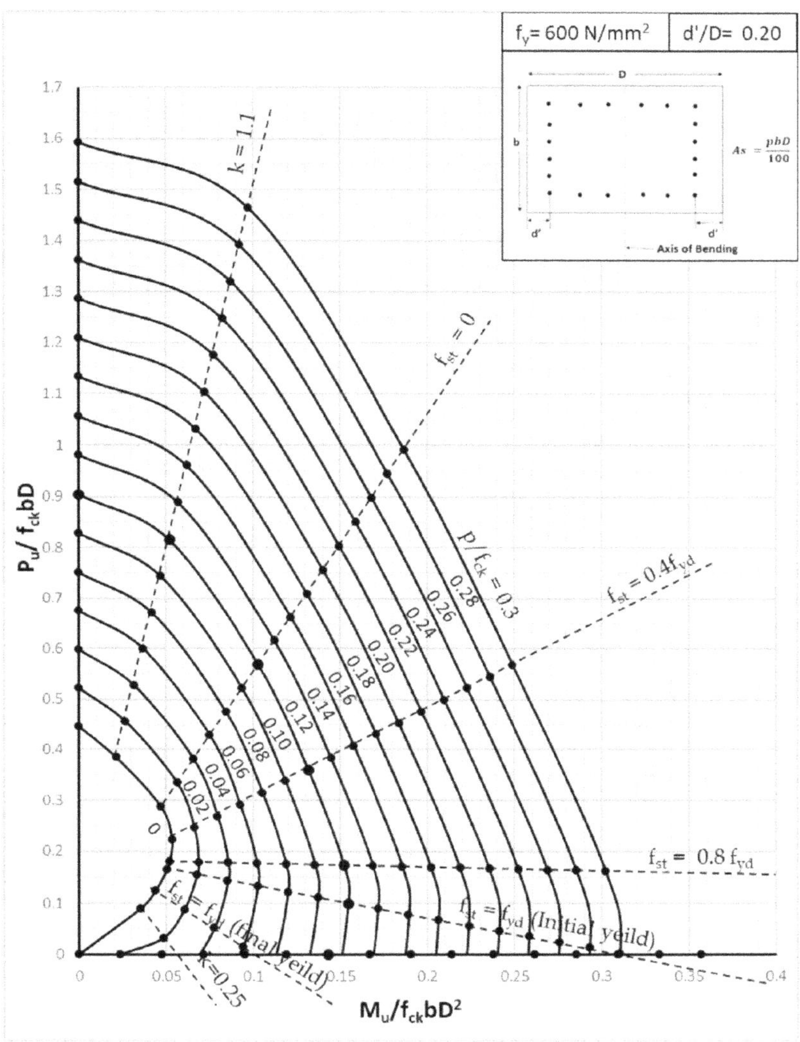

CHART 11.32 Compression with bending – rectangular section – reinforcement distributed equally on four sides.

11.24 COLUMNS UNDER AXIAL LOAD WITH BIAXIAL BENDING

The proposed charts have been developed for sections subjected to axial compression with uniaxial bending. The axis of bending can be along the minor axis or along the major axis of the column. However, in real-life conditions, the axis of bending can be oriented along any intermediate inclined axis. In order to check the safety of the sections from fundamental principles, an interaction surface is to be generated which will take the shape of an onion. This process is significantly complicated and cumbersome and requires extensive calculations which are beyond the scope of this book. Simplified methods for checking the safety of such columns based on interaction charts developed for axial load with uniaxial bending are widely used and the Indian standard follows the Bressler's load contour approach which is given in IS:456-2000.

11.25 SUMMARY

In this chapter an attempt has been made to explore the design philosophy of compression members using the fundamental principles of the limit state method of design. Generally, designers use design aids or tools for performing the design of compression members without trying to understand their basic behaviour under different types of loads. For the design of flexural members, design is primarily based on the moments acting on the section. Since for columns, axial loads and moments produce similar normal stresses on the section, design is performed to cater to the requirements of maximum axial force, maximum moments or a combination of axial force and moment acting simultaneously. Since two parameters are to be checked, interaction charts are necessary. The interaction charts are derived from axial and moment capacities of the section and not on external forces or moment acting on the section. In this chapter, detailed discussions on strain distribution and stress distribution across column sections for different failure modes have been provided as per the fundamental assumptions of the limit state of collapse in compression. Interaction charts catering to all grades of structural concrete and commonly used HYSD steel grades of Fe 415, Fe 500, Fe 550 and Fe 600 have been developed which will not only serve as excellent design tools but more importantly will give an insight into the behaviour of the columns under different types of loads.

BIBLIOGRAPHY

Bandyopadhyay, J. N. (2011). *Design of Concrete Structures*, PHI Learning, New Delhi.

Bhatt, Prab, McGinley, Thomas J. and Choo, Ban Seng (2006). *Reinforced Concrete, Design Theory and Examples*, Third Edition, E & FN Spon, Taylor and Francis, Oxon.

Ferguson, Phil M., Green, John E., Jirsa, James O. (1988). *Reinforced Concrete Fundamentals*, Fifth Edition, John Wiley, New York.

Indian Standard IS: 456-2000. *Plain and Reinforced Concrete–Code of Practice* (Fourth Revision), Tenth Reprint April 2007, Bureau of Indian Standards, New Delhi, India.

Park, R. and Paulay, T. (1975). *Reinforced Concrete Structures*, John Wiley, New York.

Paulay, T. and Priestley, M. J. N. (2018). *Seismic Design of Reinforced Concrete and Masonry Buildings*, John Wiley, New York.

Pillai, S. U. and Menon, D. (2012). *Reinforced Concrete Design*, Third Edition, Reprint 2012, Tata McGraw Hill Education, New Delhi.

Purushothaman, P. (1984). *Reinforced Concrete Structural Elements, Behaviour, Analysis, and Design*, Tata McGraw-Hill, Torsteel Research Foundation India.

Shah, V. L. and Karve, S. R. (2016). *Limit State Theory and Design of Reinforced Concrete*, Eighth Edition, Structures Publications, Parva, Pune.

SP: 16-1980. *Design Aids for Reinforced Concrete to IS: 456-1978*, Fourth Reprint March 1988, Bureau of Indian Standards, New Delhi, India.

12 Limit State of Collapse in Shear

12.1 INTRODUCTION

The cross-section of a beam and its longitudinal reinforcement are calculated from the moment of resistance required to resist the maximum bending moment to which the beam section can be subjected to. Since bending moments are not constant over the beam length, shear forces come into play. The mechanism of shear failure in reinforced concrete is complicated, and extensive theoretical research and experimentation are being carried out throughout the world. The method of design is quite empirical in nature. Failure due to a combination of shear and bending moments can be diagonal tension failure, flexural shear failure and compression failure. Instead of detailed theoretical discussions, the basic methodology of shear design of reinforced concrete sections is covered in this chapter. Shear failures are very brittle and should be avoided.

In a beam made of homogeneous material subjected to transverse loads, the neutral axis will be passing through the centroid. Due to flexure, the portion of the section above the neutral axis will be subjected to normal stresses which are compressive in nature, and the portion below the neutral axis will be subjected to normal tensile stresses. Shear forces result in shear stresses at the section orthogonal to the normal stresses. Near the supports of a simply supported beam subjected to uniformly distributed load, the shear stresses are high and bending stresses are small. At the neutral axis, the normal stresses are zero and a state of pure shear stress exists. This results in principal tensile and compressive stresses of the same magnitude as the shear stress inclined at 45° to the neutral axis. Concrete is strong in compression but weak in tension. Hence shear cracks generally occur at 45° inclination near the supports (Figure 12.1).

In a rectangular beam as per elastic analysis, the distribution of shear stress is parabolic with the maximum value as (Figure 12.2):

$$\tau_{max} = 1.5\frac{V}{bh} \tag{12.1}$$

where, V = shear force at the section

b = width of the section

h = depth of the beam

For rectangular reinforced concrete beams of uniform depth the distributions of shear stresses are considered to be uniform and the nominal shear stress is calculated as follows:

DOI: 10.1201/9781003415398-12

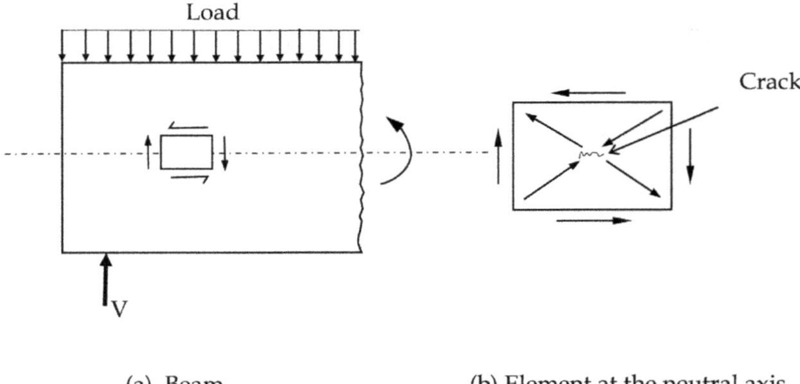

FIGURE 12.1 Shear in beam.

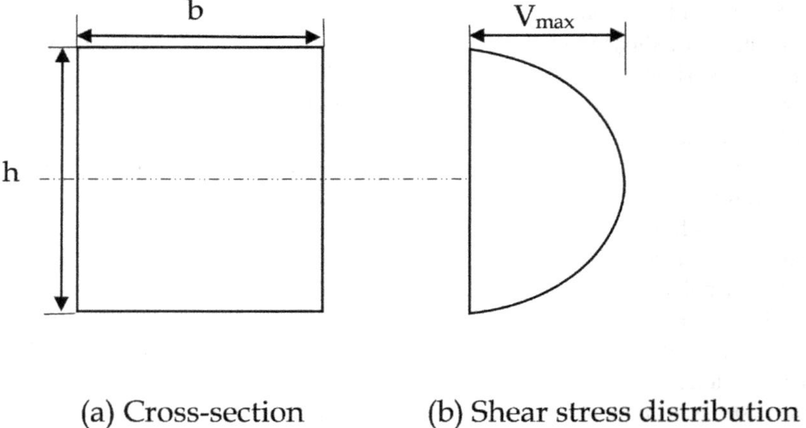

FIGURE 12.2 Shear stress in beams: (a) cross-section; (b) shear stress distribution.

$$\tau_v = \frac{V}{bd} \tag{12.2}$$

where, V = design shear force
b = breadth
d = effective depth

12.2 SHEAR STRENGTH OF RC SECTIONS

Shear in a reinforced concrete beam without shear reinforcement results in cracks along an inclined plane near the support, as shown in Figure 12.3.

The cracks are caused by the diagonal tensile stresses. The shear failure mechanism is complex and depends on the ratio of the shear span a_v (distance between the support and the major concentrated load acting on the beam) to the effective depth, d. The different mechanisms resisting shear in the beam can be stated as:

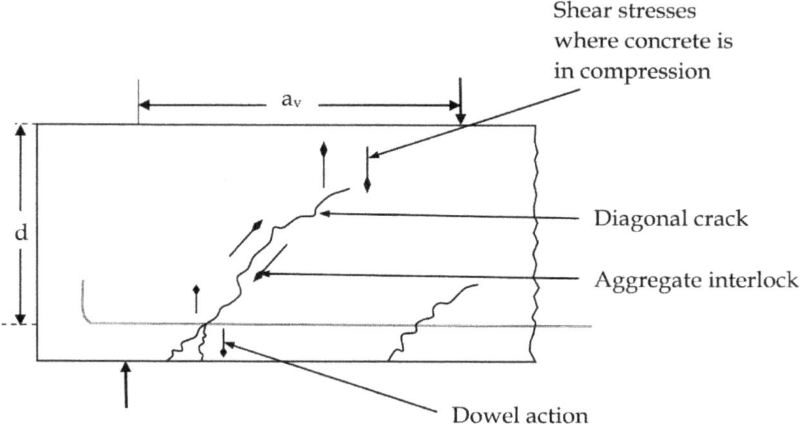

FIGURE 12.3 Different actions resulting in shear strength.

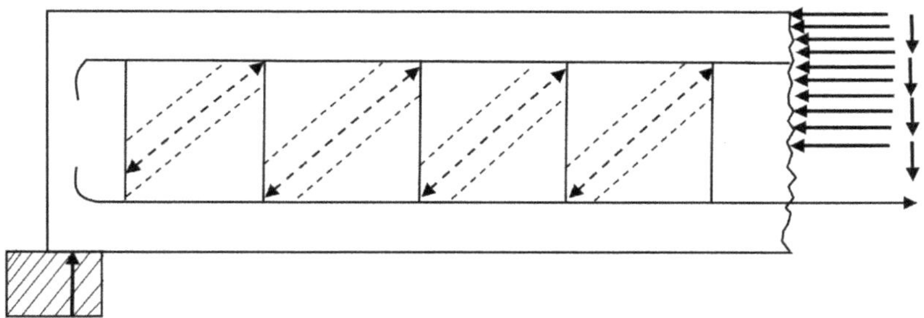

FIGURE 12.4 Diagonal compressive stresses in RC beams.

(a) Shear stresses in the compression zone resisted by uncracked concrete;
(b) Aggregate interlocking along the cracks;
(c) Dowel actions in the bars where the concrete between the cracks transmits shear forces to the bars.

An accurate analysis of shear strength is not possible and results from experimentation are generally relied upon. The shear strength of an RC beam depends on the grade of concrete and percentage of tension reinforcement in beams. For any grade of concrete, the maximum permissible shear stress is limited by the standards, which cannot be exceeded under any circumstances.

12.3 ACTION OF SHEAR REINFORCEMENTS

As explained in Figure 12.4, a cracked beam essentially acts as a truss. The provision of shear reinforcement which crosses the cracks will prevent cracking in the beam. Shear reinforcement in the form of vertical links or inclined bars can be effective in resisting failure in shear.

The behaviour of beams with shear reinforcement is analogous to a simple truss where the tension reinforcement acts as the bottom chord, the stirrups as the vertical members, the cracked concrete acts as diagonal compression concrete, and the top compression concrete as the top chord of the truss acting in compression. The action of a reinforced concrete beam with shear reinforcement is analogous to a simple truss, as shown in Figure 12.5.

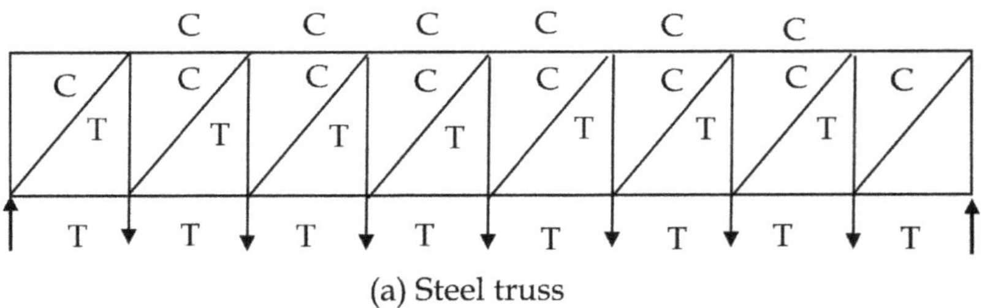

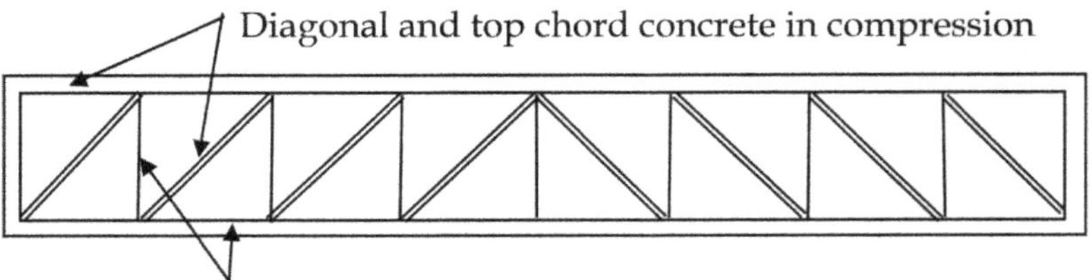

FIGURE 12.5 Truss analogy of shear resistance of a beam with shear reinforcement.

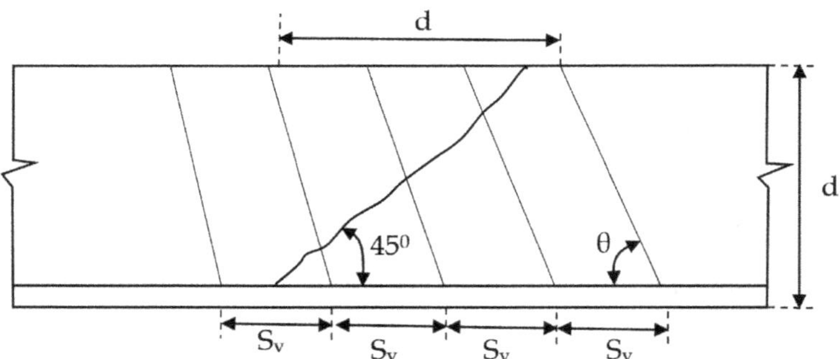

FIGURE 12.6 Shear strength of shear reinforcement.

The total shear force acting at a section is resisted by the concrete and the shear reinforcement. Depending upon the percentage of tension reinforcement, p_t, the shear strength of the concrete section can be determined. Hence shear force resisted by the shear reinforcement can be expressed as:

$$V_{US} = V_U - V_C$$

where,

V_U = transferred shear force acting at the section
V_C = shear strength of the concrete section based on τ_c depending on p_t
V_{US} = shear force to be resisted by the shear reinforcement

An expression for resisting shear force V_{US} by shear reinforcement can be illustrated by using Figure 12.6.

The forces in the various elements can be determined from statics based on the assumptions that the joints are hinged and the shear crack is inclined at 45° with the axis of the beam. It is assumed that "*n*" number of bars inclined at an angle θ and spaced at distance S_v are intersected by the crack. The vertical component of the tensile force developed in the "*n*" number of shear reinforcing bars is the shear force across the section.

$$V_{US} = n.\ \sigma_{SV}.A_{SV}.Sin\theta$$

where, A_{sv} = area of cross-section of a leg of a shear reinforcing bar

$$n = (\text{Cot}\,45° + \text{Cot}\theta) \times \frac{d}{S_V}$$

$$= (1 + \cot\theta)\frac{d}{S_V}$$

$$V_{US} = \sigma_{SV}.A_{SV}.Sin\theta.(1 + \text{Cot}\ \theta) \times \frac{d}{S_V}$$

$$= \sigma_{SV}.A_{SV}.\frac{d}{S_V}(Sin\theta + \text{Cos}\ \theta)$$

For vertical stirrups θ = 90°

$$V_{US} = \sigma_{SV}.A_{SV}.\frac{d}{S_V}$$

where, $\sigma_{SV} = 0.87 f_y$, S_V = spacing of shear reinforcement

$$V_{US} = 0.87 f_y.A_{SV}.\frac{d}{S_V}$$

S_v = spacing of shear reinforcement and the other symbols have their standard meaning. This expression is provided in IS:456-2000.

As per IS:456-2000, the characteristic strength of the stirrup shall not be taken as greater than 415 N/mm^2.

It needs to be clearly understood that shear failure, being brittle in nature, should be avoided since it is highly undesirable and can result in significant risk of loss of life and property. Hence shear design should be highly conservative in nature because of which the characteristic strength of HYSD reinforcement has been restricted to 415 N/mm^2.

It needs to be mentioned that for aseismic design only vertical links are effective under seismic forces. Shear forces can change direction as seismic forces are reversible and cyclic in nature. The inclined stirrups and bent-up bars, effective in one direction for resisting shear force, will not be effective for the opposite direction. The anchoring of the hooks should be performed using a 135° angle with an extension of eight times the diameter of the bar or 75 mm, whichever is more (Figure 12.7).

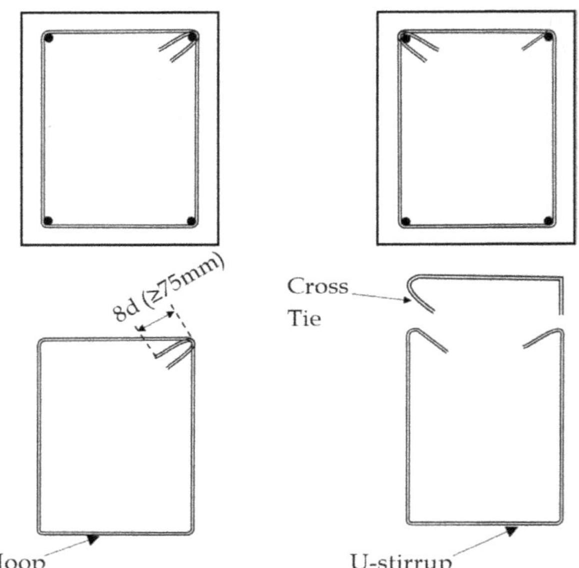

FIGURE 12.7 Transverse reinforcement in RC elements.

12.4 PRESENT-DAY CONCEPT OF SHEAR DESIGN

For reinforced concrete, the flexural mode of failure can only be made ductile. This has been discussed in detail in the chapters dealing with flexure (Chapter 10) and ductility (Chapter 17). All other modes of failure are brittle in nature, and occur suddenly without prior warning, and can result in huge loss of life and property. As per the limit state method of design, failure refers to some predetermined states governed by strain levels which, when reached, the section is considered to have failed. However, failure from the viewpoint of design never means complete collapse of the structure. Under design loads the structure can at the most reach the limiting state, but actual collapse can occur only under overloading. This can occur under seismic load, and the designer should take the necessary measures in their design so that the failure mode is ductile and not brittle. Hence, flexural failure is warranted and not shear failure. Before failure, the design moments will increase and reach their maximum or plastic moment values. The rate of change of the bending moment is shear force, and hence the shear forces will also increase significantly at that stage and the shear reinforcement provided should be capable of resisting the shear in the beams where two plastic hinges develop at the two ends. The plastic moments of resistance at the ends of the beams are to be calculated on the basis of reinforcement provided from the load design. As the seismic forces are reversible and act in the lateral direction, the shear to be resisted by the beams constitutes the sum of the shear coming from the dead and imposed loads and shear due to lateral forces which will be equal to the sum of the plastic moments at the beam ends divided by its length. From the stress–strain relationship of reinforcement, it can be observed that the maximum stress in steel as a result of strain hardening can be $1.25f_y$, where f_y is the yield stress of steel. In the limit state method of design, design stress in reinforcement is taken as $0.87f_y$. Hence it can be written that,

$$\frac{\textit{Plastic moment}}{\textit{Design moment}} = \frac{1.25\, f_y}{0.87\, f_y} = 1.437$$

which is considered as 1.4 by the code, i.e., IS:13920-2016 (Figure 12.8).

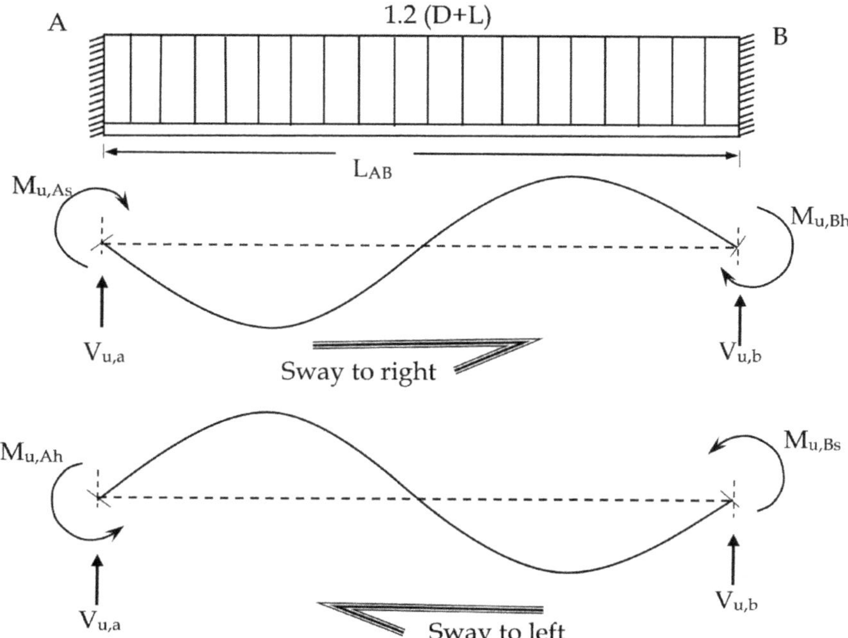

FIGURE 12.8 Calculation of design shear forces due to the formation of plastic hinges at beam ends.

12.5 EFFECT OF STIRRUP IN CONFINEMENT OF CORE CONCRETE

Confinement of concrete entirely transforms the stress–strain relationship of concrete with an increase in the failure strains and corresponding stresses. The stress–strain relationship forms the stress block of concrete and hence with an increase in the failure stresses and strains, strengths of the sections will increase. The increase in the ultimate strains will improve the ductility of the sections. Thus, closely spaced links at the ends of the structural elements help to confine the core concrete and improve the strength and ductility of the section significantly.

12.6 IS CODAL REQUIREMENTS

For the limit state of collapse in shear as per IS:456-2000, the following information is provided:

(a) Nominal shear stress;
(b) Design shear strength of concrete;
(c) Maximum permissible shear stress in sections;
(d) Minimum shear reinforcement to be provided;
(e) Design shear reinforcement.

The critical sections for shear design and enhanced shear strength of the sections close to supports are also provided. As per IS:13920-2016, transverse reinforcement is to be calculated considering the development of plastic hinges at the ends of the beams as discussed in earlier sections. In the design of the shear force capacity of RC beams, the contributions of bent-up bars, inclined links and concrete are to be neglected in the calculations. Concrete can crack severely during earthquakes, and hence its action in resisting shear forces is also to be neglected. Spacing of closely spaced links in beams has been specified. For columns, links at the ends are to be calculated considering the development of plastic hinges at the beam ends and special confining reinforcement at the ends have been specified in the standard.

12.7 SUMMARY

In this chapter, the basic features of shear design of reinforced concrete beams have been discussed. Though beams behave predominantly in flexure, and flexure design dominates, shear design plays a vital role in the strength design of beams. Reinforced concrete beams must have a minimum amount of shear reinforcement as per the standard. Nowadays, the concept of shear design has been significantly modified to suit the requirements of aseismic design. The transverse reinforcement also results in confinement of the core concrete, thereby improving the strength and ductility of beams. Shear failure should never precede flexural failure, as shear failure is highly brittle in nature.

BIBLIOGRAPHY

Bandyopadhyay, J. N. (2011). *Design of Concrete Structures*, PHI Learning, New Delhi.

Bhatt, Prab, MacGinley, Thomas J. and Choo, Ban Seng (2014). *Reinforced Concrete Design to Eurocodes, Design Theory and Examples*, Fourth Edition, CRC Press, New Jersey.

Indian Standard IS: 456-2000 (2007). *Plain and Reinforced Concrete–Code of Practice*, Fourth Revision, Tenth Reprint April 2007, Bureau of Indian Standards, Manak Bhavan, New Delhi, India.

Indian Standard IS: 13920-2016. (July 2016). *Ductile Design and Detailing of Reinforced Concrete Structures Subjected to Seismic Forces–Code of Practice*, First Revision, Bureau of Indian Standards, Manak Bhavan, New Delhi, India.

Mallick, S. K. and Gupta, A. P. (1995). *Reinforced Concrete*, Fifth Revision, Oxford & IBH, New Delhi.

Setareh, Mehdi, and Darvas, Rober (2007). *Concrete Structures*, Pearson Prentice Hall, New Jersey, Columbus, Ohio.

Sinha, S. N. (2002). *Reinforced Concrete Design*, Second revised edition, Tata McGraw-Hill, New Delhi, India.

13 Development Length of Reinforcement in RC Sections

13.1 INTRODUCTION

Reinforced concrete is a combination of two different materials, concrete and reinforcement, and there should be perfect composite action between the two. The bond between the two should be perfect so that strain compatibility is maintained between any point on the reinforcement and the surrounding concrete. This refers to the adhesion between the two materials. It is because of this bond that stress transfer between the two materials is possible and the same reinforcing bar can be subjected to different forces at different sections of the member. Bond stress is the shear stress acting at the steel concrete interface, i.e., interface shear. The success of reinforced concrete as a construction material can be attributed to the basic assumptions that the perfect bond exists between steel and concrete.

Two types of bond stresses can develop:

a) Flexural bond stresses develop due to a change in bar force along the length of the bar or due to a change in the bending moment along the length of the member.
b) Anchorage or development bond due to tension and compression in the bar.

These two types of bond stresses are illustrated in Figure 13.1.

Flexural bond stress develops due to a change of the bending moment along the length of the member and hence is a function of the shear force. The bond stress developed is a function of the total perimeter of the bars at the beam section. Hence, to increase the total perimeter of the bars, more smaller-diameter bars need to be provided than higher-diameter bars. Although for simplicity the bond stresses are considered as uniform along the bar, cracking can result in significant variation of the stress.

Let the flexural bond stresses at two adjacent sections of beam at a distance *dx* subjected to a differential moment *dM* be τ_{bf} and the differential force in steel bars be *dT* then:

$$dT = \frac{dM}{lever\ arm} = \frac{dM}{jd}$$

Assuming that the bond stress is uniform

$$dT = \tau_{bf}\left(\pi d\right)_n dx = \frac{dM}{jd}$$

DOI: 10.1201/9781003415398-13

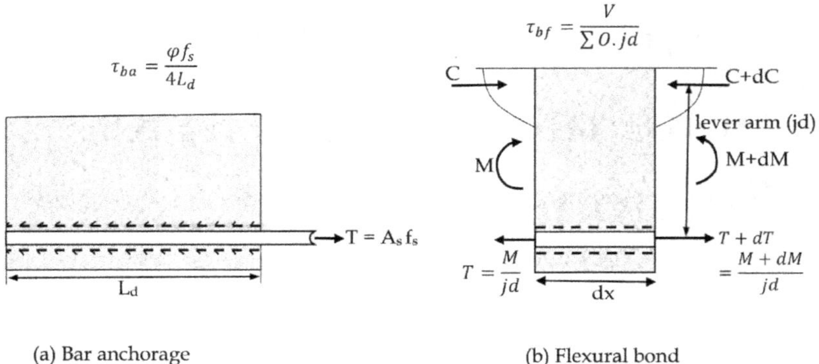

FIGURE 13.1 (a) Anchorage/development bond; (b) flexural bond.

$$\therefore \; \tau_{bf} = \frac{dM}{dx} \times \frac{1}{(\pi d)_n} \times \frac{1}{jd}$$

$$\therefore \; \tau_{bf} = \frac{V}{\sum_n O.\,jd} \tag{13.1}$$

where, n is number of bars

$\sum_n O$ = summation of the perimeter or total circumference of n bars

jd = lever arm

dM/dx = V, shear force at the section

Anchorage bond

This bond arises over the length of anchorage provided near the end/cut-off point or it resists pulling out of the bar in tension or pushing the bar into concrete.

13.2 CONCEPT OF DEVELOPMENT LENGTH

A bar must extend a length L_d beyond any section on either side at which it is required to reach its full yield strength or design strength.

Design force of the bar = full bond force developed in it

$$\frac{\pi (d_b)^2}{4} \times f_d = \tau_{ba} \times \pi d_b \times L_d$$

$$\therefore L_d = \frac{f_d \times d_b}{4\tau_{ba}} \tag{13.2}$$

where, τ_{ba} = bond stress

d_b = diameter of bar

L_d = development length

The bond stress is assumed to be uniform over this length of the bar. A bar must extend a certain minimum length L_d on either side of a point of its maximum stress to transfer the bar force to the surrounding concrete through the bond and to prevent the bar from being pulled out under tension or being pushed in under compression. This length is called the anchorage length in the case of tension or compression, and the development length in the case of flexure. Current design practices focus on development length rather than determination of bond stress. Since failure in a bond is brittle, the development of proper bond stresses between concrete and steel, i.e., providing proper development length of bars, needs to be properly ensured. Closely spaced transverse reinforcement to properly confine the concrete, proper spacing of bars, provisions of adequate cover, provision of hooks, bends or mechanical anchorages, proper design of ribs on reinforcement, staggering splices, avoiding curtailment of reinforcement in tension zone, splicing of reinforcements at proper locations and with development lengths etc., are required to prevent failure in bond. Design and detailing for bond resistance should be conservative in nature as this type of failure is highly undesirable.

13.3 BENDS, HOOKS AND MECHANICAL ANCHORAGES

Standard bends, hooks and mechanical anchorages may be used to satisfy the requirements of development length in tension. In compression these are not effective. IS:456-2000 specifies that deformed bars may be used with or without end anchorages in the form of hooks or bends. However, it seems to be more justified to consider the straight lengths before bends only in compression. In this context it needs to be mentioned that in aseismic design, shear reinforcements are to be anchored properly so that the longitudinal reinforcements are properly kept in position (Figures 13.2 and 13.3).

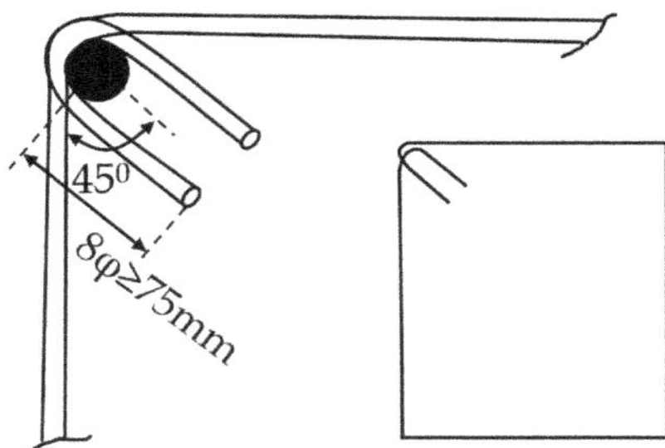

FIGURE 13.2 Details of anchorage to be provided for shear reinforcement.

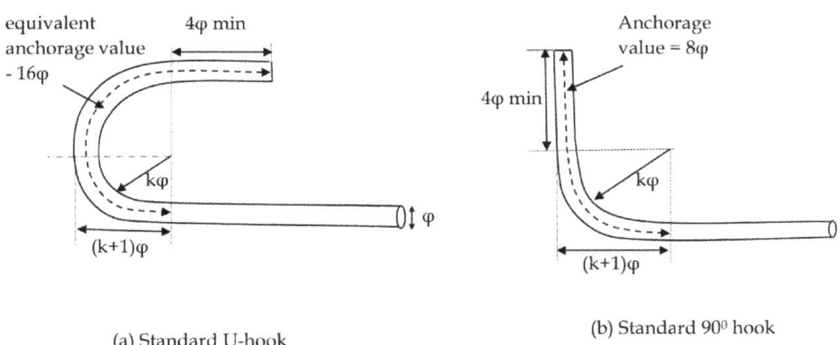

FIGURE 13.3 Standard hook and bend.

Note: k = 4 for each 45° bend

13.4 CRITICAL SECTIONS FOR CHECKING DEVELOPMENT LENGTH

It has been observed that designers are more concerned with the strength and serviceability design and mostly emphasize compliance with the different limit states. If the detailing features covered in the standards or allied documents like handbooks are properly followed, the basic requirements of development length are mostly complied with, especially with the present-day ductility detailing considerations. However, a detailed understanding of the phenomenon is generally lacking. Some basic checks when properly performed during design and detailing can prevent bond failure. Reinforcements in concrete structural elements are mostly subjected to axial forces. At locations where the design load conditions are reached, the reinforcements attain their maximum or yield stresses and full development length is required on either side of the section, in tension or compression as the case may be. Development length needs to be checked at:

I. Sections of maximum moment, i.e., where full design stresses are being developed in the bars
II. At sections, where the reinforcements are cut off or curtailed
III. At simple supports for simply supported beams
IV. At points of inflection.

At locations where bars are to be spliced, the lapping length should be at least equal to the development length.

Termination of longitudinal reinforcement in beams in the tension zone should be avoided. Nowadays, software can easily specify the positions of the points of contraflexure for different loading conditions and accordingly bar curtailment needs to be performed. The concrete cover, grade of concrete, spacing of reinforcement, effect of settling of concrete below the top bars of beams, types of bars such as plain or deformed bars, use of bundled bars, lap splices, use of lightweight concrete, etc., are all parameters that affect the development length.

Reinforced concrete is very strong in compression but very weak in tension. The lap length of bars should be at least equal to the development length and provision of lap length in tension, compression and flexure needs special attention.

Figures 13.4–13.6 demonstrate ductility detailing which is mandatory for structures in seismic-vulnerable areas and is optional but desirable in areas of low seismic vulnerability.

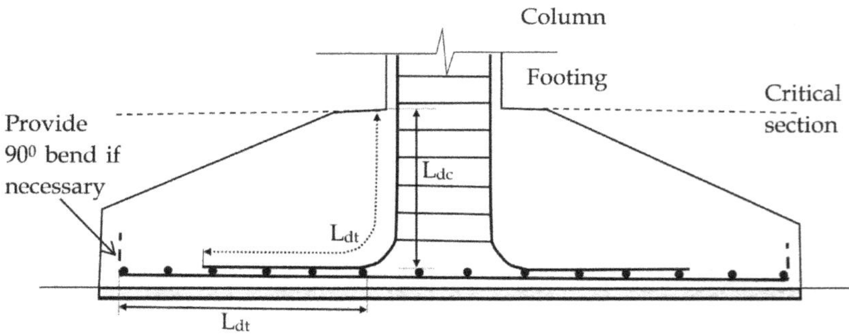

FIGURE 13.4 Critical section for checking the development length of column bars at the footing level.

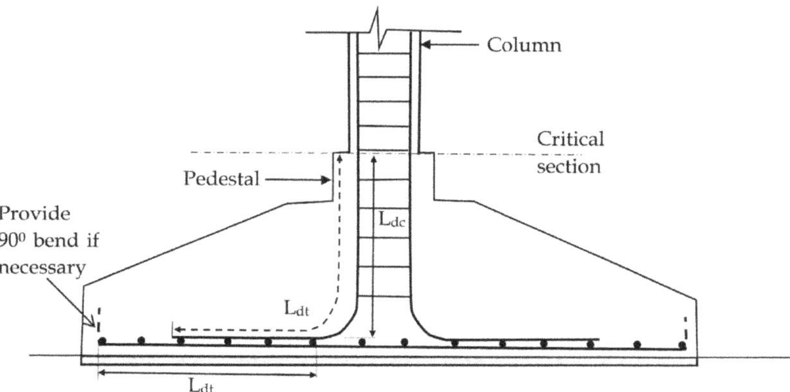

FIGURE 13.5 Critical section for checking the development length of column bars at the pedestal level.

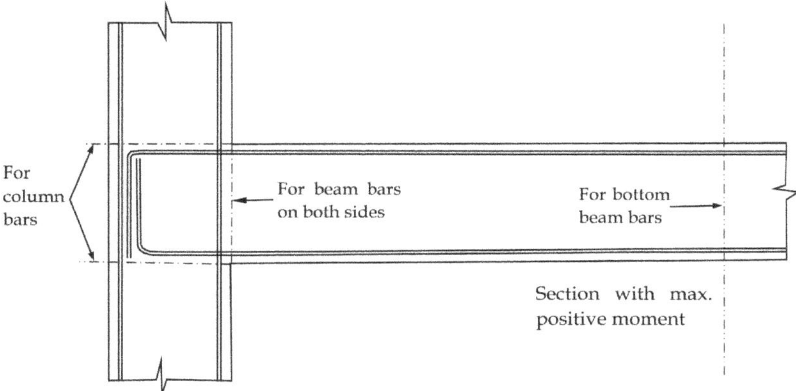

FIGURE 13.6 Critical sections for checking the development length of column and beam reinforcement.

Dotted line indicates critical section for checking development length.

13.5 IS CODAL REQUIREMENTS

Development or anchorage bond

As per IS:456-2000, the development length is given by,

$$L_d = \frac{\varphi . f_s}{4\tau_{bd}} \tag{13.3}$$

where, φ = nominal diameter of the bar

f_s = stress in bar at the section considered at design load

τ_{bd} = design bond stress

The values specified for bond stress are 1.2, 1.4, 1.5, 1.7 and 1.9 MPa for concrete grades M20, M25, M30, M35 and M40 and above, respectively, for plain bars in tension with an increase of 60%

TABLE 13.1
$\frac{L_d}{\varphi}$ values for fully stressed (HYSD) bars in tension

Grades of steel	Grade of concrete				
	M20	M25	M30	M35	M40 and above
Fe 415	47	40	38	32	30
Fe 500	57	49	45	40	36
Fe 550	62	53	49	44	39
Fe 600	68	58	54	48	43

TABLE 13.2
$\frac{L_d}{\varphi}$ values for fully stressed (HYSD) bars in compression

Grades of steel	Grade of concrete				
	M20	M25	M30	M35	M40 and above
Fe 415	38	32	30	27	24
Fe 500	45	39	36	32	29
Fe 550	50	43	40	35	31
Fe 600	54	47	44	38	34

for deformed bars in tension and a further increase of 25% for bars in compression. The development lengths expressed in terms of bar diameters are shown in Tables 13.1 and 13.2.

The development length calculated as per Equation (13.3) is for fully stressed bars ($f_s = 0.87f_y$) for different combinations of grades of concrete and steel. However, the area of steel provided is generally higher than the area of steel required. Therefore, the development length required may be reduced proportionally as per the following:

$$L_{d,modified} = L_d \times \frac{A_{st,required}}{A_{st,provided}}$$

Nowadays, generally seismic forces govern the design whereby over-loading can be a routine phenomenon and this practice of reducing the development length is not encouraged.

In the case of bundled bars, the development length needs to be increased due to the reduction in the steel concrete interface.

Flexural bond

As per IS:456-2000, at simple supports and at points of inflection, positive moment tension reinforcement shall be limited to a diameter such that L_d, computed for the design stress f_d, is expressed as follows:

$$L_d \leq \frac{M_1}{V} + L_0 \tag{13.4}$$

where, M_1 = moment of resistance of the section assuming all reinforcement at the section to be stressed to f_d

$f_d = 0.87f_y$ in the case of limit state design and the permissible stress σ_{st} in the case of working stress design

V = shear force at the section due to design loads

L_0 = sum of the anchorage beyond the centre of the support and the equivalent anchorage value of any hook or mechanical anchorage at simple support; and at a point of inflection. L_0 is limited to the effective depth of the members or 12φ, whichever is greater, and,

φ = diameter of bar

The value of $\frac{M_1}{V}$ may be increased by 30% when the ends of the reinforcement are confined by a compressive reaction.

The main significance of this clause is to check the embedded length of the reinforcement from the point where it is fully stressed. Figure 13.7 demonstrates the shear force and bending moment diagrams of a simply supported beam or the portion between points of inflection of a continuous beam. Let the bars in question be fully stressed at a moment M_1 occurring at distance from a simple support given by $a = \frac{M_1}{V}$ where V is the shear force at the support section. L_0 should be the extension of the bar beyond the support or inflection point, as the case may be, to satisfy the requirement of the development length required from the point where the bar is fully stressed. This ensures that the sum of the distances measured along the bar from the support or inflection point to its innermost extension and the point where it is fully stressed is greater than the development length of the bar.

The equivalent anchorage value of the hook and mechanical anchorage needs to be taken into consideration. For uniformly distributed load the concept is similar and even more conservative as the moment M_1 will occur at an even further distance from the support. This consideration is for positive moment reinforcement. Since M_1 and V are not directly under the control of the designer and L_0 is restricted by the code, the only way to satisfy this requirement is to reduce the bar diameter of positive moment tension reinforcements.

From the above discussions on anchorage and flexural bond, it needs to be mentioned that no bar should be terminated without providing the necessary development length L_d on either side of the point of maximum design stress, $f_d = 0.87f_y$.

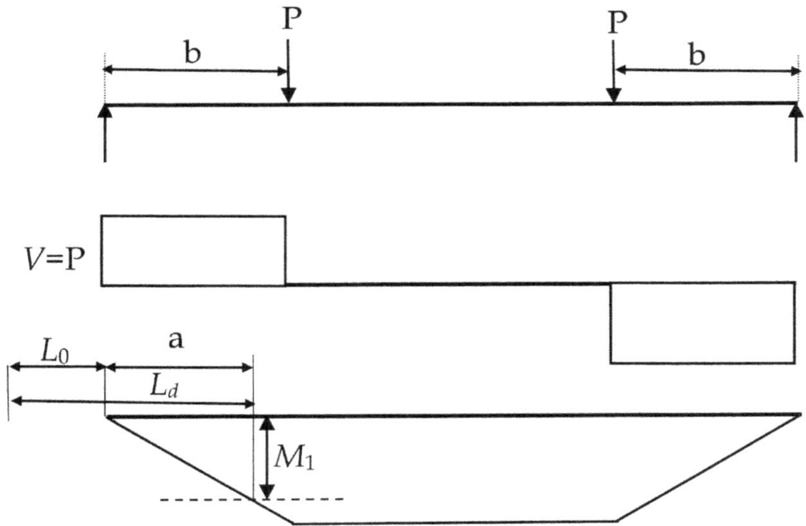

FIGURE 13.7 Checking for the development length at a simple support or at a point of inflection.

13.6 CALCULATIONS OF BREAKING STRENGTH OF CONCRETE AND ITS EFFECT

Concrete is strong in compression but very weak in tension. The basic instinct of any structure is to transmit the forces acting on it to the supporting soil through the sub-structure or foundations. Towards the end of the load transfer path, the vertical load-bearing elements transfer the forces to the foundations. Concrete footings with pedestals can act as supports to steel columns which are connected to the pedestals using base plates and anchor bolts. The function of anchor bolts is to transfer shear and tension to the pedestals. Uplift due to wind or earthquake forces is to be resisted by the anchors placed within the concrete pedestals. Anchors can be of different types, such as cast-in-place anchors or post-installed anchors, depending on whether they are placed during casting or inserted into the hardened concrete. During uplift, different types of failure can occur such as failure of steel anchor, pullout, concrete breakout, side-face blowout, bond failure, etc. which are shown in Figure 13.8.

The tensile strength of concrete is its capacity to resist cracking or breaking under tension. Determining the tensile strength is necessary to understand the extent of the possible damage due to uplift. Breaking and cracking arise when tensile forces surpass the tensile strength. Therefore, the strength of concrete without failure against tension is known as the breaking strength of concrete.

Steel failure

Steel failure occurs due to necking near the shank of the anchor. As the tensile forces increase on the anchors, the anchor will fail at its ultimate strength. ACI code prescribes an equation utilizing the ultimate strength of the steel:

$$Ns = A_s \times f_{ult}$$

where, Ns is the ultimate strength of the steel, A_s is the effective area of the steel in tension, and f_{ult} is the ultimate strength of the material

Pull-out

There is another common failure which occurs when the frictional component is surpassed, and the anchor simply slips out of the concrete. This can occur due to various reasons, development length, number of threads embedded in concrete, when the yield strength is much higher than the pull-out strength and the frictional force is surpassed by this pull-out strength.

Concrete breakout

Breakout failure occurs when the tensile load on the anchor increases more than the tensile strength of the concrete specimen and the anchor breaks out and shears out in a cone (Figure 13.9).

ACI estimates the typical breakout angle as approximately 35°, which also depends on the depth of the embedment of anchors. There are many ways to calculate the breakout capacity. ACI 318 prescribes the use of the concrete capacity design (CCD) method. The CCD also assumes a 35° failure angle and a rectangular breakout. The basic concrete breakout capacity of a single anchor in tension is predicted by ACI, where N_b is the ultimate breakout capacity, k_c is a constant based on the anchor type, f'_c is the specified compressive strength of concrete, λ_a is a modification factor based on the concrete type, and h_{ef} is the effective embedment depth of the anchor:

$$N_b = k_c \times \lambda_a \times \sqrt{f'_c} \times h_{ef}^{1.5}$$

There is another method where the ACI code assumes a 45° failure angle and a conical breakout. the equation below predicts the breakout strength using this failure angle, where P is the ultimate

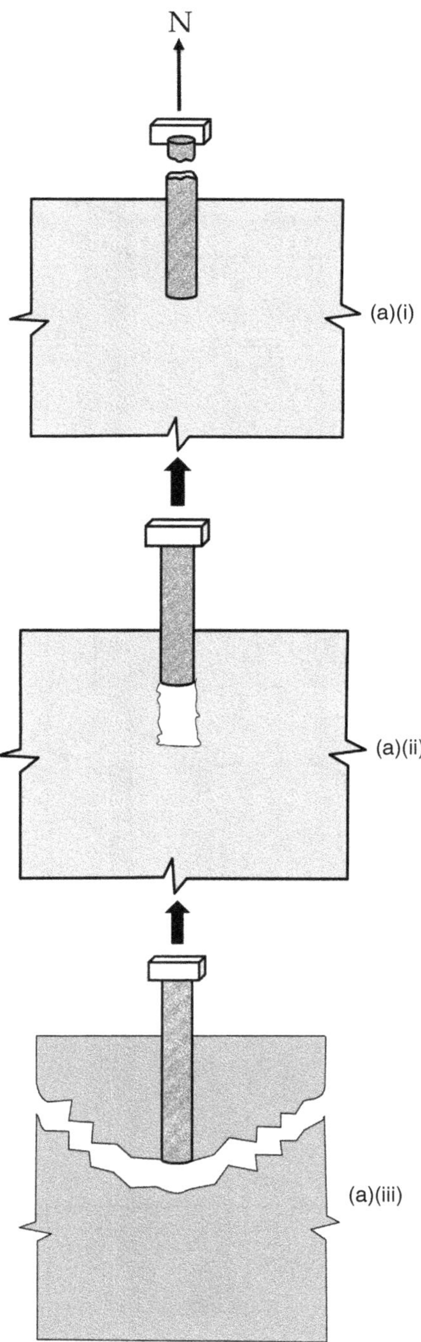

FIGURE 13.8 Failure modes for anchor: (a) under tensile loading - (i) steel failure; (ii) pullout; (iii) concrete breakout; (iv) concrete splitting; (v) side-face blowout and (b) shear loading.

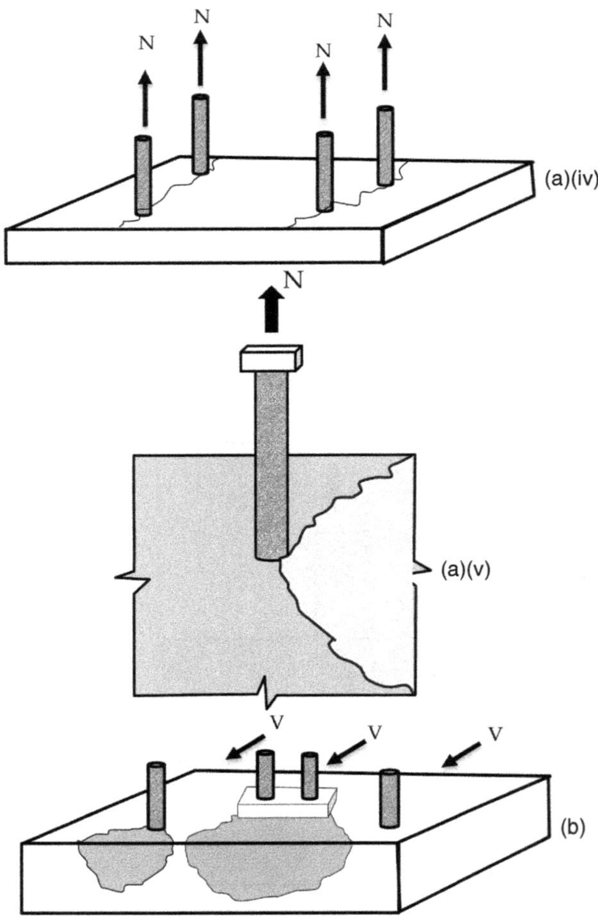

FIGURE 13.8 (Continued)

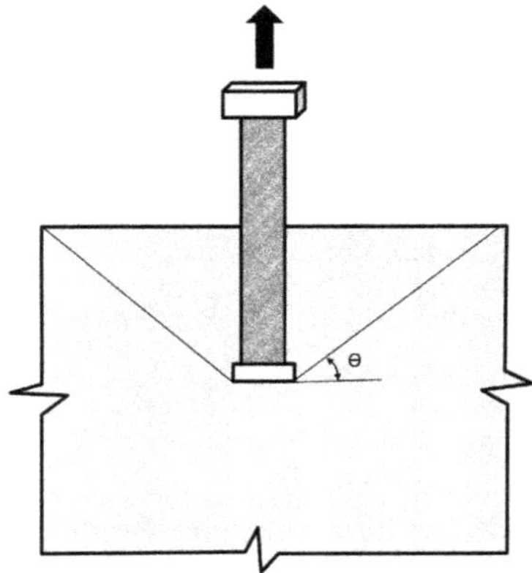

FIGURE 13.9 Failure angle θ.

breakout capacity, L is the embedment length, d is the diameter of the anchor head, and f_c' is the compressive strength of the concrete:

$$P = \pi \times L \times (L + d) \times 4 \times \sqrt{f_c'}$$

There are various methods to calculate the breakout capacity using tensile strength, changing failure angles, etc.

Concrete breakout strength of anchors in tension, N_{cb}

Nominal concrete breakout strength in tension, N_{cb} of a single anchor or N_{cbg} of an anchor group, can be calculated by (a) or (b), respectively:

(a) For a single anchor:

$$N_{cb} = \frac{A_{Nc}}{A_{Nco}} \psi_{ed,N} \psi_{c,N} \psi_{cp,N} N_b$$

(b) For an anchor group:

$$N_{cbg} = \frac{A_{Nc}}{A_{Nc0}} \psi_{ec,N} \psi_{ed,N} \psi_{c,N} \psi_{cp,N} N_b$$

where, ψ's are different factors specified in the standard

A_{Nc} is the total projected area of the concrete cone for an anchor group

A_{Nc0} is the projected area of the concrete cone for a single anchor

The spacing of anchors and edge distance affect the concrete breakout strength.

13.7 SUMMARY

This chapter explains the concept of bonding between concrete and reinforcement. The design philosophy adopted nowadays puts greater emphasis on checking the requirements of development length rather than calculating the bond stresses. On the basis of permissible bond stresses for different grades of concrete and steel, development lengths in tension and compression have been calculated for all grades of structural concrete of M20 to M60 and HYSD steel of Fe 415 to Fe 600. Using simple sketches critical sections for checking the development length of different structural elements has been demonstrated. The codal clauses dealing with bond and development length have been discussed. The current trend in reinforced concrete design and detailing is focused on ductility considerations and providing ductility detailing in elements whereby not only the bond failure can be prevented but the reserve strength of sections and their performances under load can be significantly improved also. The breaking strength of concrete and calculation of length of anchors have been discussed as per ACI specifications.

BIBLIOGRAPHY

ACI 318-19. (2019). *Building Code Requirements for Structural Concrete and Commentary.*

Bandyopadhyay, J. N. (2011). *Design of Concrete Structures*, PHI Learning, New Delhi.

Ferguson, Phil M., Green, John E. and Jirsa, James O. (1988). *Reinforced Concrete Fundamentals*, Fifth Edition, John Wiley, Chichester, UK.

Park, R. and Paulay, T. (1975). *Reinforced Concrete Structures*, John Wiley, New York.

Pillai, S. U. and Menon, D. (2012).. *Reinforced Concrete Design*, Third Edition, Reprint 2012, Tata McGraw Hill Education, New Delhi.

Mallick, S. K. and Gupta, A. P. (1989). *Reinforced Concrete* (Fifth Revision), Oxford & IBH, New Delhi.

14 Limit State of Serviceability

14.1 INTRODUCTION

The code-based prescriptive method of design is based on loads. Limit state of collapse is based on ultimate loads and checks the safety of the structure with respect to strength or resistance to rupture. Limit state of serviceability is based on working or service loads. These loads are not used for strength checking but for checking other parameters which are related to working loads. Application of forces on structures induces stresses, which in turn result in strains, deformations and deflections. It is essential to consider both the short-term and long-term deflections as for reinforced concrete structures both are significant. Long-term deflection can be three to four times the short-term effects and can cause warping in the elements. Excessive deflection can result in undesirable vibration. Deflection can be easily measured in contrast to stresses or strains, which are not that easy to measure. Control of deflection is a requirement of serviceability and is a criterion for serviceability design. Creep and shrinkage effects are also manifested in the form of deflection or deformations. Another criterion for serviceability design is to restrict the crack widths under these working loads. Reinforced concrete structures are robust, costly and require sufficient time for project completion. Hence, they should be sufficiently durable. Durability aspects are also included within the serviceability requirements. Durability aspects are controlled in the form of minimum cement content, maximum water–cement ratio, cover to reinforcement, requirements of fire resistance, etc.

As per the design philosophy of the limit state method, two independent classes of limit states need to be considered: limit state of collapse and limit state of serviceability. Whereas the former deals with safety in terms of strength, overturning, sliding, buckling, etc. the latter deals with serviceability in terms of deflection, cracking, vibration, durability, etc.

The primary serviceability limit states are comprised of the following criteria:

- The member should not undergo excessive deformation under service loads. This limit state is generally referred to as the limit state of deflection.
- The width of cracks developed on the surface of reinforced concrete members under service loads should be limited to the values prescribed in the codes of practice. This limit state is referred to as the limit state of cracking.

14.2 SERVICEABILITY DESIGN OF COMMONLY USED STRUCTURAL ELEMENTS

The design of reinforced concrete structures using the limit state method requires simultaneous checking for both these limit states. Normally, designers check for limit states of collapse only. Limit

DOI: 10.1201/9781003415398-14

states of serviceability are specifically checked in special cases as detailed serviceability checks are elaborate and cumbersome, involving a different set of serviceability load factors. However, if the designer complies with the requirements of the standards properly while performing the design, serviceability checks are almost fully complied with. The requirements of durability including fire resistance are controlled in the form of minimum and maximum cement content, maximum water–cement ratio, cover to embedded steel, proper compaction and curing, shape and size of the member, etc. The environmental exposure condition to which the structure will be exposed also plays a vital role in durability. Regarding deflection, parameters like span to effective depth ratio, effect of spans >10 metres, effect of tension and compression reinforcement, flanged beams, slenderness limits for beam to ensure lateral stability, etc. need to be checked as per the codal requirements. However, generally strength design governs for beams and serviceability design does not play a key role except for slabs, which are of much lesser depth, where design is governed by serviceability requirements and generally not by strength. Excessive deflection may result in vibration, which is highly undesirable. Regarding check for cracking, the maximum horizontal distances between parallel reinforcements in flexural members are specified in IS:456-2000, which if properly complied with means that detailed calculations for crack widths are not necessary. Side face reinforcements are necessary for beams where the depth of the web exceeds 750 mm. Good detailing requirements are also needed for avoiding cracks in concrete. If all the above-mentioned specifications are properly complied with the serviceability requirements are generally automatically indirectly satisfied.

14.3 CALCULATION OF DEFLECTION

Deflection predictions of reinforced concrete members are quite difficult, mostly because of uncertainties in calculating the flexural rigidity (EI), uncertainties in assessment of long-term effects like creep and shrinkage, etc. Deflection of reinforced concrete elements are calculated in two parts:

(i) Immediate or short-term deflection occurring on application of the load.
(ii) Long-term deflection mostly occurring from creep and shrinkage under sustained loading which is permanent in nature, i.e., dead load plus the sustained part of live load.

14.3.1 Deflection behaviour of beams

For understanding the deflection behaviour of beams, study of the moment curvature relationship is necessary. A typical moment curvature relationship as obtained from measurements on singly reinforced beam failing in tension is shown in Figure 14.1.

The relationship between moment and curvature is given by the classical theory of bending,

$$\frac{\sigma}{y} = \frac{M}{I} = \frac{E}{R} \tag{14.1}$$

where the symbols have their standard meanings.

$$\therefore \text{EI} = \text{MR} = \frac{\text{M}}{\phi}$$

where EI is the flexural rigidity of the section. With an increase in moment, cracking of the concrete reduces the flexural rigidity and decreases the slope of the curve. When the steel yields, a large increase in curvature occurs at nearly constant bending moment. Hence, the curve becomes almost

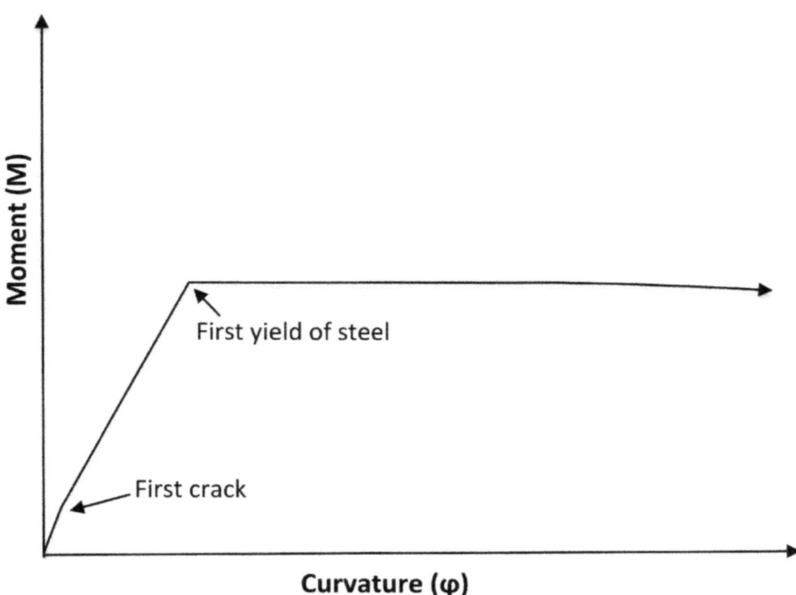

FIGURE 14.1 Moment curvature relationship for singly reinforced beam failing in tension.

horizontal. The moment curvature relationship for under-reinforced beams can be idealized to a tri-linear relationship. The first stage is up to the initiation of the first crack, the second to yield of the tension steel and the third to the limit of useful strain in concrete.

Up to the initiation of first crack in the moment curvature relationship, concrete behaves in a linear elastic manner and the value of EI is maximum. The maximum tensile stress is less than the tensile strength in flexure.

In general, the deflection calculation of RC elements can be approximately performed on the basis of gross concrete section ignoring the reinforcements. For accurate calculation of the moment of inertia for reinforced concrete sections, the effect of reinforcement is to be taken into consideration, and the transformed section is based on equivalent concrete area calculated using the modular ratio $\left(\frac{E_s}{E_C}\right)$. For a section made of uniform material, the neutral axis passes through the centroid. Hence the position of the centroid of the transformed uncracked cross-section of the concrete can be calculated and the moment of inertia needs to be determined about this axis. This is indicated in Figure 14.2.

The pre-cracking region ends at the initiation of the first flexural crack when the concrete stress reaches its modulus of rupture, f_{cr}.

Hence the cracking moment of resistance of the section can be expressed as:

$$M_{cr} = \frac{I_g f_{cr}}{y_T} \tag{14.2}$$

where, I_g = gross moment of inertia of the section
f_{cr} = flexural strength of concrete
y_T = half the depth of the beam

Beyond the first crack, that is post-cracking stress, beams are generally subjected to service loads. When flexural cracking develops, the contribution of the concrete in the tension zone reduces substantially with a reduction in flexural rigidity. For fully cracked sections the contribution of the tension concrete is completely ignored in the calculations. When the steel yields, there is a substantial loss in stiffness of the section because of extensive cracking. As the load continues to increase, strain in steel goes on increasing without an increase in stress. When the limiting strain in compression concrete is reached, the section fails. These has been described in the chapter on flexure (Chapter 10). When the beam reaches the limit state of collapse it is considered to have failed from the consideration of strength design and hence these deformations are not of much importance to designers dealing with the prescriptive method of design. However, these deformations are considered in performance-based design (PBD).

As per Indian standard, IS:456-2000, structural analysis for concrete sections can be performed on the basis of any one of the following:

(a) Gross concrete section ignoring the effect of reinforcements.
(b) Transformed section considering the effect of modular ratio.
(c) Cracked section, calculated on the basis of area of concrete in compression plus equivalent area of concrete in lieu of steel.

Short-term deflections are calculated for working loads within which concrete can be considered as a linear elastic material. Long-term deflections are time dependent and occur over a period of time mainly due to creep and shrinkage of concrete.

14.3.2 Deflection by elastic theory

Short-term deflections, due to the applied loads, are generally based on the assumption of linear elastic behaviour, whereby reinforced concrete is treated as a homogeneous material. Expression for the maximum elastic deflection, Δ, of a homogeneous beam of effective span, L, and flexural rigidity, EI, can be derived using the standard methods of structural analysis.

Typically, it can be expressed as:

$$\Delta = k_w \frac{WL^3}{EI} = k_m \frac{ML^2}{EI} \tag{14.3}$$

where, W = total load on the span, M = maximum moment, and k_w and k_m are constants which depend on the load, end conditions and flexural rigidity of the member. Generally, deflections are calculated at positions close to the mid span for simply supported and continuous beams and at the free end of cantilevers.

14.3.3 Tension stiffening effect

If the external moment is less than the cracking moment (M_{cr}) the section will remain uncracked and the full section including the concrete in the tension zone will remain effective. The moment of inertia (I_T) is calculated with respect to the uncracked transformed section, i.e., for the gross concrete section and the additional area corresponding to the reinforcement. When the applied moment exceeds M_{cr}, it is expected that the entire section will crack, and I should be calculated for I_{cr}. However, in practice this does not occur and significant portions between the cracks remain uncracked. Hence, for calculating deflection under service load an intermediate value between EI_{gr} and EI_{cr} is used, which is termed as EI_{eff}.

The tension stiffening effect may be described as the increase in stiffness over the cracked stiffness value due to the capacity of concrete to carry some tensile stresses.

The different values of EI may be expressed as:

1) Calculated on the basis of uncracked transformed section, EI_T.
2) Calculated on the basis of gross section without considering the effect of reinforcement, EI_{gr}.
3) Calculated on the basis of cracked transformed section (ignoring concrete on the tension side and only considering the equivalent concrete area calculated for steel), EI_{cr}.
4) Calculated on the basis of intermediate effective value, EI_{eff}.

$$EI_T > EI_{gr} > EI_{eff} > EI_{cr}$$

14.3.4 Calculation of flexural rigidity

A rectangular reinforced concrete beam having width "b" and depth "D", with tension reinforcement A_{st} is shown in Figure 14.2. The method of calculating flexural rigidities for the gross cross-section, uncracked transformed section and cracked section are demonstrated in the following section.

(a) Calculation of moment of inertia of the gross cross-section

In this condition, the centroid will be located at mid depth of the section and the neutral axis will pass through the centroid. Hence the moment of inertia about the neutral axis,

$I_{gross} = \frac{bd^3}{12}$. This condition is shown in Figure 14.2(b).

(b) *Calculation of moment of inertia of uncracked transformed section*

For this section, as shown in Figure 14.2(c), the effect of reinforcements is transformed into equivalent concrete area using the modular ratio, $= \frac{E_s}{E_c}$. Now, the total area of the cross-section becomes $[(b \times D) + \{(m - 1) \times A_{st}\}]$, as A_{st} area of steel is now replaced by concrete. The depth of the centroid of the section from the top fibre can be determined from basic statics as:

$$y = \frac{\Sigma A_i . y_i}{\Sigma A_i} = \frac{b.D.\frac{D}{2} + (m-1)A_{st}.d}{b.D + (m-1)A_{st}} \tag{14.4}$$

For any grade of concrete, the modulus of elasticity can be determined from codal expression as $E_c = 5000\sqrt{f_{ck}}$, Young's modulus of steel $E_s = 2 \times 10^5$ MPa. Knowing the area of reinforcement A_{st} and the modular ratio, the depth of the neutral axis can easily be calculated. The moment of inertia of the section about the neutral axis which will pass through the centroid can be evaluated as:

$$\frac{bD^3}{12} + bD\left(y - \frac{D}{2}\right)^2 + (m-1)A_{st}(d-y)^2 \tag{14.5}$$

In the above expression, the moment of inertia of the transformed area of steel about its own axis is ignored as its value is insignificant.

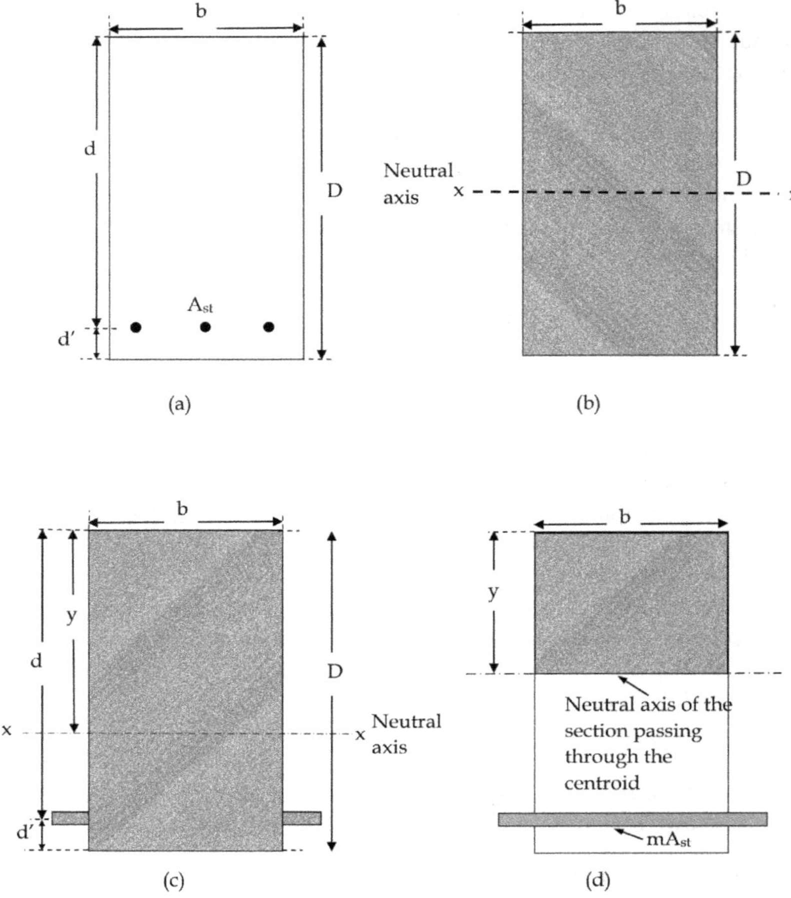

FIGURE 14.2 Different types of beam cross-section for I calculation: (a) beam cross-section; (b) gross cross-section ignoring reinforcement; (c) uncracked transformed section; (d) cracked section.

(c) *Calculation of moment of inertia of cracked section*

In this condition the effect of concrete in tension is ignored and area of steel is converted into equivalent concrete area, as shown in Figure 14.2(d). For this condition, if y is the depth of the centroid from the top fibre, equating moments of areas about the centroidal axis (which is the neutral axis) the position of neutral axis can be determined as:

$$\left(b \times y \times \frac{y}{2}\right) = m.A_{st}.(d - y) \tag{14.6}$$

The value of y can be obtained by solving the quadratic equation. The moment of inertia of cracked section I_{cr} about the neutral axis can be written as:

$$I_{cr} = b.\frac{y^3}{3} + m.A_{st}(d - y)^2 \tag{14.7}$$

Calculating the deflection of a reinforced concrete beam using the flexural rigidity on the basis of uncracked section, results in a lower value of the actual deflection, whereas calculating on the basis of the cracked section results in over-estimation of the actual deflection. Hence, generally, deflection calculations of reinforced concrete beams are based on the effective flexural rigidity, EI_{eff}.

14.3.5 Effective moment of inertia formulation

The expression of I_{eff} given in the Indian standard (IS 456: 2000, Cl. C-2.1) is based on an earlier version of the British code, which assumed an idealized trilinear moment–curvature relationship.

The expression given in the IS code for effective moment of inertia calculation has been modified on the basis of test data. The expression is of the form:

$$I_{eff} = \frac{I_{cr}}{1.2 - \left(\frac{M_{cr}}{M}\right)\frac{z}{d}\left(1 - \frac{x}{d}\right)\frac{b_w}{b}} \tag{14.8}$$

Where, M = maximum moment under service load, M_{cr} = cracking moment
z = lever arm
x = depth of neutral axis
d = effective depth
b_w = breadth of web
b = breadth of compression face

14.3.6 Long-term deflection

Creep, shrinkage and temperature effects result in deflections and deformations, and they are time dependent. The effect of creep is to produce deflections which are often two to three times as great as deflections resulting from all other effects.

14.3.6.1 Deflection due to creep

Creep deflections are dependent on the loading applied on the member. Under permanent loads the compressive strength in concrete increases with time, resulting in an increase of curvature. If compression reinforcements are present this effect of increase in curvature is reduced. Approximate methods are used to evaluate creep. As per the sustained modulus method, the modified modulus of elasticity of concrete due to creep may be expressed as:

$$E_{cc} = \frac{E_c}{1 + C_t}$$

where, C_t is the creep coefficient, i.e., the ratio of creep strain to initial strain. The values of creep coefficients are specified in IS:456 for different loading stages.

14.3.6.2 Deflection due to shrinkage

Shrinkage in concrete introduces additional stresses for determinate and indeterminate structures. The shortening of a beam due to shrinkage is resisted by the reinforcements, thereby inducing compressive stresses in steel and tensile stresses in concrete, which can result in cracking. In statically indeterminate members, shrinkage results in additional bending moments. Shrinkage strain is obtained as a product of several factors and the problem is often simplified by assigning a numerical value. As per IS:456, the approximate value of total shrinkage strain for design may be taken as 0.0003.

14.4 CONTROL OF DEFLECTION

Deflection of flexural members like beams and slabs, if excessive, causes distress to users and is also likely to cause cracking of partitions. As given in IS:456 2000, the accepted limits to permissible deflection are given below:

(a) The final deflection due to all loads including the effects of temperature, creep and shrinkage and measured from the as-cast level of the supports of floors, roofs and all other horizontal members, should not normally exceed $\frac{span}{250}$.
(b) The deflection including the effects of temperature, creep and shrinkage occurring after erection of partitions and application of finishes should not normally exceeds $\frac{span}{350}$ or 20 mm, whichever is less.

It is essential that both requirements are fulfilled.

14.5 DEFLECTION LIMITS/REQUIREMENTS AS PER INDIAN STANDARDS

As per IS:456-2000, the vertical deflection limits may generally be assumed to be satisfied provided the requirements as specified in the standard are satisfied. Checking for deflection has been primarily envisaged by restricting the span to effective depth ratios and using an additional co-efficient to account for larger spans, percentages of tension reinforcement, compression reinforcement and provision of flange. The factor depending on the areas and stress of steel for tension reinforcement is presented in the form of a graph, whereby depending on the areas of tension reinforcement and its stress, the modification factor can be determined.

This modification/multiplication factor has been calculated on the basis of an empirical formula from a work by A. W. Beeby, published in a Cement and Concrete Association publication. The equation is presented below.

$$\text{Multiplication factor} = \frac{1}{0.225 + 0.00322 f_s - 0.625 \log_{10}\left(\frac{bd}{100A_s}\right)}$$

Where, f_s = service stress on the steel, 0.58 f_y
f_y = characteristic strength of steel

The area of tension reinforcement (A_s) is taken at the centre of the span for beams and slabs and at the support for a cantilever.

However, in the graph provided in IS:456-2000, relationships are provided for HYSD bars up to Fe 500 grade. Using the above equation, relationships for steel grade for Fe 550 and Fe 600 have been drawn and shown in Figure 14.3.

The coefficient for larger spans, compression reinforcement and flanged beams will be the same as that provided in IS:456.

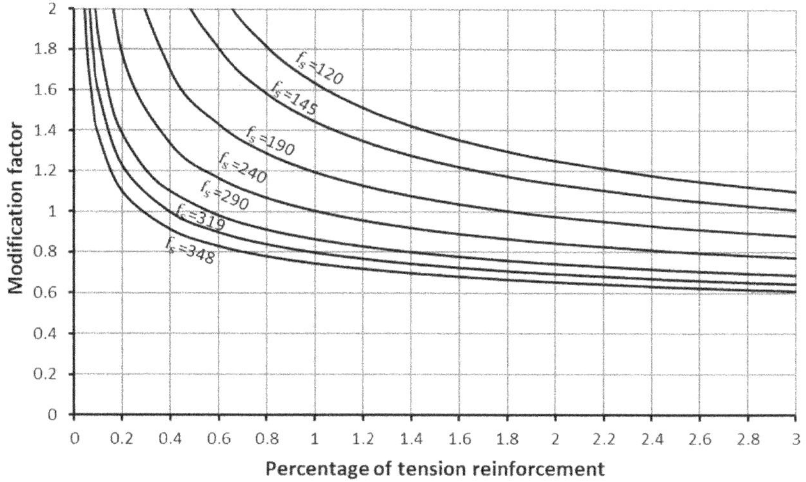

FIGURE 14.3 Modification factor for tension reinforcement.

Note: Service stress of steel in N/mm², $f_s = 0.58 f_y \dfrac{\text{Area of cross} - \text{section of steel required}}{\text{Area of cross} - \text{section of steel provided}}$.

14.6 CALCULATIONS OF CRACK WIDTH

Concrete is weak in tension and cracks at very low values of tensile strain. With the use of high-strength bars, the strains in the concrete surrounding such reinforcements are generally higher than the values of tensile strains at which cracking occurs. However, the steel becomes effective and load transfer occurs only when the surrounding concrete cracks. Fortunately, the cracks are not concentrated at a particular location but are rather distributed over the span and cross-sections and are generally too small to be noticed by the naked eye. The permissible widths of cracks are also specified by the codes of practice for different exposure conditions, and these requirements are to be strictly complied with. Since the actual calculations of crack widths are elaborate and cumbersome, the codes of practice provide necessary restrictions on bar spacings and proper detailing which, if properly followed, it is expected that the crack width will be within tolerable limits. The actual width of cracks may vary between wide limits and hence the prediction of the maximum crack width is quite complicated. Instead, it seems that the characteristic and admissible crack widths are more practical and useful.

Cracking is a complex phenomenon. Several investigators have proposed different equations for estimating the crack width. The Indian standard code of practice has included a formula to calculate the crack width based on the British Standard.

This expression takes into considerations the proximity of reinforcing bars perpendicular to the crack, proximity of the neutral axis and the average steel strain. Design surface crack width as per IS 456: 2000 is presented below:

$$W_{cr} = \frac{3a_{cr}\varepsilon_m}{1+\dfrac{2(a_{cr}-C_{min})}{h-x}} \tag{14.9}$$

Where, a_{cr} = distance from the point considered to the surface of the nearest longitudinal bar (refer to Figure 14.4)

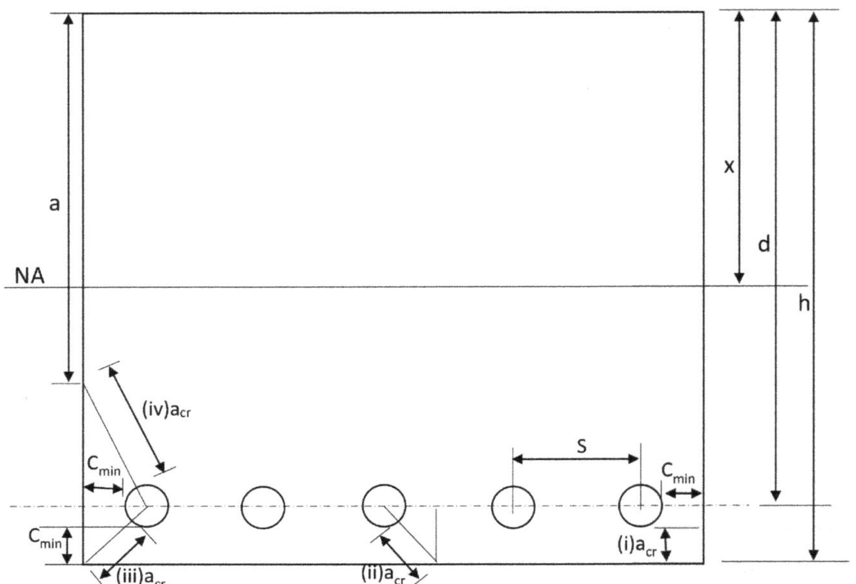

FIGURE 14.4 Calculations of parameters of crack width in beams.

s = centre to centre spacing between bars
C_{min} = minimum cover to the longitudinal bars
x = depth of the neutral axis
h = overall depth of the member
ε_m = average steel strain at the level considered

The average steel strain ε_m may be considered from the strain distribution diagram after employing some simplified assumptions. Alternatively, the average steel strain can be calculated for a rectangular tension zone by using the following expression

$$\varepsilon_m = \varepsilon_1 - \varepsilon_2$$

Where, ε_1 = strain at the level considered ignoring the stiffening of the concrete in the tension zone, considering the section to be cracked

ε_2 = reduction in strain due to the tension stiffening effect

Since Bernoulli's theorem is applicable, the strain distribution diagram will be linear. The code assumes that both concrete and steel behave in a linear elastic manner in both tension and compression, as a result of which the stress distribution diagram will also be linear. The tensile stress in concrete at the level of steel has been considered as 1.0 MPa for short-term calculations and 0.55 MPa for long-term effects. When the stiffening effect of concrete is considered, some tensile strain will be sustained by concrete as a result of which the strain in steel will be reduced.

The code has provided an empirical expression for calculation of ε_2, as:

$$\varepsilon_2 = \frac{b(h-x)(a-x)}{3E_s A_s (d-x)} \tag{14.10}$$

Where b = width of section at the centroid of tension steel

a = distance from the compression face to the point at which crack width is being calculated

d = effective depth

E_s = modulus of elasticity of steel (N/mm^2)

A_s = area of tension reinforcement

Hence,

$$\varepsilon_m = \varepsilon_1 - \left[\frac{b(h-x)(a-x)}{3E_s A_s (d-x)}\right] \tag{14.11}$$

14.7 GUIDELINES TO PROVIDE CRACK CONTROL

As per IS:456-2000, the maximum spacing of bars in flexural members has been specified. These values must be complied with if actual calculations of crack width are not performed. For beams, where no moment redistribution from the section is performed, the maximum horizontal distance between parallel main reinforcement bars in tension has been specified as 180 mm and 150 mm for Fe 415 and Fe 500, respectively. For slabs, the horizontal distance between parallel main reinforcement bars shall not be more than three times the effective depth of solid slab or 300 mm whichever is smaller, and the horizontal distance between parallel reinforcement bars provided against shrinkage and temperature shall not be more than five times the effective depth of a solid slab or 300 mm whichever is smaller. When the depth of web of a beam exceeds 750 mm side face reinforcements are to be provided.

The maximum horizontal spacing of bars in tension reinforcements of beams as specified in IS:456 does not have any background information based on which the values have been achieved. However, similar expressions available in international standards are discussed herewith. The maximum probable crack width for FRP-reinforced concrete members, as given in ACI-440.1R-06, may be calculated as

$$w = 2\frac{f_f}{E_f}\,\beta\, k_b \sqrt{d_c^2 + \left(\frac{S}{2}\right)^2} \tag{14.12}$$

Where, w = maximum crack width in mm

f_f = reinforcement stress, MPa

E_f = reinforcement modulus of elasticity, MPa

β = ratio of distance between neutral axis and tension face to distance between neutral axis and centroid of reinforcement

d_c = thickness of cover from tension face to centre of closest bar in mm

S = bar spacing in mm

and k_b is a coefficient that accounts for the degree of bond between the bar and surrounding concrete. For FRP bars having bond behaviour similar to uncoated steel bars, the bond coefficient k_b is assumed to be equal to 1. As per ACI, the crack control provision for steel reinforcement corresponds to a maximum crack width of 0.4 mm. As per Indian standards, the surface width of a crack should not, in general, exceed 0.3 mm in members where cracking is not harmful.

From the above equation, the maximum spacing between bars may be calculated as per the following expression:

$$S=2\sqrt{\left[\left(\frac{wE_f}{2f_f\beta k_b}\right)^2-d_c^2\right]} \tag{14.13}$$

The above equation can be used for both FRP and steel reinforced concrete design.

As per Indian standard, the modulus of elasticity of steel, $E_f = 2\times10^5$MPa

Stress in reinforcement, $f_f = 0.58 \times f_y$

A. J. Frosch has prescribed the following simplified expression for calculation of β which reads

$$\beta = 1+0.0031\times d_c$$

where, d_c is the effective cover in mm.

Using the above-mentioned formulation (equation 14.13) and considering the value of effective cover, d_c, the maximum horizontal spacing of main reinforcing bars in beams can be calculated.

As per the requirements of IS:456-2000, for a crack width of 0.3 mm, considering a value of nominal concrete cover of 20 mm, bar diameter of 20 mm and diameter of stirrups as 8 mm, d_c = 20 + 8 + 10 = 38 mm. Using equation (14.13), the maximum clear distance between bars (spacing between bars – bar diameter) for Fe 415, Fe 500, Fe 550 and Fe 600 steel grades are obtained as 189, 149, 130 and 114 mm, respectively. In IS:456-2000, these values for Fe 415 and Fe 500 steel grades are prescribed as 180 and 150 mm, respectively, for no redistribution of moments from the section.

14.8 LIMITS ON CRACKING AS PER INDIAN STANDARDS

As per Indian standards, cracking of concrete should not adversely affect the appearance and durability of the structure. The acceptable limits of cracking vary with the type of structure and environment. The surface width of the cracks should not in general exceed 0.3 mm in members where cracking is not harmful and does not have any serious adverse effect upon the preservation of reinforcing steel nor upon the durability of the structure. In members where cracking in the tension zone is harmful because of their exposure to the effects of weather or are continuously exposed to moisture or in contact with soil and ground water, an upper limit of 0.2 mm is suggested for the maximum width of cracks. For particularly aggressive environments, the assessed surface width of cracks should not in general exceed 0.1 mm.

14.9 SUMMARY

The limit state of serviceability forms an integral part of the limit state method. In general, designers perform strength design as a rule and design for serviceability is not commonly performed. However, if the specifications of the codes of practice are properly adhered to, the requirements of serviceability are automatically satisfied. IS:456-2000 deals with limit states of serviceability in deflection and cracking. Requirements of concrete durability and fire resistance are also provided in the standard. In this chapter these aspects have been discussed with an emphasis on deflection and cracking considerations. The methodology of calculations of short-term and long-term deflections have been outlined along with the necessary aspects that are to be followed by designers for the control of deflection. Regarding cracking, the equation for calculating crack width and guidelines for providing crack control have been discussed.

BIBLIOGRAPHY

Nawy, Edward G. (2009). *Reinforced Concrete, A Fundamental Approach*", Fifth Edition, Pearson Education. Upper Saddle River, New Jersey, 07458.

OSPINA, Carlos E. and Bakis, Charles E. (2007). *Indirect Flexural Crack Control of Concrete Beams and One-Way Slabs Reinforced with FRP Bars*, FRPRCS-8, University of Patras, Greece, July 16–18.

Pillai, S. U. and Menon, D. (2012). *Reinforced Concrete Design*, Third Edition, Reprint 2012, Tata McGraw Hill Education, New Delhi, 2012.

Purushothaman, P. (1984). *Reinforced Concrete Structural Elements, Behaviour, Analysis, and Design*, Tata McGraw-Hill, Torsteel Research Foundation, India.

Raju, Krishna N. (2016). *Design of Reinforced Concrete Structures, IS:456-2000*, Fourth Edition, CBS Publishers, New Delhi.

15 Limit State Method as per IRC:112-2020 for RC Sections

15.1 INTRODUCTION

The code of practice for concrete road bridges (IRC:112-2020 – Code of Practice for Concrete Road Bridges) is based on the limit state design philosophy. This is a unified code covering various types of concrete bridges made of plain concrete, reinforced concrete and prestressed concrete. Design with reinforced concrete as per this standard is quite similar to the requirements of Eurocode 2: "Design of concrete structures – part 1-1: General rules and rules for buildings". The Indian standard code of practice on plain and reinforced concrete (IS:456) was last revised and published in the year 2000 as a fourth revision. Unprecedented growth of knowledge in the field of concrete technology, production of steel and advancement of design philosophy have opened up new horizons in the field of concrete making and reinforced concrete design. In this chapter, the requirements of IRC:112-2020 for the design of RC sections by ultimate limit state have been presented, which covers advanced design considerations beyond IS:456-2000.

15.2 STRESS–STRAIN RELATIONSHIPS OF CONCRETE AND STEEL

Structural design is based on the stress–strain relationship of the material. Reinforced concrete being a combination of two materials, concrete and reinforcements, the stress–strain relationships of both of these elements form the basis of design. The stress–strain relationship of concrete up to M60 grade is parabolic-rectangular and identical with the relationship provided in IS:456:2000. Hence, the stress blocks for concrete as per IRC:112-2020 are identical to those of IS:456-2000. However, the design stress–strain relationships of reinforcement provided in IRC:112 for HYSD bars are significantly different from that provided in IS:456. IS:456 deals with HYSD bars made by the CTD process and those steels do not have a definite yield point because of which the concept of proof stress was applied for arriving at the yield stress. IRC:112 provides bilinear stress–strain relationships of HYSD bars. IS:456 does not limit the maximum strain in steel because of which the stress–strain diagram is open ended. In IRC:112 there are two types of bilinear stress–strain diagrams, a simplified bilinear diagram whereby the diagram is horizontal beyond the yield point and an idealized bilinear diagram which is inclined beyond the yield point with an upward slope indicating strain hardening of the material and having a maximum strain limit. Design can be performed by either of the two relationships. As per IRC:112 it has been mentioned that when the upward sloping branch of the idealized stress–strain curve is used the tensile stresses in reinforcing steel are limited to those corresponding to a strain of $0.9\varepsilon_{uk}$, where ε_{uk} values for different grades of steel are given in IRC:112-2020. As per the explanatory handbook to IRC:112-2011 "Code of practice for concrete road bridges (IRC: SP105-2015)" it has been mentioned that there is no limit on

DOI: 10.1201/9781003415398-15

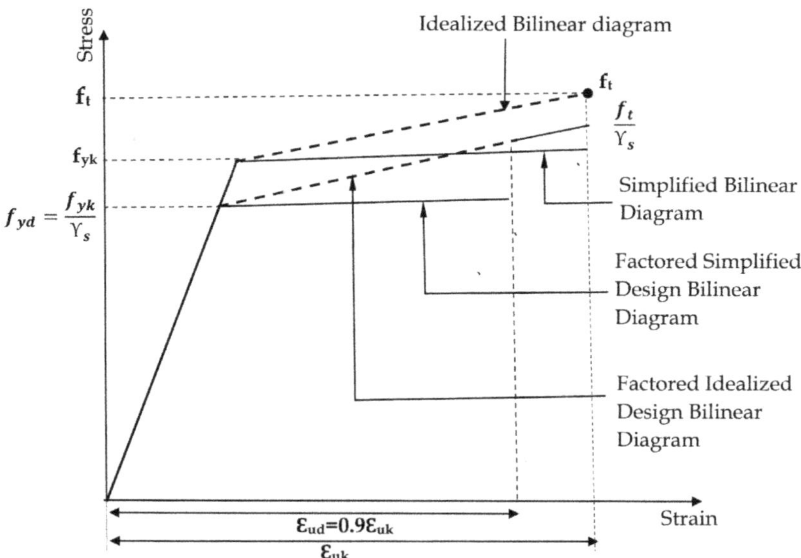

FIGURE 15.1 Stress–strain relationship for HYSD steel as per IRC:112-2020.

strain if the horizontal branch of the curve is used as the limiting ultimate stress remains constant at f_{yd}. This is similar to the requirements of Eurocode 2 (EN:1992-1-1).

As per Eurocode 2, EN:1992-1-1, two stress–strain relationships of steel have been provided, with an inclined plastic branch termed as the idealized one and another with a horizontal plastic branch. In the first case, the maximum stress is greater than f_{yd} but the maximum strain is limited to ε_{ud} taken as $0.9\varepsilon_{uk}$, where ε_{uk} values for different classes of steel are given in Eurocode 2. An even more simplified option is to limit the maximum stress to f_{yd} with no limit on the maximum strain as is presented in the second type of relationship. For simplicity, the second option is preferred for all common design situations. The material safety factor of steel is taken as 1.15 and the behaviour in tension and compression is taken to be the same. The stress–strain relationships of HYSD steel as per different codes of practice are shown in Figures 15.1–15.3. The symbols have their standard meanings as specified in the codes.

15.3 STRESS BLOCKS FOR CONCRETE

The stress–strain relationships for concrete up to M60 grade are identical for IS:456 and IRC:112. However, for higher grades of concrete beyond M60 the limiting strains in axial and flexural compression are different and the values from M65 to M90 are provided in IRC:112. Based on IRC:112 the stress–strain relationships are shown in Figure 15.4. These relationships when rotated by 90 degrees and placed across the beam cross-sections along the compression side with the origin at the neutral axis form the stress blocks for concrete.

The stress–strain relationships for concrete as per Eurocode 2 for characteristic cube strength of concrete, $f_{ck,cube}$ ranging from 20 MPa to 105 MPa are shown in Figure 15.5. Hence, for calculating design parameters as per Eurocode 2 the stress block in concrete will be as per Figure 15.5.

15.4 ASSUMPTIONS IN ULTIMATE LIMIT STATE FOR DESIGN OF MEMBERS

Design of reinforced concrete sections, as per IRC:112, is based on the assumptions of the limit state philosophy of design which consists of two basic groups of limit states, namely oltimate limit

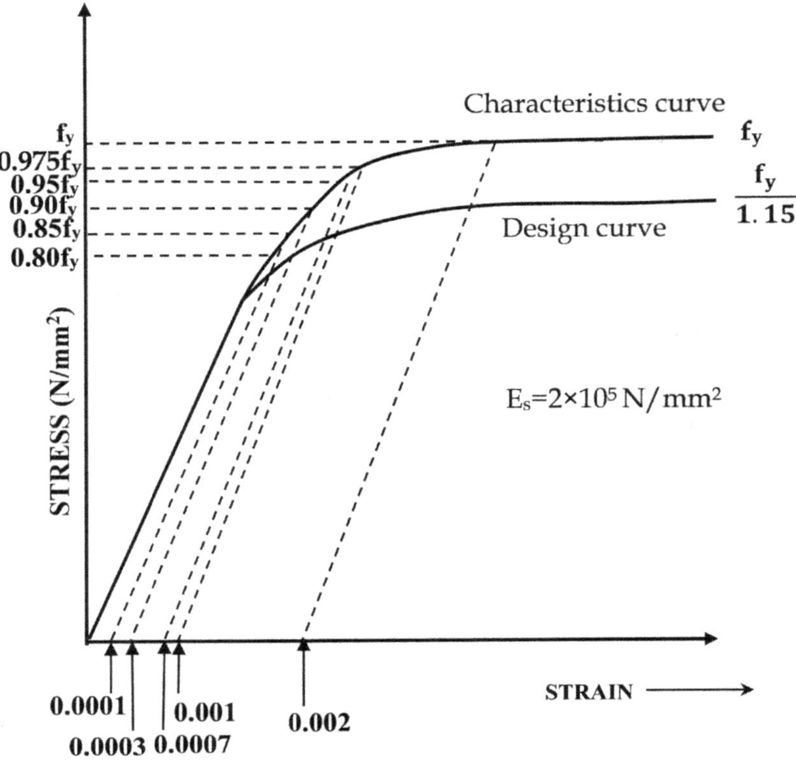

FIGURE 15.2 Stress–strain relationship for HYSD steel as per IS:456-2000.

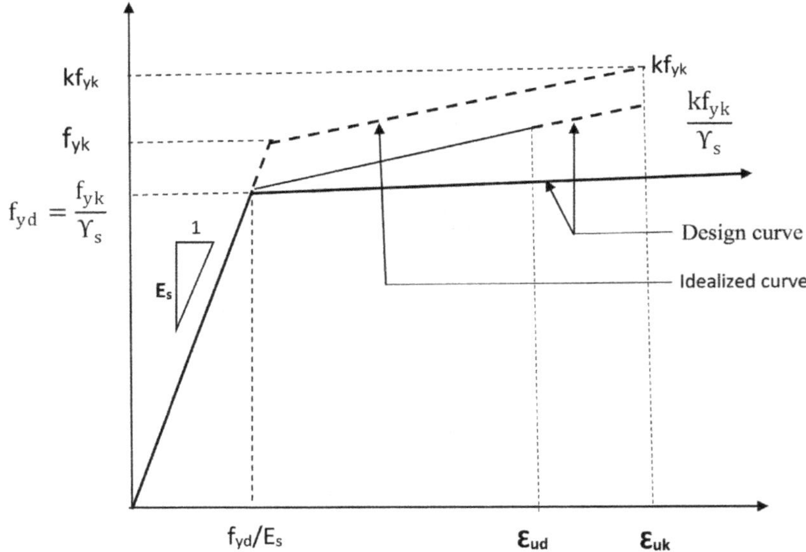

FIGURE 15.3 Idealized and design stress–strain diagrams for reinforcing steel for tension and compression as per Eurocode 2 (EN 1992-1-1).

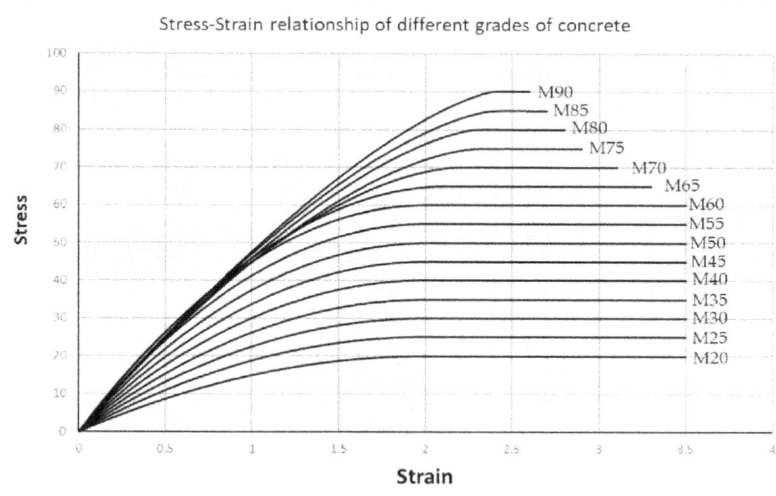

FIGURE 15.4 Stress–strain relationships for different grades of concrete from M20 to M90 as per IRC:112-2020.

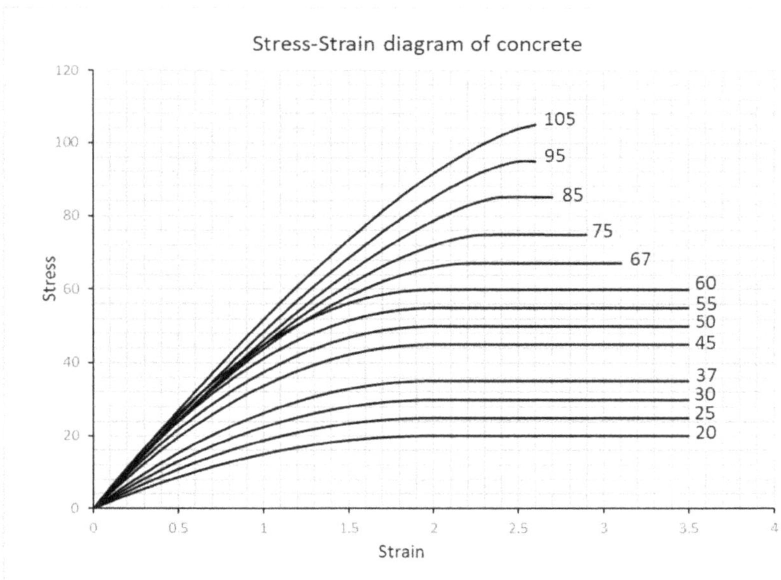

FIGURE 15.5 Stress–strain relationships for different grades of concrete from $f_{ck,\ cube}$ 20 to 105 as per Eurocode 2.

state (ULS) and serviceability limit state (SLS). ULS deals with static equilibrium and failure of a structural element or structure as a whole and SLS deals with deflection, cracking, vibration, fatigue, internal stress levels, etc. Section 8 of IRC:112-2020 deals with the ultimate limit state of linear elements for bending and axial forces. The limit state philosophy of the ultimate limit state (ULS) considers the effects of axial forces and flexure simultaneously. These clauses are applicable to both beams and columns. In the limit state method, section design requires determination of the strain distribution diagram across the section. Once the strain distribution diagram is obtained, stress blocks can be easily developed and capacities of the section in carrying axial forces and moments can be evaluated. For columns, since axial forces act along with bending, the extreme section conditions can vary between pure axial compression to pure axial tension with intermediate positions of neutral axis where the section is subjected to axial load and flexure. However, concrete being a material which is significantly strong in compression and very weak in tension, cases whereby the entire cross-sections are subjected to tension are generally not desirable in design and may lead to significant cracking in the section which can pose severe durability problems.

Stress–strain diagrams of concrete and steel are the building blocks for performing the design of sections. Using these the flexural capacity of beams, axial load and moment carrying capacities of columns can be determined. In Chapters 10 and 11 the design of beams and columns following the stress–strain relationships for concrete and steel as per IS:456 have been discussed in detail. For IRC:112, the methodology is almost identical with the following exceptions:

a) Stress–strain diagrams for concrete up to M60 grade are exactly the same as for IS:456 but for higher grades of concrete from M65 to M90, they are different.
b) Stress–strain diagrams for steel given in IS:456 are for CTD bars which do not have a definite yield point, and hence the concept of proof stress is used to evaluate the yield stress. For IRC:112 the diagrams are bilinear with definite yield points.

In IS:456 design for flexure and design for compression are specified separately, whereas for both IRC:112 and Eurocode 2 design methodology is combined, i.e., prescribed for sections in flexure with or without axial forces. The basic methodology of design has been discussed in the standards in the form of strain and stress distributions at ultimate limit states. The assumptions are quite similar to those of the limit state of collapse in compression as per IS:456, with minor variations due to modification of stress–strain relationships as mentioned above. Following these assumptions, the strain distribution diagrams are to be developed for designing the sections. In IS:456 limit state of collapse is dealt with separately for flexure and compression and there is virtually no mention of tension in the section as the code deals with the limit state of collapse in compression. However, as per IRC, designs of beams and columns are governed by the ultimate limit state of linear elements for bending and axial forces. The section conditions governed by this theory range from pure compression to pure tension along with compression with bending and tension with bending. For tension, concrete will be totally ineffective in carrying loads and all forces will be resisted by reinforcements. The detailed strain distribution diagrams along with the domains of strain distribution following both the idealized and simplified stress–strain relationships for steel as per IRC:112 are discussed in the following sections.

The strain distribution diagram corresponding to the factored idealized bilinear diagram has the limiting strain restricted to $\mathcal{E}_{ud} = 0.9\mathcal{E}_{uk}$ corresponding to the stress of f_t which is a function of f_{yk} and is provided in Table 18.1 in the code. (Refer to Figure 15.1.)

As per the idealized bilinear stress–strain relationship diagram, shown in Figure 15.6, failure in reinforced concrete sections can occur in the following conditions:

1) Failure in pure axial compression due to concrete reaching limiting axial compressive strain, ε_{c2}, where the neutral axis is at infinity. Failure criteria defined by point [C], line ④ corresponds to this condition.
2) Failure by concrete reaching the maximum strain in flexural compression, $\mathcal{E}_{cu2}$, when the neutral axis is passing through the section face. Failure criteria defined by point [B], line ③ correspond to this condition. Zone (e) denotes the intermediate conditions when the section is subjected to pure axial compression and neutral axis passing through the face of the section. For this zone, the neutral axis lies outside the section and the full section is in compression.

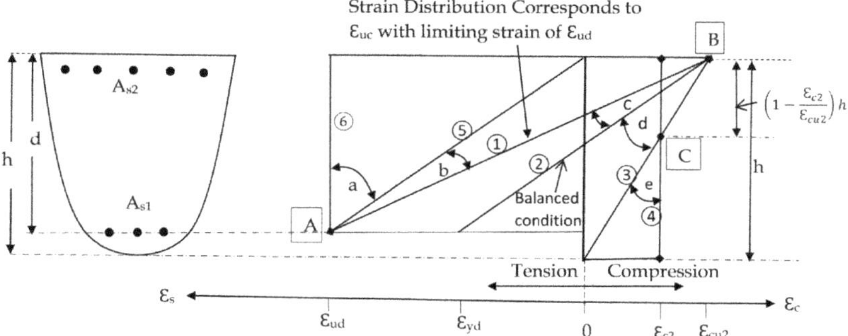

FIGURE 15.6 Possible strain distribution diagrams for section subjected to axial force and bending following the factored idealized bilinear stress–strain relationship as per IRC:112.

For intermediate conditions when the neutral axis lies outside the section the maximum strain in concrete will range from ε_{c2} to ε_{cu2}.

3) When the neutral axis is within the section, failure can occur due to concrete reaching maximum strain but maximum strain in steel may or may not be reached. Under balanced conditions, the maximum strain in concrete will be reached and steel will simultaneously reach limiting or yield strain, failure criteria defined by point [B], and this condition corresponds to line ②. Zone (d) corresponds to sections which can range from balanced to over-reinforced sections having both tension and compression in the section and neutral axis varies from the face of the section to the balanced neutral axis depth.
4) With increasing conditions of under-reinforcement, strain in steel will increase and when the maximum strain in steel is reached, failure will occur due to crushing of concrete and steel reaching maximum strain simultaneously. Failure criteria defined by points [B] and [A] and line ① correspond to this condition. Zone (c) corresponds to under-reinforced condition and neutral axis lying within the section. Strain in steel lies between the yield and maximum value. In Eurocode 2, failure up to this condition is specified and not beyond this, where the section may be subjected to higher levels of tension.
5) With a further reduction in depth of the neutral axis, failure will occur due to steel reaching the maximum strain but strain in concrete less than the maximum permitted, i.e., failure occurring from the steel point and not from concrete. Failure occurs when steel reaches the maximum strain and the neutral axis is lying within the section. Failure criteria defined by point [A], this condition comes within zone (b) and when the neutral axis passes through the face of the section the entire section is subjected to tension. This condition is denoted by line ⑤. In zone (b) the section is either under-reinforced subjected to compression with bending or subjected to tension with bending.
6) Failure occurring due to steel reaching the maximum strain and neutral axis lying outside the section resulting in tension on the entire section. Axial capacity and moment capacity contributed by steel only. Failure criteria defined by point [A], limiting conditions corresponding to line ⑤ and line ⑥. This zone is denoted as zone (a).
7) Failure occurring due to all layers of steel reaching the maximum strain, the neutral axis is at infinity. Section having only axial capacity with zero moment carrying capacity for symmetrical arrangement of reinforcement about the centroidal axis. Failure criteria defined by point [A] and this condition corresponds to line ⑥.

The strain distribution diagram for factored simplified bilinear stress–strain relationship where the maximum strain is not limited and stress is constant at f_{yd} is presented in Figure 15.7.

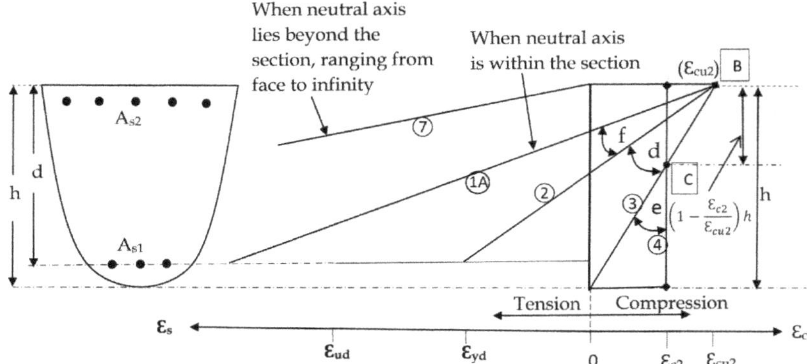

FIGURE 15.7 Domains of strain distributions for factored simplified bilinear stress-strain relationship as per IRC:112.

For the simplified bilinear stress–strain relationship of steel the plastic branch is horizontal without any strain limit and the stress in steel is constant at f_{yk}. The behaviour of the section from pure axial compression to the balanced condition is identical to that occurring due to the idealized bilinear stress–strain diagram, and lines ②, ③ and ④ depict the same conditions as mentioned above. For the depth of the neutral axis less than the balanced depth, the failure criteria will be governed by the concrete reaching the limiting strain and strain in steel will go on increasing with a reduction in the depth of the neutral axis. The failure criteria are governed by point [B] and this condition is typically shown by line 1A (Figure 15.7). However, when the depth of the neutral axis reduces significantly the capacity of the section in axial force may become tensile with flexure. When the neutral axis goes beyond the section, that is the section is fully in tension, concrete will be ineffective, and the section capacities will be governed by steel only. Strain in steel at the outermost layer will even increase and the condition is represented by line ⑦. In the final limiting case, the neutral axis will go to infinity and all the layers of steel will yield in tension. Stresses in all the layers will be constant at yield stress, $0.87f_y$ and the section will have axial tension capacity with no moment capacity for symmetrical arrangement of reinforcement about the centroidal axis. The strain distribution diagram is not well defined and hence has not been shown.

15.5 DESIGN FOR FLEXURAL MEMBERS – SINGLY REINFORCED SECTIONS

For beams, the neutral axis will always lie within the section and its position will vary from a minimum value corresponding to $A_{st,min}$ to a maximum value corresponding to $A_{st,max}$ since the minimum and maximum longitudinal reinforcements in beams are limited by the codes of practice and the effects of axial forces are either negligible or not present. As per IRC:112, the design assumptions are given for flexure along with axial load and there is no restriction on the type of section and hence all types of sections, i.e., balanced, under-reinforced and over-reinforced, can be adopted. As discussed in Chapter 10, failure of sections in the limit state method occur in two stages. Stage I is the yield stage where the material yields and Stage II is the ultimate stage when the material reaches the failure criteria or strain. For under-reinforced sections the yield stage occurs due to yielding of steel and for balanced and over-reinforced sections this stage occurs when concrete reaches the strain of 0.002. Beyond this point strain increases at constant stress that is as if concrete yields. Failure generally occurs due to concrete reaching the failure strain of 0.0035 that is by crushing of concrete for all types of sections in flexure. Theoretically, for singly reinforced beams, design can range from a maximum under-reinforced section with minimum percentage of tension reinforcement ($A_{st,min}$) to a section with maximum percentage of tension reinforcement ($A_{st,max}$). Maximum

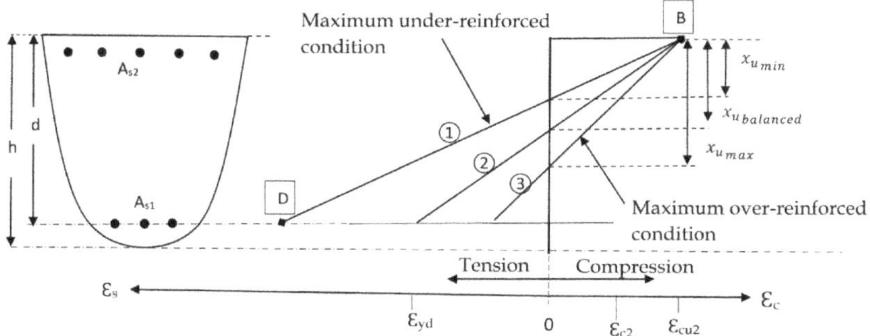

FIGURE 15.8 Possible domains of strain distributions following bilinear stress–strain diagrams of steel as per IRC:112.

Note: For an idealized stress–strain relationship the strain in steel corresponding to the maximum under reinforcement condition, *D* will be limited to ε_{ud} but for a simplified stress–strain relationship there is no limitation on strain.

strain in steel, for an idealized stress–strain relationship will be governed by ε_{ud}, hence the minimum percentage of tension reinforcement should result in strains in steel less than or equal to ε_{ud}. For a simplified stress–strain relationship there is no limitation on strain. This is demonstrated in Figure 15.8. The design parameters corresponding to different types of sections for singly reinforced beams are discussed in the subsequent sections.

15.6 FOR FACTORED SIMPLIFIED BILINEAR STRESS–STRAIN RELATIONSHIP OF STEEL (HORIZONTAL PLASTIC BRANCH)

15.6.1 Computation of Design Parameters for Concrete Grades from M20 to M90 and Grades of Steel from Fe 415 to Fe 600 for Balanced Condition

The methodology of calculating the basic design parameters has already been presented in detail in Chapter 10 for IS:456. In IS:456, the stress–strain relationship for HYSD bars has been provided for CTD bars which do not have a definite yield point and hence the concept of proof stress that is stress at an inelastic strain of 0.002 is considered as yield stress. However, as per IRC:112, the stress–strain relationships of steel are bilinear in nature having the design stress of $0.87f_y$. Accordingly, the inelastic strain of 0.002 has not been considered in the calculation of design parameters. The limiting or balanced values of strains are 0.001805, 0.002175, 0.002392 and 0.00262, for Fe 415, Fe 500, Fe 550 and Fe 600, respectively. This has been demonstrated in Figure 15.9. However, the methodology of calculation remains identical. The design parameters are presented in the following tables.

The values of ε_{c2}, ε_{cu2}, ε_{st} and balanced depth of neutral axis factors for different grades of concrete and steel have been presented in Table 15.1. Using strain distribution diagrams (Figure 15.9), the values of balanced depth of neutral axis have been calculated. The balanced percentages of steel corresponding to different grades of concrete and steel are presented in Table 15.2 using the force equilibrium equation and areas of stress blocks in balanced conditions for different grades of concrete.

$$x_u = \frac{0.87 f_y A_{st}}{N f_{ck} b}$$

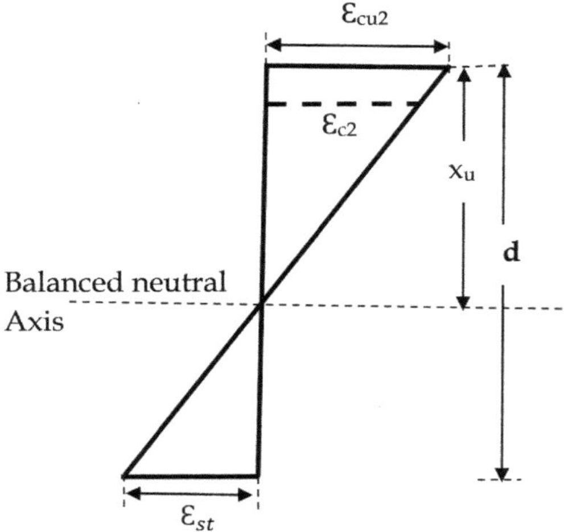

FIGURE 15.9 Strain distribution diagram for limiting/balanced condition.

TABLE 15.1

Table showing limiting/balanced depth $\left(\frac{x_u}{d}\right)$ for different grades of steel and concrete

Balanced Depth of Neutral Axis $(x_u/d)_{Balanced}$						
			Grades of Steel			
Grades of Concrete	ε_{c2}	ε_{cu2}	Fe 415, ε_{st} = 0.001805	Fe 500, ε_{st} = 0.002175	Fe 550, ε_{st} = 0.002392	Fe 600, ε_{st} = 0.00261
M20 to M60	0.002	0.0035	0.66	0.62	0.59	0.57
M65	0.0021	0.0033	0.6463	0.6027	0.5797	0.5583
M70	0.0022	0.0031	0.6319	0.5876	0.5644	0.5429
M75	0.0023	0.0029	0.6163	0.5714	0.5479	0.5263
M80	0.0023	0.0028	0.608	0.5628	0.5392	0.5175
M85	0.0024	0.0027	0.5993	0.5538	0.5301	0.5084
M90	0.0024	0.0026	0.5902	0.5445	0.5207	0.4990

$$\frac{x_{u,max}}{d} = \frac{0.87 f_y}{N f_{ck}} \times \frac{p_{t,lim}}{100}$$

$$\text{or, } \frac{p_{t,lim}}{100} = \left(\frac{N f_{ck}}{0.87 f_y}\right)\left(\frac{x_{u,max}}{d}\right)$$

The values of N for different grades of concrete are given in Table 15.3.

Since the calculations are similar to those performed for IS:456 and shown in Chapter 10, they have not been repeated here.

15.6.1.1 Compressive force

Compressive force in concrete is determined from the stress block which is directly derived from the stress–strain relationship of concrete. Since the stress–strain relationships of concrete up to

TABLE 15.2
Limiting percentage of steel in balanced condition

Grades of Concrete	Fe415		Fe500		Fe550		Fe600	
	Tension reinforcement ratio ($\rho_{balanced}$) and percentage p_{tlim} or $p_{tbalanced}$ in balanced condition — Grades of Steel							
	$\rho_{balanced}$	$p_{tbalanced}$	$\rho_{balanced}$	$p_{tbalanced}$	$\rho_{balanced}$	$p_{tbalanced}$	$\rho_{balanced}$	$p_{tbalanced}$
M20	0.01315	1.32	0.0102	1.03	0.0089	0.89	0.0079	0.79
M25	0.01644	1.65	0.0127	1.28	0.01117	1.11	0.00987	0.99
M30	0.01973	1.98	0.0153	1.54	0.0134	1.34	0.01185	1.18
M35	0.02302	2.31	0.01786	1.80	0.01564	1.56	0.01382	1.38
M40	0.0263	2.64	0.02041	2.06	0.01787	1.78	0.0158	1.58
M45	0.02960	2.97	0.02296	2.32	0.0201	2.00	0.0177	1.77
M50	0.03289	3.30	0.0255	2.57	0.02234	2.23	0.01975	1.97
M55	0.03617	3.62	0.0280	2.83	0.02457	2.45	0.02172	2.17
M60	0.03946	3.95	0.03062	3.06	0.02681	2.67	0.0237	2.37
M65	0.04189	4.189	0.0324	3.242	0.02834	2.834	0.02503	2.503
M70	0.0441	4.41	0.03404	3.404	0.02972	2.972	0.0262	2.62
M75	0.04609	4.609	0.0354	3.546	0.03091	3.091	0.0272	2.722
M80	0.0435	4.35	0.03726	3.726	0.03245	3.245	0.0285	2.855
M85	0.0441	4.410	0.03896	3.896	0.0339	3.390	0.0298	2.98
M90	0.0457	4.57	0.04055	4.055	0.03526	3.526	0.03097	3.097

$p_{t,balanced}$ = Percentage area of steel in balanced condition = $100A_{stbalanced}/bd$.

TABLE 15.3
Compressive forces in concrete for different grades as per IRC:112

Grade of Concrete	$C_u = N.f_{ck}.x_u.b$
M20 to M60	$0.36 f_{ck} x_u b$
M65	$0.351 f_{ck} x_u b$
M70	$0.341 f_{ck} x_u b$
M75	$0.328 f_{ck} x_u b$
M80	$0.323 f_{ck} x_u b$.
M85	$0.313 f_{ck} x_u b$
M90	$0.311 f_{ck} x_u b$

M60 grade are similar for IS:456 and IRC:112, the compressive force is the same for both codes of practice. For higher grades of concrete the stress blocks are different due to differences in the stress–strain relationship. The areas of the stress blocks have been calculated for M20 to M90 grades of concrete and the corresponding forces are shown in Table 15.3.

15.6.1.2 Line of action of the compressive force

The compressive forces in concrete act at the centroid of the stress blocks which have been calculated for different grades of concrete and are presented in Table 15.4. The compressive force in concrete and its point of application are shown in Figure 15.10.

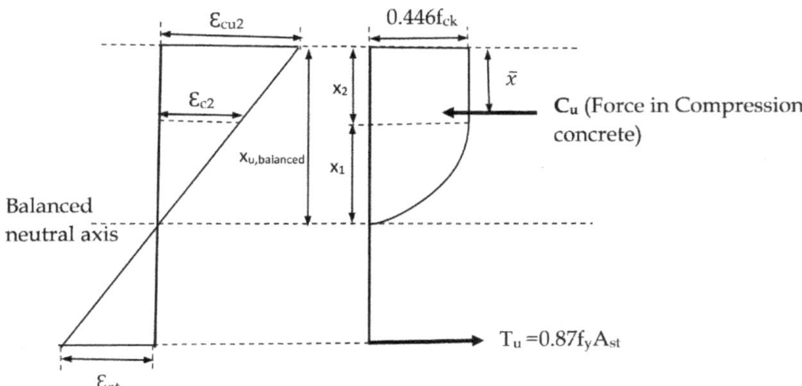

FIGURE 15.10 Strain distribution diagram and stress block in flexure as per IRC:112 for balanced condition.

TABLE 15.4
Line of action of compressive force for concrete as per IRC:112 for balanced condition

Grade of Concrete	$\bar{x}$	Lever arm (z) $z = d - \bar{x}$
M20 to M60	$0.416x_u$	$d - 0.416x_u$
M65	$0.407x_u$	$d - 0.407x_u$
M70	$0.400x_u$	$d - 0.400x_u$
M75	$0.391x_u$	$d - 0.391x_u$
M80	$0.388x_u$	$d - 0.388x_u$
M85	$0.383x_u$	$d - 0.383x_u$
M90	$0.381x_u$	$d - 0.381x_u$

15.6.1.3 Tensile force

The tensile force acting on the section, T_u, is given by the product of the stress on the reinforcing steel and its cross-sectional area.

Thus,
$$T_u = f_{yd} A_{st} = \frac{f_{yk}}{1.15} \times A_{st} = 0.87 f_{yk} A_{st} \tag{15.1}$$

Where, f_{yk} is the characteristic yield strength of reinforcing steel.

15.6.1.4 Balanced/limiting moment of resistance

15.6.1.4.1 Moment of resistance from compression concrete ($M_{ulim,concrete}$) or tension steel ($M_{ulim,steel}$)

The moment of resistance of the section can be determined by multiplying the compressive force with the lever arm or the tensile force with the lever arm. For balanced condition the moment of resistance is generally calculated from the concrete viewpoint, whereas for under-reinforced sections the moment of resistance is calculated from the steel side. The limiting moment of resistance factors corresponding to the balanced condition for different grades of concrete are shown in Table 15.5.

TABLE 15.5
Moment of resistance from compression concrete and tension steel for different grades of concrete as per IRC:112 under balanced condition

Grade of Concrete	$M_{ulim,concrete}$ (*compressive force × lever arm*)	$M_{ulim,tension}$ (*tensile force × lever arm*)
M20 to M60	$0.36 f_{ck} x_u b \times (d - 0.416 x_u)$	$0.87 f_y A_{st} \times (d - 0.416 x_u)$
M65	$0.351 f_{ck} x_u b \times (d - 0.407 x_u)$	$0.87 f_y A_{st} \times (d - 0.407 x_u)$
M70	$0.341 f_{ck} x_u b \times (d - 0.400 x_u)$	$0.87 f_y A_{st} \times (d - 0.400 x_u)$
M75	$0.328 f_{ck} x_u b \times (d - 0.391 x_u)$	$0.87 f_y A_{st} \times (d - 0.391 x_u)$
M80	$0.323 f_{ck} x_u b \times (d - 0.388 x_u)$	$0.87 f_y A_{st} \times (d - 0.388 x_u)$
M85	$0.313 f_{ck} x_u b \times (d - 0.383 x_u)$	$0.87 f_y A_{st} \times (d - 0.383 x_u)$
M90	$0.311 f_{ck} x_u b \times (d - 0.381 x_u)$	$0.87 f_y A_{st} \times (d - 0.381 x_u)$

TABLE 15.6
Limiting moment of resistance factor $\frac{M_{ulim}}{bd^2}$ for different grades of concrete and steel as per IRC:112

Grades of Concrete	Fe 415	Fe 500	Fe 550	Fe 600
M20	3.47	3.29	3.21	3.14
M25	4.33	4.11	4.02	3.92
M30	5.20	4.93	4.82	4.71
M35	6.07	5.75	5.62	5.49
M40	6.93	6.57	6.43	6.28
M45	7.80	7.39	7.23	7.06
M50	8.67	8.22	8.04	7.85
M55	9.53	9.04	8.84	8.63
M60	10.40	9.86	9.64	9.42
M65	10.93	10.44	10.17	9.91
M70	11.27	10.73	10.43	10.14
M75	11.49	10.89	10.57	10.26
M80	11.94	11.30	10.95	10.62
M85	12.28	11.60	11.23	10.88
M90	12.86	12.14	11.75	11.38

The limiting or balanced moment of resistance of the section for different grades of concrete and steel have been calculated and the values are presented in Table 15.6

15.6.2 Computation of Design Parameters for Concrete Grades from M20 to M90 and Grades of Steel from Fe 415 to Fe 600 for Maximum Under-Reinforced Condition

When the area of steel provided in a section is less than that required for balanced condition the section is termed as an under-reinforced section. The codes of practice limit the minimum amount of tension reinforcement to be provided in flexural members so that the moment of resistance of the section is greater than that of the cracking moment of resistance. Under-reinforced sections are always preferred in reinforced concrete because they possess higher levels of ductility, and the level of ductility depends upon the level of under-reinforcement. This has been discussed in detail in Chapter 17 following the requirements of IS:456 and IS:13920. For IRC:112, the minimum percentage of tension reinforcement in flexural members should not be less than

$$A_{st,min} = 0.26\left(\frac{f_{ctm}}{f_{yk}}\right)b_t \cdot d \text{ or } 0.0013\, b_t \cdot d \tag{15.2}$$

$$= \rho_{min} \cdot b_t \cdot d \text{ (calculated value)}$$

subject to a minimum of $0.0013 b_t.d$ (adopted value)

where $\rho = p/100$, where p is the percentage of tension reinforcement, b_t denotes the mean width of the tension zone and f_{ctm} = mean value of axial tensile strength of concrete whose values for different strength classes are provided in Table 6.5 of IRC:112, f_{yk} = characteristic yield strength of reinforcement.

Values of minimum area of tension reinforcement calculated from the above equation for different grades of concrete and steel are presented in Table 15.7. This condition refers to the maximum under-reinforced condition with the minimum depth of neutral axis, and is shown in Figure 15.11.

As already discussed in earlier chapters, the flexural strength of sections will be reached when the strain in concrete reaches the ultimate value of ε_{cu2}. For the minimum area of tension reinforcement, the depth of neutral axis will also be the minimum and joining this point with the ultimate concrete strain the maximum strain in tension steel can be determined. Using the force equilibrium condition in the section with minimum area of tension reinforcement, the minimum value of neutral axis depth factor $\left(\frac{x_u}{d}\right)_{min}$ can easily be calculated and the corresponding maximum strain in steel

TABLE 15.7
Minimum area of tension reinforcement ratios in beams (ρ_{min}) as per IRC:112 for different grades of concrete and steel

		ρ_{min}							
		Grades of Steel							
		Fe415		Fe500		Fe550		Fe600	
Grades of Concrete	f_{ctm} (MPa)	Calculated	Adopted	Calculated	Adopted	Calculated	Adopted	Calculated	Adopted
M20	1.9	0.00119	0.0013	0.00098	0.0013	0.0008	0.0013	0.00082	0.0013
M25	2.2	0.00137	0.00137	0.00114	0.0013	0.0010	0.0013	0.00095	0.0013
M30	2.5	0.00156	0.00156	0.0013	0.0013	0.0011	0.0013	0.00108	0.0013
M35	2.8	0.00175	0.00175	0.00145	0.00145	0.0013	0.00132	0.00121	0.0013
M40	3	0.00187	0.00187	0.00156	0.00156	0.0014	0.00141	0.0013	0.0013
M45	3.3	0.00206	0.00206	0.00171	0.00171	0.0015	0.00156	0.00143	0.00143
M50	3.5	0.00219	0.00219	0.00182	0.00182	0.0016	0.00165	0.0015	0.0015
M55	3.7	0.00231	0.00231	0.00192	0.001924	0.0017	0.00174	0.00160	0.00160
M60	4	0.00250	0.00250	0.00208	0.00208	0.0018	0.00189	0.00173	0.00173
M65	4.4	0.00275	0.00275	0.00228	0.00228	0.0020	0.00208	0.0019	0.0019
M70	4.5	0.00281	0.00281	0.00234	0.00234	0.0021	0.002127	0.0019	0.0019
M75	4.7	0.00294	0.00294	0.0024	0.00244	0.0022	0.00222	0.00203	0.00203
M80	4.8	0.0030	0.0030	0.00249	0.00249	0.0022	0.002269	0.00208	0.00208
M85	4.9	0.00306	0.00306	0.0025	0.00254	0.0023	0.00231	0.00212	0.00212
M90	5	0.0031	0.0031	0.0026	0.0026	0.0023	0.00236	0.00216	0.00216

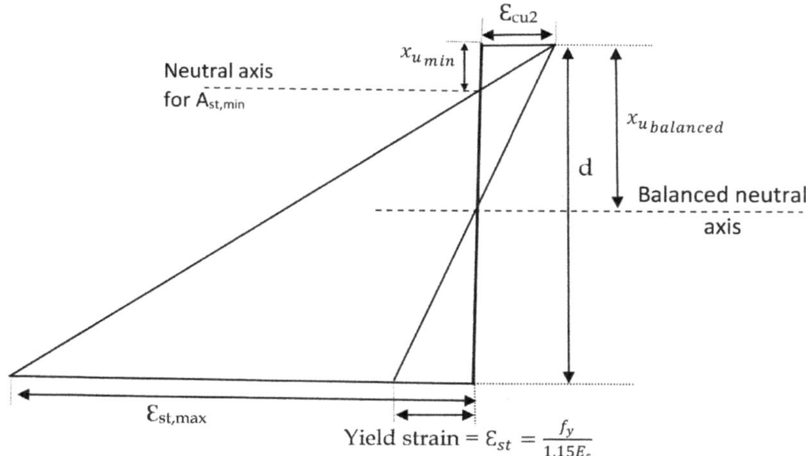

FIGURE 15.11 Strain distribution diagram for maximum under-reinforced section showing minimum depth of neutral axis and maximum strain in steel with respect to balanced condition.

TABLE 15.8

Minimum depth of neutral axis factor $\left(\frac{X_u}{d}\right)_{min}$ and maximum strains in steel corresponding to minimum area of tension reinforcement that is maximum under-reinforced condition

Grades of steel	$\left(\frac{X_u}{d}\right)_{min}$ Corresponding to $A_{st,min}$	$\varepsilon_{st,max}$
M20	0.05969	0.05513
M25	0.05529	0.05979
M30	0.05236	0.06334
M35	0.05026	0.06612
M40	0.04712	0.07077
M45	0.04607	0.07245
M50	0.04398	0.07607
M55	0.04226	0.07930
M60	0.04188	0.08005
M65	0.04253	0.07428
M70	0.04039	0.07364
M75	0.03937	0.07074
M80	0.0377	0.07147
M85	0.03622	0.07184
M90	0.03490	0.07188

can be evaluated. These have been demonstrated in Chapter 17 as per the requirements of IS:456 and hence the methodology is not repeated here. In flexure, failure will occur when concrete reaches the ultimate strain of ε_{cu2}. These values calculated as per requirements of IRC:112 are presented in Table 15.8.

TABLE 15.9
Strain in steel under different limiting conditions and strain ductility values

IRC:112												
	Strain in Reinforcing steel (ε_{st})											
	Grades of Steel											
	Fe415			Fe500			Fe550			Fe600		
Grades of concrete	Balanced Strain	Maximum Under-reinforced Max. strain	Ratio (Strain Ductility)	Balanced Strain	Maximum Under-reinforced Max. strain	Ratio (Strain Ductility)	Balanced Strain	Maximum Under-reinforced Max. strain	Ratio (Strain Ductility)	Balanced Strain	Maximum Under-reinforced Max. strain	Ratio (Strain Ductility)
M20	0.0018	0.0551	30.54	0.0022	0.0551	25.35	0.002392	0.0551	23.04	0.0026	0.0551	21.12
M25	0.0018	0.0598	33.12	0.0022	0.0598	27.49	0.002392	0.0598	24.99	0.0026	0.0598	22.91
M30	0.0018	0.0633	35.09	0.0022	0.0633	29.12	0.002392	0.0633	26.48	0.0026	0.0633	24.26
M35	0.0018	0.0661	36.63	0.0022	0.0661	30.4	0.002392	0.0661	27.64	0.0026	0.0661	25.33
M40	0.0018	0.071	39.20	0.0022	0.071	32.54	0.002392	0.071	29.58	0.0026	0.071	27.11
M45	0.0018	0.0725	40.14	0.0022	0.0725	33.31	0.002392	0.0725	30.29	0.0026	0.0725	27.76
M50	0.0018	0.076	42.14	0.0022	0.076	34.98	0.002392	0.076	31.8	0.0026	0.076	29.14
M55	0.0018	0.079	43.93	0.0022	0.079	36.46	0.002392	0.079	33.15	0.0026	0.079	30.38
M60	0.0018	0.08	44.34	0.0022	0.08	36.81	0.002392	0.08	33.46	0.0026	0.08	30.67
M65	0.0018	0.074	41.15	0.0022	0.074	34.154	0.002392	0.074	31.05	0.0026	0.0742	28.46
M70	0.0018	0.0736	40.8	0.0022	0.0736	33.86	0.002392	0.0736	30.78	0.0026	0.0736	28.21
M75	0.0018	0.0707	39.195	0.0022	0.0707	32.52	0.002392	0.0707	29.575	0.0026	0.0707	27.10
M80	0.0018	0.0715	39.59	0.0022	0.0715	32.86	0.002392	0.0715	29.872	0.0026	0.0715	27.38
M85	0.0018	0.0718	39.8	0.0022	0.0718	33.03	0.002392	0.0718	30.027	0.0026	0.0718	27.52
M90	0.0018	0.0719	39.819	0.0022	0.0719	33.049	0.002392	0.0718	30.04	0.0026	0.0718	27.54

After yielding of steel, inelastic deformation occurs. The ratio of strain in tension steel at the ultimate stage to the strain in steel corresponding to the yield stage is termed as strain ductility.

$$\text{Strain ductility} = \frac{\text{Strain at ultimate}}{\text{Strain at yield}}$$

The strain ductility values calculated as per IS:456 have been presented in Chapter 17. Calculations as per IRC:112-2020 are shown in Table 15.9.

15.6.3 Computation of Design Parameters for Concrete Grades from M20 to M90 and Grades of Steel from Fe 415 to Fe 600 for Maximum Percentage of Tension Reinforcement

For over-reinforced sections, the strain in steel will not reach the yield value either in Stage 1, the yield stage, or in Stage II, which is the ultimate stage. Since steel does not reach the yield point the stress in steel will not be constant and will vary with strain. Hence, for over-reinforced sections strain in steel will be less than yield. It will be in the elastic range that is up to f_{yk}. The partial safety factor for steel, i.e., 1.15, is applicable only for the inelastic portion of the stress–strain curve. The strain in steel will depend upon the level of over-reinforcement. Hence, for an over-reinforced section only one point in the strain distribution diagram is known, i.e., strain at the outermost fibre in concrete is reached that is ε_{cu2}. Neither the depth of the neutral axis nor the strain in steel is known. In order to solve the unknowns, two equations i.e., force equilibrium equation and strain distribution diagram across the cross-section, are used. The process is described below.

As per force equilibrium condition, $C = T$

$$N.\, f_{ck}.x_u.\, b = f_{st}.\, A_{st} \tag{15.3}$$

where, x_u is the depth of neutral axis for over-reinforced conditions, f_{st} is the stress in steel and other symbols have their standard meanings.

$$N.f_{ck}.x_u.b = f_{st}.\, \frac{pbd}{100}$$

where, $A_{st} = pbd/100$, p is the percentage of reinforcement.

$$f_{st} = \frac{N.\, f_{ck}..x_u.b.100}{pbd}$$

$$f_{st} = \frac{N.\, f_{ck}.100}{p}\left(\frac{x_u}{d}\right) \tag{15.4}$$

From the strain distribution diagram,

$$\frac{\varepsilon_{cu2}}{x_u} = \frac{\varepsilon_s}{d - x_u}$$

or, $$\frac{f_{st}}{E_s\left(d - x_u\right)} = \frac{\varepsilon_{cu2}}{x_u}$$ (as $\varepsilon_s = f_{st}/E_s$ for over-reinforced condition)

$$\text{or,} \quad f_{st} = \left\{ \frac{E_s \times \varepsilon_{cu2}}{\frac{x_u}{d}} \right\} \cdot \left(1 - \frac{x_u}{d} \right) \tag{15.5}$$

Now equating these two equations,

$$\frac{N.f_{ck}.100}{p} \left(\frac{x_u}{d} \right) = \left\{ \frac{E_s \times \varepsilon_{cu2}}{\frac{x_u}{d}} \right\} \cdot \left(1 - \frac{x_u}{d} \right)$$

$$\text{let,} \left(\frac{x_u}{d} \right) = X$$

$$\frac{\{N \times f_{ck} \times 100\}}{p} \times X = \left\{ \frac{E_s \times \varepsilon_{cu2}}{X} \right\} \cdot (1 - X)$$

$$\text{Or,} \quad 1 - X = \left[\frac{N \times f_{ck} \times 100}{p \times \varepsilon_{cu2} \times E_s} \right] . X^2$$

$$\text{Or,} \quad \left[\frac{N \times f_{ck} \times 100}{p \times \varepsilon_{cu2} \times E_s} \right] . X^2 + X - 1 = 0 \tag{15.6}$$

As per IRC:112, the maximum area of tension reinforcement for a beam is 2.5% of the area of concrete, *i.e.*, $A_{st,max}$ = 2.5% of A_c.

The maximum reinforcement is converted in terms of the effective cross-section, i.e., width "*b*" and effective depth "*d*" as per the following.

$$A_{st,max} = (2.5/100) \times A_c$$

$$\text{or,} \ A_s = \left(\frac{2.5}{100} \right) \times (A_g - A_s)$$

$$\text{Now,} \ A_s = 0.025 A_g - 0.025 A_s$$

$$\text{or,} \ A_s (1 + 0.025) = 0.025 bD$$

$$\text{or,} \ A_s = \frac{0.025}{(1 + 0.025)} bD$$

$$\text{or,} \ A_s = 0.02439 bD$$

$A_{effective} = b \times d$, and assuming $d = 0.9D$

$$A_s = (0.02439/0.9) bd$$

$$A_s = 0.0271 \times bd$$

Hence, the area of steel expressed in terms of effective cross-sectional area (*bd*) is 2.71 percent. Substituting this value in Equation (15.6), the maximum depth of neutral axis $\left(\frac{x_u}{d} \right)_{\max}$

TABLE 15.10
$\left(\frac{x_u}{d}\right)_{max}$ for different grades of concrete

Grade of concrete	$\left(\frac{x_u}{d}\right)_{max}$	Grade of concrete	$\left(\frac{x_u}{d}\right)_{max}$
M20	0.7731	M65	0.5753
M25	0.7401	M70	0.5578
M30	0.7116	M75	0.5413
M35	0.6867	M80	0.5270
M40	0.6646	M85	0.5159
M45	0.6448	M90	0.5011
M50	0.6269		
M55	0.6107		
M60	0.5958		

can be calculated. Up to M60 grade concrete the values of maximum flexural compressive strength of concrete ε_{cu2} are considered as 0.0035, however for higher grades of concrete the corresponding values of ε_{cu2} are to be used for calculating $\left(\frac{x_u}{d}\right)_{max}$. This value is a function of grade of concrete only and in Table 15.10 different values of $\left(\frac{x_u}{d}\right)_{max}$ for different grades of concrete are presented.

15.7 FOR FACTORED IDEALIZED BILINEAR STRESS–STRAIN DIAGRAM OF STEEL WITH INCLINED PLASTIC BRANCH

For an idealized bilinear diagram as provided by IRC:112 there is a capping on the strain limit for steel, ε_{uk}. The design limiting strain ε_{ud} is restricted to $0.9\varepsilon_{uk}$. Hence the design when performed by an idealized bilinear stress–strain diagram will differ from that performed by a simplified bilinear stress–strain diagram for only the plastic branch of the curve. The calculations up to the yield point will remain similar for both the stress–strain relationships, i.e., idealized or simplified. Thus, for balanced and over-reinforced condition, the design calculations will remain identical for both relationships. However, for under-reinforced conditions when the strains in steel will be more than the yield strain, the corresponding stress values are to be read from the idealized stress–strain diagram and the maximum or limiting design strain ε_{ud} will be restricted to $0.9\varepsilon_{uk}$. As per IRC:112 for grades Fe 415D, Fe 500D and Fe 550D, ε_{uk} shall be taken as 5% (max.), for grades Fe 415S and Fe 500S, ε_{uk} shall be taken as 8% (max.) and for other grades it shall be taken as 2.5% (max.).

15.7.1 COMPUTATION OF DESIGN PARAMETERS FOR CONCRETE GRADES FROM M20 TO M90 AND GRADES OF STEEL FROM FE415 TO FE600

The strain distribution diagram at the ultimate limit state for beams using idealized bilinear stress–strain relationship is shown in Figure 15.12, which is a modification derived from Figure 15.6 dealing with a section subjected to flexure along with axial forces.

15.7.1.1 Balanced condition

The design parameters corresponding to balanced condition like x_u/d, $A_{st,balanced}$, limiting or balanced moment of resistance factor, etc. will remain identical with that derived for a factored simplified stress–strain relationship.

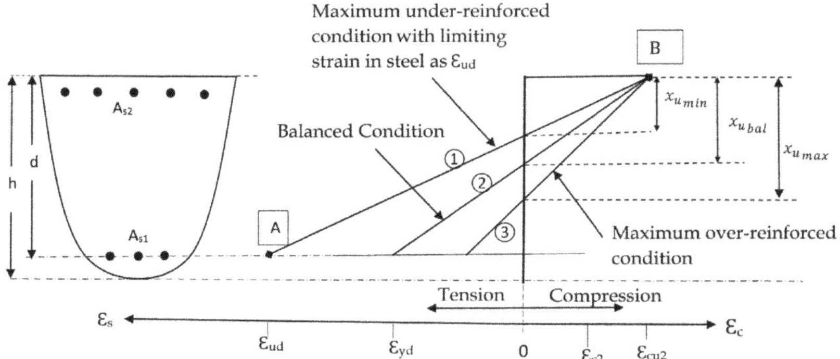

FIGURE 15.12 Possible domains of strain distributions following factored idealized bilinear stress–strain diagram of steel as per IRC:112.

15.7.1.2 Under-reinforced condition

Since in an under-reinforced condition the strain in steel is higher than the yield strain, calculations will differ from those performed for factored simplified bilinear diagram. For factored simplified bilinear stress–strain relationship of steel, stress beyond the yield point remains constant with an increase in strain and there is no capping on the maximum strains in steel. However, for factored idealized bilinear stress–strain relationship, stress in steel increases with an increase in strain up to the point having coordinates (ε_{uk}, f_t/Υ_s), where ε_{uk} is the characteristic strain of reinforcement at maximum load, f_t is the tensile strength of reinforcement and Υ_s is the partial safety factor for reinforcing steel.

Under-reinforced condition is specified by the strain domains between lines ① and ② as shown in Figure 15.12. Failure of the section can occur due to concrete reaching the strain of ε_{cu2} but strain in steel can range from the yield strain, ε_{yd}, to the maximum strain permitted, i.e., ε_{ud}. The stress blocks of concrete, i.e., the compressive force, remain similar to those presented in Table 15.3.

15.7.1.3 Maximum under-reinforced condition

For the factored simplified bilinear relationship, there is no capping on the maximum strain in steel, the calculations are straightforward and comparatively simple. However, for factored idealized stress–strain relationship of steel the maximum strain in steel cannot exceed the specified maximum strain value ε_{ud} as mentioned in the code. Hence $\varepsilon_{st,max}$ corresponding to the minimum amount of tension reinforcement ($A_{st,min}$) is to be compared with the maximum permitted design strain ε_{ud} as specified in IRC:112. If the values of $\varepsilon_{st,max}$ for different grades of concrete are less than ε_{ud} the condition is tenable and permitted by the code. However, if $\varepsilon_{st,max}$ exceeds ε_{ud}, the condition is not permitted by the code and the modified values of $A_{st,min}$ are to be calculated so that $\varepsilon_{st,max}$ is equal to ε_{ud}. From Table 15.9, it is observed that the maximum strains in steel can be very high and can exceed the values of ε_{ud} permitted by the code. In flexure, the limiting strength of the section is generally reached when concrete reaches the ultimate strain ε_{cu2} and the full stress block of concrete develops. Hence, as discussed in Section 15.5, the value of A_{stmin} needs to be modified so that the maximum strain in steel is restricted to ε_{ud} when ε_{cu2} is reached in concrete. This strain distribution matches with an extreme condition of the range of possible strain distribution as per Eurocode 2.

For the idealized bilinear stress–strain diagram of steel, strain hardening of steel has been taken into consideration, along with an increase in the stress level beyond the yield value. The plastic branch of the curve is obtained by joining the two points, corresponding to the yield level (f_{yk}) and the ultimate tensile stress level (f_t). This makes the design process somewhat complicated. For the simplified bilinear stress–strain relationship, stresses in steel beyond the yield point are constant and independent of the strain values. However, for the idealized relationship, stress values are not constant and will depend upon the values of strain in the reinforcement. This significantly complicates the problem and the methodology of determination of section parameters is quite cumbersome. The solution can be performed by assumptions and successive iterations involving a lengthy process (similar to those shown in Sections 10.21 and 10.22). Hence, for simplicity the factored simplified bilinear stress–strain relationship is preferred in all common design situations.

15.7.1.4 Maximum over-reinforced condition or section with maximum percentage of tension reinforcement

The over-reinforced condition deals with the initial branch of the stress–strain relationship of steel, which is identical for both relationships, idealized and simplified, and hence previous calculations performed for factored simplified bilinear relationships are also applicable for this condition.

15.8 DOUBLY REINFORCED SECTIONS

Doubly reinforced sections are those in which, in addition to tension reinforcement, compression reinforcements are also provided. Since a certain percentage of support or span reinforcements are to be extended all through the section length as per requirements of the standards, all sections in flexure have both tensile and compression reinforcements. Conventionally, designers opt for a doubly reinforced section when the external moment on the section exceeds the balanced moment of resistance capacity of the section. However, singly reinforced sections are only hypothetical as reinforcements are provided at both the tension and compression sides of flexural members. The design of doubly reinforced sections is similar to that of singly reinforced sections. The only difference being that due to the presence of reinforcement on the compression side, forces in compression steel are calculated considering the strain compatibility, that is the strain in steel is equal to the strain in the surrounding concrete. Thus, in the force equilibrium equation, the force in the tension steel should be equal to the sum of the forces in the concrete and compression steel and the moment of resistance of the section is to be calculated accordingly. This process has already been covered following the requirements IS 456 and presented in Chapter 10.

15.9 DESIGN FOR COMPRESSION MEMBERS

As per IS:456 limit state philosophy, the design for limit state of collapse was dealt with separately for flexure and compression. In IRC:112 the limit state design philosophy deals with flexure along with axial load, which is similar to Eurocode 2. Stress–strain relationships, stress blocks of concrete, strain distribution diagrams across cross-sections subjected to flexure along with axial loads and design assumptions have been dealt with in detail in earlier sections. These are primarily applicable to columns which are subjected to flexure along with axial loads and to beams which are subjected to flexure only as the axial load is either zero or negligibly small. The design of compression members has been dealt with in detail following the provisions of IS:456 in Chapter 11, and hence the process is not repeated here. For the design of compression members as per IRC:112 the methodology will differ from IS:456 primarily because of the fact that the stress–strain relation of HYSD steel has a bilinear relationship, which is different from that prescribed in IS:456. Simplified bilinear relationship which is commonly used has been considered in the present section for compression design.

The philosophy of compression design as per IRC:112 is similar to that of IS:456 as envisaged in Chapter 11. The process of development of safety profiles for RC sections, with different percentages of reinforcement with different combinations of grade of concrete and steel, i.e., interaction charts are similar for both standards, with minor modifications which will be demonstrated using numerical examples.

15.10 COLUMNS UNDER AXIAL LOAD AND UNIAXIAL BENDING

As per IRC:112, the assumptions of the ultimate limit state are provided in a combined manner for elements subjected to axial forces and bending. This is similar to the design provisions of Eurocode 2. Typical details of strain distribution diagrams and stress blocks for sections subjected to compression with bending have been discussed in earlier sections.

15.11 DEVELOPMENT OF STRAIN DISTRIBUTION DIAGRAMS AND STRESS BLOCKS

Similar to Eurocode 2, IRC:112 provides a bilinear stress–strain relationship for steel reinforcements which have already been discussed in Section 15.2. In idealized bilinear diagram, strain hardening

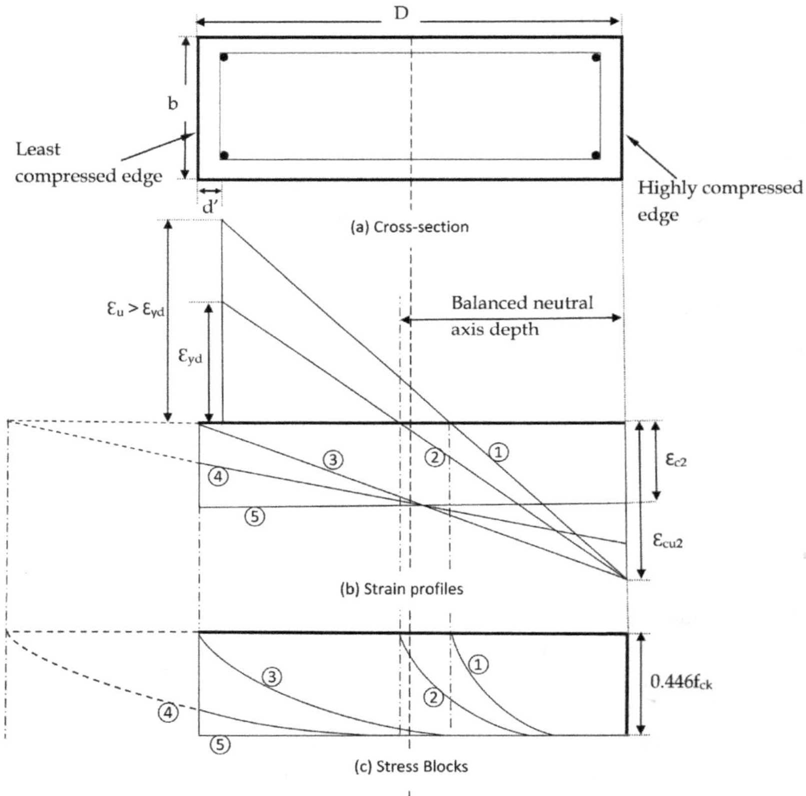

FIGURE 15.13 Diagram showing strain distribution and concrete stress blocks on section subjected to axial load with bending: (a) cross-section; (b) strain profiles; (c) stress blocks.

Note: The numerals indicate different section conditions which have been defined by the strain and stress distribution diagrams.

① Neutral axis within the section – strain in steel exceeds ε_{yd} and strain in concrete reaches ε_{cu2}.

② Strain in steel reaches ε_{yd}, strain in concrete reaches ε_{cu2} and the section is in balanced condition.

③ Neutral axis is at the face of the section, i.e., least compressed edge.

④ Strain in concrete is between ε_{c2} and ε_{cu2} and neutral axis lies outside the section.

⑤ Neutral axis is at infinity.

is considered and the plastic branch of the curve slopes upward, whereas in the simplified bilinear curve the plastic branch is horizontal without considering the strain hardening effect. Hence, the design yield stress will be $f_y/1.15$ and yield strain will be $(f_y/1.15\times E_s)$. The material safety factor Υ_s has been taken as 1.15 as per the standards. The strain distribution diagrams for different types of failure conditions will be similar to those presented in Chapter 11 with the only exception being that instead of proof stress, the yield stress of steel has been considered, as a result of which the inelastic strain of 0.002 has been deleted.

In Figure 15.13, strain distributions in the section for different positions of the neutral axis have been shown in detail. The corresponding stress blocks in concrete have also been shown. Since the diagrams have been developed in detail in Chapter 11 the procedure has not been repeated here.

In the same manner as discussed in Chapter 11 safety profiles of sections with constant percentages of reinforcements have been developed considering the stress–strain relationship of concrete and the simplified bilinear stress–strain diagram of steel. The stress blocks in concrete will be similar to those developed for IS:456 for concrete grades up to M60 and for higher grades of concrete up to M90 the modified strain values as per IRC:112 will be used for developing the stress blocks. Typical interaction charts are shown in Figures 15.16 and 15.17 involving grades of steel Fe500 and Fe550, $d'/D = 0.10$ which are applicable for all grades of concrete from M20 toM60.

15.12 NUMERICAL EXAMPLE (FOR CONCRETE GRADE UP TO M60)

Problem 15.1: Prepare an axial force–moment safety profile for a rectangular column section [area =$(b\times D)$] with 3 percent of longitudinal steel using grade of concrete M40 and grade of steel Fe 550 and considering $d'/D = 0.10$ (Table 15.11).

Given data: Grade of concrete $(f_{ck}) = 40$ N/mm^2

Grade of reinforcement $(f_y) = 550$ N/mm^2

$d'/D = 0.10$

Percentage of steel $(p) = 3\%$

Considered number of rows of reinforcement = six along each face (Figure 15.14)

Total number of reinforcements = 20

Percentage of steel (p_i) in each row of reinforcement and corresponding distance from centroidal axis (y_i/D). Spacing between bars $= \frac{1-d'-d'}{6-1}D = \frac{1-0.1-0.1}{6-1}D = 0.16D$

TABLE 15.11
Detailing of reinforcement distribution within the rectangular section

	Number of bars	Percentage of steel (p_i)	Distance from centroidal axis (y_i/D)
Row 1	6	$p_1=\left(\frac{3}{20}\right)\times 6= 0.9\%$	$\frac{y_1}{D}=\frac{1-0.1-0.1}{2}= 0.4$
Row 2	2	$p_2=\left(\frac{3}{20}\right)\times 2= 0.3\%$	$\frac{y_2}{D}= 0.4 - 0.16 = 0.24$
Row 3	2	$p_3=\left(\frac{3}{20}\right)\times 2= 0.3\%$	$\frac{y_3}{D}= 0.24 - 0.16 = 0.08$
Row 4	2	$p_4=\left(\frac{3}{20}\right)\times 2= 0.3\%$	$\frac{y_4}{D}= 0.08 - 0.16 = -0.08$
Row 5	2	$p_5=\left(\frac{3}{20}\right)\times 2= 0.3\%$	$\frac{y_5}{D}= -0.08 - 0.16 = -0.24$
Row 6	6	$p_6=\left(\frac{3}{20}\right)\times 6= 0.9\%$	$\frac{y_6}{D}= -0.24 - 0.16 = -0.4$

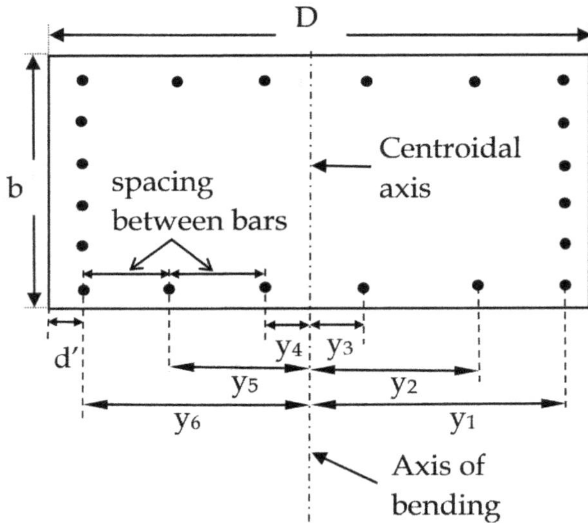

FIGURE 15.14 Reinforcement equally distributed on four sides for a rectangular column section.

Since the strain distribution diagrams and stress blocks are similar to Problem 11.19.2, the same have not been repeated here.

When neutral axis is outside the sections $\left(\frac{x_u}{D} > 1\right)$

Case 1: When neutral axis is at infinity

$$\therefore \frac{x_u}{D} = \infty$$

Hence, the whole column section is subjected to pure axial load.

When neutral axis is at infinity, strain is 0.002 throughout the section and the whole section is in compression.

Strain in concrete = $\varepsilon_c = 0.002$

Strain in concrete at the position of reinforcement = 0.002 = strain in steel

$f_{sc} = 0.002 \times 200{,}000$(stress–strain relationship is elastic up to $0.8f_y$)

$= 400$ N/mm^2

As the load is purely axial, the section is in constant stress of $0.446f_{ck}$ throughout the sections. Therefore, the total compressive force in concrete is:

$$C_c = 0.446 f_{ck} \times b \times D$$

Now, the steel bars of area $\left(\frac{pbD}{100}\right)$ are subjected to constant stress of f_{sc} when the strain is 0.002.

Hence, force in steel can be stated as follows:

$$C_s = \left(\frac{pbD}{100}\right) \times \left(f_{sc} - 0.446 f_{ck}\right)$$

$$P_u = C_c + C_s = (0.446 f_{ck} \times bD) + \left(\frac{pbD}{100}\right) \times (f_{sc} - 0.446 f_{ck})$$

$$\frac{P_u}{f_{ck} \times bD} = 0.446 + \left(\frac{p}{100 f_{ck}}\right) \times (f_{sc} - 0.446 f_{ck})$$

For the present problem $\frac{P_u}{f_{ck} \times bD} = 0.446 + [\left(\frac{3}{100 \times 40}\right) \times (400 - 0.446 \times 40)]$

$$= 0.446 + 0.28662$$
$$= 0.73262$$

and $\frac{M_u}{f_{ck} bD^2} = 0$

Case 2: $\frac{x_u}{D} = 1.1$

For this case, strain is maximum at the left edge which is $\varepsilon_{c,max} = 0.0035 - 0.75\, \varepsilon_{c,min}$ and $\varepsilon_{c,min}$ is the strain in concrete at the left edge of the section.

Therefore, now from similar triangles $\varepsilon_{c,min}$ can be calculated.

$$\frac{\varepsilon_{c,min}}{(1.1-1)D} = \frac{\varepsilon_{c,max}}{1.1D}$$

or, $\frac{\varepsilon_{c,min}}{0.1} = \frac{0.0035 - 0.75\varepsilon_{c,min}}{1.1}$

or, $\varepsilon_{c,min} = \frac{(0.0035 \times 0.1)}{1.1 + (0.75 \times 0.1)} = 0.000297$

$$\varepsilon_{c,max} = 0.0035 - (0.75 \times \varepsilon_{c,min}) = 0.0035 - (0.75 \times 0.000297)$$

$$= 0.003276$$

Again, from similar triangles strain in concrete in each row $\varepsilon_{c,1}$, $\varepsilon_{c,2}$, $\varepsilon_{c,3}$, $\varepsilon_{c,4}$, $\varepsilon_{c,5}$, $\varepsilon_{c,6}$ can be calculated.

$$\varepsilon_{c,1} = \frac{[0.003276 \times (1.1 - 0.1)]}{1.1} = 0.00298$$

$$\varepsilon_{c,2} = \frac{[0.003276 \times (1.1 - 0.1 - 0.16)]}{1.1} = 0.00250$$

$$\varepsilon_{c,3} = \frac{[0.003276 \times (1.1 - 0.1 - 2 \times 0.16)]}{1.1} = 0.00203$$

$$\varepsilon_{c,4} = \frac{\left[0.003276 \times (1.1 - 0.1 - 3 \times 0.16)\right]}{1.1} = 0.00155$$

$$\varepsilon_{c,5} = \frac{\left[0.003276 \times (1.1 - 0.1 - 4 \times 0.16)\right]}{1.1} = 0.00107$$

$$\varepsilon_{c,6} = \frac{\left[0.003276 \times (1.1 - 0.1 - 5 \times 0.16)\right]}{1.1} = 0.00060$$

From strain compatibility, $\varepsilon_{c,1} = \varepsilon_{s,1}$, $\varepsilon_{c,2} = \varepsilon_{s,2}$, $\varepsilon_{c,3} = \varepsilon_{s,3}$, $\varepsilon_{c,4} = \varepsilon_{s,4}$, $\varepsilon_{c,5} = \varepsilon_{s,5}$, $\varepsilon_{c,6} = \varepsilon_{s,6}$.

From $\varepsilon_{s,1}$, $\varepsilon_{s,2}$, $\varepsilon_{s,3}$, $\varepsilon_{s,4}$, $\varepsilon_{s,5}$, $\varepsilon_{c,6}$ one can calculate $f_{s1}, f_{s2}, f_{s3}, f_{s4}, f_{s5}, f_{s6}$ from the stress–strain diagram of steel as per IRC:112.

$$f_{s1} = 478.2608 \text{ N/mm}^2$$
$$f_{s2} = 478.2608 \text{ N/mm}^2$$
$$f_{s3} = 405.1063 \text{ N/mm}^2$$
$$f_{s4} = 309.7872 \text{ N/mm}^2$$
$$f_{s5} = 214.468 \text{ N/mm}^2$$
$$f_{s6} = 119.1489 \text{ N/mm}^2$$

Hence,

$$f_{c1} = 0.446 f_{ck} = 0.446 \times 40 = 17.84 \text{ N/mm}^2$$
$$f_{c2} = 0.446 f_{ck} = 0.446 \times 40 = 17.84 \text{ N/mm}^2$$
$$f_{c3} = 0.446 f_{ck} = 0.446 \times 40 = 17.84 \text{ N/mm}^2$$

$$f_{c4} = f_{cd} \times \left[1 - \left(1 - \frac{\varepsilon_{c,4}}{\varepsilon_{c2}}\right)^n\right]$$

As per IRC:112-2020, values of n depend upon the grades of concrete and are as follows:

Up to M60, $n = 2$ and for M65, M70, M75, M80, M85 and M90 the values of n are 1.9, 1.7, 1.6, 1.5, 1.5 and 1.4, respectively.

$$f_{c4} = 0.446 \times 40 \times \left[1 - \left(1 - \frac{0.00155}{0.002}\right)^2\right] = 16.932574 \text{ N/mm}^2$$

$$f_{c5} = f_{cd} \times \left[1 - \left(1 - \frac{\varepsilon_{c,5}}{\varepsilon_{c2}}\right)^n\right]$$

$$= 0.446 \times 40 \times \left[1 - \left(1 - \frac{0.00107}{0.002}\right)^2\right] = 14.001936 \text{ N/mm}^2$$

$$f_{c6} = f_{cd} \times \left[1 - \left(1 - \frac{\varepsilon_{c,6}}{\varepsilon_{c2}}\right)^n\right]$$

$$= 0.446 \times 40 \times \left[1 - \left(1 - \frac{0.00060}{0.002}\right)^2\right] = 9.045178 \text{ N/mm}^2$$

Calculation of area of stress blocks of concrete:

C_1 = coefficient of area of the compressive stress block

C_2= coefficient of distance of the centroid of the compressive stress block of concrete measured from the right edge

From Chapter 11, as per Table 11.1,

For $\frac{x_u}{D}$ = 1.1, C_1 = 0.384 and C_2 = 0.443

$$\text{Now } \frac{P_u}{f_{ck}bD} = C_1 + \sum_{i=1}^{n} \frac{p_i}{100 f_{ck}} \times \left(f_{si} - f_{ci}\right)$$

$$= 0.384 + [\frac{1}{100 \times 40} \times \{0.9 \times (478.2608 - 17.84) + 0.3 \times (478.2608 - 17.84)$$
$$+ 0.3 \times (405.106383 - 17.84) + 0.3 \times (309.787234 - 16.932574)$$
$$+ 0.3 \times (214.468085 - 14.001936) + 0.9 \times (119.148936 - 9.045178)$$
$$= 0.384 + 0.228943$$
$$= 0.612943$$

$$\frac{M_u}{f_{ck}bD^2} = C_1(0.5 - C_2) + \sum_{i=1}^{n} \frac{p_i}{100 f_{ck}} \times \left(f_{si} - f_{ci}\right) \times (y_i / D)$$

$$=\{0.384 \times (0.5 - 0.443)\} + [\frac{1}{100 \times 40} \times \{0.9 \times \{(478.2608 - 17.84) \times 0.4$$
$$+ 0.3 \times (478.2608 - 17.84) \times 0.24 + 0.3 \times (405.106383 - 17.84) \times 0.08$$
$$+ 0.3 \times (309.787234 - 16.932574) \times (-0.08) + 0.3 \times (214.468085 - 14.001936)$$
$$\times (-0.24) + 0.9 \times (119.148936 - 9.045178) \times (-0.4)\}]$$
$$= 0.02189 + 0.036774$$
$$= 0.058662$$

Neutral axis within the section $\left(\frac{x_u}{D} < 1\right)$

Case 3: when tensile stress at longitudinal steel at outermost edge is zero, $f_{st} = 0$

$x_u = D - d'$

Hence, $\frac{x_u}{D} = 1 - 0.1 = 0.90$

Therefore, $\varepsilon_{c,6} = \varepsilon_{s,6} = 0$

At the right edge strain in concrete is maximum which is,

$$\varepsilon_{c,max} = 0.0035$$

Therefore, from a similar triangle we can get the value of strain at the position of reinforcement.

$$\frac{\varepsilon_{c,1}}{(1-0.1-0.1)D} = \frac{\varepsilon_{c,max}}{(1-0.1)D}$$

Hence, $$\varepsilon_{c,1} = \frac{\left[0.003276 \times (0.90 - 0.1)\right]}{0.90} = 0.003111$$

$$\varepsilon_{c,2} = \frac{\left[0.003276 \times (0.90 - 0.1 - 0.16)\right]}{0.90} = 0.002489$$

$$\varepsilon_{c,3} = \frac{[0.003276 \times (0.90 - 0.1 - 2 \times 0.16)]}{0.90} = 0.001867$$

$$\varepsilon_{c,4} = \frac{[0.003276 \times (0.90 - 0.1 - 3 \times 0.16)]}{0.90} = 0.001244$$

$$\varepsilon_{c,5} = \frac{[0.003276 \times (0.90 - 0.1 - 4 \times 0.16)]}{0.90} = 0.000622$$

$$\varepsilon_{c,6} = \frac{[0.003276 \times (0.90 - 0.1 - 5 \times 0.16)]}{0.90} = 0$$

From strain compatibility, $\varepsilon_{c,1} = \varepsilon_{s,1}$, $\varepsilon_{c,2} = \varepsilon_{s,2}$, $\varepsilon_{c,3} = \varepsilon_{s,3}$, $\varepsilon_{c,4} = \varepsilon_{s,4}$, $\varepsilon_{c,5} = \varepsilon_{s,5}$, $\varepsilon_{c,6} = \varepsilon_{s,6}$.

From $\varepsilon_{s,1}$, $\varepsilon_{s,2}$, $\varepsilon_{s,3}$, $\varepsilon_{s,4}$, $\varepsilon_{s,5}$, $\varepsilon_{s,6}$ one can calculate $f_{s1}, f_{s2}, f_{s3}, f_{s4}, f_{s5}, f_{s6}$ from the stress–strain diagram of steel as per IRC:112-2020,

$$f_{s1} = 478.2608 \text{ N/mm}^2$$
$$f_{s2} = 478.2608 \text{ N/mm}^2$$
$$f_{s3} = 373.3333 \text{ N/mm}^2$$
$$f_{s4} = 248.8888 \text{ N/mm}^2$$
$$f_{s5} = 124.4444 \text{ N/mm}^2$$
$$f_{s6} = 0 \text{ N/mm}^2$$

Stress in concrete at the position of steel can be calculated.

From $\varepsilon_{c,1}$ to $\varepsilon_{c,5}$ one can calculate stress in concrete f_{c1} to f_{c5}.

$$f_{c1} = 0.446 f_{ck} = 0.446 \times 40 = 17.84 \text{ N/mm}^2$$

$$f_{c2} = 0.446 f_{ck} = 0.446 \times 40 = 17.84 \text{ N/mm}^2$$

$$f_{c3} = f_{cd} \times \left[1 - \left(1 - \frac{\varepsilon_{c,3}}{\varepsilon_{c2}}\right)^n\right]$$

$$= 0.446 \times 40 \times \left[1 - \left(1 - \frac{0.001867}{0.002}\right)^2\right]$$

$$= 17.760711 \text{ N/mm}^2$$

$$f_{c4} = f_{cd} \times \left[1 - \left(1 - \frac{\varepsilon_{c,4}}{\varepsilon_{c2}}\right)^n\right] = 15.293 \text{ N/mm}^2$$

$$f_{c5} = f_{cd} \times \left[1 - \left(1 - \frac{\varepsilon_{c,5}}{\varepsilon_{c2}}\right)^n\right] = 9.3737 \text{ N/mm}^2$$

$$f_{c6} = 0$$

$$\text{now} \quad \frac{P_u}{f_{ck}bD} = 0.36 \times \left(\frac{x_u}{D}\right) + \sum_{i=1}^{n} \frac{p_i}{100 f_{ck}} \times \left(f_{si} - f_{ci}\right)$$

$$= (0.36 \times 0.9) + \left(\frac{1}{40 \times 100}\right) \times [\{0.9 \times (478.2608 - 17.84)\}$$
$$+ \{0.3 \times (478.2608 - 17.84)\} + \{0.3 \times (373.333 - 17.7607)\}$$
$$+ \{0.3 \times (248.89 - 15.293)\} + \{0.3 \times (124.444 - 9.373717\}$$
$$+ \{0.9 \times (0 - 0)\}]$$
$$= 0.324 + 0.190944$$
$$= 0.514944$$

$$\frac{M_u}{f_{ck}bD^2} = 0.36 \times \left(\frac{x_u}{D}\right) \times (0.5 - 0.42 \times \left(\frac{x_u}{D}\right)) + \sum_{i=1}^{n} \frac{p_i}{100 f_{ck}} \times \left(f_{si} - f_{ci}\right) \times (y_i / D)$$

$$= 0.36 \times 0.9 \times (0.5 - 0.42 \times 0.9)\left(\frac{1}{40 \times 100}\right) \times [\{0.9 \times (478.2608 - 17.84) \times 0.4\}$$
$$+ \{0.3 \times (478.2608 - 17.84) \times 0.24\} + \{0.3 \times (373.333 - 17.7607) \times 0.08\}$$
$$+ \{0.3 \times (248.89 - 15.293) \times (-0.08)\} + \{0.3 \times (124.444 - 9.373717 \times (-0.24)\}$$
$$+ \{0.9 \times (0 - 0)\} \times (-0.4)]$$
$$= 0.087914$$

Case 4: when the tensile stress in longitudinal steel at outermost edge is $-0.4f_{yd} = -0.4 \times 0.87 \times f_y$

$$f_{s6} = (-0.4 \times 0.87 \times 550) = (-191.4) \text{ N/mm}^2 \text{ (tensile)}$$

Now from the stress–strain relation one can find strain at the position of longitudinal steel at the outermost edge.

$$\varepsilon_{s,6} = -\left(\frac{0.87 \times 550}{200000}\right) \times 0.4 = -0.000957$$

At the right edge of the section strain in concrete is maximum,

$$\varepsilon_{c,max} = 0.0035$$

From similar triangles $\frac{x_u}{D}$ can be calculated.

$$\frac{0.0035}{\frac{x_u}{D}} = \frac{0.000957}{\left\{1 - 0.10 - \left(\frac{x_u}{D}\right)\right\}}$$

$$\text{or,} \quad \frac{x_u}{D} \times 0.000957 = 0.0035 \times (1 - 0.10) - 0.0035 \times \left(\frac{x_u}{D}\right)$$

$$\text{or,} \quad \frac{x_u}{D} = \frac{0.0035 \times 0.90}{0.000957 + 0.0035}$$

$$\text{or,} \quad \frac{x_u}{D} = 0.706753$$

Again, from a similar triangle one can get the value of strain at the position of reinforcement.

$$\frac{\varepsilon_{c,1}}{0.706753-0.10} = \frac{\varepsilon_{c,max}}{0.706753}$$

Hence,

$$\varepsilon_{c,1} = \frac{\left[0.0035\times(0.706753-0.1)\right]}{0.706753} = 0.003004$$

$$\varepsilon_{c,2} = \frac{\left[0.0035\times(0.706753-0.1-0.16)\right]}{0.706753} = 0.002212$$

$$\varepsilon_{c,3} = \frac{\left[0.0035\times(0.706753-0.1-2\times0.16)\right]}{0.706753} = 0.001420$$

$$\varepsilon_{c,4} = \frac{\left[0.0035\times(0.706753-0.1-3\times0.16)\right]}{0.706753} = 0.000627$$

$$\varepsilon_{c,5} = \frac{\left[0.0035\times(0.706753-0.1-4\times0.16)\right]}{0.706753} = -0.000164$$

$$\varepsilon_{c,6} = \frac{\left[0.00355\times(0.706753-0.1-5\times0.16)\right]}{0.706753} = -0.000957$$

From strain compatibility, $\varepsilon_{c,1} = \varepsilon_{s,1}$, $\varepsilon_{c,2} = \varepsilon_{s,2}$, $\varepsilon_{c,3} = \varepsilon_{s,3}$, $\varepsilon_{c,4} = \varepsilon_{s,4}$, $\varepsilon_{c,5} = \varepsilon_{s,5}$, $\varepsilon_{c,6} = \varepsilon_{s,6}$.

From $\varepsilon_{s,1}$, $\varepsilon_{s,2}$, $\varepsilon_{s,3}$, $\varepsilon_{s,4}$, $\varepsilon_{s,5}$, $\varepsilon_{s,6}$ one can calculate $f_{s1}, f_{s2}, f_{s3}, f_{s4}, f_{s5}, f_{s6}$ from the stress–strain diagram of steel as per IRC:112-2020,

$$f_{s1} = 478.2608 \text{ N/mm}^2$$
$$f_{s2} = 442.4844 \text{ N/mm}^2$$
$$f_{s3} = 284.0133 \text{ N/mm}^2$$
$$f_{s4} = 125.5422 \text{ N/mm}^2$$
$$f_{s5} = -32.92889 \text{ N/mm}^2$$
$$f_{s6} = -191.4 \text{ N/mm}^2$$

Stress in concrete at the position of steel can be calculated.

From $\varepsilon_{c,1}$ to $\varepsilon_{c,5}$ one can calculate stress in concrete f_{c1} to f_{c5}.

$$f_{c1} = 0.446f_{ck} = 0.446 \times 40 = 17.84 \text{ N/mm}^2$$
$$f_{c2} = 0.446f_{ck} = 0.446 \times 40 = 17.84 \text{ N/mm}^2$$

$$f_{c3} = f_{cd} \times \left[1-\left(1-\frac{\varepsilon_{c,3}}{\varepsilon_{c2}}\right)^n\right]$$

$$= 0.446 \times 40 \times \left[1-\left(1-\frac{0.001420}{0.002}\right)^2\right] = 16.34 \text{ N/mm}^2$$

$$f_{c4} = f_{cd} \times \left[1-\left(1-\frac{\varepsilon_{c,4}}{\varepsilon_{c2}}\right)^n\right] = 9.441 \text{ N/mm}^2$$

$$f_{c5} = 0$$

$$f_{c6} = 0$$

$$\frac{P_u}{f_{ck}bD} = 0.36 \times \left(\frac{x_u}{D}\right) + \sum_{i=1}^{n} \frac{p_i}{100 f_{ck}} \times \left(f_{si} - f_{ci}\right)$$
$$= 0.373122$$

$$\frac{M_u}{f_{ck}bD^2} = 0.36 \times \left(\frac{x_u}{D}\right) \times (0.5\text{-}0.42 \times \left(\frac{x_u}{D}\right)) + \sum_{i=1}^{n} \frac{p_i}{100 f_{ck}} \times \left(f_{si} - f_{ci}\right) \times (y_i / D)$$
$$= 0.119500$$

Case 5: When the tensile stress in longitudinal steel at outermost edge is $-0.8f_{yd} = -0.8 \times 0.87 \times f_y$

$$f_{s6} = (-0.8 \times 0.87 \times 550) = (-382.8) \text{ N/mm}^2 \text{ (tensile)}$$

Now from the stress–strain relation one can find strain at the position of longitudinal steel.

$$\varepsilon_{s,6} = -\left(\frac{0.87 \times 550}{200000}\right) \times 0.8$$
$$= -0.001914$$

At the right edge of the section strain in concrete is maximum, $\varepsilon_{c,max} = 0.0035$

From similar triangle $\frac{x_u}{D}$ can be calculated.

$$\frac{0.0035}{\frac{x_u}{D}} = \frac{0.001914}{\left\{1-0.10-\left(\frac{x_u}{D}\right)\right\}}$$

$$\text{or, } \frac{x_u}{D} \times 0.001914 = 0.0035 \times (1\text{-}0.10)\text{-}0.0035 \times \left(\frac{x_u}{D}\right)$$

$$\text{or, } \frac{x_u}{D} = \frac{0.0035 \times 0.90}{0.001914 + 0.0035}$$

$$\text{or, } \frac{x_u}{D} = 0.58182$$

Again, from similar triangles one can get the value of strain at the position of reinforcement.

$$\frac{\varepsilon_{c,1}}{0.58182 - 0.10} = \frac{\varepsilon_{c,max}}{0.58182}$$

Hence,
$$\varepsilon_{c,1} = \frac{\{0.0035\times(0.58182-0.10)\}}{0.58182} = 0.002898$$

$$\varepsilon_{c,2} = \frac{[0.0035\times(0.58182-0.1-0.16)]}{0.58182} = 0.001935$$

$$\varepsilon_{c,3} = \frac{[0.0035\times(0.58182-0.1-2\times0.16)]}{0.58182} = 0.000973$$

$$\varepsilon_{c,4} = \frac{[0.0035\times(0.58182-0.1-3\times0.16)]}{0.58182} = 0.000010$$

$$\varepsilon_{c,5} = \frac{[0.0035\times(0.58182-0.1-4\times0.16)]}{0.58182} = -0.000951$$

$$\varepsilon_{c,6} = \frac{[0.00355\times(0.58182-0.1-5\times0.16)]}{0.58182} = -0.001914$$

From strain compatibility, $\varepsilon_{c,1} = \varepsilon_{s,1}$, $\varepsilon_{c,2} = \varepsilon_{s,2}$, $\varepsilon_{c,3} = \varepsilon_{s,3}$, $\varepsilon_{c,4} = \varepsilon_{s,4}$, $\varepsilon_{c,5} = \varepsilon_{s,5}$, $\varepsilon_{c,6} = \varepsilon_{s,6}$.

From $\varepsilon_{s,1}$, $\varepsilon_{s,2}$, $\varepsilon_{s,3}$, $\varepsilon_{s,4}$, $\varepsilon_{s,5}$, $\varepsilon_{s,6}$ one can calculate $f_{s1}, f_{s2}, f_{s3}, f_{s4}, f_{s5}, f_{s6}$ from the stress–strain diagram,

$$f_{s1} = 478.2608 \text{ N/mm}^2$$
$$f_{s2} = 387.1911 \text{ N/mm}^2$$
$$f_{s3} = 194.6933 \text{ N/mm}^2$$
$$f_{s4} = 2.195555 \text{ N/mm}^2$$
$$f_{s5} = -190.3022 \text{ N/mm}^2$$
$$f_{s6} = -382.8 \text{ N/mm}^2$$

Stress in concrete at the position of steel can be calculated.

From $\varepsilon_{c,1}$ to $\varepsilon_{c,6}$ one can calculate stress in concrete f_{c1} to f_{c6}.

$$f_{c1} = 0.446 f_{ck} = 0.446 \times 40 = 17.84 \text{ N/mm}^2$$

$$f_{c2} = f_{cd} \times \left[1-\left(1-\frac{\varepsilon_{c,2}}{\varepsilon_{c2}}\right)^n\right]$$

$$= 0.446 \times 40 \times \left[1-\left(1-\frac{0.001935}{0.002}\right)^2\right] = 17.8217 \text{ N/mm}^2$$

$$f_{c3} = f_{cd} \times \left[1-\left(1-\frac{\varepsilon_{c,3}}{\varepsilon_{c2}}\right)^n\right]$$

$$= 0.446 \times 40 \times \left[1-\left(1-\frac{0.000973}{0.002}\right)^2\right] = 13.14018 \text{ N/mm}^2$$

$$f_{c4} = f_{cd} \times \left[1-\left(1-\frac{\varepsilon_{c,4}}{\varepsilon_{c2}}\right)^n\right] = 0.195306 \text{ N/mm}^2$$

$$f_{c5} = 0$$
$$f_{c6} = 0$$

$$\frac{P_u}{f_{ck}bD} = 0.36 \times \left(\frac{x_u}{D}\right) + \sum_{i=1}^{n} \frac{p_i}{100 f_{ck}} \times \left(f_{si} - f_{ci}\right)$$

$$= 0.254118$$

$$\frac{M_u}{f_{ck}bD^2} = 0.36 \times \left(\frac{x_u}{D}\right) \times (0.5\text{-}0.42 \times \left(\frac{x_u}{D}\right)) + \sum_{i=1}^{n} \frac{p_i}{100 f_{ck}} \times \left(f_{si} - f_{ci}\right) \times (y_i / D)$$

$$= 0.140585$$

Case 6: When the tensile stress in longitudinal steel is $-f_{yd} = -0.87 \times f_y$ and strain $= -0.87 f_y/E_s$

Hence, $f_{s6} = (-0.87 \times 550) = (-478.5)$ N/mm^2 (tensile)
Now from the stress–strain relation we can find strain at the position of longitudinal steel.

$$\varepsilon_{s,6} = -\left(\frac{0.87 \times 550}{200000}\right) = -0.002392$$

At the right edge of the section strain in concrete is maximum.

$$\varepsilon_{c,max} = 0.0035$$

From similar triangles, $\frac{x_u}{D}$ can be calculated.

$$\frac{0.0035}{\frac{x_u}{D}} = \frac{0.002392}{\left\{1 - 0.10 - \left(\frac{x_u}{D}\right)\right\}}$$

$$\text{or,} \quad \frac{x_u}{D} \times 0.002392 = 0.0035 \times (1 - 0.10) - 0.0035 \times \left(\frac{x_u}{D}\right)$$

$$\text{or,} \quad \frac{x_u}{D} = \frac{0.0035 \times 0.90}{0.002392 + 0.0035}$$

$$\text{or,} \quad \frac{x_u}{D} = 0.534577$$

Again, from similar triangles one can get the value of strain at the position of reinforcement.

$$\frac{\varepsilon_{c,1}}{0.534577 - 0.10} = \frac{\varepsilon_{c,max}}{0.534577}$$

Hence,

$$\varepsilon_{c,1} = \frac{\{0.0035 \times (0.534577 - 0.10)\}}{0.534577} = 0.002845$$

$$\varepsilon_{c,2} = \frac{\left[0.0035 \times (0.534577 - 0.1 - 0.16)\right]}{0.534577} = 0.001797$$

$$\varepsilon_{c,3} = \frac{[0.0035 \times (0.534577 - 0.1 - 2 \times 0.16)]}{0.534577} = 0.000750$$

$$\varepsilon_{c,4} = \frac{[0.0035 \times (0.534577 - 0.1 - 3 \times 0.16)]}{0.534577} = -0.000297$$

$$\varepsilon_{c,5} = \frac{[0.0035 \times (0.534577 - 0.1 - 4 \times 0.16)]}{0.534577} = -0.001344$$

$$\varepsilon_{c,6} = \frac{[0.00355 \times (0.534577 - 0.1 - 5 \times 0.16)]}{0.534577} = -0.002392$$

From strain compatibility, $\varepsilon_{c,1} = \varepsilon_{s,1}$, $\varepsilon_{c,2} = \varepsilon_{s,2}$, $\varepsilon_{c,3} = \varepsilon_{s,3}$, $\varepsilon_{c,4} = \varepsilon_{s,4}$, $\varepsilon_{c,5} = \varepsilon_{s,5}$, $\varepsilon_{c,6} = \varepsilon_{s,6}$.

From $\varepsilon_{s,1}$, $\varepsilon_{s,2}$, $\varepsilon_{s,3}$, $\varepsilon_{s,4}$, $\varepsilon_{s,5}$, $\varepsilon_{s,6}$ one can calculate $f_{s1}, f_{s2}, f_{s3}, f_{s4}, f_{s5}, f_{s6}$ from the stress–strain diagram,

$$f_{s1} = 478.2608 \text{ N/mm}^2$$
$$f_{s2} = 359.5444 \text{ N/mm}^2$$
$$f_{s3} = 150.0333 \text{ N/mm}^2$$
$$f_{s4} = -59.4778 \text{ N/mm}^2$$
$$f_{s5} = -268.989 \text{ N/mm}^2$$
$$f_{s6} = -478.5 \text{ N/mm}^2$$

Stress in concrete at the position of steel can be calculated.

From $\varepsilon_{c,1}$ to $\varepsilon_{c,6}$ one can calculate stress in concrete f_{c1} to f_{c6}.

$$f_{c1} = 0.446 f_{ck} = 0.446 \times 40 = 17.84 \text{ N/mm}^2$$

$$f_{c2} = f_{cd} \times \left[1 - \left(1 - \frac{\varepsilon_{c,2}}{\varepsilon_{c2}}\right)^n\right]$$

$$= 0.446 \times 40 \times \left[1 - \left(1 - \frac{0.001797}{0.002}\right)^2\right] = 17.65751 \text{ N/mm}^2$$

$$f_{c3} = f_{cd} \times \left[1 - \left(1 - \frac{\varepsilon_{c,3}}{\varepsilon_{c2}}\right)^n\right].$$

$$= 0.446 \times 40 \times \left[1 - \left(1 - \frac{0.000750}{0.002}\right)^2\right] = 10.8731 \text{ N/mm}^2$$

$$f_{c4} = f_{c5} = f_{c6} = 0$$

$$\frac{P_u}{f_{ck} bD} = 0.36 \times \left(\frac{x_u}{D}\right) + \sum_{i=1}^{n} \frac{p_i}{100 f_{ck}} \times (f_{si} - f_{ci})$$

$$= 0.199823$$

$$\frac{M_u}{f_{ck}bD^2} = 0.36 \times \left(\frac{x_u}{D}\right) \times (0.5\text{-}0.42 \times \left(\frac{x_u}{D}\right)) + \sum_{i=1}^{n} \frac{p_i}{100 f_{ck}} \times (f_{si} - f_{ci}) \times (y_i / D)$$

$$= 0.149705$$

Case 7: When the depth of neutral axis $= 0.25D$

$$\therefore \frac{x_u}{D} = 0.25$$

At the right edge, strain in concrete is maximum.

$$\varepsilon_{c,max} = 0.0035$$

From similar triangle one can get the value of strain at the position of reinforcement.

$$\frac{\varepsilon_{c,1}}{0.25 - 0.10} = \frac{\varepsilon_{c,max}}{0.25}$$

Hence,

$$\varepsilon_{c,1} = \frac{\{0.0035 \times (0.25 - 0.10)\}}{0.25} = 0.0021$$

$$\varepsilon_{c,2} = \frac{[0.0035 \times (0.25 - 0.1 - 0.16)]}{0.25} = -0.00014$$

$$\varepsilon_{c,3} = \frac{[0.0035 \times (0.25 - 0.1 - 2 \times 0.16)]}{0.25} = -0.00238$$

$$\varepsilon_{c,4} = \frac{[0.0035 \times (0.25 - 0.1 - 3 \times 0.16)]}{0.25} = -0.00462$$

$$\varepsilon_{c,5} = \frac{[0.0035 \times (0.25 - 0.1 - 4 \times 0.16)]}{0.25} = -0.00686$$

$$\varepsilon_{c,6} = \frac{[0.0035 \times (0.25 - 0.1 - 5 \times 0.16)]}{0.25} = -0.0091$$

From strain compatibility, $\varepsilon_{c,1} = \varepsilon_{s,1}$, $\varepsilon_{c,2} = \varepsilon_{s,2}$, $\varepsilon_{c,3} = \varepsilon_{s,3}$, $\varepsilon_{c,4} = \varepsilon_{s,4}$, $\varepsilon_{c,5} = \varepsilon_{s,5}$, $\varepsilon_{c,6} = \varepsilon_{s,6}$.

From $\varepsilon_{s,1}$, $\varepsilon_{s,2}$, $\varepsilon_{s,3}$, $\varepsilon_{s,4}$, $\varepsilon_{s,5}$, $\varepsilon_{s,6}$ one can calculate $f_{s1}, f_{s2}, f_{s3}, f_{s4}, f_{s5}, f_{s6}$ from the stress–strain diagram of steel as per IRC:112-2020,

$$f_{s1} = 420 \text{ N/mm}^2$$
$$f_{s2} = -28 \text{ N/mm}^2$$
$$f_{s3} = -476 \text{ N/mm}^2$$
$$f_{s4} = -478.2608 \text{ N/mm}^2$$
$$f_{s5} = -478.2608 \text{ N/mm}^2$$
$$f_{s6} = -478.2608 \text{ N/mm}^2$$

Stress in concrete at the position of steel can be calculated. From strain in concrete, $\varepsilon_{c,1}$ to $\varepsilon_{c,6}$ one can calculate stress in concrete f_{c1} to f_{c6}.

$$f_{c1} = 0.446f_{ck} = 0.446 \times 40 = 17.84 \text{ N/mm}^2$$

$$f_{c2} = f_{c3} = f_{c4} = f_{c5} = f_{c6} = 0$$

$$\frac{P_u}{f_{ck}bD} = 0.36 \times \left(\frac{x_u}{D}\right) + \sum_{i=1}^{n} \frac{p_i}{100 f_{ck}} \times \left(f_{si} - f_{ci}\right)$$

$$= -0.036661$$

$$\frac{M_u}{f_{ck}bD^2} = 0.36 \times \left(\frac{x_u}{D}\right) \times (0.5 - 0.42 \times \left(\frac{x_u}{D}\right)) + \sum_{i=1}^{n} \frac{p_i}{100 f_{ck}} \times \left(f_{si} - f_{ci}\right) \times (y_i / D)$$

$$= 0.122906$$

Case 8: When a column behaves like a steel beam

This is the specific situation when it is subjected to pure moment only.

The column has symmetrical longitudinal steel on both sides of the centroidal axis, hence the column will have pure moment capacity, by yielding of both tensile and compressive steel bars.

$$M_u = \sum_{i=1}^{n} \left(\frac{p_i bD}{100}\right) \left(0.87 f_y\right) \left(\frac{y_i}{D}\right)$$

$$\frac{M_u}{f_{ck}bD^2} = \sum_{i=1}^{n} \left(\frac{p_i}{100 f_{ck}}\right) \left(0.87 f_y\right) \left(\frac{y_i}{D}\right)$$

$$= \frac{1}{100} \times (0.87 \times 550) \times [\{\frac{0.9}{40} \times 0.4\} + \{\frac{0.3}{40} \times 0.24\}$$

$$+ \{\frac{0.3}{40} \times 0.08\} + \{\frac{0.3}{40} \times 0.08\} + \{\frac{0.3}{40} \times 0.24\} + \{\frac{0.9}{40} \times 0.4\}]$$

$$= 0.109098$$

$$\frac{P_u}{f_{ck}bD} = 0$$

Preparation of graphical chart

The coordinates of the eight points $\left(\frac{P_u}{f_{ck}bD}, \frac{M_u}{f_{ck}bD^2}\right)$ can be plotted to generate the safety profile (Table 15.12).

These 8 points are now plotted in a graph to prepare a graphical chart.

TABLE 15.12
Values of non-dimensional parameters generated from different positions of neutral axis of the section

Cases	Illustrations of case	$\frac{P_u}{f_{ck}bD}$	$\frac{M_u}{f_{ck}bD^2}$
Case 1	When neutral axis is at infinity	0.73262	0
Case 2	When neutral axis is outside the sections ($x_u/D > 1$), i.e., ($x_u/D = 1.1$)	0.61294	0.05866
Case 3	When tensile stress at longitudinal steel is zero, $f_{st} = 0$	0.51495	0.08791
Case 4	When the tensile stress in longitudinal steel is $-0.4f_{yd} = -0.4 \times 0.87 \times f_y$	0.37313	0.11950
Case 5	When the tensile stress in longitudinal steel is $-0.8f_{yd} = -0.8 \times 0.87 \times f_y$	0.25412	0.14058
Case 6	When the tensile stress in longitudinal steel is $-f_{yd} = -0.87 \times f_y$ and strain $= -0.87f_y/E_s$	0.19987	0.14969
Case 7	When the depth of neutral axis $x_u = 0.25D$	−0.0367	0.1229
Case 8	When column behaves like a steel beam	0	0.10909

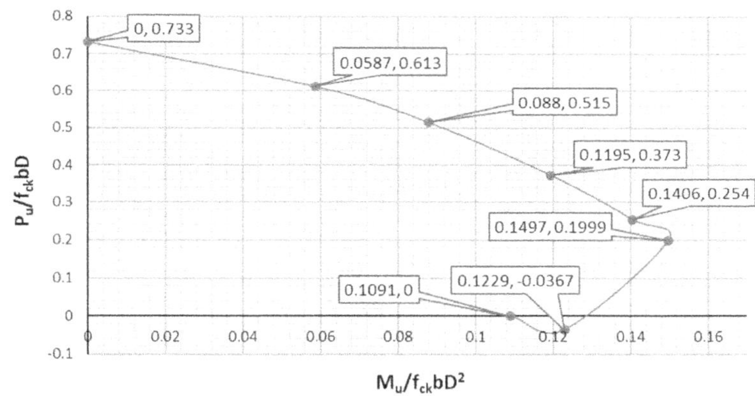

FIGURE 15.15 Safety profile corresponding to Problem 15.1.

15.13 SAMPLE DESIGN AIDS FOR COMPRESSION DESIGN

P–M interaction charts needed for column design are prepared by plotting a number of safety profiles with different percentages of tension reinforcement up to the maximum percentage permitted in the standard in a single diagram. Similar to IS:456-2000, IRC:112-2020 prescribes similar stress–strain diagrams of concrete up to M60. From M65 to M90 the concrete stress strain diagrams are to be provided as per stress–strain values provided in the standard. The factored simplified design diagram is generally used in design. Values of d'/D can range from 0.05 to 0.2, with 0.05 as the common interval as is observed in the commonly available design charts. For grades of concrete up to M60 using the similar assumptions as mentioned in Chapter 11, P–M interaction charts have been drawn using M40 grade of concrete whereby the reinforcement percentages are expressed in terms of p/f_{ck}. The variations in the values obtained from this chart with respect to individual charts developed for

M20 to M60 concrete are almost insignificant, as has been obtained from actual calculations. Four grades of steel and four d'/D ratios need to be used, resulting in the development of 16 interaction charts that can cater to all grades of structural concrete up to M60 and a particular distribution of longitudinal reinforcements, following the stress–strain relationship of concrete and steel as per IRC:112-2020. Reinforcement has been expressed as p/f_{ck}. Since in Chapter 11 all 32 charts for reinforcement distributed on two and four sides have been provided following the stipulations of IS:456-2000, in this chapter only two sample charts prepared for Fe 500 with d'/D equal to 0.1 have been included and reinforcement distributed on two and four sides, following the stipulations of IRC:112-2020, to restrict the volume of the book. This is presented in Charts 15.1 and 15.2. Using the same considerations, designers can develop interaction charts for any grade of steel with any value of d'/D.

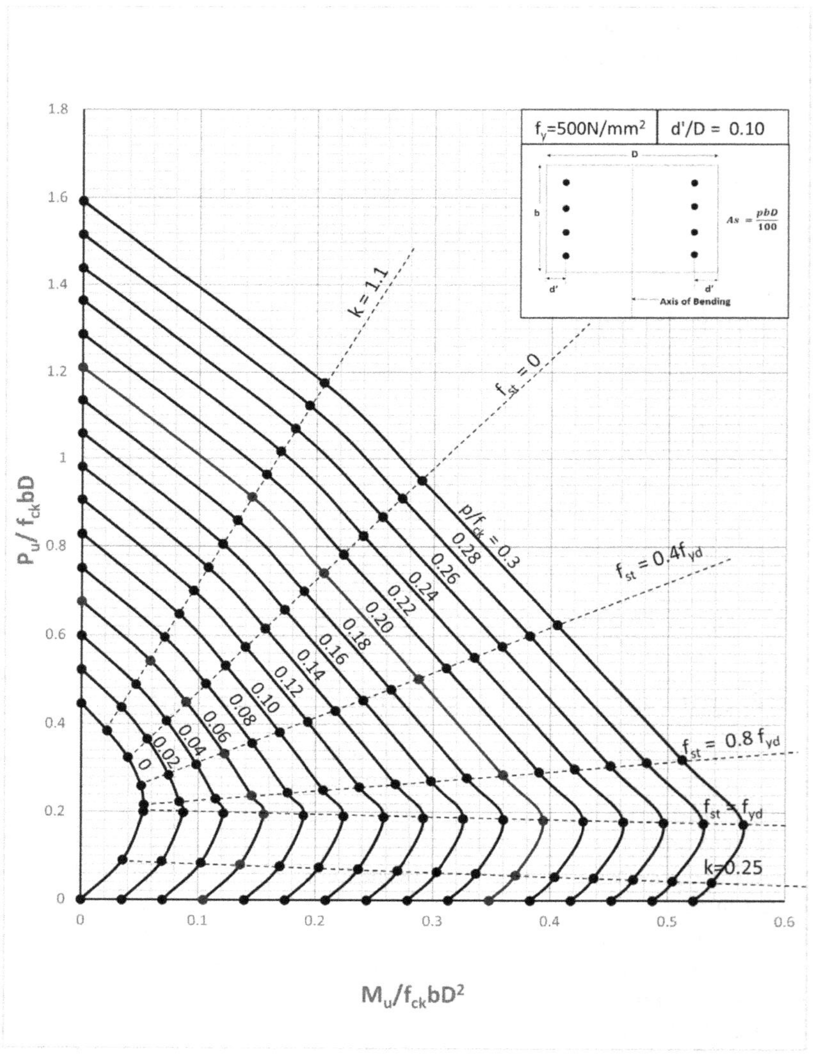

CHART 15.1 Compression with bending – rectangular section reinforcement equally distributed on two sides.

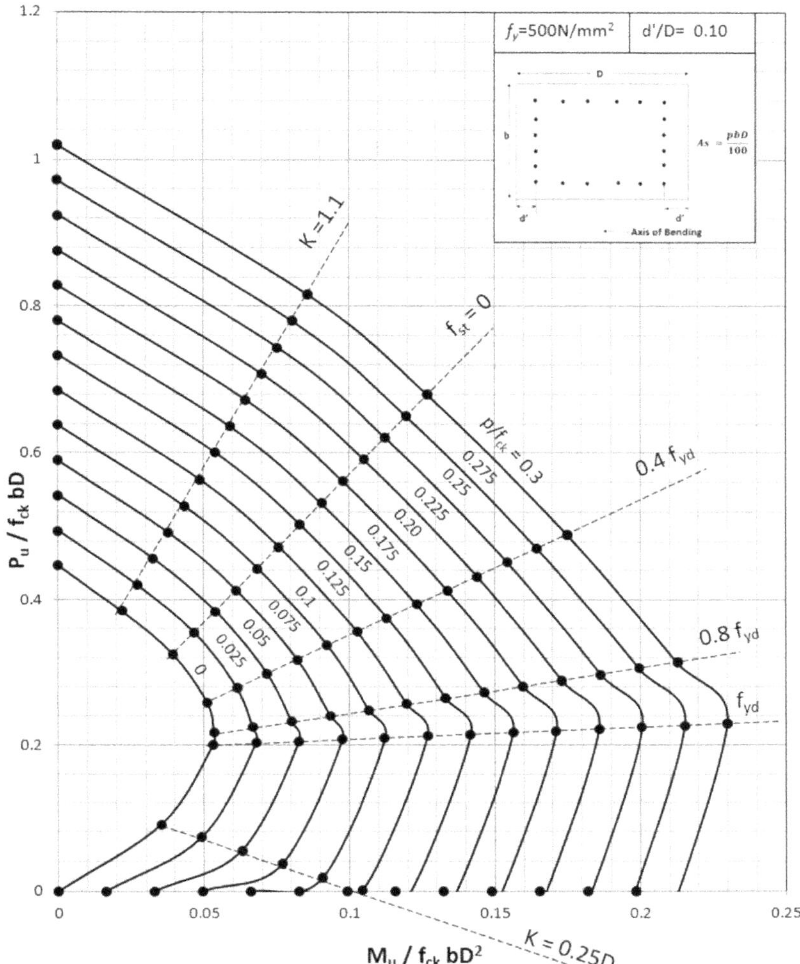

CHART 15.2 Compression with bending – rectangular section reinforcement equally distributed on four sides.

15.14 NUMERICAL EXAMPLE (HIGHER GRADES OF CONCRETE BEYOND M60)

Problem 15.2: Prepare an axial force–moment safety profile for a rectangular column section [area = ($b \times D$)] with 3.5 percent of longitudinal steel using grade of concrete M75 and grade of steel Fe 500 and considering $d'/D = 0.15$.

Solution

Given data: Grade of concrete (f_{ck}) = 75 N/mm²

Grade of reinforcement (f_y) = 500 N/mm²

$d'/D = 0.15$

Percentage of steel (p) = 3.5%

Consider number of rows of reinforcement = 6 (Figure 15.16)

Total number of reinforcements = 20

Percentage of steel (p_i) in each row of reinforcement and corresponding distance from centroidal

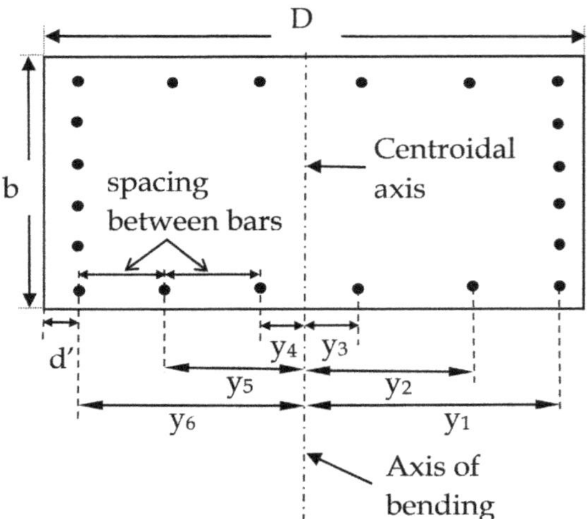

FIGURE 15.16 Reinforcement equally distributed on four sides for a rectangular column section.

TABLE 15.13
Detailing of reinforcement distribution within the rectangular section

	Number of bars	Percentage of steel (p_i)	Distance from centroidal axis (y_i/D)
Row 1	6	$p_1=\left(\frac{3.5}{20}\right)\times 6= 1.05\%$	$\frac{y_1}{D}=\frac{1-0.15-0.15}{2}=0.35$
Row 2	2	$p_2=\left(\frac{3.5}{20}\right)\times 2 = 0.35\%$	$\frac{y_2}{D}=0.35-0.14=0.21$
Row 3	2	$p_3=\left(\frac{3.5}{20}\right)\times 2 = 0.35\%$	$\frac{y_3}{D}=0.21-0.14=0.07$
Row 4	2	$p_4=\left(\frac{3.5}{20}\right)\times 2 = 0.35\%$	$\frac{y_4}{D}=0.07-0.14=-0.07$
Row 5	2	$p_5=\left(\frac{3.5}{20}\right)\times 2 = 0.35\%$	$\frac{y_5}{D}=-0.07-0.14=-0.21$
Row 6	6	$p_6=\left(\frac{3.5}{20}\right)\times 6 = 1.05\%$	$\frac{y_6}{D}=-0.21-0.14=-0.35$

axis $\left(y_i / D\right)$. Spacing between bars $=\frac{1-d'-d'}{6-1}D=\frac{1-0.15-0.15}{6-1}D=0.14D$ (Table 15.13)

When the neutral axis is outside the sections ($x_u/D > 1$)

Case 1: When neutral axis is at infinity:

$$x_u/D=\infty$$

In this case the whole column section is subjected to pure axial load.

As per IRC:112-2020, table 6.5, for M75 grade of concrete, $\varepsilon_{c2}=0.0023$.

Strain is 0.0023 throughout the section and the whole section is in compression. Hence this is strain in steel. This is higher than yield strain.

Stress in steel, $f_{sc}=0.87\times 500=435\ \text{N/mm}^2$.

As the load is purely axial, the section is in constant stress of $0.446f_{ck}$ throughout the section. Therefore, the total compressive force in concrete is $C_c = 0.446 f_{ck} \times b \times D$.

Now, the steel bars of area $\left(\dfrac{pbD}{100}\right)$ are subjected to constant stress of f_{sc} when the strain is 0.0023 in steel.

$$C_s = \left(\frac{pbD}{100}\right) \times \left(f_{sc} - 0.446 f_{ck}\right)$$

$$P_u = C_c + C_s = \left(0.446 f_{ck} \times bD\right) + \left(\frac{pbD}{100}\right) \times \left(f_{sc} - 0.446 f_{ck}\right)$$

$$\therefore \ \frac{P_u}{f_{ck} bD} = 0.446 + \left(\frac{p}{100 f_{ck}}\right) \times \left(f_{sc} - 0.446 f_{ck}\right)$$

For the present problem $\dfrac{P_u}{f_{ck} bD} = 0.446 + [\left(\dfrac{3.5}{100 \times 75}\right) \times (435 - 0.446 \times 75)]$

$= 0.446 + 0.1874$

$= 0.6334$

and $\dfrac{M_u}{f_{ck} bD^2} = 0$

Case 2: Neutral axis depth factor, $\dfrac{x_u}{D} = 1.1$

$\varepsilon_{c,max} = \varepsilon_{cu2} - 0.75\varepsilon_{c,min}$, and $\varepsilon_{c,min}$ in the left edge, as shown in Figure 11.6.
As per IRC:112-2020, for M75 grade of concrete, ultimate strain, $\varepsilon_{cu2} = 0.0029$.
Therefore, from similar triangles one can now calculate, $\varepsilon_{c,min}$.

$$\frac{\varepsilon_{c,min}}{(1.1 - 1)D} = \frac{\varepsilon_{c,max}}{1.1D}$$

$$\text{or,} \quad \frac{\varepsilon_{c,min}}{0.1} = \frac{0.0029 - 0.75\varepsilon_{c,min}}{1.1}$$

$$\text{or,} \quad \varepsilon_{c,min} = \frac{(0.0029 \times 0.1)}{1.1 + (0.75 \times 0.1)} = 0.000246$$

Hence, $\varepsilon_{c,max} = 0.0029 - \left(0.75 \times \varepsilon_{c,min}\right) = 0.0029 - (0.75 \times 0.000246)$

or, $\varepsilon_{c,max} = 0.00271$

Again, from similar triangles, strains in concrete can be calculated at each row as $\varepsilon_{c,1}$, $\varepsilon_{c,2}$, $\varepsilon_{c,3}$, $\varepsilon_{c,4}$, $\varepsilon_{c,5}$, $\varepsilon_{c,6}$.

$$\varepsilon_{c,1} = \frac{\left[0.00271 \times (1.1 - 0.15)\right]}{1.1} = 0.00234$$

$$\varepsilon_{c,2} = \frac{[0.00271 \times (1.1 - 0.15 - 0.14)]}{1.1} = 0.00200$$

$$\varepsilon_{c,3} = \frac{[0.00271 \times (1.1 - 0.15 - 2 \times 0.14)]}{1.1} = 0.00165$$

$$\varepsilon_{c,4} = \frac{[0.00271 \times (1.1 - 0.15 - 3 \times 0.14)]}{1.1} = 0.00131$$

$$\varepsilon_{c,5} = \frac{[0.00271 \times (1.1 - 0.15 - 4 \times 0.14)]}{1.1} = 0.00096$$

$$\varepsilon_{c,6} = \frac{[0.00271 \times (1.1 - 0.15 - 5 \times 0.14)]}{1.1} = 0.00062$$

From strain compatibility, $\varepsilon_{c,1} = \varepsilon_{s,1}$, $\varepsilon_{c,2} = \varepsilon_{s,2}$, $\varepsilon_{c,3} = \varepsilon_{s,3}$, $\varepsilon_{c,4} = \varepsilon_{s,4}$, $\varepsilon_{c,5} = \varepsilon_{s,5}$, $\varepsilon_{c,6} = \varepsilon_{s,6}$.

From $\varepsilon_{s,1}$, $\varepsilon_{s,2}$, $\varepsilon_{s,3}$, $\varepsilon_{s,4}$, $\varepsilon_{s,5}$, and $\varepsilon_{s,6}$ one can calculate $f_{s1}, f_{s2}, f_{s3}, f_{s4}, f_{s5}$, and f_{s6} respectively from the stress–strain diagram.

$f_{s1} = 435$ N/mm^2
$f_{s2} = 399.14894$ N/mm^2
$f_{s3} = 330.72340$ N/mm^2
$f_{s4} = 261.617021$ N/mm^2
$f_{s5} = 192.510638$ N/mm^2
$f_{s6} = 123.404255$ N/mm^2

Also, from $\varepsilon_{c,1}$ to $\varepsilon_{c,6}$ one can calculate stress in concrete f_{c1} to f_{c6}.

$$f_{c1} = 0.446 f_{ck} = 0.446 \times 75 = 33.45 \text{ N/mm}^2$$

$$f_{c2} = 0.446 \times f_{ck} \times \left[1 - \left(1 - \frac{\varepsilon_{c,2}}{0.0023}\right)^{1.6}\right] = 32.158798 \text{ N/mm}^2$$

$$f_{c3} = 0.446 \times f_{ck} \times \left[1 - \left(1 - \frac{\varepsilon_{c,3}}{0.0023}\right)^{1.6}\right] = 29.060496 \text{ N/mm}^2$$

$$f_{c4} = 0.446 \times f_{ck} \times \left[1 - \left(1 - \frac{\varepsilon_{c,4}}{0.0023}\right)^{1.6}\right] = 24.740551 \text{ N/mm}^2$$

$$f_{c5} = 0.446 \times f_{ck} \times \left[1 - \left(1 - \frac{\varepsilon_{c,5}}{0.0023}\right)^{1.6}\right] = 19.400074 \text{ N/mm}^2$$

$$f_{c6} = 0.446 \times f_{ck} \times \left[1 - \left(1 - \frac{\varepsilon_{c,6}}{0.0023}\right)^{1.6}\right] = 13.156479 \text{ N/mm}^2$$

Calculation of area of stress block of concrete:

C_1 = Coefficient of area of the compressive stress block

C_2 = Coefficient of distance of the centroid of the compressive stress block of concrete measured from the maximum compressed edge

When, $\frac{x_u}{D} = 1.1$, C_1 and C_2 have been calculated as per the similar methodology adopted for deter mination of these coefficients in case of M20 to M60 grade of concrete.

$$C_1 = 0.353018 \text{ and } C_2 = 0.420528$$

$$\text{Now } \frac{P_u}{f_{ck}bD} = \text{C}_1 + \sum_{i=1}^{n} \frac{p_i}{100 f_{ck}} \times \left(f_{si} - f_{ci}\right)$$

$$= 0.353 + [\frac{1}{100 \times 75} \times \{1.05 \times (435 - 33.45)\}$$
$$+ \{0.35 \times (399.14894 - 32.1588)\}$$
$$+ \{0.35 \times (330.723 - 29.06)\}$$
$$+ \{0.35 \times (261.617 - 24.74)\} + \{0.35 \times (192.510 - 19.40)\}$$
$$+ \{1.05 \times (123.404 - 13.15)\}]$$

$$= 0.353 + 0.122007$$

$$= 0.475007$$

$$\frac{M_u}{f_{ck}bD^2} = C_1(0.5 - C_2) + \sum_{i=1}^{n} [\frac{p_i}{100 f_{ck}} \times \left(f_{si} - f_{ci}\right) \times y_i / D]$$

$$= 0.353 \times (0.5 - 0.4205)\}$$

$$+ \frac{1}{100 \times 75} \times [\{1.05 \times (435 - 33.45) \times 0.35\}$$
$$+ \{0.35 \times (399.14894 - 32.1588) \times 0.21\}$$
$$+ \{0.35 \times (330.723 - 29.06) \times 0.07\}$$
$$+ \{0.35 \times (261.617 - 24.74) \times (-0.07)\}$$
$$+ \{0.35 \times (192.510 - 19.40) \times (-0.21)\}$$
$$+ \{1.05 \times (123.404 - 13.15) \times (-0.35)\}]$$

$$= 0.02806 + 0.01638$$
$$= 0.044436$$

When neutral axis is inside the section

For this case, compressive stress block parameters should be calculated for the considered grade of concrete which is M75. Design parameters for all grades of concrete are already calculated and presented in Section 15.6. Calculations for M75 are shown below.

The strain and stress diagrams are shown in Figure 15.17. For M75 grade of concrete the limiting strain for the parabolic part of the stress block of concrete in compression is $\varepsilon_{c2} = 0.0023$ and that of the rectangular part is $\varepsilon_{cu2} = 0.0029$. Let x_2 denote the length of the rectangular part of the stress block and x_1 denote the length of the parabolic part of the stress block. Form the concept of similar triangles it can be stated that,

$$\frac{x_1}{0.0023} = \frac{x_u}{0.0029}$$

$$or,\ x_1 = \frac{0.0023}{0.0029}$$

$$or,\ x_1 = \frac{23}{29} x_u$$

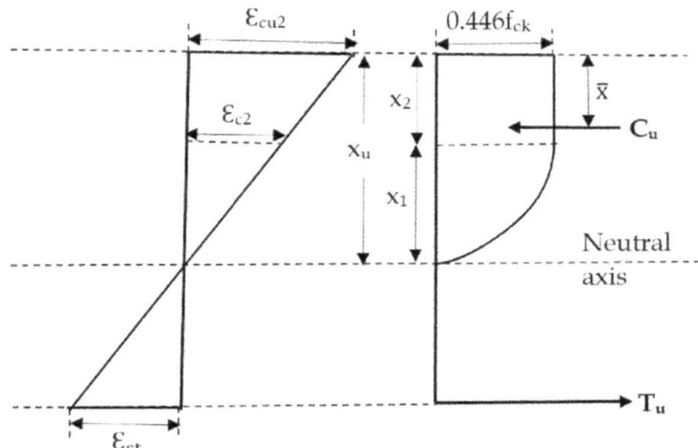

FIGURE 15.17 Strain distribution diagram and stress block.

$$\therefore\ x_2 = \frac{6}{29}x_u$$

From the stress block, one can determine the compressive force on the section for M75 grade of concrete.

Let C_u represent the compressive force. Thus, for a single lamina, the compressive force is given by:

$$C_u = 0.446f_{ck} \times \frac{6}{29}x_u + \frac{2}{3} \times 0.446f_{ck} \times \frac{23}{29}x_u = 0.3281x_u.f_{ck}$$

Thus, for a section having cross-sectional width b, the compressive force will be:

$$C_u = 0.3281f_{ck}x_u b$$

Line of action of the compressive force

The line of action of C_u is to be determined by the centroid of the stress block located at a distance $\bar{x}$ from the concrete fibre subjected to maximum compressive strain. Let C_1 and C_2 be the force acting on the parabolic and rectangular part of the compressive stress block, respectively. Force C_1, which is the force due to the parabolic part only, will be acting through the centroid of the parabola which is located at a distance of $\frac{5}{8}x_1$ from the neutral axis. Similarly, force C_2, which is the force due to the rectangular part of the stress block, will be acting through its centroid which is located at a distance of $\frac{x_2}{2}$ from the most compressed fibre.

Let us determine the line of action of compressive force for the M75 grade of concrete. Considering moments of the compressive forces C_1 and C_2 about the concrete fibre under maximum compressive strain, it can be stated that,

$$0.3281f_{ck}x_u \times \bar{x} = 0.446f_{ck} \times \frac{6}{29}x_u \times \frac{6}{58}x_u + \frac{2}{3} \times 0.446f_{ck} \times \frac{23}{29}x_u \times \left(x_u - \frac{5}{8} \times \frac{23}{29}x_u\right)$$

$$\therefore\ \bar{x} = 0.392x_u$$

Now, for M75 grade of concrete,

$$\frac{P_u}{f_{ck}bD} = 0.328 \times \left(\frac{x_u}{D}\right) + \sum_{i=1}^{n} \frac{p_i}{100 f_{ck}} \times \left(f_{si} - f_{ci}\right)$$

$$\frac{M_u}{f_{ck}bD^2} = [0.328 \times \left(\frac{x_u}{D}\right) \times \{0.5\text{-}0.392 \times \left(\frac{x_u}{D}\right)\}] + \sum_{i=1}^{n} \frac{p_i}{100 f_{ck}} \times \left(f_{si} - f_{ci}\right) \times (y_i/D)$$

Case 3: when tensile stress at the outermost layer of longitudinal steel is zero, $f_{st} = 0$

$$x_u = D - d'$$

$$\therefore \frac{x_u}{D} = 1 - 0.15 = 0.85$$

$$\text{So,} \quad \varepsilon_{c,6} = \varepsilon_{s,6} = 0$$

Strain in concrete is maximum at the right edge, $\varepsilon_{c,max} = 0.0029$.

From similar triangles, values of strain at the position of reinforcement can be calculated.

$$\frac{\varepsilon_{c,1}}{(0.85 - 0.15)D} = \frac{\varepsilon_{c,max}}{(0.85)D}$$

$$\text{Hence,} \quad \varepsilon_{c,1} = \frac{\{0.0029 \times (0.85 - 0.15)\}}{0.85} \quad \text{or,} \quad \varepsilon_{c,1} = 0.002388$$

$$\varepsilon_{c,2} = \frac{\{0.0029 \times (0.85 - 0.15 - 0.14)\}}{0.85} = 0.001910$$

$$\varepsilon_{c,3} = \frac{\{0.0029 \times (0.85 - 0.15 - 2 \times 0.14)\}}{0.85} = 0.001432$$

$$\varepsilon_{c,4} = \frac{\{0.0029 \times (0.85 - 0.15 - 3 \times 0.14)\}}{0.85} = 0.000955$$

$$\varepsilon_{c,5} = \frac{\{0.0029 \times (0.85 - 0.15 - 4 \times 0.14)\}}{0.85} = 0.000477$$

$$\varepsilon_{c,6} = \frac{\{0.0029 \times (0.85 - 0.15 - 5 \times 0.14)\}}{0.85} = 0$$

From strain compatibility, $\varepsilon_{c,1} = \varepsilon_{s,1}$, $\varepsilon_{c,2} = \varepsilon_{s,2}$, $\varepsilon_{c,3} = \varepsilon_{s,3}$, $\varepsilon_{c,4} = \varepsilon_{s,4}$, $\varepsilon_{c,5} = \varepsilon_{s,5}$, $\varepsilon_{c,6} = \varepsilon_{s,6}$.
From $\varepsilon_{s,1}$, $\varepsilon_{s,2}$, $\varepsilon_{s,3}$, $\varepsilon_{s,4}$, $\varepsilon_{s,5}$, and $\varepsilon_{s,6}$, stress in steel f_{s1}, f_{s2}, f_{s3}, f_{s4}, f_{s5}, and f_{s6} can be calculated from the stress–strain diagram of steel as per IRC:112.

$f_{s1} = 434.782608$ N/mm²
$f_{s2} = 382.117647$ N/mm²
$f_{s3} = 286.588235$ N/mm²
$f_{s4} = 191.058823$ N/mm²
$f_{s5} = 95.529411$ N/mm²
$f_{s6} = 0$

Also, one can calculate stress in concrete at the position of steel.

From $\varepsilon_{c,1}$ to $\varepsilon_{c,6}$ one can calculate stress in concrete f_{c1} to f_{c6}.

$$f_{c1} = 0.446 \times f_{ck} = 33.45 \text{ N/mm}^2$$

$$f_{c2} = 0.446 \times f_{ck} \times \left[1 - \left(1 - \frac{\varepsilon_{c,2}}{0.0023}\right)^{1.6}\right] = 31.498870 \text{ N/mm}^2$$

$$f_{c3} = 0.446 \times f_{ck} \times \left[1 - \left(1 - \frac{\varepsilon_{c,3}}{0.0023}\right)^{1.6}\right] = 26.427216 \text{ N/mm}^2$$

$$f_{c4} = 0.446 \times f_{ck} \times \left[1 - \left(1 - \frac{\varepsilon_{c,4}}{0.0023}\right)^{1.6}\right] = 19.277865 \text{ N/mm}^2$$

$$f_{c5} = 0.446 \times f_{ck} \times \left[1 - \left(1 - \frac{\varepsilon_{c,5}}{0.0023}\right)^{1.6}\right] = 10.401451 \text{ N/mm}^2$$

$$f_{c6} = 0$$

$$\text{now } \frac{P_u}{f_{ck}bD} = 0.328 \times \left(\frac{x_u}{D}\right) + \sum_{i=1}^{n} \frac{p_i}{100 f_{ck}} \times \left(f_{si} - f_{ci}\right)$$

$$= (0.328 \times 0.85) + \left(\frac{1}{100 \times 75}\right) \times [\{1.05 \times (434.782 - 33.45)\}$$
$$+ \{0.35 \times (382.117 - 31.498)\} + \{0.35 \times (286.588 - 26.427)\}$$
$$+ \{0.35 \times (191.058 - 19.277)\} + \{0.35 \times (95.529 - 10.401)\}$$
$$+ \{1.05 \times 0\}]$$
$$= 0.2788 + 0.096675$$
$$= 0.375556$$

$$\frac{M_u}{f_{ck}bD^2} = 0.328 \times \left(\frac{x_u}{D}\right) \times (0.5 - 0.391 \times \left(\frac{x_u}{D}\right)) + \sum_{i=1}^{n} \frac{p_i}{100 f_{ck}} \times \left(f_{si} - f_{ci}\right) \times \left(\frac{y_i}{D}\right)$$

$$= 0.328 \times 0.85 \times \{0.5 - (0.391 \times 0.85)\}$$
$$+ \left(\frac{1}{100 \times 75}\right) \times [\{1.05 \times (434.782 - 33.45) \times 0.35\}$$
$$+ \{0.35 \times (382.117 - 31.498) \times 0.21\}$$
$$+ \{0.35 \times (286.588 - 26.427) \times 0.07\}$$
$$+ \{0.35 \times (191.058 - 19.277) \times (-0.07)\}$$
$$+ \{0.35 \times (95.529 - 10.401) \times (-0.21) + \{1.05 \times 0 \times (-0.35)\}]$$
$$= 0.04697 + 0.02255$$
$$= 0.069175$$

Case 4: when the tensile stress in the outermost layer of longitudinal steel is $-0.4 f_{yd} = -0.4 \times 0.87 \times f_y$

Hence, stress in steel at the sixth row, $f_{s6} = (-0.4 \times 0.87 \times 500) = (-174)$ N/mm² (tensile)

From the stress–strain relation, strain at the position of longitudinal steel at the outermost edge will be,

$$\varepsilon_{s,6} = \left(-\frac{174}{200000}\right) = -0.00087 \text{ (tensile)}$$

At the right edge, strain in concrete is maximum, $\varepsilon_{c,max} = 0.0029$. From similar triangle one can calculate x_u/D.

$$\frac{0.0029}{\frac{x_u}{D}} = \frac{0.00087}{\left\{1-0.15-\left(\frac{x_u}{D}\right)\right\}}$$

$$\text{or,}\quad \frac{x_u}{D}\times(0.00087) = \{0.0029\times(1-0.15)\} - \left\{0.0029\times\left(\frac{x_u}{D}\right)\right\}$$

$$\text{or,}\ \frac{x_u}{D} = \frac{0.0029\times 0.85}{0.00087+0.0029}$$

$$\text{or,}\ \frac{x_u}{D} = 0.653846$$

Again, from similar triangles the value of strain at the position of reinforcement can be calculated.

$$\frac{\varepsilon_{c,1}}{(0.653846-0.15)D} = \frac{\varepsilon_{c,max}}{(0.653846)D}$$

$$\text{Therefore, } \varepsilon_{c,1} = \frac{\{0.0029\times(0.503846)\}}{0.653846} \quad \text{or, } \varepsilon_{c,1} = 0.002234$$

$$\varepsilon_{c,2} = \frac{\{0.0029\times(0.653846-0.15-0.14)\}}{0.653846} = 0.001613$$

$$\varepsilon_{c,3} = \frac{\{0.0029\times(0.653846-0.15-2\times0.14)\}}{0.653846} = 0.000992$$

$$\varepsilon_{c,4} = \frac{\{0.0029\times(0.653846-0.15-3\times0.14)\}}{0.653846} = 0.000371$$

$$\varepsilon_{c,5} = \frac{\{0.0029\times(0.653846-0.15-4\times0.14)\}}{0.653846} = -0.000249$$

$$\varepsilon_{c,6} = \frac{\{0.0029\times(0.653846-0.15-5\times0.14)\}}{0.653846} = -0.00087$$

From strain compatibility, $\varepsilon_{c,1} = \varepsilon_{s,1}$, $\varepsilon_{c,2} = \varepsilon_{s,2}$, $\varepsilon_{c,3} = \varepsilon_{s,3}$, $\varepsilon_{c,4} = \varepsilon_{s,4}$, $\varepsilon_{c,5} = \varepsilon_{s,5}$, $\varepsilon_{c,6} = \varepsilon_{s,6}$.

From $\varepsilon_{s,1}$, $\varepsilon_{s,2}$, $\varepsilon_{s,3}$, $\varepsilon_{s,4}$, $\varepsilon_{s,5}$, and $\varepsilon_{s,6,}$ stress in steel f_{s1}, f_{s2}, f_{s3}, f_{s4}, $f_{s5,}$ and f_{s6} can be calculated from the stress–strain diagram of steel as per IRC:112-2020.

$$f_{s1} = 434.782608 \text{ N/mm}^2$$

$$f_{s2} = 322.752941 \text{ N/mm}^2$$

$$f_{s3} = 198.564705 \text{ N/mm}^2$$

$$f_{s4} = 74.376470 \text{ N/mm}^2$$

$$f_{s5} = -49.811764 \text{ N/mm}^2$$

$$f_{s6} = -174 \text{ N/mm}^2$$

Also, one can calculate stress in concrete at the position of steel.

From $\varepsilon_{c,1}$ to $\varepsilon_{c,6}$ stress in concrete f_{c1} to f_{c6} can be calculated.

$$f_{c1} = 0.446 \times f_{ck} \times \left[1 - \left(1 - \frac{\varepsilon_{c,1}}{0.0023}\right)^{1.6}\right] = 33.337945 \text{ N/mm}^2$$

$$f_{c2} = 0.446 \times f_{ck} \times \left[1 - \left(1 - \frac{\varepsilon_{c,2}}{0.0023}\right)^{1.6}\right] = 28.619540 \text{ N/mm}^2$$

$$f_{c3} = 0.446 \times f_{ck} \times \left[1 - \left(1 - \frac{\varepsilon_{c,3}}{0.0023}\right)^{1.6}\right] = 19.905395 \text{ N/mm}^2$$

$$f_{c4} = 0.446 \times f_{ck} \times \left[1 - \left(1 - \frac{\varepsilon_{c,4}}{0.0023}\right)^{1.6}\right] = 8.224183 \text{ N/mm}^2$$

$$f_{c5} = 0$$

$$f_{c6} = 0$$

$$\begin{aligned}
\frac{P_u}{f_{ck}bD} &= 0.328 \times \left(\frac{x_u}{D}\right) + \sum_{i=1}^{n} \frac{p_i}{100 f_{ck}} \times \left(f_{si} - f_{ci}\right) \\
&= (0.328 \times 0.653846) + \left(\frac{1}{100 \times 75}\right) \times [\{1.05 \times (434.782 - 33.337)\} \\
&\quad + \{0.35 \times (322.753 - 28.619)\} \\
&\quad + \{0.35 \times (198.564 - 19.905)\} \\
&\quad + \{0.35 \times (74.376 - 8.224)\} \\
&\quad + \{0.35 \times (-49.811 - 0)\} + \{1.05 \times (-174 - 0)\}] \\
&= 0.214521 + 0.054947 \\
&= 0.269469
\end{aligned}$$

$$\begin{aligned}
\frac{M_u}{f_{ck}bD^2} &= 0.328 \times \left(\frac{x_u}{D}\right) \times (0.5 - 0.391 \times \left(\frac{x_u}{D}\right)) \\
&\quad + \sum_{i=1}^{n} \frac{p_i}{100 f_{ck}} \times \left(f_{si} - f_{ci}\right) \times (y_i / D) \\
&= [0.328 \times 0.653846 \times \{0.5 - (0.391 \times 0.653846)\}] \\
&\quad + \left(\frac{1}{100 \times 75}\right) \times [\{1.05 \times (434.782 - 33.337) \times 0.35\} \\
&\quad + \{0.35 \times (322.753 - 28.619) \times 0.21\} \\
&\quad + \{0.35 \times (198.564 - 19.905) \times 0.07\} \\
&\quad + \{0.35 \times (74.376 - 8.224) \times (-0.07)\} \\
&\quad + \{0.35 \times (-49.811 - 0)\} \times (-0.21) \\
&\quad + \{1.05 \times (-174 - 0)\} \times (-0.35)]
\end{aligned}$$

$= 0.052337 + 0.031876$
$= 0.084214$

Case 5: When the tensile stress at the outermost layer of longitudinal steel is $-0.8f_{yd} = -0.8 \times 0.87 \times f_y$

Hence, stress in steel at the sixth row, $f_{s6} = (-0.8 \times 0.87 \times 500) = (-348)$ N/mm^2 (tensile)

From the stress–strain relation, strain at the position of longitudinal steel at the outermost edge will be

$$\varepsilon_{s,6} = \left(-\frac{348}{200000}\right) = -0.00174 \text{ (tensile)}$$

At the right edge, strain in concrete is maximum, $\varepsilon_{c,max} = 0.0029$. From similar triangles we can calculate x_u/D.

$$\frac{0.0029}{\frac{x_u}{D}} = \frac{0.00174}{\left\{1 - 0.15 - \left(\frac{x_u}{D}\right)\right\}}$$

$$\text{or,}\quad \frac{x_u}{D} \times (0.00174) = \{0.0029 \times (1 - 0.15)\} - \left\{0.0029 \times \left(\frac{x_u}{D}\right)\right\}$$

$$\text{or,}\quad \frac{x_u}{D} = \frac{0.0029 \times 0.85}{0.00174 + 0.0029}$$

$$\text{or,}\quad \frac{x_u}{D} = 0.531$$

Again, from similar triangles one can get the value of strain at the position of reinforcement.

$$\frac{\varepsilon_{c,1}}{(0.531 - 0.15)D} = \frac{\varepsilon_{c,max}}{(0.531)D}$$

$$\text{Therefore,}\quad \varepsilon_{c,1} = \frac{\{0.0029 \times (0.381)\}}{0.531} \text{ or, } \varepsilon_{c,1} = 0.002081$$

$$\varepsilon_{c,2} = \frac{\{0.0029 \times (0.381 - 0.14)\}}{0.531} = 0.001316$$

$$\varepsilon_{c,3} = \frac{\{0.0029 \times (0.381 - 2 \times 0.14)\}}{0.531} = 0.000552$$

$$\varepsilon_{c,4} = \frac{\{0.0029 \times (0.381 - 3 \times 0.14)\}}{0.531} = -0.000211$$

$$\varepsilon_{c,5} = \frac{\{0.0029 \times (0.381 - 4 \times 0.14)\}}{0.531} = -0.000975$$

$$\varepsilon_{c,6} = \frac{\{0.0029 \times (0.381 - 5 \times 0.14)\}}{0.531} = -0.00174$$

From strain compatibility, $\varepsilon_{c,1} = \varepsilon_{s,1}$, $\varepsilon_{c,2} = \varepsilon_{s,2}$, $\varepsilon_{c,3} = \varepsilon_{s,3}$, $\varepsilon_{c,4} = \varepsilon_{s,4}$, $\varepsilon_{c,5} = \varepsilon_{s,5}$, $\varepsilon_{c,6} = \varepsilon_{s,6}$.

From $\varepsilon_{s,1}$, $\varepsilon_{s,2}$, $\varepsilon_{s,3}$, $\varepsilon_{s,4}$, $\varepsilon_{s,5}$, and $\varepsilon_{s,6,}$ stress in steel, f_{s1}, f_{s2}, f_{s3}, f_{s4}, f_{s5}, and f_{s6} can be calculated from the stress–strain diagram of steel as per IRC:112.

$$f_{s1} = 416.235294 \text{ N/mm}^2$$
$$f_{s2} = 263.388235 \text{ N/mm}^2$$
$$f_{s3} = 110.541176 \text{ N/mm}^2$$
$$f_{s4} = -42.305882 \text{ N/mm}^2$$
$$f_{s5} = -195.152941 \text{ N/mm}^2$$
$$f_{s6} = -348 \text{ N/mm}^2$$

Also, one can calculate stress in concrete at the position of steel.

From $\varepsilon_{c,1}$ to $\varepsilon_{c,6}$ stress in concrete f_{c1} to $f_{c,6}$ can be calculate.

$$f_{c1} = 0.446 \times f_{ck} \times \left[1 - \left(1 - \frac{\varepsilon_{c,1}}{0.0023}\right)^{1.6}\right] = 32.674142 \text{ N/mm}^2$$

$$f_{c2} = 0.446 \times f_{ck} \times \left[1 - \left(1 - \frac{\varepsilon_{c,2}}{0.0023}\right)^{1.6}\right] = 24.864634 \text{ N/mm}^2$$

$$f_{c3} = 0.446 \times f_{ck} \times \left[1 - \left(1 - \frac{\varepsilon_{c,3}}{0.0023}\right)^{1.6}\right] = 11.901491 \text{ N/mm}^2$$

$$f_{c4} = 0$$
$$f_{c5} = 0$$
$$f_{c6} = 0$$

$$\text{Now } \frac{P_u}{f_{ck} bD} = 0.328 \times \left(\frac{x_u}{D}\right) + \sum_{i=1}^{n} \frac{p_i}{100 f_{ck}} \times \left(f_{si} - f_{ci}\right)$$

$$= (0.328 \times 0.531) + \left(\frac{1}{100 \times 75}\right) \times [\{1.05 \times (416.235294 - 32.674142)\}$$
$$+ \{0.35 \times (263.388235 - 24.864634)\}$$
$$+ \{0.35 \times (110.54117 - 11.901491)\}$$
$$+ \{0.35 \times (-42.305882 - 0)\}$$
$$+ \{0.35 \times (-195.152941 - 0)\} + \{1.05 \times (-348 - 0)\}]$$
$$= 0.174298 + 0.009631$$
$$= 0.183930$$

$$\frac{M_u}{f_{ck} bD^2} = 0.328 \times \left(\frac{x_u}{D}\right) \times (0.5 - 0.391 \times \left(\frac{x_u}{D}\right)) + \sum_{i=1}^{n} \frac{p_i}{100 f_{ck}} \times \left(f_{si} - f_{ci}\right) \times (y_i / D)$$

$$= [0.328 \times 0.531 \times \{0.5 - (0.391 \times 0.531)\}]$$
$$+ \left(\frac{1}{100 \times 75}\right) \times [\{1.05 \times (416.235294 - 32.674142) \times 0.35\}$$
$$+ \{0.35 \times (263.388235 - 24.864634) \times 0.21\}$$
$$+ \{0.35 \times (110.54117 - 11.901491) \times 0.07\}$$
$$+ \{0.35 \times (-42.305882 - 0) \times (-0.07)\}$$

$$+ \{0.35 \times (-195.152941 - 0)\} \times (-0.21)$$
$$+ \{1.05 \times (-348 - 0) \times (-0.35)\}]$$
$$= 0.050891 + 0.040556$$
$$= 0.091448$$

Case 6: When the tensile stress at the outermost layer of longitudinal steel is $-f_{yd} = -0.87 \times f_y$ and strain $= -0.87f_y/E_s$ (i.e., at the yield point)

Hence, stress in steel at the sixth row, $f_{s6} = (-0.87 \times 500) = -435$ N/mm^2 (tensile)

From the stress–strain relation, strain at the position of longitudinal steel at the outermost edge will be

$$\varepsilon_{s,6} = \left(-\frac{435}{200000}\right) = -0.002175 \text{ (tensile)}$$

At the right edge, strain in concrete is maximum, $\varepsilon_{c,max} = 0.0029$. From similar triangles one can calculate $\frac{x_u}{D}$.

$$\frac{0.0029}{\frac{x_u}{D}} = \frac{0.002175}{\left\{1 - 0.15 - \left(\frac{x_u}{D}\right)\right\}}$$

$$\text{or,} \quad \frac{x_u}{D} \times (0.002175) = \{0.0029 \times (1 - 0.15)\} - \left\{0.0029 \times \left(\frac{x_u}{D}\right)\right\}$$

$$\text{or,} \quad \frac{x_u}{D} = \frac{0.0029 \times 0.85}{0.002175 + 0.0029}$$

$$\text{or,} \quad \frac{x_u}{D} = 0.485714$$

Again, from similar triangles one can get the value of strain at position of reinforcement.

$$\frac{\varepsilon_{c,1}}{(0.4857 - 0.15)D} = \frac{\varepsilon_{c,max}}{(0.4857)D}$$

$$\text{Hence, } \varepsilon_{c,1} = \frac{\{0.0029 \times (0.3357)\}}{0.4857} \text{ or, } \varepsilon_{c,1} = 0.002004$$

$$\varepsilon_{c,2} = \frac{\{0.0029 \times (0.485714 - 0.15 - 0.14)\}}{0.4857} = 0.001168$$

$$\varepsilon_{c,3} = \frac{\{0.0029 \times (0.485714 - 0.15 - 2 \times 0.14)\}}{0.4857} = 0.000332$$

$$\varepsilon_{c,4} = \frac{\{0.0029 \times (0.485714 - 0.15 - 3 \times 0.14)\}}{0.4857} = -0.000503$$

$$\varepsilon_{c,5} = \frac{\{0.0029 \times (0.485714 - 0.15 - 4 \times 0.14)\}}{0.4857} = -0.001339$$

$$\varepsilon_{c,6} = \frac{\{0.0029 \times (0.485714 - 0.15 - 5 \times 0.14)\}}{0.4857} = -0.002175$$

From strain compatibility, $\varepsilon_{c,1} = \varepsilon_{s,1}$, $\varepsilon_{c,2} = \varepsilon_{s,2}$, $\varepsilon_{c,3} = \varepsilon_{s,3}$, $\varepsilon_{c,4} = \varepsilon_{s,4}$, $\varepsilon_{c,5} = \varepsilon_{s,5}$, $\varepsilon_{c,6} = \varepsilon_{s,6}$.

From $\varepsilon_{s,1}$, $\varepsilon_{s,2}$, $\varepsilon_{s,3}$, $\varepsilon_{s,4}$, $\varepsilon_{s,5}$, and $\varepsilon_{s,6,}$ stress in steel $f_{s1}, f_{s2}, f_{s3}, f_{s4}, f_{s5}$, and f_{s6} can be calculated from the stress–strain diagram of steel as per IRC:112.

$$f_{s1} = 400.882352 \text{ N/mm}^2$$
$$f_{s2} = 233.705882 \text{ N/mm}^2$$
$$f_{s3} = 66.529411 \text{ N/mm}^2$$
$$f_{s4} = -100.647058 \text{ N/mm}^2$$
$$f_{s5} = -267.823529 \text{ N/mm}^2$$
$$f_{s6} = -435 \text{ N/mm}^2$$

Also, one can calculate stress in concrete at the position of steel.

From $\varepsilon_{c,1}$ to $\varepsilon_{c,6}$ we can calculate stress in concrete f_{c1} to f_{c6}.

$$f_{c1} = 0.446 \times f_{ck} \times \left[1-\left(1-\frac{\varepsilon_{c,1}}{0.0023}\right)^{1.6}\right] = 32.194747 \text{ N/mm}^2$$

$$f_{c2} = 0.446 \times f_{ck} \times \left[1-\left(1-\frac{\varepsilon_{c,2}}{0.0023}\right)^{1.6}\right] = 22.698704 \text{ N/mm}^2$$

$$f_{c3} = 0.446 \times f_{ck} \times \left[1-\left(1-\frac{\varepsilon_{c,3}}{0.0023}\right)^{1.6}\right] = 7.397870 \text{ N/mm}^2$$

$$f_{c4} = 0$$

$$f_{c5} = 0$$

$$f_{c6} = 0$$

$$\text{Now } \frac{P_u}{f_{ck}bD} = 0.328 \times \left(\frac{x_u}{D}\right) + \sum_{i=1}^{n} \frac{p_i}{100 f_{ck}} \times \left(f_{si} - f_{ci}\right)$$

$$= (0.328 \times 0.485714) + \left(\frac{1}{100 \times 75}\right) \times [\{1.05 \times (400.882352 - 32.19475)\}$$
$$+ \{0.35 \times (233.70588 - 22.6987)\} + \{0.35 \times (66.5294 - 7.3978)\}$$
$$+ \{0.35 \times (-100.647 - 0)\} + \{0.35 \times (-267.8235 - 0)\}$$
$$+ \{1.05 \times (-435 - 0)\}]$$
$$= 0.159358 + (-0.013872)$$

$$= 0.145486$$

$$\frac{M_u}{f_{ck}bD^2} = 0.328 \times \left(\frac{x_u}{D}\right) \times (0.5 - 0.391 \times \left(\frac{x_u}{D}\right)) + \sum_{i=1}^{n} \frac{p_i}{100 f_{ck}} \times \left(f_{si} - f_{ci}\right) \times (y_i / D)$$
$$= [0.328 \times 0.485714 \times \{0.5 - (0.391 \times 0.485714)\}]$$
$$+ \left(\frac{1}{100 \times 75}\right) \times [\{1.05 \times (400.882352 - 32.19475) \times 0.35\}$$
$$+\{0.35 \times (233.70588 - 22.6987) \times 0.21\}$$
$$+\{0.35 \times (66.5294 - 7.3978) \times 0.07\}$$

$$+ \{0.35 \times (-100.647 - 0) \times (-0.07)\}$$
$$+ \{0.35 \times (-267.8235 - 0) \times (-0.21)\}$$
$$+ \{1.05 \times (-435 - 0) \times (-0.35)\}]$$
$$= 0.049370 + 0.044595$$
$$= 0.093966$$

Case 7: When the depth of neutral axis $x_u = 0.25D$

$$\frac{x_u}{D} = 0.25$$

Again, from similar triangles one can get the value of strain at the position of reinforcement.

$$\frac{\varepsilon_{c,1}}{(0.25-0.15)D} = \frac{\varepsilon_{c,max}}{(0.25)D}$$

$$\text{Therefore, } \varepsilon_{c,1} = \frac{\{0.0029\times(0.25-0.15)\}}{0.25} \text{ or, } \varepsilon_{c,1} = 0.00116$$

$$\varepsilon_{c,2} = \frac{\{0.0029\times(0.25-0.15-0.14)\}}{0.25} = -0.000464$$

$$\varepsilon_{c,3} = \frac{\{0.0029\times(0.25-0.15-2\times0.14)\}}{0.25} = -0.002088$$

$$\varepsilon_{c,4} = \frac{\{0.0029\times(0.25-0.15-3\times0.14)\}}{0.25} = -0.003712$$

$$\varepsilon_{c,5} = \frac{\{0.0029\times(0.25-0.15-4\times0.14)\}}{0.25} = -0.005336$$

$$\varepsilon_{c,6} = \frac{\{0.0029\times(0.25-0.15-5\times0.14)\}}{0.25} = -0.00696$$

From strain compatibility, $\varepsilon_{c,1} = \varepsilon_{s,1}$, $\varepsilon_{c,2} = \varepsilon_{s,2}$, $\varepsilon_{c,3} = \varepsilon_{s,3}$, $\varepsilon_{c,4} = \varepsilon_{s,4}$, $\varepsilon_{c,5} = \varepsilon_{s,5}$, $\varepsilon_{c,6} = \varepsilon_{s,6}$.

From $\varepsilon_{s,1}$, $\varepsilon_{s,2}$, $\varepsilon_{s,3}$, $\varepsilon_{s,4}$, $\varepsilon_{s,5}$, and $\varepsilon_{s,6}$, stress in steel f_{s1}, f_{s2}, f_{s3}, f_{s4}, f_{s5}, and f_{s6} can be calculated from the stress–strain diagram of steel as per IRC:112.

$$f_{s1} = 232 \text{ N/mm}^2$$
$$f_{s2} = -92.8 \text{ N/mm}^2$$
$$f_{s3} = -417.6 \text{ N/mm}^2$$
$$f_{s4} = -435 \text{ N/mm}^2$$
$$f_{s5} = -435 \text{ N/mm}^2$$
$$f_{s6} = -435 \text{ N/mm}^2$$

Also, one can calculate stress in concrete at the position of steel.

From $\varepsilon_{c,1}$ to $\varepsilon_{c,6}$ stress in concrete f_{c1} to f_{c6} can be calculated.

$$f_{c1} = 0.446 \times f_{ck} \times \left[1-\left(1-\frac{\varepsilon_{c,1}}{0.0023}\right)^{1.6}\right] = 22.568736 \text{ N/mm}^2$$

$$f_{c2} = 0$$
$$f_{c3} = 0$$
$$f_{c4} = 0$$
$$f_{c5} = 0$$
$$f_{c6} = 0$$

$$\begin{aligned}\frac{P_u}{f_{ck}bD} &= 0.328 \times \left(\frac{x_u}{D}\right) + \sum_{i=1}^{n} \frac{p_i}{100 f_{ck}} \times \left(f_{si} - f_{ci}\right) \\ &= (0.328 \times 0.25) + \left(\frac{1}{100 \times 75}\right) \times [\{1.05 \times (232 - 22.56873)\} \\ &\quad + \{0.35 \times (-92.8 - 0)\} + \{0.35 \times (-417.6 - 0)\} \\ &\quad + \{0.35 \times (-435 - 0)\} + \{0.35 \times (-435 - 0)\} + \{1.05 \times (-435 - 0)\}] \\ &= 0.082022 + (-0.095947) \\ &= -0.013924\end{aligned}$$

$$\begin{aligned}\frac{M_u}{f_{ck}bD^2} &= 0.328 \times \left(\frac{x_u}{D}\right) \times (0.5\text{-}0.391 \times \left(\frac{x_u}{D}\right)) + \sum_{i=1}^{n} \frac{p_i}{100 f_{ck}} \times \left(f_{si} - f_{ci}\right) \times (y_i / D) \\ &= [0.328 \times 0.25 \times \{0.5 - (0.391 \times 0.25)\}] \\ &\quad + \left(\frac{1}{100 \times 75}\right) \times [\{1.05(232 - 22.56873) \times 0.35\} \\ &\quad + \{0.35 \times (-92.8 - 0) \times 0.21\} + \{0.35 \times (-417.6 - 0) \times 0.07\} \\ &\quad + \{0.35 \times (-435 - 0) \times (-0.07)\} + \{0.35 \times (-435 - 0) \times (-0.21)\} \\ &\quad + \{1.05 \times (-435 - 0) \times (-0.35)\}] \\ &= 0.034974 + 0.032982 \\ &= 0.067956\end{aligned}$$

Case 8: When a column behaves like a steel beam

This is the specific situation when the section has only moment capacity with zero axial load capacity.

The column has symmetrical longitudinal steel on both sides of the centroidal axis. Therefore, the column will have pure moment capacity due to yielding of both tensile and compressive steel bars.

$$\text{Therefore, } M_u = \sum_{i=1}^{n} \left(\frac{p_i bD}{100}\right)\left(0.87 f_y\right)\left(\frac{y_i}{D}\right)$$

$$\begin{aligned}\frac{M_u}{f_{ck}bD^2} &= \sum_{i=1}^{n} \left(\frac{p_i}{100 f_{ck}}\right)\left(0.87 f_y\right)\left(\frac{y_i}{D}\right) \\ &= \frac{1}{100} \times (0.87 \times 500) \times [\{\frac{1.05}{75} \times 0.35\} \\ &\quad + \{\frac{0.35}{75} \times 0.21\} + \{\frac{0.35}{75} \times 0.07\} + \{\frac{0.35}{75} \times 0.07\} \\ &\quad + \{\frac{0.35}{75} \times 0.21\} + \{\frac{1.05}{75} \times 0.35\}] \\ &= 0.053998\end{aligned}$$

$$\frac{P_u}{f_{ck}bD} = 0$$

Preparation of graphical chart

Hence these eight pairs of $\dfrac{P_u}{f_{ck}bD}$ and $\dfrac{M_u}{f_{ck}bD^2}$ are now plotted to obtain the safety profile of this section (Table 15.14).

These eight coordinates are now plotted in a graph to prepare a graphical chart as shown in Figure 15.18.

TABLE 15.14
Values of nondimensional parameters generated from different conditions

Cases	Illustrations of case	$\dfrac{P_u}{f_{ck}bD}$	$\dfrac{M_u}{f_{ck}bD^2}$
Case 1	When neutral axis is at infinity	0.6334	0
Case 2	When neutral axis is outside the sections ($x_u/D > 1$), i.e., ($x_u/D = 1.1$)	0.475007	0.04443
Case 3	When tensile stress at longitudinal steel is zero, $f_{st} = 0$	0.375556	0.069175
Case 4	When the tensile stress in longitudinal steel is $-0.4f_{yd} = -0.4 \times 0.87 \times f_y$	0.269469	0.08421
Case 5	When the tensile stress in longitudinal steel is $-0.8f_{yd} = -0.8 \times 0.87 \times f_y$	0.18393	0.091448
Case 6	When the tensile stress in longitudinal steel is $-f_{yd} = -0.87 \times f_y$ and strain $= -0.87f_y/E_s$	0.145486	0.093966
Case 7	When the depth of neutral axis $x_u = 0.25/D$	−0.013924	0.067956
Case 8	When column behaves like a steel beam	0	0.053998

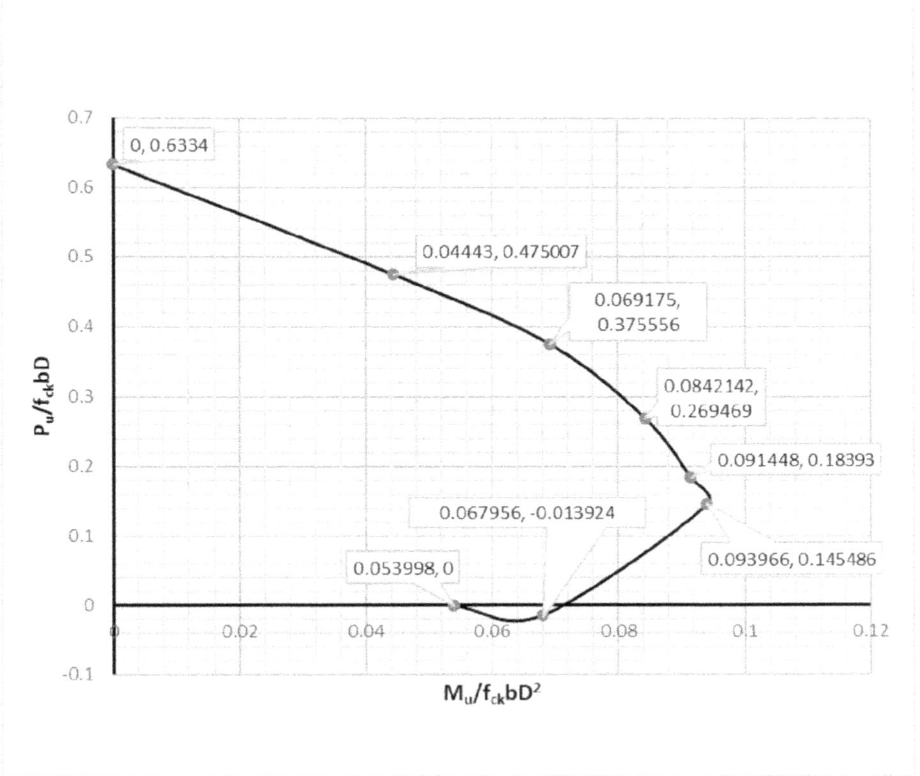

FIGURE 15.18 Safety profile corresponding to Problem 15.2.

15.15 SAMPLE DESIGN AIDS FOR COMPRESSION MEMBERS BEYOND M60 GRADE OF CONCRETE

As discussed in Problem 15.2 a number of safety profiles can be developed in a similar manner for M75 grade concrete and the *d'/D* ratio is equal to 0.15 using different percentages of longitudinal reinforcement. These safety profiles when plotted in the same diagram will lead to the force–moment interaction chart valid for the grade of concrete and steel considered and *d'/D* value adopted. This has been shown in Chart 15.3. For grades of concrete beyond M60, since the stress–strain diagrams of concrete are unique for each concrete grade, interaction charts are to be developed for different combinations of concrete and steel and different *d'/D* ratios. Hence for six grades of concrete ranging from M65 to M90, using four grades of steel, namely Fe 415 to Fe 600 and four values of *d'/D* ratios ranging from 0.05 to 0.2 a total of 96 interaction charts (for each set of distribution of longitudinal reinforcements) are necessary. However, in this chapter a typical interaction chart using M75

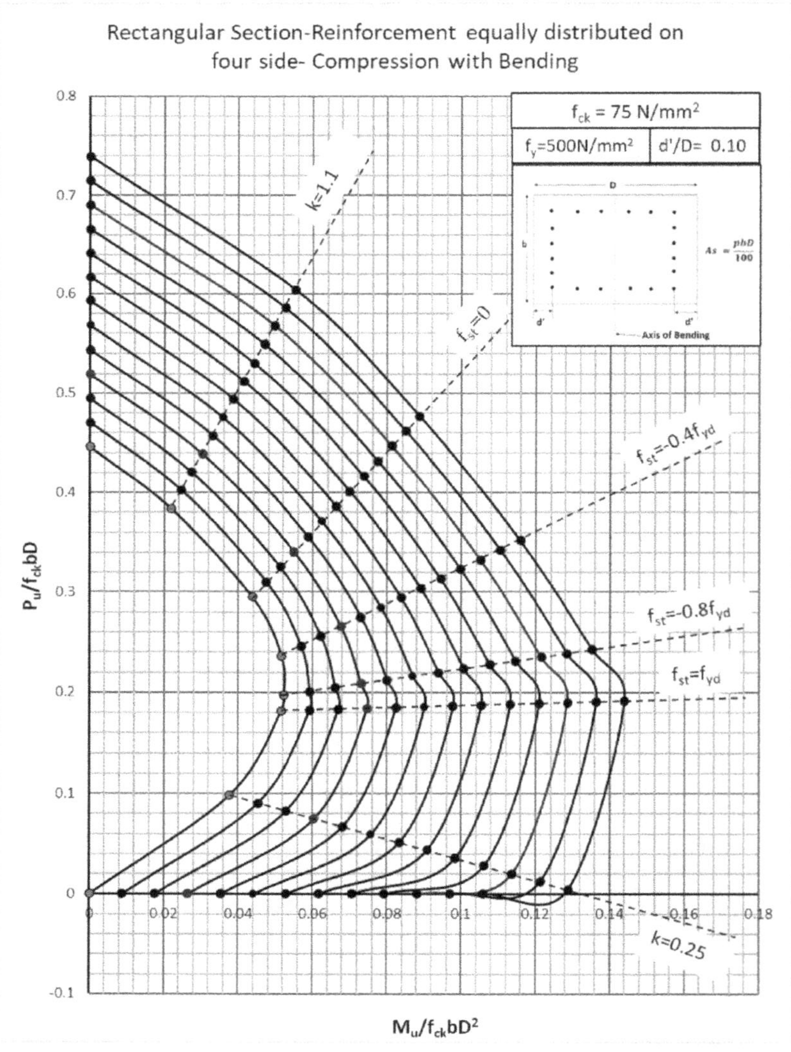

CHART 15.3 Compression with bending – rectangular section reinforcement equally distributed on four sides.

grade concrete, Fe 500 grade steel, d'/D equal to 0.1 with reinforcement distributed uniformly along the four sides has been presented in Chart 15.3.

15.16 SUMMARY

The present version of IS:456 has been taken up for revision and a revised version is expected soon. In the revised version, significant modifications to the material properties and specifications are expected to be enforced as a result of which the design considerations will undergo significant modifications. These modifications are expected to be in line with the relevant Indian and international standards such as IRC:112-2020 and Eurocode 2. In this chapter the modified material properties and assumptions of design have been discussed as per the requirements of IRC:112-2020 indicating the related requirements of Eurocode 2. The methodology of flexure design and interaction charts for compression have been developed in this chapter along with the basic requirements of stress and strain distributions across the sections. Typical interaction chart for higher grade of concrete also have been presented in this chapter.

BIBLIOGRAPHY

Bandyopadhyay, J. N. (2011). *Design of Concrete Structures*, PHI Learning, New Delhi.

Bhatt, P., MacGinley, T. J. and Choo, B. S. (2005). *Reinforced Concrete Design: Design Theory and Examples.* CRC Press, London.

EN 1992-1-1. (2004). *Eurocode 2: Design of Concrete Structures Part 1-1: General Rules and Rules for Buildings*, European Committee For Standardization, Brussels.

Indian Standard IS: 456-2000 (2017). *Plain and Reinforced Concrete–Code of Practice,* Fourth Revision, Tenth Reprint April 2007, Bureau of Indian Standards, New Delhi.

IRC:112-2020 (July, 2020). *Code of Practice for Concrete Road Bridges*, First Revision, Indian Road Congress, New Delhi.

Park, R. and Paulay, T. (1975). 'Ductile reinforced concrete frames some comments on the special provisions for seismic design of ACI 318-71 And on capacity design', *Bulletin of the New Zealand National Society for Earthquake Engineering*, Vol. 8, No. 1, March.

Paulay, T. and Priestley, M. N. (1992). *Seismic Design of Reinforced Concrete and Masonry Buildings* (Vol. 768), Wiley, New York.

SP 16-1980. (1980). *Design Aids for Reinforced Concrete to IS 456:1978*, Special Publication SP 16, Bureau of Indian Standards, New Delhi.

16 Design beyond Limit State

16.1 INTRODUCTION

The standards or codes of practice are based on prescriptive methods of design and deal with design loads for both strength and serviceability design. The design methodologies are valid up to design loads. Once the design load is reached, the section is considered to have failed and the design process comes to an end.

16.2 LIMIT STATE CORRESPONDS TO FAILURE UNDER DESIGN LOAD

Reinforced concrete design is generally performed by the limit state method. Sections are to be checked for the limit state of collapse and limit state of serviceability. The design loads for the limit state of collapse and limit state of serviceability are ultimate and service loads, respectively. When the design loads are reached, the sections are considered to have failed. Load, or to be more specific design load, is the criterion for design. Hence the design method is based on the design load.

16.3 OVERLOADING BEYOND DESIGN LOADS

If the design loads are exceeded, the sections are considered to be subjected to overloading. Under such conditions the design methodology fails as it is based on design loads and is valid up to the design load only. The behaviour of a structural element under loads is mostly governed by the stress–strain relationship. For design, an idealized stress–strain relationship is considered using some safety factors and a portion of the stress–strain relationship is adopted. For RC design, the stress–strain relationships of both concrete and steel are used. Concrete, being a brittle material, fails at a low value of total strain, whereas steel, being ductile, has a total strain at failure that is quite high. In the limit state method, strain is the criterion for failure. In order to ensure ductile failure, initiation of failure should occur in steel and designers should always aim to adopt under-reinforced sections. In the limit state method, a section fails in two stages: yielding occurs in Stage I and failure occurs in Stage II. The more the section is under-reinforced, the more will be the gap between Stage I and Stage II and the more will be the ductility of the section. This failure does not mean collapse of the section, instead it signifies that the section has reached the predetermined failure criteria. This is the condition under design load.

DOI: 10.1201/9781003415398-16

16.4 LIMIT STATE METHOD FAILS BEYOND DESIGN LOADS

Designers need to appreciate the fact that the sections are inert and cannot understand how they have been designed. It will behave as per its own natural characteristics. It is the duty of the designer to understand the natural behaviour of the sections and to try to interpret it following a design principle. However, any design principle will have its own limitations and cannot be expected to be valid over the entire load–deformation behaviour right from initiation of load to its actual collapse. The prescriptive methods of design as per the standards or codes of practice are based on loads and are valid up to design loads.

16.5 BEHAVIOUR OF STRUCTURAL ELEMENTS BEYOND LIMIT STATE, I.E., UNDER OVERLOADING

If the section is under-reinforced, failure will be ductile, and if the section is over-reinforced, failure will be brittle. Under both conditions failure will occur by crushing of concrete. Failure occurs at the limiting load when the strain in concrete reaches the limiting strain. Actual failure strain of steel is about two orders of magnitude higher than its yield strain, hence in reinforced concrete, failure generally occurs in flexure by crushing of concrete. Thus, under overloading strain in concrete will always exceed the limiting strain. Concrete actually fails at a strain varying from 0.003 to 0.005 and depends on the grade of concrete. IS:456 considers a conservative value of 0.0035 which is considered as the criterion of failure. This strain value is for unconfined concrete. For confined concrete, the failure strain will increase significantly. For structural concrete, transverse reinforcements provide confinement of concrete and hence result in higher failure strain values. The amount of increase will depend on a number of parameters such as the grade of concrete, level of confinement, etc. Hence there is a gap between code-specified failure and actual collapse and under loading the section will move from the designated failure to actual collapse (refer to Figures 16.1–16.3).

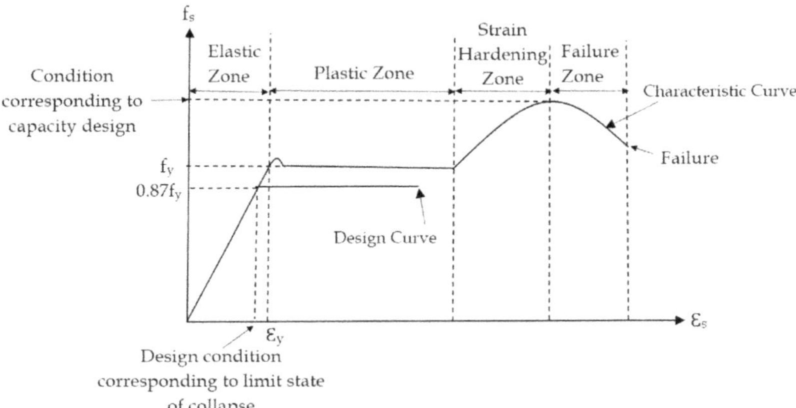

FIGURE 16.1 Stress–strain diagram for steel showing the conditions corresponding to the limit state of collapse (failure initiation stage in steel) and the maximum capacity of the section.

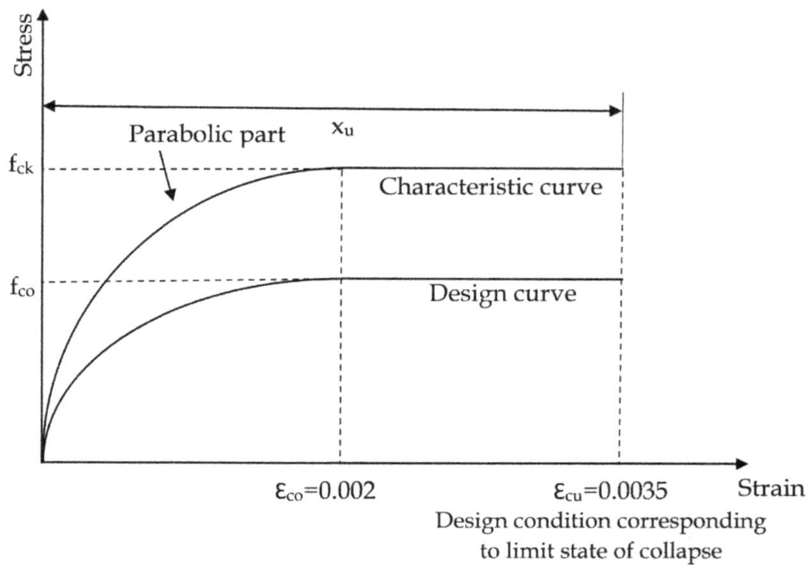

FIGURE 16.2 Stress–strain diagram for concrete showing the condition corresponding to the limit state of collapse.

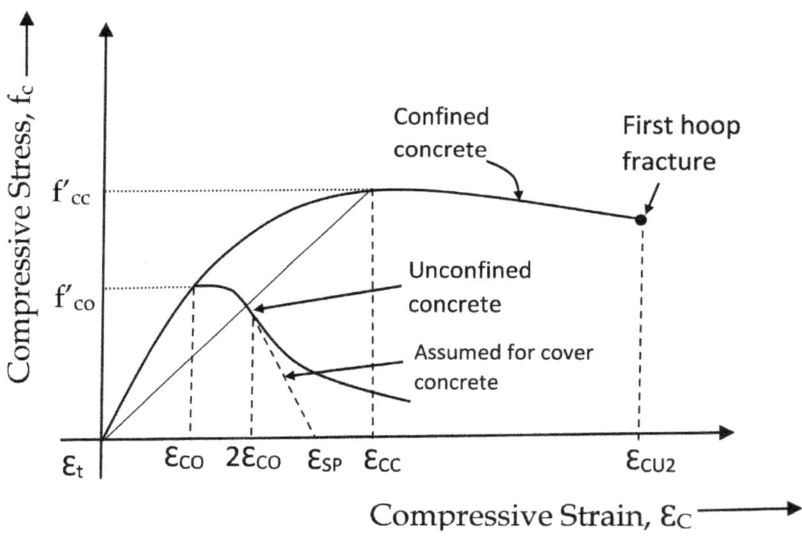

FIGURE 16.3 Stress–strain diagram of confined concrete along with that for unconfined concrete.

16.6 LOAD-BASED DESIGN FAILURES

The range between design load and collapse load is beyond the scope of load-based design. For gravity loads like dead load, imposed load and lateral loads like wind load it is expected that the design loads have been estimated with considerable accuracy and they are not expected to be exceeded within the design life of the structure. Hence the prescriptive method of design is well suited for design under dead, imposed and wind loads whereby the chances of exceeding design loads are minimal. As per IS:456-2000, characteristic load is that load which has a 95% probability of not being exceeded during the life of the structure. Design load is obtained from the characteristic load by multiplying with the partial safety factor appropriate to the nature of loading and the limit state being considered. However, the codes dealing with earthquake-resistant design categorically mention that under actual seismic conditions the earthquake forces can be much higher than the design forces and the chances of overloading are quite possible. Hence, for seismic forces, load-based design fails.

16.7 DISPLACEMENT-BASED DESIGN

In force-based systems due to the application of loads, strains and stresses develop in the elements. Deformations or displacements occur because of loads. Hence, load or force is the cause and displacement is the effect. In a displacement-based system, displacement is the cause and force is the effect. Dead, imposed and wind loads are load-based systems, whereas earthquake load is a displacement-based system. During earthquakes, ground vibrations result in vibrations of the superstructure whereby displacements, velocities and accelerations are induced in the structure and because of which forces are generated as per Newton's laws of motion. In a load-based system, when the failure strain is reached, the design load or limiting load is attained.

In displacement-based design, different levels of displacements are defined, and these levels are related to the different conditions or performance levels of the structure. Hence displacement-based design deals with different levels of displacements and corresponding conditions of the structure at those levels.

In load-based design, only the failure load is specified with respect to certain predefined failure strain. The process ends over that point and no further information about the actual condition of the structure under such strain is provided. Obviously, a failure condition is not an actual collapse but no further information about the actual behaviour or condition is available. However, in displacement-based design, the strain or displacement levels at different stages are defined with respect to different damage levels. These damage states provide more information about the structural behaviour or condition of the structure at that level.

In displacement-based design, displacements result in the generation/development of forces, which is the case in seismic design.

Damage is better correlated to displacements rather than forces. In force-based methods, displacements are treated in a somewhat cursory manner and are checked only in serviceability design and not in strength design. It is expected that if the displacement levels satisfy the serviceability requirements, then the structural element will be safe in strength design also. Hence, checking of displacements or the condition of a structure are not performed in strength design.

16.8 DUCTILITY DESIGN

Both methods of structural design for RC sections as prescribed by IS:456-2000, i.e., the working stress method and limit state method, are performed for strength and serviceability. Strength design ensures safety against failure and serviceability design requires checking for deflections, crack width, durability requirements, etc. so that the sections are properly functional and fit for use. Both of these designs are based on loads, and it may be clearly stated that the prescriptive methods of design are load-based, or to be more precise, design load-based. It is well accepted that the actual seismic forces can be much higher than the design seismic forces but if the structures are to be designed for the actual forces their cost would be prohibitive. Hence, the design for seismic forces is performed for the design loads only; keeping some alternative arrangements in the structure so that even if the actual forces exceed the design forces damage or distress can occur in the structural elements but actual collapse will be prevented so that the safety of life and property can be ensured. Here appears the role of ductility design, whereby a portion of a significant amount of energy imparted on the structure during earthquakes can be dissipated in the form of a significant amount of work done by the structure due to inelastic deformations. The basic definition of ductility can be mentioned as the ratio of ultimate deformation to yield deformation, which signifies that for incorporating ductility into elements they must cross the yield point. This is not possible in the working stress method of design as the concept of yielding is totally absent and single-stage failure occurs with the attainment of permissible stresses either in concrete or steel, or both simultaneously. However, in the limit state method of design, yielding is a pre-requirement of failure. Here failure occurs in multiple stages, the first stage is the yield stage (Stage I) and the second stage is the ultimate stage (Stage II), where the predetermined failure criteria are reached. For steel, yielding can be very well conceived, however for concrete it becomes a bit more difficult to understand. One can relate yielding to an increase in strain without an increase in stress. It is observed from the stress–strain diagram of concrete that, beyond the strain level of 0.002, an increase in strain occurs up to a limiting value of 0.0035 without an increase in stress. The strain at which actual snapping of steel occurs may be about two orders of magnitude higher than the yield strain. However, for concrete the difference between the failure strain and yield strain is very low. Thus, the designer should always aim at initiating failure in steel so that the ductility can be significantly high.

As per IS:456-2000 the failure strain in concrete in axial compression is 0.002, which may be considered as the yield point of concrete, hence for pure axial compression failure is a single stage one. Both stages, i.e., Stage I and Stage II, are reached simultaneously. Hence ductility cannot be incorporated into the section and the failure will be a brittle one. However, for all other cases such as flexure only or flexure with compression, the failure will occur in two stages allowing the designer to incorporate ductility into the sections. With the incorporation of axial compressive forces, ductility will be reduced. Thus, beams will have greater ductility than columns. The concept of ductility has been discussed in detail in Chapter 17.

16.9 INTRODUCTION TO PERFORMANCE-BASED DESIGN (PBD)

In the prescriptive method of design as specified in the standards or codes of practice, the design methodology is prescribed, and the designer has to follow those guidelines strictly. The entire design process is almost completely defined, and the designer has to follow the process rigorously with little freedom to exercise in the design process. In performance-based design, the design process is not shackled by mere do's and don'ts and the designer can incorporate his or her innovations. On the other hand, performance-based design specifies the performance that is expected of the given structural element as well as the structural system. This method can be used for the design of a new structure as well as for rehabilitation of an old one. New building codes have detailed stipulations for PBD such as EN codes and the Japan Code on concrete structures. The PBD concept makes use of demand and capacity of various structural elements. Initially, the basic design of the structural elements is performed as per prescriptive load-based design. The next level of design is capacity design, whereby, based on the section and reinforcement provided, the designer determines which elements will yield earlier and which will yield later. PBD is used for seismic design and verifies the performance of the structure for various design limit states. A major requirement of PBD is to determine the seismic demand and corresponding capacity of the structure. In PBD, different performance-based limit states are defined with corresponding expected damage levels. Design is performed to ensure that the desired performance-based limit state is reached at the expected seismic demand of that area. PBD is a process whereby the designer can tailor-make the design to suit the seismic demand and financial capacity of the owners. The target performance level is divided into two parts, non-structural damage and structural damage, and the combination of these gives the combined performance level. Seismic performance is defined by designating the maximum allowable damage state

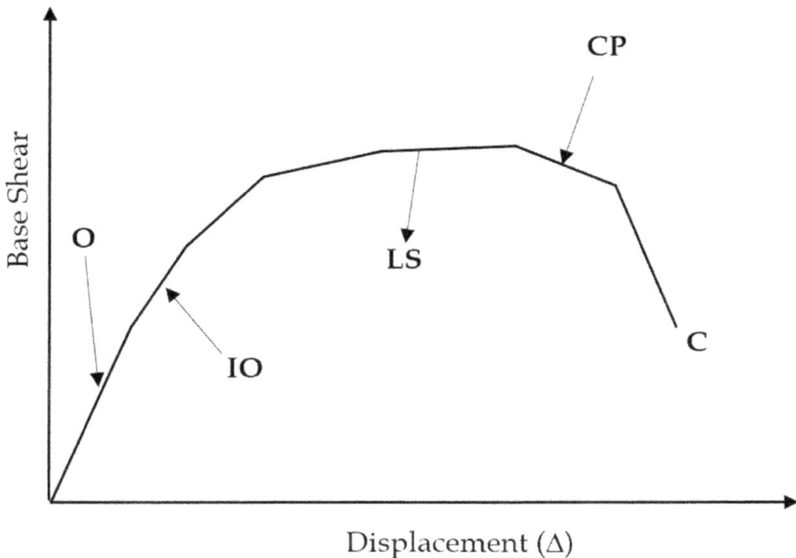

FIGURE 16.4 Different performance limits.

(performance level) for an identified seismic exposure (earthquake ground motion). The damage states or performance levels can be termed as operational (O), immediate occupancy (IO), life safety (LS) and collapse prevention (CP). Finally, the structure will collapse (C). These performance levels are shown in Figure 16.4. Performance objectives are defined in different hazard levels in seismic documents such as ATC40 and FEMA356. Both ATC 40 and FEMA 356 prescribe a basic objective with respect to the performance level and expected hazard levels and include the details and acceptance criteria. The goal of PBD is to design a structure which will reach a target displacement profile, i.e., response, when subjected to a specified earthquake motion, i.e., demand. The performance levels of a structure are governed by suitable values of maximum displacements and maximum inter-storey drift.

16.10 SUMMARY

From this chapter the following conclusions can be drawn.

The standards or codes of practice deal with prescriptive methods of design based on loads. Load-based designs are valid up to design loads.

For dead, imposed and wind loads, load-based designs are quite acceptable and result in safe and serviceable design. For seismic forces, exceeding of design loads is quite common and hence load-based design fails and displacement-based design is adopted.

Under seismic conditions, forces are not actually present, they are induced or developed in structures due to vibrations. Because of vibrations, displacements, velocities and accelerations are induced which result in the development of inertial forces. Hence this is a displacement-based system and not a force-based one.

For the design of earthquake-resistant structures, performance-based design is adopted as force-based design fails.

The aim of performance-based design is to design a structure which will reach a target displacement profile when subjected to a specified earthquake motion.

BIBLIOGRAPHY

ATC-40. (1996). *Seismic Evaluation and Retrofit of Concrete Buildings*, Volume 1, Applied Technology Council, California.

Bakhaswala, Harsh, Vasu Shah, Anuj Chandiwala, and Dimple Desai, (2017). 'Capacity based earthquake resistant design of multi-storied R.C. building: A review', *International Journal of Advance Engineering and Research Development*, Vol. 4, No. 11, pp. 327–333.

Choudhari, D. J. and Dhoot, G. O. (2016), 'Performance based seismic design of reinforced concrete building', *Open Journal of Civil Engineering*, Vol. 6, pp. 188–194.

FEMA 356. (2000). *Pre-standard and Commentary for the Seismic Rehabilitation of Buildings*, Federal Emergency Management Agency, Washington DC.

Jha, B. and Bhanja, S. (2021). Optimum design of R.C. Section and shortcomings of prescriptive method of design', *Asian Journal of Civil Engineering*, Springer, Vol. 22, No. 4, pp. 769–787. Doi: https://doi.org/10.1007/s42107-020-00346-9

Paulay, T. and Priestley, M. J. N. (2018). *Seismic Design of Reinforced Concrete and Masonry Buildings*, John Wiley, New York.

Priestley, M. J. N, Calvi, G. M. and Kowalsky. (2007). *Displacement Based Seismic Design of Structures*, IUSS Press, Italy.

Vinod, Hosur. (2013). *Earthquake-Resistant Design of Building Structures*, Wiley, New Delhi.

17 Ductility Design

17.1 INTRODUCTION

Design of reinforced concrete sections is performed by the limit state method throughout the world and in India also the basic design code advocates the limit state method although the working stress method is still allowed for structures where the limit state method cannot be conveniently applied. A structural element is expected to reach the limit state only if the design load is reached. The aim of the limit state method of design is that the structure should not reach the impending state of failure, i.e., the limit state. The code-based prescriptive method of design deals with strength and serviceability and is based on load design only. Once an element reaches the limit state it is considered to have failed. Failure does not mean collapse of the structure, it only indicates that the predetermined state defining failure has been reached.

17.2 NEED FOR DUCTILITY DESIGN

The load-based prescriptive method of design seems to be adequate for all types of loads except seismic loads. The codes dealing with seismic forces categorically mention that the actual seismic forces can be much higher than the design seismic forces. Structures under design loads, other than seismic effects, are not pushed into their ultimate states. However, under seismic effects they may be pushed up to and beyond their ultimate states. Hence, to prevent collapse, energy dissipation should occur through large inelastic deformations and therefore adequate ductility is essential to avoid brittle failure modes. This leads the designer to a very difficult situation as our basic code based on prescriptive design can only ensure safety of the sections up to and at the limiting states. However, performance of the structure or assessment of its state or condition is never checked at the ultimate state or collapse level and is only checked at the limit state of serviceability. It is expected or implied that if a structure satisfies the serviceability requirements at working loads/service level loads, its condition at ultimate or collapse level need not be checked and may be considered to be safe. The standards or codes of practice deal with load-based design and are silent about the behaviour of sections beyond the limit state, whereas there is every possibility of the sections crossing limit states under seismic forces. The designer needs to understand clearly that failure and actual collapse are not the same. A section will collapse only when all its reserved strength has been mobilized. Concrete is a brittle material and reinforced concrete elements are mostly massive and robust, whose failure may result in huge loss of life and property. Hence, a great deal of uncertainly looms over the designer as he or she has no clue how the elements will behave beyond the limit state and up to the actual collapse level. During severe earthquakes there is every possibility that the limit state may be exceeded. Earthquake-resistant design of reinforced concrete structures is gaining increased

DOI: 10.1201/9781003415398-17

importance and has immense prospects in the future to emanate as an offshoot of normal reinforced concrete design. However, the limit state may also be exceeded due to a significant amount of plastic rotation of sections for accommodating a huge amount of yielding of the tension reinforcement required for highly under-reinforced sections. However, unfortunately, this provision is generally never considered as a criterion for failure. In the standards, the maximum permitted strains in steel or the maximum plastic rotations of reinforced concrete sections are not provided and the designers are unable to determine whether the huge amounts of strains or plastic rotations in beams will lead to partial or complete collapse of the sections. In India, reinforced concrete structures are designed and detailed as per IS:456-2000. However, structures located in high seismic regions require ductile design and detailing. The standard entitled "Ductile Design and Detailing of Reinforced Concrete Structures Subjected to Seismic Forces – Code of Practice" (First Revision) – IS 13920:2016 covers the requirements for design and detailing of reinforced concrete structures for earthquake resistance so as to give them adequate stiffness, strength and ductility. It clearly mentions that the provisions of this standard are specifically applied to monolithic reinforced concrete construction and should be mandatorily adopted to all lateral load resisting systems of reinforced concrete structures located in all seismic zones other than zone II, where it is optional. Designers should keep in mind that reinforced concrete can be made ductile only in flexure, and hence flexural failure should occur before any other mode of failure.

17.3 SIGNIFICANCE OF DUCTILITY

Ductility plays a significant role in statically indeterminate structures by allowing redistribution of overstresses from one critical section to another, thus delaying local failure. Ductility demand is also very high for beams designed for nominal bending moment requirement with a minimum or very low amount of tension reinforcement. In the philosophy of limit state design of reinforced concrete structures, the concept of ductility is seamlessly incorporated in the flexure design by ensuring that at failure steel should reach at least the yield strain, i.e., only under-reinforced and balanced sections are permitted for design in flexure. Although over-reinforced sections are prohibited, the Indian Code of Practice does not emphasize the fact that under-reinforced sections are much more desirable than balanced sections. Only regarding redistribution of moments some restriction on the depth of neutral axis has been imposed by the standard. Adequate ductility must be built into the structure and a hierarchy must be formed amongst the different failure modes by virtue of which the ductile failure modes occur before the brittle ones. However, in many of the international standards, balanced section design is not permitted and designers have to mandatorily opt for under-reinforced sections. In this chapter ductility and plastic rotations of flexural sections for the two extreme or limiting cases, i.e., balanced and maximum under-reinforced sections for different combinations of concrete and steel will be evaluated. These will be initially performed for singly reinforced sections and then the values will be evaluated for doubly reinforced sections also. Determination of ductility and plastic rotations of under-reinforced sections is quite simple but evaluation of these values for balanced sections is not commonly available, and these values have been derived from fundamental principles. These values can provide necessary information in quantitative terms to the designers regarding the maximum and minimum values of ductility and plastic rotations. Higher ductility also requires higher plastic rotations, which might be difficult to be mobilized especially for deeper sections. Ductility and plastic rotation values of maximum under-reinforced sections are significantly higher than their balanced counterparts.

17.4 ROLE OF DUCTILITY IN DESIGN

The present standards do not provide any kind of guidelines regarding the desirable values of ductility and plastic rotations which are adequate for reinforced concrete structures. The prescriptive method of design does not provide any information about the performance/deformation of the section at the limit state of collapse level. Even within the domain of the limit state method of design very low values of tension reinforcement can lead to very high values of plastic rotations that can result in failure of the section. The minimum percentage of tension reinforcement is generally governed by the cracking moment consideration of the section. For performance assessment of a section up to the actual collapse or failure level, performance-based design is to be referred. However, prescriptive and performance-based methods are two distinctly different methods having different perspectives. Therefore, a designer dealing with the design of earthquake-resistant structures should compare some important features of these two methods and study their implications in design.

17.5 MOMENT CURVATURE RELATIONSHIP IN RC MEMBERS

The theory of pure bending states

$$\frac{f}{y}=\frac{M}{I}=\frac{E}{R} \tag{17.1}$$

This is a well-known flexure formula and is also referred to as Bernoulli–Euler bending equation. From the equation

$$\frac{f}{y}=\frac{E}{R}$$

$$\text{Or,}\quad \frac{1}{R}=\frac{f}{yE}$$

$$\text{Or,}\quad \frac{1}{R}=\frac{\varepsilon}{y}$$

where f is the bending stress, E is the Young's modulus of elasticity and $1/R$ or φ is called the curvature of the beam and is given by:

$$\varphi=\frac{\varepsilon}{y} \tag{17.2}$$

$$=\frac{\text{strain in the fibre under consideration}}{\text{distance of the fibre under consideration from neutral axis}}$$

$$\text{Again,}\quad \varphi=\frac{M}{EI}=\frac{\text{Moment of Resistance}}{\text{Flexural Rigidity}}$$

Elaborate analysis shows that normal stresses due to bending are not altered due to the presence of shear stresses in general. Thus, it is justifiable to use the theory of pure bending even if the bending moment is not constant.

17.6 DETERMINATION OF YIELD AND ULTIMATE CURVATURE

An initially straight element of an RC member with equal moments and axial force is shown in Figure 17.1. Considering only small length dx of the member and using the standard notations, the relationship of the ends of the element is given by

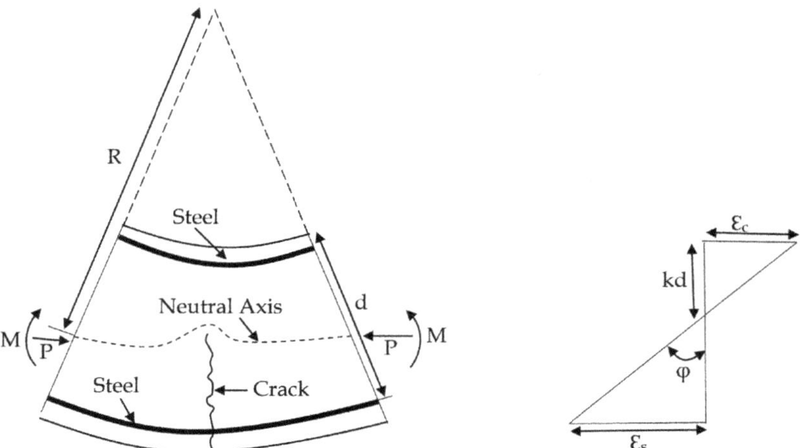

FIGURE 17.1 Deformation of flexural member.

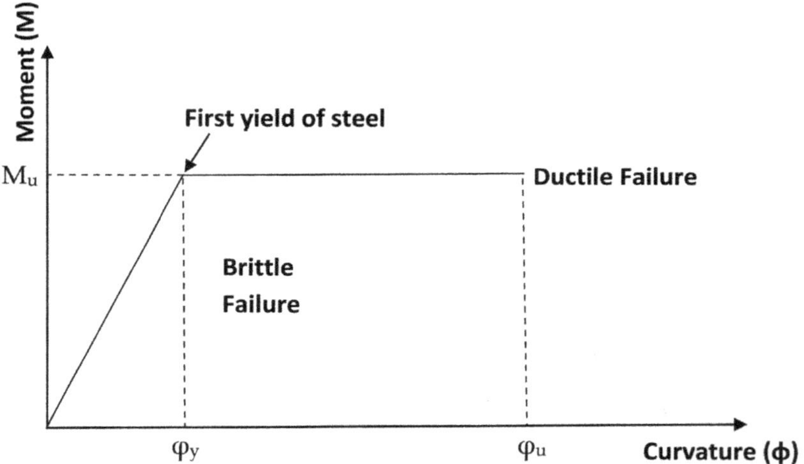

FIGURE 17.2 Idealized moment curvature relationship of a singly reinforced section in tension.

$$\frac{dx}{R} = \frac{\epsilon_c dx}{kd} = \frac{\epsilon_s dx}{d(1-k)}$$

$$\frac{1}{R} = \frac{\epsilon_c}{kd} = \frac{\epsilon_s}{d(1-k)}$$

Now $1/R$ is the curvature of the element (the rotation per unit length of the member) and thus,

$$\varphi = \frac{\epsilon_c}{kd} = \frac{\epsilon_s}{d(1-k)} \tag{17.3}$$

In many cases, the moment curvature relationship of the beam can be idealized to a bilinear relationship which gives successive degrees of approximation as shown in Figure 17.2, where φ_y is the respective yield curvature, when the extreme tension reinforcements (i.e., the farthest from the neutral axis) first attains yield strain and φ_u is the ultimate curvature of the section when outermost compression fibre of concrete reaches the limiting strain.

17.7 LIMIT STATE METHOD AND MAXIMUM UNDER-REINFORCED SECTIONS

The limit state method of design is based on the stress–strain characteristics of concrete and steel, and strain is considered as the criterion for failure. In the limit state method of design since over-reinforced sections are not permitted in flexure, as per IS:456 there are only two most critical or limiting sections, namely:

a) Balanced section – A section is called balanced when the area of tension steel is such that at the ultimate limit state, the limiting strains in concrete and steel are reached simultaneously.
b) Maximum under-reinforced section – This is another critical section corresponding to minimum percentage of tension reinforcement. As per IS:456-2000, the minimum percentage of tension reinforcement is $\frac{0.85bd}{f_y}$, where b is the width and d is the effective depth of section and the limiting strain in steel at failure should not be less than ($0.87f_y/E_s$ +0.002) where f_y is the grade of steel and E_s is the modulus of elasticity of steel. The minimum tension reinforcement is a function of grade of steel only and is presented in Table 17.1. Using this minimum tension reinforcement, the minimum depth of neutral axis has been determined from the force equilibrium ($C = T$) equation and has been obtained as $\frac{2.054}{f_{ck}}$. The maximum strain in concrete is 0.0035, irrespective of the grade of concrete (up to M60), and the maximum strain in steel is obtained by joining the maximum concrete strain with the minimum depth of neutral axis by a straight line, since as per Bernoulli's theorem the strain distribution diagram is linear. The depth of neutral axis and the maximum strain at the level of tensile steel are functions of concrete grade only. These values are presented in Table 17.2 for different grades of concrete and steel. As per IS:13920 the minimum longitudinal reinforcement ratio required on any face at

TABLE 17.1
Minimum reinforcement (%) as per different standards

	Minimum Reinforcement (%) as per IS: 456-2000			
	Grade of steel			
Grade of Concrete	**Fe 415**	**Fe 500**	**Fe 550**	**Fe 600**
M20 to M60	0.2048	0.1700	0.1545	0.1416
	Minimum Reinforcement (%) as per IS:13920-2016			
M20	0.287	0.239	0.217	0.199
M25	0.321	0.267	0.242	0.222
M30	0.352	0.292	0.266	0.243
M35	0.380	0.316	0.287	0.263
M40	0.406	0.337	0.307	0.281
M45	0.431	0.358	0.325	0.298
M50	0.454	0.377	0.343	0.314
M55	0.477	0.396	0.360	0.330
M60	0.498	0.413	0.376	0.344

TABLE 17.2
Minimum neutral axis depth and maximum strain in steel with minimum tension reinforcement as per Indian Standards

Grade of Concrete	Minimum tension reinforcement provided as per IS:456-2000		Minimum tension reinforcement provided as per IS:13920-2016	
	Minimum neutral axis depth $(x_{u,min})\ \frac{2.054}{f_{ck}}$	Maximum Strain in Steel $(\epsilon_{st,max})$	Minimum neutral axis depth $(x_{u,min})\ \frac{0.644}{\sqrt{f_{ck}}}$	Maximum Strain in Steel $(\epsilon_{st,max})$
M20	0.1027d	0.0305	0.1441d	0.0208
M25	0.08216d	0.0391	0.1289d	0.0237
M30	0.06846d	0.0476	0.1177d	0.0262
M35	0.05868d	0.0561	0.1089d	0.0286
M40	0.05135d	0.0647	0.1019d	0.0308
M45	0.0456d	0.0732	0.0961d	0.0329
M50	0.04108d	0.0817	0.0911d	0.0349
M55	0.0373d	0.0902	0.0869d	0.0368
M60	0.0342d	0.0987	0.0832d	0.0386

any beam section is $0.24\frac{\sqrt{f_{ck}}}{f_y}bD$, which is a function of both the grades of concrete and steel, and is presented in Table 17.1. The effective depth of the section can be taken as $d = 0.9D$, where D is the overall depth. Using this minimum tension reinforcement, the minimum depth of neutral axis can be determined from the force equilibrium ($C = T$) equation and has been obtained as $\frac{0.644}{\sqrt{f_{ck}}}$. Hence for IS:13920, both the maximum strains and minimum neutral axes depths are functions of the characteristic strength of concrete only. These values have been calculated and presented in Table 17.2.

17.8 CALCULATION AND INTERPRETATION OF STRAIN DUCTILITY VALUES

The maximum and minimum (yield) values of strains in steel for maximum under-reinforced sections and balanced sections and the ratios between maximum strain to minimum or yield strain in steel i.e.,$\frac{\epsilon_{stmax}}{\epsilon_y}$, referred to as strain ductility in steel, are presented in Table 17.3 for different grades of concrete and steel. These ratios will help in understanding how the section needs to actually rotate to accommodate the strain increments as shown in Figure 17.3. It is observed from Table 17.3 that with higher strengths of concrete and lower grades of steel the strain ductility keeps increasing. The maximum strain values in steel are not specified in the standards however, the minimum or limiting strain values are specified. There is no limit on the maximum strain values, which are open-ended. This implies that a significant amount of yielding or deformation is required for development of concrete stress block, resulting in an increase in curvature of the section which may not be feasible or very easy for RCC sections, especially in the case of deep girders. These rotations, with minimum percentage of steel, might seem to be hypothetical. However, since the actual percentage of steel is almost always more than $A_{st,min}$, the actual values will be less. The codes should provide a capping on the maximum strain in steel, otherwise associated curvatures required to generate

TABLE 17.3
Strains in reinforcing steel

Grade Of Concrete	Fe 415 Balanced Min. Strain	Fe 415 Max. under reinforced Max. Strain	Fe 415 Ratio (Strain ductility)	Fe 500 Balanced Min. Strain	Fe 500 Max. under reinforced Max. Strain	Fe 500 Ratio (Strain ductility)	Fe 550 Balanced Min. Strain	Fe 550 Max. under reinforced Max. Strain	Fe 550 Ratio (Strain ductility)	Fe 600 Balanced Min. Strain	Fe 600 Max. under reinforced Max. Strain	Fe 600 Ratio (Strain ductility)
						IS:456-2000						
						STRAINS IN REINFORCING STEEL (ϵ_{stmax})						
						Grade of Steel						
M 20	0.0038	0.0305	8.03	0.0042	0.0305	7.31	0.0044	0.0305	6.95	0.0046	0.0305	6.62
M 25	0.0038	0.0391	10.29	0.0042	0.0391	9.37	0.0044	0.0391	8.91	0.0046	0.0391	8.48
M 30	0.0038	0.0476	12.53	0.0042	0.0476	11.41	0.0044	0.0476	10.85	0.0046	0.0476	10.33
M 35	0.0038	0.0561	14.77	0.0042	0.0561	13.45	0.0044	0.0561	12.79	0.0046	0.0561	12.18
M 40	0.0038	0.0647	17.02	0.0042	0.0647	15.49	0.0044	0.0647	14.73	0.0046	0.0647	14.03
M 45	0.0038	0.0732	19.26	0.0042	0.0732	17.53	0.0044	0.0732	16.67	0.0046	0.0732	15.87
M 50	0.0038	0.0817	21.50	0.0042	0.0817	19.57	0.0044	0.0817	18.67	0.0046	0.0817	17.72
M 55	0.0038	0.0902	23.74	0.0042	0.0902	21.61	0.0044	0.0902	20.55	0.0046	0.0902	19.57
M 60	0.0038	0.0987	25.98	0.0042	0.0987	23.65	0.0044	0.0987	22.49	0.0046	0.0987	21.42
						IS: 13920-2016						
M 20	0.0038	0.0208	5.47	0.0042	0.0208	4.98	0.0044	0.0208	4.74	0.0046	0.0208	4.51
M 25	0.0038	0.0237	6.23	0.0042	0.0237	5.67	0.0044	0.0237	5.39	0.0046	0.0237	5.13
M 30	0.0038	0.0262	6.91	0.0042	0.0262	6.29	0.0044	0.0262	5.98	0.0046	0.0262	5.69
M 35	0.0038	0.0286	7.53	0.0042	0.0286	6.86	0.0044	0.0286	6.52	0.0046	0.0286	6.21
M 40	0.0038	0.0308	8.12	0.0042	0.0308	7.39	0.0044	0.0308	7.03	0.0046	0.0308	6.69
M 45	0.0038	0.0329	8.67	0.0042	0.0329	7.89	0.0044	0.0329	7.50	0.0046	0.0329	7.14
M 50	0.0038	0.0349	9.19	0.0042	0.0349	8.36	0.0044	0.0349	7.95	0.0046	0.0349	7.57
M 55	0.0038	0.0368	9.68	0.0042	0.0368	8.81	0.0044	0.0368	8.38	0.0046	0.0368	7.98
M 60	0.0038	0.0386	10.12	0.0042	0.0386	9.24	0.0044	0.0386	8.79	0.0046	0.0386	8.37

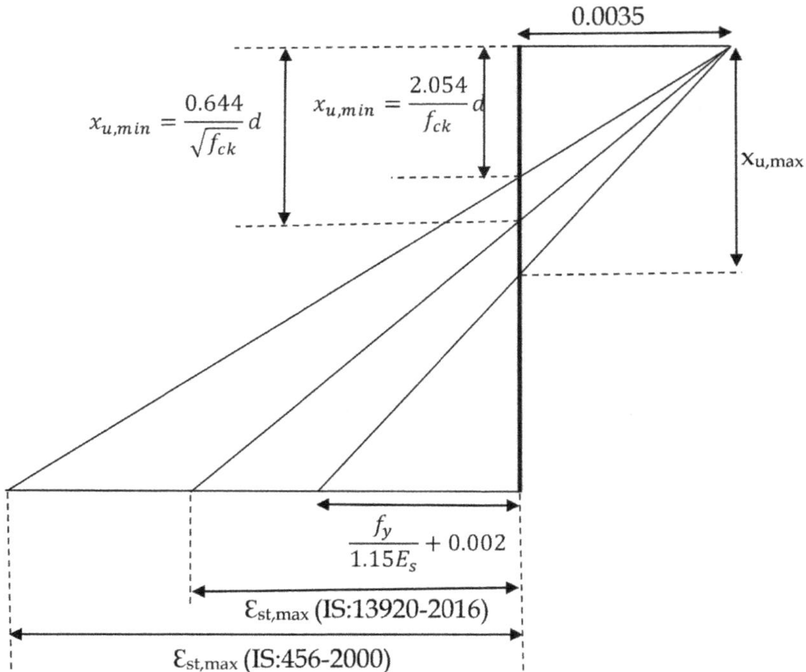

FIGURE 17.3 Strain distribution diagrams for balanced condition and with a minimum amount of tension reinforcement as per IS:456 and IS:13920.

these strains in steel can be very large and not even realizable, particularly in deeper girders of greater depths which are built integrally with structural frames. The standards may consider specifying an upper limit for the strain in the extreme layer steel or an upper limit for the maximum curvatures that can be generated in RC beams. Ductility plays a significant role in statically indeterminate structures by allowing redistribution of overstresses from one critical section to another, thus delaying local failure. Ductility demand is also very high for beams designed for nominal bending moment requirement with a minimum or very low amount of tension reinforcement. However, no specific recommendations are available in the code of practice regarding the choice of the section. Some designers are of the opinion that in balanced condition both materials are optimally used and hence it is better but the performance of the under-reinforced section in terms of ductility is always better than balanced section which is categorically emphasized. The greater is the level of under-reinforcement, the greater is the level of ductility but the greater also will be the plastic rotations. Hence, it can be summarized that for design of earthquake-resistant structures under-reinforced sections are preferred and not balanced sections.

17.9 DUCTILITY OF UNDER-REINFORCED SECTIONS

For maximum under-reinforced sections, design strains in steel are reached first, whereas the strain in concrete is significantly less than its limiting strain, and under such conditions the concrete stress block cannot fully develop. The stress block must develop fully and for this the inclined arm of the strain distribution diagram should rotate to accommodate higher strains in steel until the limiting strain in concrete (0.0035) is reached. At the initial stage when the strain in the steel reaches yield, the strain in the extreme fibre of the concrete is somewhat lower and for that low level of strain, the stress–strain relationship of concrete can be taken as linear where the working stress method is

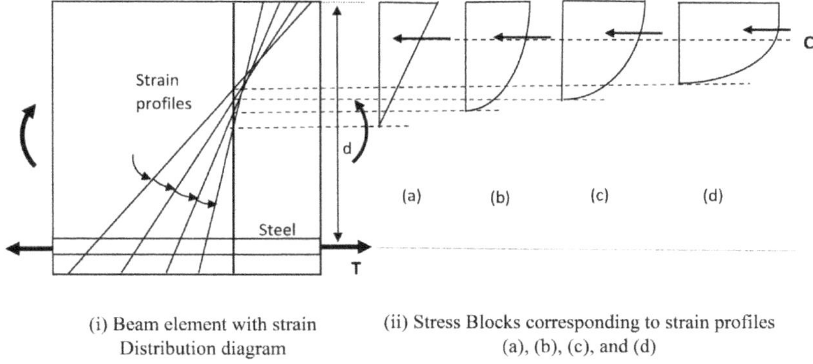

FIGURE 17.4 Accumulation of strain and development of stress block.

found to be acceptable, as shown in Figure 17.4. In that case, the concrete stresses may be assumed to be in the linear elastic range and the neutral axis depth factor k may be calculated using elastic (straight line) theory. Figure 17.4 also presents the rotation of the strain distribution diagram and accumulation of strain in steel with the changing shape of stress block as the bending moment at the section is increased. This rotation in the strain distribution diagram will necessitate a significant amount of rotation in beams resulting in ductile failure. This rotation will be minimum for a balanced section and maximum for an under-reinforced section with a minimum percentage of tension steel. Since the minimum amount of tension reinforcement is higher as per IS:13920 than the corresponding value of IS:456, the rotation of the strain distribution diagram or the yielding of the section will be somewhat less for sections designed as per IS:13920. When reinforced concrete structures are subjected to ultimate loads, inelastic analyses are required to take into account the non-linearity of materials because at ultimate loads, i.e., at stress levels of $0.87f_y$ in steel and $0.45f_{ck}$ for concrete, the corresponding strains are not linearly related. Unfortunately, till now there is no such analysis, which takes into account the complex behaviour of reinforced concrete just before collapse and hence all the RC structures are generally analysed by the conventional linear elastic theory as prescribed by the code. However, ductility arising from inelastic behaviour of concrete and steel, proper design and detailing of reinforcement, over-strength due to reserve strength, etc. are to be relied upon to take care of this difference between the actual and the design lateral load in the case of overloading. It needs to be ensured that brittle failure should be avoided. Thus, ductility of reinforced concrete structures is very important so that they can perform satisfactorily when subjected to overloading.

17.10 MEASURES OF DUCTILITY

The measure of ductility has to be made with reference to a load deformation response. The ratio of the ultimate deformation to the deformation at the yield can give the measure of ductility. However, each choice of deformation (strain, rotation, curvature or deflection) may give different values of ductility.

Displacement Ductility – The displacement ductility factor is defined as the ratio of Δ_u and Δ_y, where Δ_u and Δ_y are the respective deflection at the end of the post-elastic range and when the yield is first reached. The displacement ductility factor should range between 3–5 for beams.

Curvature Ductility – Based on the moment curvature relationship, curvature ductility may be defined as the ratio of $\frac{\varphi_u}{\varphi_y}$ where φ_u and φ_y are the curvatures at the ultimate and yield stage, respectively, as shown in Figure 17.5.

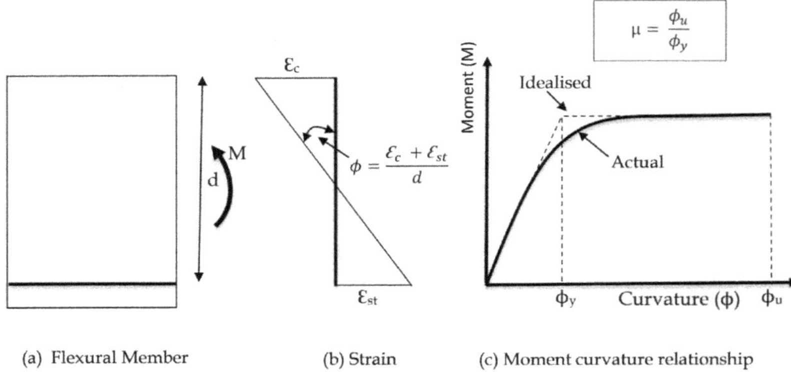

FIGURE 17.5 Curvature ductility of RC sections.

Rotational Ductility – The rotational ductility factor is defined as the ratio of θ_u and θ_y where θ_u is the rotation at the end of the post-elastic range and θ_y is the rotation at the first yield point of the tension steel.

Greater emphasis and importance are given to measuring the curvature ductility since it is the most common and desirable source of inelastic structural deformations in potential plastic hinges. Therefore, it is useful to relate section rotations per unit length (i.e., curvature) to causative bending moments.

17.11 PLASTIC HINGE IN REINFORCED CONCRETE SECTIONS

In reinforced concrete members the plastic hinge is defined as that section of the beam where plastification of concrete in compression and yielding of steel in tension zone has occurred causing rotation of the section under constant ultimate moment. In a reinforced concrete member, a plastic hinge takes place in the critical section, where tensile reinforcing bars reach yielding due to external load. The bending moment at which a plastic hinge is formed at a section of a structure is called the section plastic moment, M_p. When the moment at any section exceeds its plastic moment, M_p, a plastic hinge is formed at the same section location. Since the hinge is formed on account of plastification of the materials, it is called a plastic hinge. Several researchers have proposed various lengths of plastic hinges. As per ATC40, $l_p = d/2$, which is considered as an acceptable value that usually gives conservative results, where d is the section depth in the direction of loading and l_p is the length of the plastic hinge. The formation of plastic hinges decreases the structure stiffness and, thus, causes the deflections to increase significantly. The formation of plastic hinges also has a significant effect on the behaviour and failure mechanism of flexural members.

17.12 EXPRESSION OF PLASTIC ROTATION AND CURVATURE DUCTILITY OF BALANCED SECTIONS

For a balanced section, the limiting strains in steel and concrete are reached simultaneously and steel strain need not increase beyond the yield level as the stress block in concrete fully develops under this condition. For this type of section, the yield and the ultimate stages are defined from a concrete view point and the corresponding strains in compression concrete are considered as 0.002 and 0.0035, respectively. In the case of a balanced section, the extreme concrete compression fibre reaches a strain of 0.002 prior to yielding of tension reinforcement. On reaching the strain of 0.002, the stress in concrete becomes constant which is considered as stage 1 or the yield stage. Finally, on

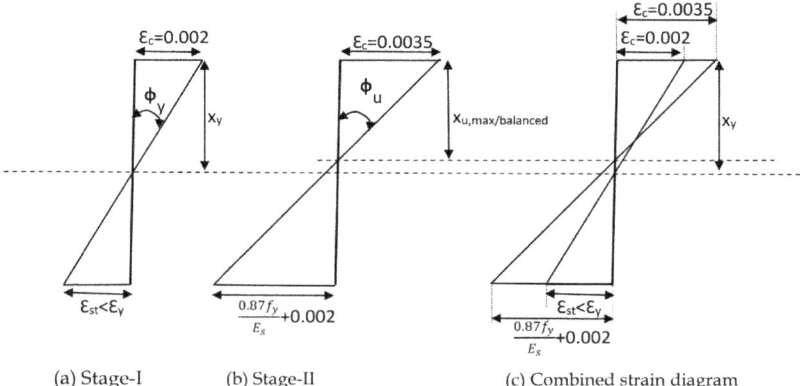

FIGURE 17.6 Strain diagram for balanced section.

reaching the value of limiting strain of 0.0035 at the outermost compression fibre failure of concrete occurs, which is termed as stage 2, as depicted in Figure 17.6.

The plastic curvature of a reinforced concrete section:

$$\varphi_p = (\text{Ultimate Curvature} - \text{Yield Curvature}) = (\varphi_u - \varphi_y)$$

$$\varphi_p = \left(\frac{\varepsilon_{cu}}{x_{u,max}} - \frac{\varepsilon_{cy}}{x_y}\right)$$

$$\text{Or, } \varphi_p = \left(\frac{0.0035}{x_{u,max}} - \frac{0.002}{x_y}\right) \tag{17.4}$$

Again, curvature (φ) = rotation (θ)/length (l), thus plastic curvature (φ_p) = plastic rotation capacity (θ_p)/length of plastic hinge (l_p), which gives $\theta_p = \varphi_p \times l_p$

$$\theta_p = \left(\frac{0.0035}{k_{umax}} - \frac{0.002}{k_y}\right) \times \frac{l_p}{d} \tag{17.5}$$

Now, curvature ductility for a balanced section can be calculated as shown below:

$$\mu = \frac{\varphi_u}{\varphi_y} \tag{17.6}$$

$$\text{Or, } \mu = \frac{0.0035}{x_{u,max/bal}} \times \frac{x_y}{0.002}$$

$$\mu = \left(\frac{0.0035}{k_{umax/bal}}\right) \times \left(\frac{k_y}{0.002}\right) \tag{17.7}$$

17.12.1 Determination of x_Y

When the strain in concrete reaches 0.002, only the parabolic part of the stress–strain diagram is formed, as per the stress–strain diagram of IS:456. Hence the stress block, which is the stress–strain diagram rotated at 90 degrees, is as shown in Figure 17.7.

From Figure 17.7:

$$\frac{0.002}{x_Y} = \frac{\epsilon_{st}}{d - x_y}$$

Or

$$\epsilon_{st} = \frac{d - x_y}{x_y} \times 0.002 \tag{17.8}$$

Considering force equilibrium, forces in steel and concrete are equated

$$\frac{2}{3} \times 0.446 f_{ck} x_y b = f_{st} A_{st,balanced} \tag{17.9}$$

$$\text{or,}\ f_{st} = \frac{0.29733 f_{ck} x_y b}{A_{st,balanced}}$$

$$\text{or,}\ \varepsilon_{st} = \frac{0.29733 f_{ck} k_y bd}{A_{st,balanced}} \times E_s$$

$$\text{or,}\ \varepsilon_{st} = \frac{0.29733 f_{ck} \mathrm{k}_y}{p_{t,balanced}/100} \times E_s \tag{17.10}$$

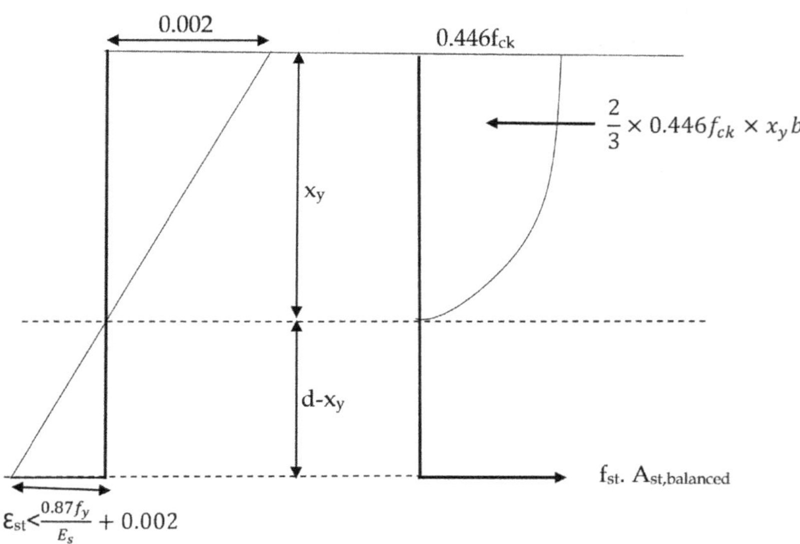

FIGURE 17.7 Strain distribution diagram and stress block when strain in outermost compression fibre of concrete is 0.002.

TABLE 17.4
Neutral axis depth factor (k_y) for different combinations of grades of concrete and steel when strain in extreme compression fibre of concrete is 0.002

	Neutral axis depth factor (k_y)			
	Grade of steel			
Grade of concrete	**Fe 415**	**Fe 500**	**Fe 550**	**Fe 600**
M20 to M60	0.5421	0.5023	0.4819	0.4633

where,

$$\epsilon_{st} = \text{total strain} = \text{elastic strain} + \text{inelastic strain}$$

$f_{st} = \text{stress in steel}$, which needs to be calculated from the strain (ϵ_{st}) using the stress–strain diagram. The stress–strain diagram of steel (HYSD bars) is linear up to $0.8f_y$, thereafter it becomes curvilinear and after reaching f_y it becomes horizontal. A factor of safety of 1.15 is applied for the design curve. For a balanced section, steel reaches the yield strain when concrete strain reaches 0.0035. Hence, when strain in concrete is 0.002, strain in steel will be less than yield but generally lies within the curvilinear part of the stress–strain diagram, i.e., between the stress levels of $0.8f_y$ to f_y. This makes the problem more complicated as the governing equation relating stress and strain is to be determined for this curvilinear part for different grades of steel. As per SP:16-1980 four intermediate points between $0.8f_y$ and f_y are specified whereby the inelastic strains are mentioned with a note that linear interpolation may be performed for intermediate values. Hence it may be considered that for cold-worked bars, which do not have a definite yield point, the relationship between stress and strain which is curvilinear between elastic and yield levels, may be approximated as an agglomeration of five straight lines joining the six cardinal points. In the above Equations (17.8) and (17.10) there are two unknowns, namely x_y and ϵ_{st} or f_{st}.

Equating the strain values from Equations (17.8) and (17.10), it may be stated that,

$$\frac{0.002}{x_y}\left(1-k_y\right)=\frac{0.29733 f_{ck} k_y}{p_{t,balanced}/100}\times E_s$$

$$0.29733 f_{ck} E_s \left(k_y\right)^2+\left(0.002\times\frac{p_{t,\,balanced}}{100}\right)k_y-\left(0.002\times\frac{p_{t,\,balanced}}{100}\right)=0 \tag{17.11}$$

By solving this quadratic equation, the depth of neutral axis factor $k_y = x_y/d$ can be evaluated for different grades of concrete and steel.

The values of k_y have been calculated for different combinations of concrete and steel grades and are presented in Table 17.4.

17.12.2 Determination of $X_{U,MAX}$ or $X_{U,BALANCED}$

The position of neutral axis at the ultimate stage ($x_{u,max/bal}$) can be determined from the strain distribution diagram shown in Figure 17.8. The values of $\frac{x_{u,max}}{d}$ for different grades of steel are presented

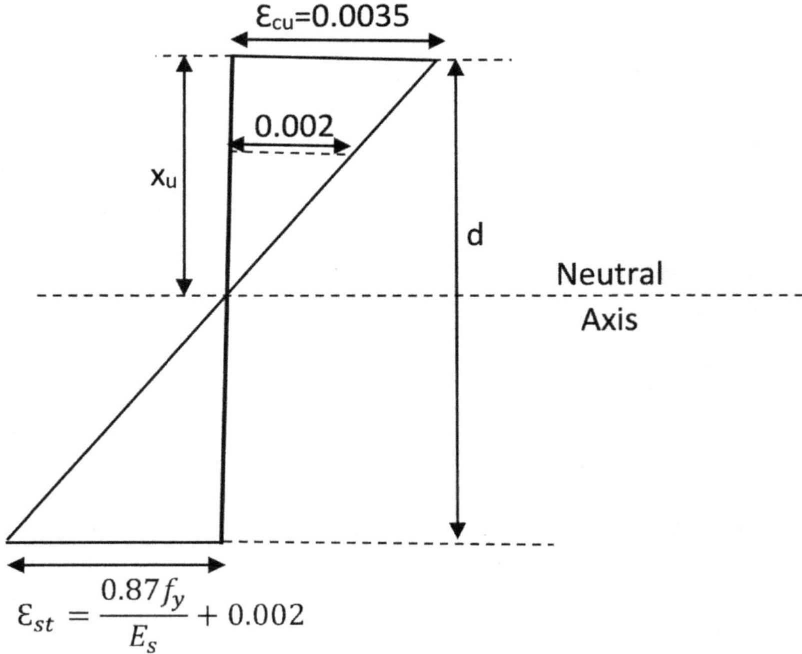

FIGURE 17.8 Strain diagram at failure for a balanced section.

TABLE 17.5
Values of limiting steel strains and depths of neutral axis for balanced sections

Grade of Steel	E_s (MPa)	f_y (Mpa)	$\varepsilon_{st} = \frac{0.87f_y}{E_s} + 0.002$	$\frac{x_{u,max}}{d}$
Fe 415	2×10^5	415	0.0038	0.48
Fe 500	2×10^5	500	0.0042	0.46
Fe 550	2×10^5	550	0.0044	0.44
Fe 600	2×10^5	600	0.0046	0.43

TABLE 17.6
Plastic rotations of balanced section for different grades of concrete and steel

	Plastic Rotation (θ_p) in Radian for l_p=0.5d for balanced section			
	IS: 456-2000			
	Grade of steel			
Grade of Concrete	**Fe 415**	**Fe 500**	**Fe 550**	**Fe 600**
M 20 to M60	0.001807	0.001846	0.001871	0.001896

TABLE 17.7
Curvature ductility values of balanced sections for different grades of concrete and steel

	Curvature Ductility (μ) for balanced sections IS:456-2000			
	Grade of steel			
Grade of Concrete	**Fe 415**	**Fe 500**	**Fe 550**	**Fe 600**
M 20 to M 60	1.98011	1.92769	1.90171	1.87856

in Table 17.5. The values of plastic rotations and curvature ductility of balanced sections have been calculated using Equations (17.4) and (17.6) and are presented in Tables 17.6 and 17.7, respectively.

17.13 EXPRESSIONS FOR PLASTIC ROTATION AND CURVATURE DUCTILITY OF MAXIMUM UNDER-REINFORCED SECTIONS

17.13.1 Determination of $x_{u,min}$

The stress–strain distribution of a maximum under-reinforced section (with minimum tension reinforcement) is presented in Figure 17.9. Maximum strain or failure strain in concrete at the ultimate curvature is ϵ_{cu} and that in steel is ϵ_{stmax}. Stage 1 indicates the end of the elastic phase, also known as the yield curvature, while stage 2 indicates the ultimate or collapse stage, also known as the ultimate curvature. In stage 1, the strain in concrete is ϵ_{cy} which is much less than ϵ_{cu} and the corresponding stress is less than $0.45f_{ck}$. The strain in steel at stage 1 is defined as the design yield strain and is given by $\epsilon_y = 0.002 + \dfrac{0.87f_y}{E_s}$, whereas in stage 2, i.e., while approaching the collapse stage, the strain in steel goes on increasing until strain in concrete at the outermost fibre reaches 0.0035 and there is development of the entire stress block of concrete. After yielding of tension steel, its stress remains constant but strain keeps increasing until compressive strain in extreme fibre of concrete reaches the strain value of ε_{cu} at the maximum stress in concrete f_c. The rotation capacity (θ_p) of an RC plastic hinge is the rotation which the section undergoes from the stage when the strain in steel is ϵ_y (stage 1) to that when the strain in concrete reaches its limiting strain, i.e. ϵ_{cu} (stage 2). Therefore, the rotation capacity of an RC section is basically the rotation that takes place between stage 1 and stage 2, i.e., from the moment when steel first yields to the moment when concrete reaches its limiting or failure strain.

From Figure 17.9 it may be stated:

$$\varphi_p = \left(\varphi_u - \varphi_y\right) = \left(\frac{\varepsilon_{cu}}{x_{u,min}} - \frac{\varepsilon_y}{d - x_{y,min}}\right)$$

$$\varphi_p = \left(\frac{0.0035}{x_{u,min}} - \frac{0.002 + \dfrac{0.87f_y}{E_s}}{d - x_{y,min}}\right)$$

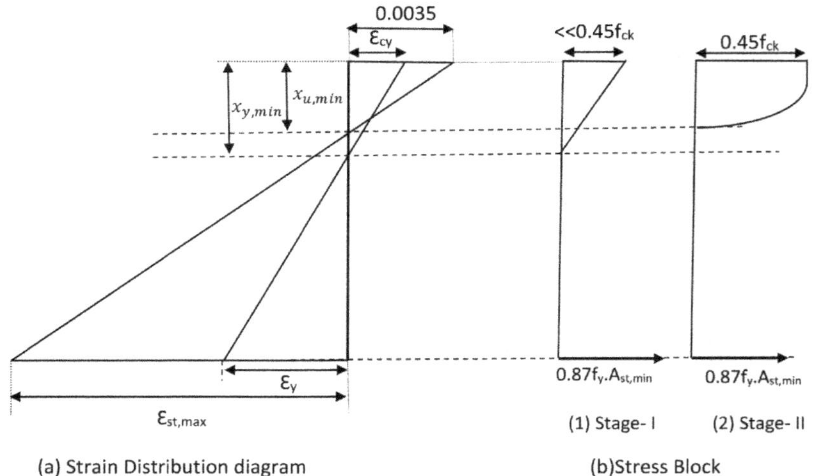

FIGURE 17.9 Strain and stress distribution diagrams of maximum under-reinforced section up to failure.

$$\text{Or, } \theta_p = \left(\frac{0.0035}{k_{u,min}d} - \frac{0.002 + \frac{0.87f_y}{E_s}}{d - k_{y,min}d} \right) \times l_p$$

$$\text{Or, } \theta_p = \left(\frac{0.0035}{k_{u,min}} - \frac{0.002 + \frac{0.87f_y}{E_s}}{1 - k_{y,min}} \right) \times \frac{l_p}{d} \tag{17.12}$$

The values of the minimum percentage of steel are different as per IS:456-2000 and IS:13920-2016.

For maximum under – reinforced condition,

$$\mu = \frac{\varphi_u}{\varphi_y} = \left(\frac{0.0035}{x_{u,min}} \right) \times \left(\frac{d - x_{y,min}}{0.002 + \frac{0.87f_y}{E_s}} \right) \tag{17.13}$$

Now, for IS : 456 – 2000 (as per Table 17.2)

$$\frac{x_{u,min}}{d} = \frac{2.05}{f_{ck}}$$

and as per IS:13920-2016 (as per Table 17.2), $\frac{x_{u,min}}{d} = \frac{0.644}{\sqrt{f_{ck}}}$

The values of $\frac{x_{u,min}}{d}$ or $k_{u,min}$ for different combinations of grades of concrete and steel as obtained from the force equilibrium equations have been presented in Table 17.2.

Thus, Equation (17.13) is modified as, $\mu = \left(\frac{0.0035}{\frac{x_{u,min}}{d}}\right) \times \left(\frac{1-\frac{x_{y,min}}{d}}{0.002+\frac{0.87f_y}{E_s}}\right)$

$$\mu = \left(\frac{0.0035}{k_{u,min}}\right) \times \left(\frac{1-k_{y,min}}{0.002+\frac{0.87f_y}{E_s}}\right) \quad (17.14)$$

The above equation gives the ductility of a maximum under-reinforced section.

17.13.2 Determination of $x_{y,min}$

For calculation of $k_{y,min}$ (minimum neutral axis depth factor at yield stage) the working stress or elastic method is considered at the yield level where the expression of $k_{y,min}$ is obtained by taking

TABLE 17.8
Neutral axis depth factor ($k_{y,min}$) when steel has reached the yield stage, Stage 1 for maximum under-reinforced sections

	Neutral axis depth factor ($k_{y,min}$)			
	IS: 456-2000			
	Grade of steel			
Grade of concrete	**Fe 415**	**Fe500**	**Fe 550**	**Fe 600**
M20	0.208	0.191	0.183	0.176
M25	0.190	0.175	0.168	0.161
M30	0.177	0.163	0.156	0.149
M35	0.166	0.153	0.146	0.140
M40	0.157	0.145	0.138	0.132
M45	0.147	0.135	0.129	0.123
M50	0.141	0.129	0.124	0.118
M55	0.134	0.123	0.117	0.112
M60	0.129	0.118	0.113	0.108
		IS: 13920-2016		
M20	0.241	0.222	0.213	0.205
M25	0.233	0.214	0.206	0.197
M30	0.226	0.208	0.199	0.191
M35	0.219	0.202	0.194	0.186
M40	0.214	0.197	0.189	0.181
M45	0.206	0.190	0.182	0.174
M50	0.202	0.186	0.179	0.171
M55	0.197	0.181	0.173	0.166
M60	0.194	0.178	0.171	0.163

TABLE 17.9
Plastic rotations of maximum under reinforced sections for different grades of concrete and steel

	Plastic Rotations (θ_p) in Radian for l_p=0.5d			
	IS:456-2000			
	Grade of steel			
Grade of Concrete	**Fe 415**	**Fe 500**	**Fe 550**	**Fe 600**
M 20	0.0146	0.0145	0.0145	0.0142
M 25	0.0190	0.0188	0.0188	0.0186
M 30	0.0233	0.0231	0.0231	0.0228
M 35	0.0275	0.0274	0.0274	0.0271
M 40	0.0318	0.0316	0.0316	0.0314
M 45	0.0361	0.0359	0.0359	0.0357
M 50	0.0404	0.0402	0.0402	0.0400
M 55	0.0447	0.0445	0.0445	0.0443
M 60	0.0489	0.0488	0.0488	0.0485
	IS:13920-2016			
M 20	0.0096	0.0095	0.0094	0.0092
M 25	0.0111	0.0109	0.0108	0.0107
M 30	0.0124	0.0122	0.0121	0.0120
M 35	0.0136	0.0134	0.0133	0.0132
M 40	0.0148	0.0146	0.0145	0.0144
M 45	0.0158	0.0156	0.0155	0.0154
M 50	0.0168	0.0166	0.0165	0.0164
M 55	0.0178	0.0176	0.0175	0.0174
M 60	0.0187	0.0185	0.0184	0.0183

the moment of compression and tension areas about the neutral axis of equivalent or transformed concrete section in terms of concrete.

$$k_{y,min} = -\left(\frac{mp}{100}\right) + \left[\left(\frac{mp}{100}\right)^2 + 2\left(\frac{mp}{100}\right)\right]^{1/2} \tag{17.15}$$

$$m = \frac{280}{3\sigma_{cbc}}$$

where σ_{cbc} = permissible stress of concrete in bending
p = % of tension reinforcement
m = modular ratio as per IS:456-2000, which is dimensionless

For the maximum under-reinforced condition, the minimum percentage of tensile reinforcement as per IS:456-2000 is $A_{stmin} = \dfrac{0.85bd}{f_y}$, and $0.24\dfrac{\sqrt{f_{ck}}}{f_y}bD$ as per IS:13920-2016, where d may be considered as 0.9D. The values of $k_{y,min}$ for different combinations of concrete and steel are presented in Table 17.8. The plastic rotations and curvature ductility of maximum under-reinforced sections have been calculated using Equations (17.12) and (17.14) and are presented in Tables 17.9 and 17.10, respectively, for different combinations of grades of concrete and steel.

TABLE 17.10
Curvature ductility of maximum under reinforced (UR) sections for different grades of concrete and steel

	Curvature Ductility (μ) for max UR section			
	IS: 456-2000			
	Grade of steel			
Grade of Concrete	Fe 415	Fe 500	Fe 550	Fe 600
M20	7.11	6.60	6.34	6.09
M25	9.08	8.41	8.08	7.75
M30	11.07	10.25	9.83	9.42
M35	13.09	12.10	11.60	11.12
M40	15.12	13.97	13.38	12.82
M45	17.21	15.88	15.21	14.57
M50	19.26	17.77	17.01	16.29
M55	21.37	19.69	18.85	18.04
M60	23.44	21.59	20.66	19.77
		IS: 13920-2016		
M20	4.85	4.52	4.35	4.19
M25	5.48	5.11	4.91	4.73
M30	6.06	5.64	5.43	5.22
M35	6.60	6.14	5.90	5.67
M40	7.10	6.60	6.35	6.10
M45	7.61	7.07	6.79	6.52
M50	8.06	7.48	7.19	6.90
M55	8.52	7.90	7.59	7.28
M60	8.93	8.28	7.95	7.63

17.14 EXPRESSIONS FOR PLASTIC ROTATION AND CURVATURE DUCTILITY OF DOUBLY REINFORCED SECTIONS

Doubly reinforced sections are those in which in addition to tension reinforcement, compression reinforcement is also provided, as shown in Figure 17.10. The designer generally opts for a doubly reinforced section only when the external bending moment exceeds the limiting or balanced moment of resistance as a singly reinforced section. Thus, when designers venture for doubly reinforced sections, the sections are generally designed as balanced ones with the depth of the neutral axis equal to its balanced value. The ductility of balanced sections has already been presented in earlier sections. Since a certain percentage of support or span reinforcement is to be extended all through the section length as per the requirement of the standards, all sections in flexure have both tensile and compression reinforcements. In this section the ductility of a doubly reinforced section has been calculated. For a typical doubly reinforced section, the area of tension reinforcement has been considered as the average of the sum of the minimum and balanced reinforcement for a typical combination of concrete and steel, and the compression reinforcement as half of the tension reinforcement value.

17.14.1 CALCULATION OF NEUTRAL AXIS AT YIELD

The value of the neutral axis at yield is calculated using the elastic method with the concept of equal moment area axis.

Equating the moments of the compressive and equivalent tensile areas about the neutral axis, it may be stated that,

$$b.x.\frac{x}{2}+\left(1.5m-1\right)A_{sc}.\left(x-d'\right)=mA_{st}\left(d-x\right) \quad (17.16)$$

Substituting $x=k_y d$, d'=0.1D = 0.1 × (d/0.9) = 0.11d, $A_{st}=\frac{p_t bd}{100}$, $A_{sc}=\frac{p_c bd}{100}$ in Equation (17.16),

where p_c and p_t are the values of percentage of area of steel provided in compression and tension and p_c is half of p_t where the symbols have their standard meanings.

Solving Equation (17.16), the following equation is obtained

$$k_y^2+(0.03mp_c-0.02p_c+0.02mp_t)k_y-(0.0033mp_c-0.0022p_c+0.02mp_t)=0 \quad (17.17)$$

This is a quadratic equation which can be solved for calculating the value of $\boldsymbol{k_y}$.

17.14.2 Calculation of depth of neutral axis at ultimate stage (x_u)

The depth of the neutral axis at the ultimate stage is determined by considering the equilibrium of forces. In a doubly reinforced section the total compressive force is due to compression steel ($f_{sc}A_{sc}-f_{cc}A_{sc}$)and concrete $\left(0.36f_{ck}\,b\,x_u\right)$. The total tensile force in a doubly reinforced section is $T=0.87f_yA_{st}$. Equating the total compressive force and tensile force:

$$\mathrm{C=T}$$

$$\left(0.36\mathrm{f_{ck}}\,\mathrm{b}\,\mathrm{x_u}+(\mathrm{f_{sc}}-\mathrm{f_{cc}})\mathrm{A_{sc}}\right)=0.87\mathrm{f_y}\mathrm{A_{st}}$$

$$x_u=\frac{0.87f_yA_{st}-\left(\mathrm{f_{sc}}-\mathrm{f_{cc}}\right)\mathrm{A_{sc}}}{0.36f_{ck}b} \quad (17.18)$$

The value of the neutral axis at ultimate stage can be determined from Equation (17.18), where f_{cc} is the stress in concrete at the level of compression steel, which can be taken as $0.45f_{ck}$ (approximately) and f_{sc} is the stress in compression steel which can be determined from the stress–strain diagram of steel after obtaining the strain at the level of compression steel, ϵ_{sc} from Equation (17.19). This assumption is valid if the strain at the level of compression steel is nearing 0.002. However, if this strain is very low then strain in concrete is obtained from the stress–strain curve of concrete.

From the strain distribution diagram shown in Figure 17.10 it can be stated that,

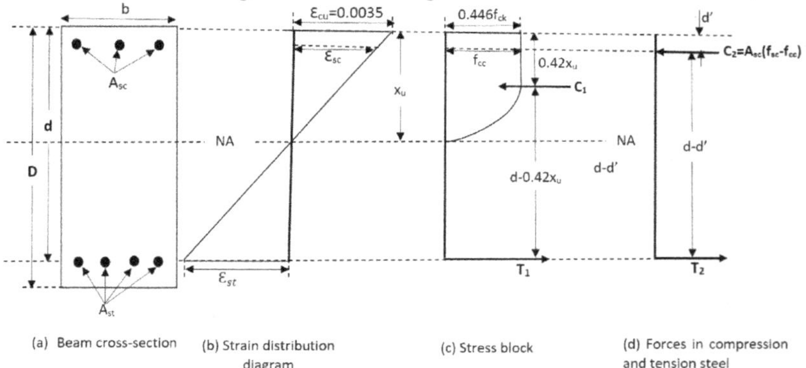

FIGURE 17.10 Doubly reinforced beam.

TABLE 17.11
Plastic rotations of doubly reinforced section

	Plastic Rotation (Radian) of Doubly Reinforced Under- Reinforced Section at l_p =0.5d											
	IS: 456-2000											
	Grade of Steel											
	Fe 415			Fe 500			Fe 550			Fe 600		
Grade of Concrete	p_t (%)	p_c (%)	θ_p radian	p_t (%)	p_c (%)	θ_p radian	p_t (%)	p_c (%)	θ_p radian	p_t (%)	p_c (%)	θ_p radian
20	0.578	0.289	0.00731	0.465	0.2325	0.00675	0.407	0.2035	0.00655	0.366	0.183	0.00634
25	0.695	0.3475	0.00754	0.56	0.28	0.00691	0.492	0.246	0.00676	0.441	0.2205	0.00654
30	0.815	0.4075	0.00767	0.655	0.3275	0.00708	0.576	0.288	0.00687	0.516	0.258	0.00665
35	0.94	0.47	0.00766	0.75	0.375	0.00714	0.657	0.3285	0.00699	0.591	0.2955	0.00676
40	1.06	0.53	0.00772	0.845	0.4225	0.00719	0.742	0.371	0.00704	0.666	0.333	0.00688
45	1.18	0.59	0.00772	0.94	0.47	0.00725	0.822	0.411	0.00716	0.741	0.3705	0.00688
50	1.3	0.65	0.00779	1.035	0.5175	0.00731	0.907	0.4535	0.00716	0.811	0.4055	0.00699
55	1.42	0.71	0.0078	1.13	0.565	0.00733	0.987	0.4935	0.00723	0.886	0.443	0.00701
60	1.54	0.77	0.0078	1.225	0.6125	0.00733	1.072	0.536	0.00724	0.961	0.4805	0.00707

TABLE 17.12
Curvature ductility of doubly reinforced section

Grade of Concrete	Curvature Ductility of Doubly Reinforced Under- Reinforced Section IS: 456-2000 Grade of Steel											
	Fe 415			Fe 500			Fe 550			Fe 600		
	p_t (%)	p_c (%)	μ	p_t (%)	p_c (%)	μ	p_t (%)	p_c (%)	μ	p_t (%)	p_c (%)	μ
20	0.578	0.289	3.71	0.465	0.2325	3.35	0.407	0.2035	3.21	0.366	0.183	3.06
25	0.695	0.3475	3.8	0.56	0.28	3.41	0.492	0.246	3.28	0.441	0.2205	3.13
30	0.815	0.4075	3.85	0.655	0.3275	3.47	0.576	0.288	3.32	0.516	0.258	3.17
35	0.94	0.47	3.84	0.75	0.375	3.49	0.657	0.3285	3.36	0.591	0.2955	3.2
40	1.06	0.53	3.86	0.845	0.4225	3.51	0.742	0.371	3.38	0.666	0.333	3.24
45	1.18	0.59	3.86	0.94	0.47	3.53	0.822	0.411	3.42	0.741	0.3705	3.24
50	1.3	0.65	3.89	1.035	0.5175	3.55	0.907	0.4535	3.42	0.811	0.4055	3.28
55	1.42	0.71	3.91	1.13	0.565	3.57	0.987	0.4935	3.46	0.886	0.443	3.3
60	1.54	0.77	3.91	1.225	0.6125	3.57	1.072	0.536	3.46	0.961	0.4805	3.32

$$\epsilon_{sc} = 0.0035\left(1 - \frac{d'}{x_u}\right)$$

$$\text{Or } \epsilon_{sc} = 0.0035\left(1 - \frac{d'}{k_u d}\right) \tag{17.19}$$

The value of $\frac{d'}{d}$ can be taken as 0.11, where d' is the effective cover to compression steel.

Now the values of plastic rotations and curvature ductility can be calculated in the same manner as mentioned in the previous sections using Equations (17.20) and (17.21). The values of plastic rotations and ductility are presented in Tables 17.11 and 17.12, respectively.

$$\theta_p = \left(\frac{0.0035}{k_u} - \frac{0.002 + \frac{0.87 f_y}{E_s}}{1 - k_y}\right) \times \frac{l_p}{d} \tag{17.20}$$

$$\mu = \left(\frac{0.0035}{k_u}\right) \times \left(\frac{1 - k_y}{0.002 + \frac{0.87 f_y}{E_s}}\right) \tag{17.21}$$

where k_u and k_y are the neutral axis depth factor at the ultimate and yield stages, respectively.

17.15 DISCUSSIONS ON DUCTILITY AND PLASTIC ROTATIONS

17.15.1 Prescriptive Design as per Indian Standards

From the above sections it may be stated that, for maximum under-reinforced sections as per IS:456, with an increase in the grade of steel, curvature ductility decreases, whereas with an increase in the grade of concrete, ductility increases. Hence higher grades of concrete are beneficial, whereas higher grades of steel are detrimental for concrete. For balanced sections, the values of ductility are almost constant and independent of the grades of concrete or steel. It has been reported that at the same tension steel ratio, a concrete section cast in high-strength concrete actually has a higher flexural ductility than a similar concrete section cast in normal concrete, despite the fact that a higher-strength concrete should have a lower ductility. However, this statement is strongly valid with a higher degree of under-reinforcement and invalid for sections close to balanced ones. The lower the tension to balanced steel ratio is, the higher is the flexural ductility, and vice versa, and this is the main factor determining the flexural ductility. With M60 and Fe 415 the values of curvature ductility for maximum under-reinforced and balanced sections are 23.44 and 1.98, respectively. For maximum under-reinforced sections as per IS:456, for any constant grade of concrete, the reduction in ductility is about 15% as the grade of steel increases from Fe 415 to Fe 600, whereas for any constant grade of steel the increase in ductility is about 230% as the grade of concrete increases from M20 to M60. Nowadays, high-strength concrete is almost routinely used in all major projects. Designers generally try to design the sections close to the balanced ones so that the full-strength potential of the high-strength concrete can be exploited and thus make the sections heavily reinforced. A heavily reinforced section cast in high-strength concrete can be very brittle and may fail quite explosively. Thus, for a section cast in high-strength concrete, it may be necessary to limit the tension to balanced steel ratio to a relatively lower value than that used with normal concrete. Designers should also not venture for maximum under-reinforced sections which though having high ductility require very high plastic rotations which might be difficult to mobilize. In this context, the codes of practice should limit the minimum reinforcement percentage in beams not only from the cracking moment consideration only but also should ensure that significant damage does not occur while trying to accommodate high plastic rotations occurring in the sections. Designers need to opt for under-reinforced sections and not balanced ones. The greater is the level of under-reinforcement, the greater is the level of ductility but also the greater will be the plastic rotations. Hence, it can be summarized that for design of earthquake-resistant structures under-reinforced sections are preferred and not balanced sections. For highly under-reinforced sections ductility demand is also very high and those sections require very high plastic rotations. As per the conventional design method, the designer opts for doubly reinforced sections only when the external moment is greater than the limiting or balanced moment of resistance of the section considering it a singly reinforced one. Thus, as per normal practice, all doubly reinforced sections are generally balanced. But for balanced sections plastic rotations and curvature ductility are quite small. Doubly reinforced sections should also preferably be under-reinforced, which significantly improves the performance of the section which has already been presented in the previous sections.

17.16 DESIRABLE DUCTILITY VALUES OF REINFORCED CONCRETE SECTIONS

A curvature ductility of at least 5 is considered to be adequate for reinforced concrete. As per another reference, a curvature ductility factor of 8 will ensure that premature spalling of concrete does not occur in moderate earthquakes. As per Park and Paulay, the curvature ductility, $\frac{\varphi_u}{\varphi_y}$, of well-designed frames should be at least 4μ, where μ is the displacement ductility factor which should range between 3–5. Hence, well-designed frames should have a curvature ductility ranging between 12–20. As can be observed from Table 17.10, the values of curvature ductility of maximum under-reinforced sections range approximately from 6 to 23 and 4 to 9, as per IS:456 and IS:13920, respectively. These sections, with minimum tension reinforcement as per the standards, lie within the purview or domain of the limit state method of design and hence are permitted as per the standards. However, the corresponding plastic rotations which are presented in Table 17.9 seem to be extremely high for a rigid, brittle and robust material like concrete, especially for deeper sections. However, designers have no information whatsoever whether these high values of rotations can actually be mobilized in reinforced concrete sections or if these can lead to actual collapse of the section. Since the prescriptive method of design as per the standards does not provide any information in this regard, different performance levels of performance-based design are compared with the plastic rotations at the limit state of collapse level as presented in Section 17.19.

17.17 EVALUATION OF DUCTILITY FOR COLUMNS

It has been discussed in Chapter 10 that failure in the limit state of collapse in flexure occurs in two stages, Stage I is the yield stage and Stage II is the ultimate stage. For under-reinforced sections, tension reinforcement yield in Stage I and strain at the outermost layer in concrete reaches the limiting strain of 0.0035 at Stage II. For balanced and over-reinforced sections, the outermost fibre of concrete reaches a strain of 0.002 at Stage I and in Stage II concrete strain reaches the value of 0.0035. The greater is the gap between Stage I and Stage II, the higher is the ductility of the section. All these conditions are valid for beams or flexural members.

For columns, the treatment is similar but slightly different since along with flexure the section is subjected to axial force. A column with pure compression will fail at a strain of 0.002. There is no separate existence of Stage I and Stage II, and hence the section has a ductility value of 1, and the failure will be brittle in nature. When the neutral axis of the section is beyond the depth of the section, section failure will occur in over-reinforced condition. Under this situation, Stage I will be reached when maximum compressive strain in concrete is 0.002 and Stage II will be reached when this strain reaches the failure strain in concrete varying from 0.002 to 0.0035. The limiting conditions will vary from the position of neutral axis lying at infinity causing uniform compressive strain on the section as 0.002 to the neutral axis at the section face causing strain in the maximum compressed concrete fibre equal to 0.0035. When the neutral axis lies within the section, the treatment is similar to beams and failure will occur under over-reinforced, balanced or under-reinforced conditions. Hence, these sections will have ductility. For balanced and under-reinforced sections, the values will be significantly small, whereas under-reinforced sections will have higher ductility values. This has been clearly demonstrated for beams, whereas for columns the nature will be very similar with even lower ductility values due to the incorporation of axial compressive forces.

17.18 EFFECT OF AXIAL COMPRESSIVE FORCE ON CURVATURE DUCTILITY

Curvature ductility is a function of the curvature values at Stage I and Stage II. The greater is the difference between these two stages, the greater will be the ductility, and vice-versa. The effects of axial force on the curvatures of different types of sections at Stages I and II are presented in Figures 17.11 and 17.12. Curvature ductility is defined as $\mu=\frac{\varphi_u}{\varphi_y}$. For over-reinforced sections, the curvatures at yield (Stage I) and ultimate (Stage II) both will decrease with the incorporation of axial compressive force. This is presented in Figure 17.11, whereby it is shown that the depth of the compression zone is increasing in both cases. Since both φ_u and φ_y are decreasing, the effect of axial compressive force will not significantly affect the ductility values. Moreover, it has been already proved that for beams, ductility values for balanced sections are quite less and almost constant irrespective of the grade of concrete and steel. The ductility values will be even less for over-reinforced sections. However, for under-reinforced sections, curvature at yield (Stage I) is increased and that at ultimate (Stage II) is reduced with incorporation of axial compressive force. As φ_u decreases and φ_y increases the curvature ductility of under-reinforced sections will be reduced due to the effect of axial compressive force. This reduction will depend on the magnitude of the compressive force. The

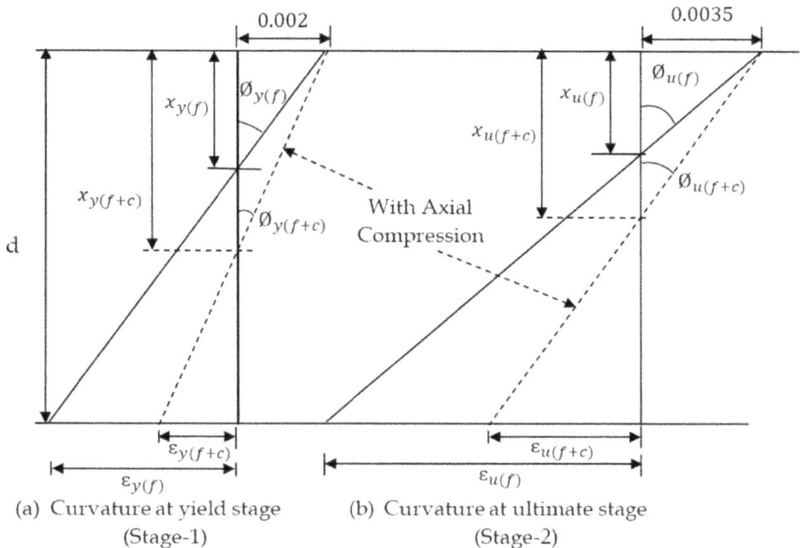

FIGURE 17.11 Effect of axial force on curvatures of balanced and over-reinforced sections at Stage I and Stage II.

Note: (f) stands for concrete flexure and (f+c) stands for flexure with compression.

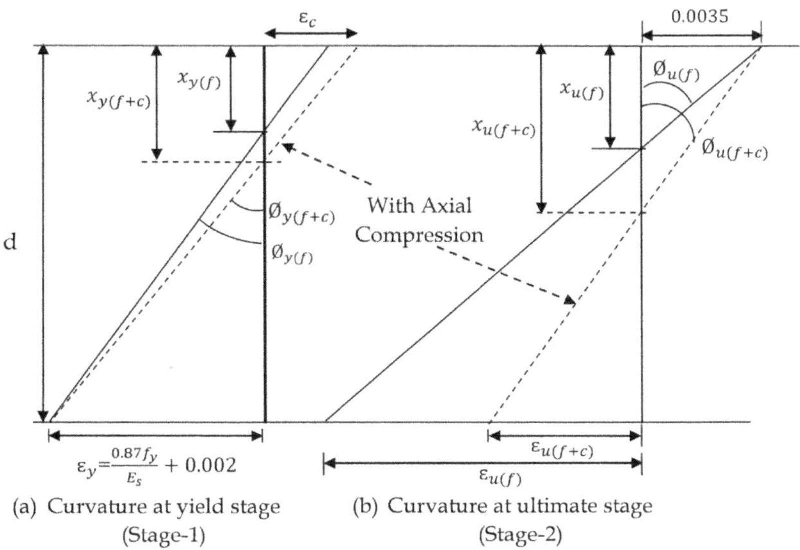

FIGURE 17.12 Effect of axial force on curvatures of under-reinforced sections at Stage I and Stage II.

Note: (f) stands for concrete flexure and (f+c) stands for flexure with compression.

greater is the compressive force, the greater will be the reduction in the ductility and the lower the compressive force, the less will be the reduction in ductility. However, the presence of axial tension force will increase the ductility values of the sections.

17.19 PERFORMANCE-BASED DESIGN

Performance-based design (PBD), which deals with structural performance levels beyond the limit state, is a process of designing new buildings or seismic upgradation of existing buildings which includes a specific intent to achieve a defined performance objective. Seismic performance is defined by designating the maximum allowable damage state (performance level) for an identified seismic exposure (earthquake ground motion). The different performance levels or performance objectives are generally specified by using the moment–curvature relationship as immediate occupancy (IO), life safety (LS) and collapse prevention (CP) states. Specialist literature and international guidelines such as ATC40 and FEMA 356 deal with performance-based design and provide numerical values of plastic rotations of structural elements at different states, namely at IO, LS and CP levels. The plastic rotations of a flexural member depend on the percentage of tension reinforcement, compression reinforcement, balanced reinforcement, concrete confinement, grades of concrete and steel, etc. The maximum values of plastic rotations for beams for IO, LS and CP states are 0.01, 0.02 and 0.025 radians, respectively, for all possible combinations of these parameters. The maximum values of plastic rotations will always yield a conservative design. The values of plastic rotations for maximum under-reinforced singly reinforced sections are presented in Table 17.9. The rotations for maximum under-reinforced sections using the limit state method are compared with the values as per the requirements of performance-based design. The values of plastic rotations corresponding to maximum under-reinforced sections as per IS:456 range from 0.0142 to 0.0489 radians (for M 60 concrete and Fe 415 steel), whereas the value for the CP or collapse prevention stage as per PBD is 0.025 for unconfined concrete. Thus, although the ductility of the sections is significantly high, the damage level will reach a much higher state than for the CP state, which can cause a significant amount of damage to the element. This seems to be a major shortcoming of IS:456 and necessary modification in the standard needs to be incorporated to reduce the amounts of plastic rotations in flexural members by increasing the minimum percentage of tension reinforcement in beams. However, if the minimum tension reinforcements are provided as per IS-13920, the values of plastic rotations for maximum under-reinforced sections range from 0.0092 to 0.0187 radians (for M60 concrete and Fe 415 steel). These values of curvature ductility are quite acceptable as the corresponding plastic rotations required in flexural members range between immediate occupancy (IO) and life safety states (LS), i.e., from 0.01 to 0.02 radians, i.e., the section lies within the limit state of collapse as mentioned in the next paragraph.

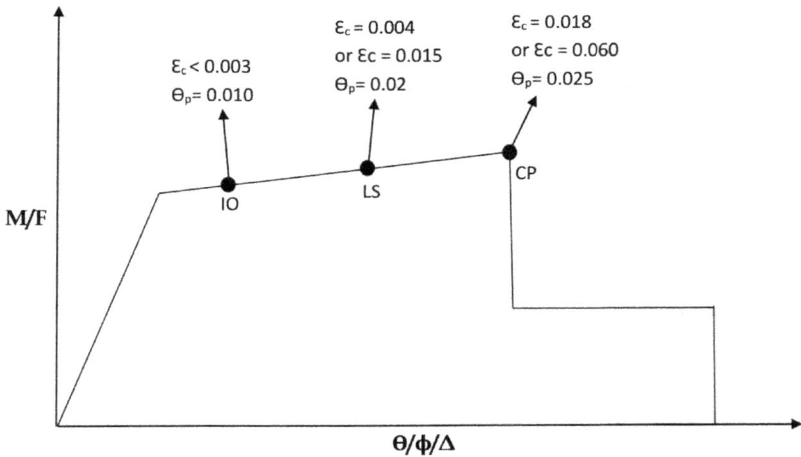

FIGURE 17.13 Different performance-based limit states and respective strain levels.

Although performance-based design cannot be directly correlated to the prescriptive method of design, some form of comparison between the limit state of collapse of the limit state method (LSM) and performance levels of performance-based design (PBD) may be performed with respect to limiting strains in concrete and steel. The limiting conditions at the different performance-based limit states are indicated in Figure 17.13, where ϵ_c, ϵ_s and θ_p are strains in concrete, strains in steel and plastic rotations, respectively. In this context it appears that the life safety (LS) state of PBD, where the limiting strain in steel is 0.004, is marginally beyond the limit state of collapse where the maximum strain in concrete is 0.0035. The theoretical moment curvature relationship has been developed following the guidelines of Park and Paulay, and is presented in Figure 17.14 for the two limiting grades of structural concrete (M20 and M60) as permitted by IS:456 using Fe415,

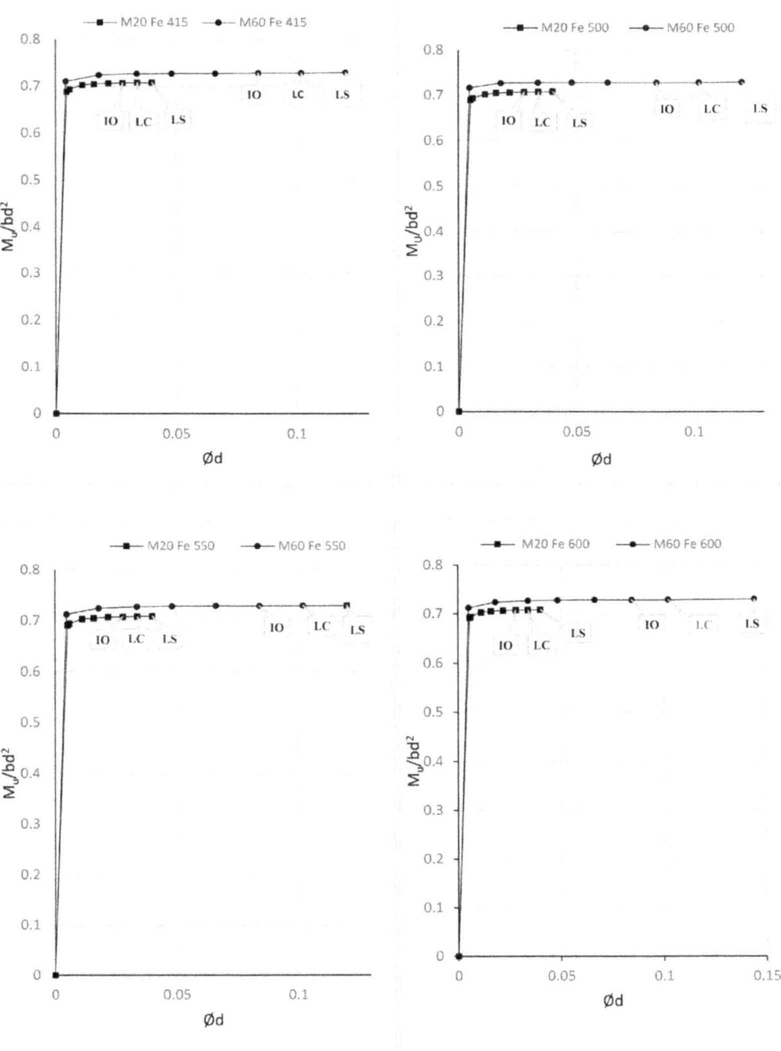

FIGURE 17.14 Theoretical moment curvature relationships of maximum under-reinforced sections.

Note: IO, immediate occupancy (PBD); limit state of collapse (LSM); life safety (LS).

Fe500, Fe550 and Fe600 steel grades. Figure 17.14 shows that the point corresponding to the limit state of collapse as per LSM is always before the LS state of PBD. This is valid for all grades of concrete and steel. However, for calculating the moment corresponding to life safety state of PBD in the moment curvature relationship, it has been assumed that the limiting or failure strain is 0.004 instead of 0.0035 for the same stress–strain diagram of concrete as per IS:456.

17.20 SUMMARY

From this chapter the following conclusions can be drawn:

1. Reinforced concrete sections can be made ductile only in flexural mode of failure, and hence failure should be initiated in flexural members. While designing beams, the designer should preferably avoid balanced sections and aim for under-reinforced sections. The value of ductility is minimum for balanced sections and highest for maximum under-reinforced sections. A lower level of ductility implies that the sections will not provide ample warning to inhabitants and may cause a risk to the life and property of inhabitants.
2. The maximum values of ductility of under-reinforced sections are very high, which is beneficial, but for ensuring those high levels of ductility, huge plastic rotations are required that may push the sections to higher limit states or to actual collapse of the structure. The minimum reinforcement specified in IS:456, being significantly less, leads to very high values of plastic rotations which are quite difficult to actually mobilize.
3. Higher grades of concrete with lower grades of steel enhance the performance of sections in terms of ductility. With an increase in the grade of steel, curvature ductility decreases, whereas with an increase in the grade of concrete ductility increases. Hence higher grades of concrete are beneficial, whereas higher grades of steel are detrimental for reinforced concrete.
4. Though performance-based design cannot be directly correlated with the prescriptive method of design, some form of comparison between the limit state of collapse and performance levels has been performed with respect to limiting strains in concrete with the help of moment curvature relationships. It appears that the life safety (LS) state, where the maximum strain in concrete is 0.004, is marginally beyond the limit state of collapse where the maximum strain in concrete is 0.0035.
5. Though the ductility of the sections will significantly increase with a minimum amount of tension reinforcement as per IS:456, the damage level will reach a higher state than the CP state, which can cause a significant amount of damage to the section. This seems to be a major shortcoming of IS:456 and necessary modification in the standard needs to be incorporated to reduce the amount of plastic rotations in flexural members.
6. The minimum tension reinforcement specified in IS:13920-2016 seems to be quite rational and leads to a plastic rotation close to the life safety (LS) state as per PBD, which is close to the limit state of collapse state as per the LSM of design.

BIBLIOGRAPHY

ATC-40. (1996). *Seismic Evaluation and Retrofit of Concrete Buildings*, Volume 1, Applied Technology Council, California.

Bandyopadhyay, J. N. (2011). *Design of Concrete Structures*, PHI Learning, New Delhi.

FEMA 356 (2000). *Pre-Standard and Commentary for the Seismic Rehabilitation of Buildings*, Federal Emergency Management Agency, Washington, DC.

Indian Standard IS: 456-2000 (April 2007). *Plain and Reinforced Concrete–Code of Practice*, Fourth Revision, Tenth Reprint April 2007, Bureau of Indian Standards, New Delhi.

Indian Standard IS: 13920-2016 (July 2016). *Ductile Design and Detailing of Reinforced Concrete Structures Subjected to Seismic Forces–Code of Practice*, First Revision, Bureau of Indian Standards, New Delhi.

Jha, B. (2020). *RCC Design – Prescriptive Versus Performance Based Design*, M. Tech Thesis, Department of Civil Engineering, NITTTR, Kolkata MAKAUT, West Bengal, India.

Jha, B. and Bhanja, S. (2021). 'Optimum design of R.C. Section and shortcomings of prescriptive method of design', *Asian Journal of Civil Engineering*, Vol. 22, No. 4, pp. 769–787. Doi: https://doi.org/10.1007/s42107-020-00346-9

Medhekar, M. S. and Jain, S. K. (1993). *Proposed Minimum Tension Reinforcement for Flexural Members*, Reprinted from the Bridge & Structural Engineer, Vol. XXIII, No. 2, pp. 78–79.

Murty, C.V.R. (2001). 'Shortcomings in structural design provisions of IS 456: 2000', *Indian Concrete Journal*, Vol. 75, No. 2, pp. 150–157.

Park, R. (1977). *Behaviour as Related to Design Criteria*, Proceedings of sixth World Conference on Earthquake Engineering, New Delhi.

Park, R. and Paulay, T. (1975). 'Ductile reinforced concrete frames some comments on the special provisions for seismic design of ACI 318-71 and on capacity design', *Bulletin of the New Zealand National Society for Earthquake Engineering*, Vol. 8, No. 1, pp. 70–90.

Park, R. and Paulay, T.(2018). *Reinforced Concrete Structures*, John Wiley, New York.

Paulay, T. and Priestley, M. J. N. (2018). *Seismic Design of Reinforced Concrete and Masonry Buildings*, John Wiley, New York.

Pillai, S. U. and Menon, D. (2012). *Reinforced Concrete Design*, Third Edition, Reprint 2012, Tata McGraw Hill Education, New Delhi.

Priestley, M. J. N, Calvi, G. M and Kowalsky (2007). *Displacement Based Seismic Design of Structures*, IUSS Press, Italy.

Shah, V. L. and Karve, S. R. (2016). *Limit State Theory and Design of Reinforced Concrete*, Eighth Edition, Structures Publications, Parva, Pune.

SP 16-1980 (1980). *Design Aids for Reinforced Concrete to IS 456:1978*, Special Publication SP 16, Bureau of Indian Standards, New Delhi.

Timoshenko, S. P. and Young, D. H. (2009). *Elements of Strength of Materials*, Fifth Edition, Reprint 2009, Original Publisher: Litton Educational Publishing, New York, Published in India by East–West Press, New Delhi.

18 Earthquake-Resistant Design of Structures

18.1 INTRODUCTION

In earthquake-resistant design of structures, the lateral load-resisting elements are chosen and suitably designed and detailed for energy dissipation. Under severe earthquakes they undergo a significant amount of inelastic deformations. Other structural elements are provided with sufficient strength. Concrete is a brittle material, whereas the reinforcements have ample ductility. The art of earthquake-resistant design needs the skilful combination of these two materials to incorporate sufficient ductility in the sections. This is the basic aspect of earthquake-resistant design and detailing. Since over-loading is almost a routine during earthquakes, aseismic design requires another attribute, ductility apart from the requirements of strength and stiffness. Incorporation of ductility in reinforced concrete sections is a major challenge as the bulk of the cross-section is made of concrete, which is a brittle material and only a small portion of the cross-section which is made of steel is ductile. The behaviour under loads of brittle and ductile materials is demonstrated in Figure 18.1, whereby it is observed that a brittle material will fail suddenly at a low value of total strain, whereas ductile material will fail after ample warnings under a high value of total strain. Ductile behaviour is always desirable as it will result in a reduction in the risk of loss of life and property. It is the duty of the designer to design the RC sections in such a manner that the behaviour of steel dominates the failure mechanism even though failure occurs generally by crushing of concrete. Hence earthquake-resistant design requires incorporation of sufficient ductility in sections and is very fast developing as an important branch of reinforced concrete design.

Safety against earthquakes can be ensured by providing energy-absorbing and energy-dissipating capability to the structure, known as ductility-based design (DBD); DBD has been adopted by the seismic codes of many countries.

It would be highly uneconomical to design a structure to withstand severe earthquakes without damage. Actual forces that can act on structures during earthquakes can be much higher than the design forces specified in the standards or codes of practice. Ductility arising from inelastic material behaviour with appropriate design and detailing and reserve strength in structures are relied upon for the deficit in actual and design forces. Hence earthquake-resistant design relies on inelastic behaviour of structures. The attributes of an earthquake resistant design can be listed as:

a) Good configuration
b) Sufficient strength
c) Sufficient stiffness
d) Sufficient ductility
e) Sufficient redundancy

DOI: 10.1201/9781003415398-18

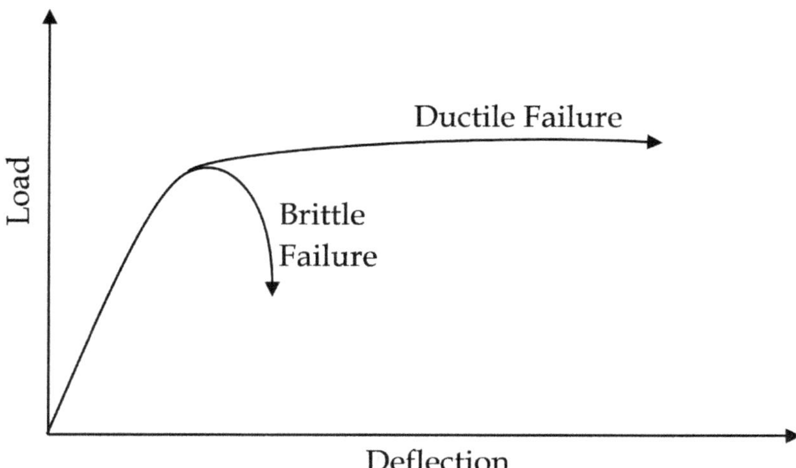

FIGURE 18.1 Load deflection curve for brittle and ductile materials.

Hence, proper planning, analysis and design are necessary for an optimum earthquake-resistant structure.

The term ductility continues to be an ambiguous parameter because of its different definitions and associated problems in its quantification. The definition and quantification of ductility have been presented in the chapter on ductility design.

Maximum ductility can be incorporated into reinforced concrete in flexural failure only. Other modes of failure, namely compression, shear, bond and torsion, are brittle in nature.

18.2 PHILOSOPHY OF DESIGN

The philosophy of earthquake-resistant design is as follows:

- Minor and frequent earthquakes should not cause any damage to the structure.
- Moderate earthquakes should not cause significant structural damage but can cause some non-structural damage.
- Major and infrequent earthquakes can result in structural damage but should not result in collapse of the structure.

18.3 ELEMENTS WHICH NEED DUCTILITY DESIGN

The elements which take part in the load path of the seismic forces need ductility to resist severe earthquake shaking without failure. The lateral load-resisting structural elements for reinforced concrete buildings which are mostly used can be listed as moment-resisting frames, moment-resisting frames with structural walls and structural walls. Seismic forces are inertial forces. In buildings, masses are assumed to be concentrated at the floor levels, as a result of which the seismic forces are generated at the floor levels. The structure should transmit the forces back to the ground safely without affecting its geometry and integrity. Thus from the centre of mass of the floor levels, the seismic forces need to move in the horizontal direction to be distributed to the moment-resisting frames or structural walls. Hence the floor slab acts as a structural element in transmitting the loads in the horizontal direction. The floor slabs are immensely stiff in their own planes and hence act as rigid diaphragms. The concept of ductility is related to the yield and ultimate deformations. Thus, for incorporation of ductility, the material must yield. The floor slabs acting as rigid diaphragms do not require any ductile design or detailing.

The moment-resisting frames consist of beams and columns. The beams are flexural members which are predominantly subjected to flexure. The columns are compression members which predominantly behave in compression, though they are also almost invariably subjected to flexure.

As discussed in earlier sections, failure in the limit state method of design in flexure or compression with flexure occurs after yielding of the materials, either steel or concrete. Failure occurs in two stages and the greater is the gap between the yield and the ultimate stage, the greater is the ductility of the section. Only in pure axial compression is the value of curvature ductility equal to one, since the yield and ultimate stages are merging failure of the section will be brittle in nature. For all other conditions, when the section is subjected to flexure or compression along with flexure curvature ductility values will be greater than one, i.e., they will have some ductility. However, when axial compressive forces are introduced into the section along with flexure, the ductility values will generally decrease and hence columns have lower ductility than beams. For increasing ductility of flexural members, the area of tension reinforcement is reduced or compression reinforcement is introduced, so that the gap between Stage I and Stage II can be increased. Since the amount of yielding of steel is significantly higher than that of concrete, over-reinforced sections (which have very low values of ductility) are banned in flexure as per IS:456-2000. These have been discussed in detail in Chapter 17.

It is a common notion among designers that reinforced concrete can be made ductile only in the flexural mode of failure. However, one needs to understand that a section subjected to flexure with axial compression has ductility but obviously the level is significantly reduced. However, ductility increases for sections subjected to flexure along with axial tension but since concrete is very weak in tension these conditions are not very desirable as they may lead to durability problems. The modes of failure like those in pure axial compression, shear and bond are brittle in nature and hence need to be avoided. Ductility depends on the grades of concrete and steel also. As per Indian Standards, design in the limit state of collapse in flexure is limited to balanced and under-reinforced sections only as over-reinforced sections are not permitted. However, for design in compression all types of sections are permitted. Failure should be initiated in the beams as then the failure will be more ductile. Failure in columns will be mostly less ductile and brittle in nature. From here the concept of strong columns and weak beams has emerged in earthquake-resistant design.

It is a common notion amongst designers that similar to beams, columns should also fail in under-reinforced condition to ensure that the mode of failure can be made ductile. With incorporation of axial compression curvature ductility of sections decreases. The variations of curvature ductility values of columns failing in highly under-reinforced, balanced and highly over-reinforced conditions are not significant. However, variations of curvature ductility values of beams are remarkably higher than those of columns. Thus, beams can be made considerably more ductile than columns. Failure of column sections in under-reinforced condition can be considered to be marginally better than failure in over-reinforced condition. Hence, the myth amongst designers that column sections should preferably fail in under-reinforced condition does not seem to be justified. Rather columns should be designed to be stronger than beams so that the beams, which are weaker, fail before the columns.

Curvature ductility values of columns failing in over-reinforced, balanced and under-reinforced conditions for concrete grades from M20 to M60 and HYSD steel grades from Fe 415 to Fe 600, calculated as per the limit state method of IS-456:2000, illustrate that variations are quite small, ranging from 1.78 to 2.56 only. The values of curvature ductility of similar sections, with identical percentage and arrangement of reinforcement, acting as doubly reinforced beams have yielded values ranging from 2.58 to 13.35. However, curvature ductility values of singly reinforced beams (from balanced to maximum under-reinforced condition) range from 1.88 to 23.44 for similar concrete and steel grades. Hence, it may be concluded that beams can be made significantly more ductile than columns.

Since a significant amount of ductility cannot be incorporated in columns, they are made sufficiently strong, and their failure should occur after that of beams. As the amount of compression increases ductility decreases, because of which many of the standards put a limit on compressive force on columns. As per IS:13920-2016, the factored axial compressive stress, i.e., the axial compressive stress corresponding to the ultimate or design load considering all load combinations relating to seismic load, shall be limited to $0.4f_{ck}$ for columns. As per Eurocode 8 the values of axial compressive force for primary seismic columns should not exceed 0.65 times the design values of compressive strength for axial force obtained from the analysis of seismic design situations. In addition, columns are mostly over-reinforced as a major part of the interaction charts for columns lie in the over-reinforced condition. This has been covered in Chapters 11 and 17.

Confining reinforcement with an appropriate amount and configuration will transform the brittle response of concrete to a ductile one. The compressive strength of confined concrete is significantly higher than that of unconfined concrete. In rectangular columns proper confinement of the vertical reinforcements will result in improved performance of the columns.

Two approaches to safety of structures against earthquakes are generally adopted: (i) to provide energy-absorbing and energy-dissipating capability to the structure known as ductility-based design (DBD) and (ii) to provide base isolation between the structure and foundation to limit the forces and accelerations experienced by the structure known as seismic base isolation (SBI). The advantage of designing ductile structures in earthquake-resistant design was highlighted in early 1930s and significant experimental and analytical studies have established that for earthquake-resistant design along with strength and serviceability, ductility should be given due importance. Unfortunately, there are different definitions for ductility and the quantifications of all these forms are somewhat cumbersome because of which some basic do's and don'ts of ductility design are generally followed by designers instead of performing in-depth theoretical analysis. Ductility depends upon several parameters, thus quantifications are quite problematic. Curvature ductility is mostly used by designers. The ductility of structural concrete members can be significantly improved by confinement. Closely spaced transverse reinforcement in the form of spirals or hoops with adequate cross-ties can result in good confinement of the core concrete and significantly improve the strength of the section.

The upper limit of strain in steel is two orders of magnitude higher than that of concrete. Hence it may be considered that an under-reinforced section may keep accumulating strain in tension steel much beyond the limiting steel strain, until eventually the extreme fibre of concrete reaches a strain of 0.0035 and is crushed. However, the associated curvature required to generate this strain of 0.0035 in concrete can be very large and may not even be realizable, particularly in deeper girders.

Adequate ductility must be built into the structure and a hierarchy must be formed among the various failure modes by virtue of which the ductile failure mode occurs before brittle ones.

Structural walls behave in a manner quite similar to columns as they are also subjected to compression with flexure. However, structural walls are mostly highly under-reinforced, because of which they can be designed and detailed as ductile sections. Since the amount of axial compression force is less with respect to moment, the flexure action will be predominant and hence the sections can be made more ductile. Therefore, a significant amount of ductility can be instilled into shear wall buildings.

18.4 PROVISIONS FOR DUCTILITY DESIGN AND DETAILING AS PER IS:13920-2016

In India, basic reinforced concrete design is covered in IS:456-2000, which deals with strength and serviceability requirements. However, earthquake-resistant structures need another attribute, i.e., ductility, in addition to strength and stiffness. It is also not necessary for every designer to calculate ductility values while performing design. However, the aim of design should be not only to make the structure safe and serviceable, but also to ensure that it exhibits improved performance up to and beyond design loads and even if actual failure occurs it should always aim at the preservation of life and property. Structures are inert and will behave as per their own natural characteristics. It is the duty of designers to understand their natural behaviour and design in such a manner that the structure will tread the path up to failure as conceived by the designer, ensuring minimum risk of loss of life and property.

An add-on standard to IS:456-2000, i.e., IS:13920-2016, has been published by the Bureau of Indian Standards whereby instead of actually estimating the values of ductility, some tips or guidelines in the form of do's and don'ts are provided which will improve the ductility of the sections.

IS:13920-2016 is primarily aimed at designing and detailing RC lateral load-resisting structural elements to give them sufficient strength, stiffness and ductility to resist severe earthquake shaking without collapse. India is divided into four seismic zones and the provisions of this standard are applicable to all three seismic zones except the least vulnerable one (Zone II), where it is not mandatory but optional.

The basic design for reinforced concrete structural elements is to be performed as per IS:456-2000, whereby the requirements of safety and serviceability are satisfied. For earthquake-resistant design along with safety and serviceability, ductility requirements need to be incorporated. IS:13920-2016, which is used as an add-on code to IS:456-2000, deals with ductility design and detailing of reinforced concrete structures. Some general specifications regarding concrete and steel are provided along with some general guidelines on analysis and building configurations. Provisions for beams include restrictions on beam dimensions, maximum and minimum longitudinal steel ratios, development length requirements, splicing, lapping of reinforcements, etc. Transverse reinforcement design has been prescribed corresponding to the development of plastic hinges at beam ends so that flexural failure will always occur before shear failure. Regarding columns, the maximum axial compressive stress has been restricted to ensure that section ductility is not remarkably reduced. Additional provisions regarding column dimensions, relative strength of beams and columns at joints, longitudinal and transverse reinforcement, special confining reinforcement, and shear design have also been specified. Experimental and analytical studies have proved that the shear stresses are the most dominant in the behaviour of RC frame joints. Design provisions of beam–column joints of moment-resisting frames have also been specified in the standard. Regarding provisions for special shear walls, detailed discussions are provided in Chapter 19.

18.5 SUMMARY

Earthquake-resistant design is nowadays emerging as an important offshoot of conventional reinforced concrete design. In order to ensure that structural elements can effectively resist earthquake effects, apart from strength and stiffness, ductility needs to be incorporated into the sections. Ductility design is a part and parcel of the limit state method as the sections go beyond their yield points before failure. In this chapter, in addition to the basic concepts of the structural design, the requirements of ductility design have been highlighted. Failure under pure axial compression is brittle. Ductility of flexural members is always higher than that of compression members as the inclusion of compressive force will reduce the ductility. Ductility can be increased by reducing the tension reinforcement and increasing the compression reinforcement into the sections. Requirements of seismic-resistant design as per IS:13920-2016 have been highlighted.

BIBLIOGRAPHY

Chowdhury, J. N., and Bhanja, S. (2023). "Justification of strong column weak beam theory by evaluation of ductility of column and beam sections using first principles", The Indian Concrete Journal, Accepted for publication.

Chowdhury, J. N., and Bhanja, S. (2024). "Columns Should Preferably Fail In Under-Reinforced Condition – Myth or Reality", The Indian Concrete Journal, Communicated.

EN 1998-1(2004) (English): *Eurocode-8: Design of Structures for Earthquake Resistance- Part-1, general rules, seismic action, and rules for buildings* (2011) [The European Union per regulation].

Jai Krishna, (1968). *Philosophy of Aseismic Design*, The second annual meeting of the Indian Society of Earthquake Engineering, University of Roorkee, Roorkee.

Jain, Ashok K. (2016). *Dynamics of Structures with MATLAB Applications*, Pearson India.

Murthy, C. V. R. (2001) 'Shortcomings in structural design provisions of IS:456-2000', *Indian Concrete Journal*, February, pp. 150–157.

Park, R. (1988). *Ductility Evaluation from Laboratory and Analytical Testing*, Vol. VIII, Proceeding to Ninth World Conference on Earthquake Engineering, August 2–9, Koyoto, Japan.

Park, R. and Paulay, T. (1975). 'Ductile reinforced concrete frames – Some comments on the special provisions for seismic design of ACI 318-71 and on capacity design', *Bulletin of the New Zealand National Society for Earthquake Engineering*, Vol. 8, No. 1, pp. 70–90.

Paulay, Thomas. (1988).*Seismic Design in Reinforced Concrete the State of The Art in New Zealand*, Vol. VIII, Proceedings of Ninth World Conference on Earthquake Engineering, Tokyo-Kyoto, Japan.

Paulay, T. and Priestley, M. J. N. (2018). *Seismic Design of Reinforced Concrete and Masonry Buildings*, John Wiley, New York.

Thakkar, S. K. (June 1990). *Seismic Design of Reinforced Concrete Framed Structures*, Bulletin Indian Society for Earthquake, technical paper, Paper no. 283, pp. 30–45.

19 Design of Shear Walls Following the Fundamental Principles of Limit State Method as per Indian and International Standards

19.1 INTRODUCTION

Shear walls can act as an excellent lateral load-resisting system in multi-storeyed buildings, in addition to providing sound and fire insulation. They have high in-plane stiffness and strength and are primarily lateral load-resisting elements while simultaneously carrying gravity loads. This chapter discusses the behaviour of single walls which are connected by floor slabs. The slabs act as rigid diaphragms with negligible bending resistance, so that only horizontal forces are transmitted. A single cantilever shear wall can be expected to behave essentially in the same way as a reinforced concrete element subjected primarily to flexure and axial forces. Shear wall, as a large cantilever, is subjected to bending moments and shear forces. Hence, the strength of the critical section across the wall can be evaluated from the moment–axial force interaction relationship. A single cantilever shear wall, as in Figure 19.1, can be expected to behave essentially as a reinforced concrete beam. The floor slabs of a multi-storeyed building act as horizontal diaphragms and will provide lateral support.

As per IS: 456-2000, reinforced concrete walls subjected to direct compression or combined flexure and direct compression may be designed by the limit state or working stress methods. Structural walls are designed to be in a high-ductility category according to modern international codes or designed with high-ductility requirements. They are expected to present extensive tensile deformations, especially in the plastic hinge region of their base, which can be up to 30% in the plastic hinge region. The compression failure of concrete and the instability of longitudinal steel are prevented by the provision of special confining reinforcement. In IS:456-2000, the stress–strain relationship of HYSD steel has been shown for CTD bars to have no definite yield point. The curve is non-linear beyond the elastic limit and is horizontal after the proof stress. This stress–strain relationship has to be used for developing stress blocks in steel.

19.2 DEFINITION AND CLASSIFICATIONS

The term "shear wall" is a misnomer because the walls behave predominantly in flexure and rarely in shear. Shear walls are primarily classified with respect to their height to length ratios but are also classified with respect to their shape. Shear walls may be planer but are often of barbell, L, T, U or I shape. The minimum ratio of length of wall to its thickness shall be 4, otherwise the element should be treated as a column. As per IS:13920-2016. "Ductile Design and Detailing of Reinforced Concrete Structures Subjected to Seismic Forces – Code of Practice" special shear wall should be classified as squat, intermediate and slender depending on the overall height, h_w to length, L_w ratio as follows:

DOI: 10.1201/9781003415398-19

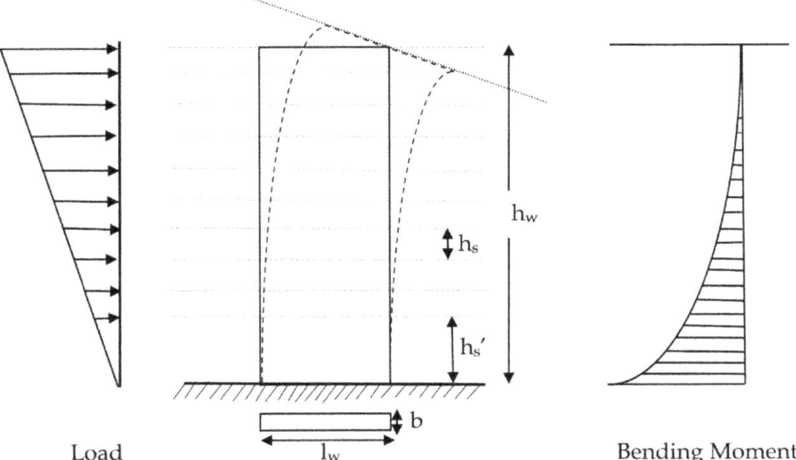

FIGURE 19.1 Behaviour of a cantilever shear wall.

(a) Squat walls: $\frac{h_w}{L_w} < 1$

(b) Intermediate walls: $1 \leq \frac{h_w}{L_w} \leq 2$

(c) Slender walls: $\frac{h_w}{L_w} > 2$

19.3 ADVANTAGES OF SHEAR WALLS

Reinforced concrete moment-resisting frames are commonly used for low-rise structures but when large lateral loads must be resisted, such as seismic loads or very high wind loads, shear walls are used. In such buildings, shear walls may be constructed between column lines or may be incorporated into stair walls, elevator, shafts, etc. In most structures, the lateral stiffness of the columns is much lower than that of the walls. Incorporation of shear walls significantly increases the stiffness and reduces the deformation of structures. When properly proportioned, shear walls possess adequate lateral stiffness to reduce interstorey distortions due to earthquake-induced motions and reduce the likelihood of damage to the non-structural elements of a building.

19.4 BRACED AND UNBRACED WALLS

A braced wall is a wall where reactions to lateral forces are provided by lateral supports such as floors or cross-walls. The wall can be considered as braced if all the lateral forces on the wall can be borne by the walls constructed at right angles to the wall being considered. Otherwise, it is considered as unbraced. An unbraced wall is a wall providing its own lateral stability such as a cantilever wall. Unbraced walls are designed for lateral loads. Braced walls subjected to vertical compression and horizontal forces can be designed by empirical equations given in the IS:456-2000. In-plane bending may be neglected if a horizontal cross-section of the wall is always under compression due to the combined effect of horizontal and vertical loads. Walls, which are part of the lateral load-resisting system of earthquake-resistant RC buildings, are termed as special shear walls or special RC structural walls and should be designed as per IS:13920-2016. These walls behave as unbraced walls.

19.5 STIFFNESS OF SHEAR WALLS

Shear walls have significant in-plane stiffness due to their greater depth. The stiffness of a shear wall may be calculated as the force required to cause unit displacement and for calculating the displacement, effects of both flexure and shear are to be considered. A typical calculation of stiffness is shown below. Refer to Figure 19.2.

For a rectangular cantilever wall,

$$\Delta_{flexure} = \frac{Hh^3}{3EI}$$

$$\Delta_{shear} = \frac{1.2Hh}{AG}$$

where, $\Delta_{flexure}$ = flexural deformation

Δ_{shear} = shear deformation

H = horizontal force

h = height of wall

I = Moment of inertia of wall $= \frac{t_w L_w^3}{12}$

A = cross-sectional area of wall $= t_w L_w$

E = Modulus of elasticity

G = Modulus of rigidity = 0.4E

$$\Delta = \frac{Hh^3}{3EI} + \frac{1.2\,Hh}{AG}$$

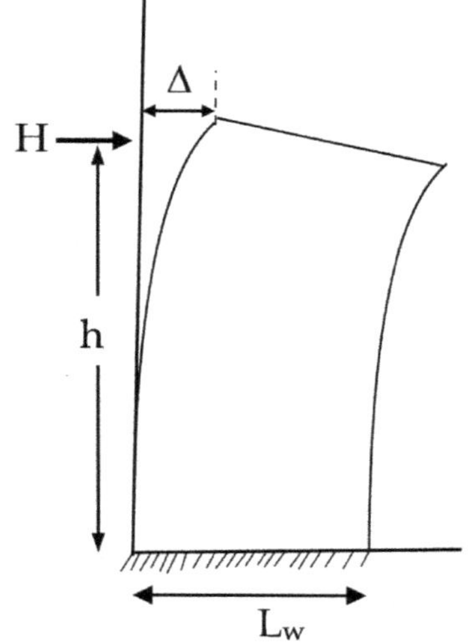

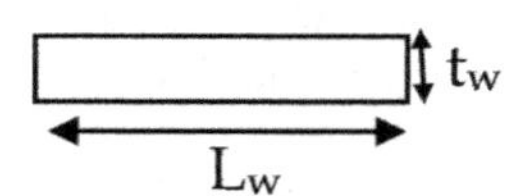

FIGURE 19.2 Lateral stiffness of walls.

$$\Delta = \frac{Hh^3}{3EI} + \frac{1.2Hh}{A(0.4E)}$$

$$= \frac{4Hh^3}{Et_w L_w{}^3} + \frac{3Hh}{t_w L_w E}$$

$$\Delta = \frac{H}{t_w E}\left(\frac{h}{L_w}\right)\left[4\left(\frac{h}{L_w}\right)^2 + 3\right]$$

$$k = \text{lateral stiffness} = \frac{H}{\Delta} = \frac{t_w E}{\left(\frac{h}{L_w}\right)\left[4\left(\frac{h}{L_w}\right)^2 + 3\right]}$$

The above expression does not consider the effect of foundation rocking.

19.6 STRESS RESULTANTS IN SHEAR WALLS

Shear walls act as cantilever beams fixed at the base. They are subjected to:

(i) Varying shear, which is transmitted from the floors, and is maximum at the base.
(ii) Varying bending moment that is maximum at the base and causes compression at one end and tension at the other.
(iii) Gravity loads from floors that produce compression.

Shear walls are to be designed as sections subjected to moment along with compressive forces. Single shear walls act as cantilever elements with maximum axial forces, bending moment, and shear forces acting at the base. The axial forces and bending moments result in normal stresses, whereas shear forces result in shear stresses. Since the effects of axial forces and bending moment are to be checked simultaneously, shear walls should be designed with the help of P–M interaction charts similar to columns.

19.7 MODES OF FAILURE IN WALLS

It has been reported that long shear walls with low height and having height to horizontal length ratios ($\frac{h_w}{l_w}$) <1 fail in shear. Squat or low-rise shear walls behave like deep beams and the flexural theories applied to beams are not applicable to them. The gravity loads acting on them are generally very low and the flexural demand is also small due to the large depth of the wall. Failure generally occurs due to the formation of inclined shear cracks. The components of the shear forces in the vertical and horizontal directions are resisted by the vertical and horizontal reinforcements. Shear walls with a height to depth ratio of about 3 behave essentially as vertical cantilever beams and should therefore be designed as flexural members with their strength governed by flexure rather than by shear. For slender shear walls the flexural theory is applicable and needs to be designed as sections subjected to flexure along with axial forces, i.e., in a manner similar to columns. The modes of failure of slender shear walls are presented in Figure 19.3.

If the vertical reinforcement in the wall is very low, failure may occur due to snapping of steel at the base. Hence minimum vertical reinforcement is to be provided in wall sections so that the moment of resistance of the section is greater than the cracking moment of the section. The second

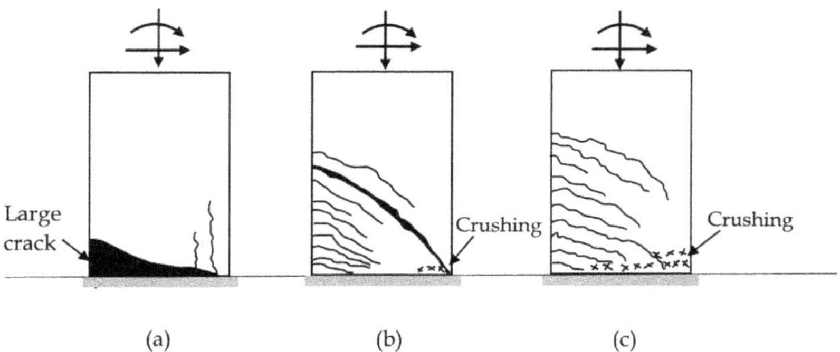

FIGURE 19.3 Failure modes for slender cantilever walls: (a) fracture of steel; (b) flexure shear failure due to diagonal tension or compression caused by shear; (c) failure due to yielding of flexural reinforcement in plastic hinge region along with crushing of concrete.

type of failure as shown in Figure 19.3(b) indicates shear failure with the formation of flexural cracks. It is a combination of flexure–shear failure, and the major flexural crack is oriented at about 45° at the base. The horizontal component of the shear force might be greater than the force resisted by the horizontal reinforcement at that section, leading to its failure along with crushing of concrete at the root of the crack. The third type of failure occurs after considerable yielding of flexural reinforcement with crushing of concrete at the outer compression fibres. The amount of flexural, shear reinforcement and the area of concrete in compression govern the failure mode. Shear walls need to be designed and detailed in such a manner that failure is ductile and the amount of plastic rotation at the bottom of the wall is significant. Generally, compressive forces in shear walls are comparatively less and moments are quite high, hence incorporation of ductility is comparatively easier than that for column sections.

As per IS:13920-2016, slender walls are defined as walls having overall height/horizontal length of wall (h_w/L_w) > 2. This ensures that the section will not behave as a deep beam, Bernoulli's theorem will be applicable and hence the assumptions for limit state of collapse in compression are applicable to the sections. When enlarged boundary elements are used along the wall edges or when the steel is concentrated at the edges, the onset of crushing is delayed.

As per IS:13920-2016, two types of flexural failure may take place in a slender rectangular shear wall, one is flexural tension failure and the other one is flexural compression failure. A flexural tension failure may happen when the tension steel yields before crushing of concrete in the extreme compression fibre. The flexural compression failure takes place when the concrete crushes at the extreme compression fibre before the tensile steel yields.

19.8 DESIGN OF SHEAR WALL SUBJECTED TO AXIAL LOAD AND UNIAXIAL BENDING

In many cases, the distributed reinforcement provided in walls has sufficient moment capacity without any additional reinforcement at the ends or in boundary elements. The vertical reinforcement may be presented by a plate of length equal to the length of wall (L_w) and a thickness such that the area of steel (A_s) is the same as that provided by uniformly distributed vertical reinforcing bars across the wall section. The strength of the critical section across the wall needs to be evaluated from the moments–axial force interaction relationship. In a reinforced concrete section under flexure or flexure with compression, the flexural strength of the section is attained when the extreme concrete compression fibre reaches the ultimate strain. Failure occurs by crushing of concrete and stress block in concrete in compression fully develops; this is valid for all positions of neutral axis within the section but for neutral axis beyond the section failure will occur in concrete at intermediate strain levels between 0.002 and 0.0035 as per IS:456-2000. This has been explained in detail in Chapter 11. Walls are similar to columns and are subjected to compression or flexure in isolation or in combination. Hence the axial and flexural strength of walls are to be derived by using the assumptions for the limit state of collapse in compression. For a section subjected to axial compression and flexure simultaneously, both the axial compression and moment result in normal stresses and hence the section can fail in axial compression or moment alone or due to a combination of both. Therefore, while checking the strength of the section, since two parameters are to be checked simultaneously, interaction charts are necessary. The addition of axial force and a distribution of reinforcement that cannot readily be represented by lumped compression and tension areas makes the calculations of flexural strength of columns and walls more tedious and complicated than for beams. Hence flexural strengths of columns and walls are estimated using design charts and computer programs.

19.9 DEVELOPMENT OF STRAIN DISTRIBUTION DIAGRAMS AND STRESS BLOCKS

Reinforced concrete design is based on a number of assumptions, of which Bernoulli's theorem and a perfect bond between concrete and steel can be considered as the basic ones. Bernoulli's principle implies that strain distribution in the concrete and steel at the various points across the cross-section is proportional to the distance of those points from the neutral axis, i.e., strain profiles are linear. Bernoulli's theorem has been found to be valid both in the compression and tension zones of a reinforced concrete section except in the case of deep beams.

In the case of compression steel, the maximum strain should not exceed the value of limiting strain for concrete to satisfy the condition of strain compatibility. Since as per IS:456-2000 the maximum strain in concrete at the outermost fibre has been considered as 0.0035, in the case of compression steel, the maximum value of strain has to be considered the same. Under balanced condition, the strain at the outermost fibre in compression reaches the ultimate or limiting strain in concrete, and the outermost fibre of steel in tension reaches the yield strain in steel simultaneously.

From the previous discussions it may be stated that strain distribution diagrams form the basis of the design of a section. For balanced condition, the strain distribution diagram is well defined as the critical or maximum values at the outermost compressed fibre in concrete and tension steel are specified. The value of limiting compressive strain in concrete in flexural compression is 0.0035 for all grades of structural concrete up to M60 as per IS:456-2000 and the limiting strain in steel is $f_y/1.15E_s + 0.002$. Only one straight line can pass through two points and hence the strain distribution diagram is well defined in a balanced condition. For conditions other than balanced, that is under-reinforced or over-reinforced, the limiting strain in concrete is known but a second point is to be defined on the strain distribution diagram so that a line joining these two points can define the strain distribution. Similar to columns, arbitrary values of stresses at the outermost tension steel are assumed wherefrom the corresponding strain in steel can be obtained from the stress–strain diagram of steel to plot the strain distribution diagram. Once the strain distribution diagram is plotted the position of the neutral axis is defined, the stress blocks in concrete and steel can be drawn and the corresponding capacities of the section in axial force and moment can be determined. This is discussed in detail in Chapter 11.

The strain distribution diagram across a wall cross-section is shown in Figure 19.4. The stress block for concrete is drawn on the compression side of the neutral axis as per the requirements of IS:456-2000. For the wall the uniformly distributed vertical reinforcement is represented by a steel plate placed at the centre of the wall spanning the entire wall length. The steel plate is subjected to compression on one side of the neutral axis and tension on the other. On the compression side, due to strain compatibility, strains in concrete and steel should be identical. Limiting strain in concrete in flexural compression is 0.0035, which should be the strain in steel at that level. For HYSD bars, the limiting/yield strains in steel are 0.0038, 0.00418, 0.0044 and 0.0046 for steel grades of Fe415, Fe500, Fe550 and Fe600, respectively. Hence, none of the HYSD bars will yield in compression. Accordingly, the stress blocks for different grades of steel are to be developed on the compression side of the neutral axis.

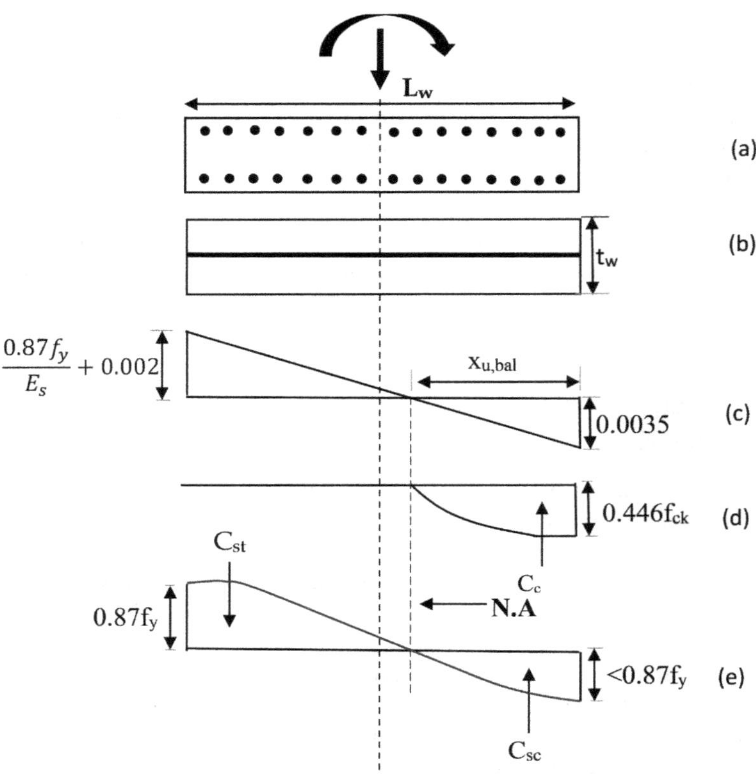

FIGURE 19.4 Strain and stress distribution across a shear wall section: (a) cross-section of a rectangular shear wall; (b) cross-section showing vertical reinforcement as an equivalent plate; (c) strain distribution diagram as per IS:456-2000 in balanced condition; (d) stress block for concrete; (e) stress block for steel as per IS:456-2000 in balanced condition.

Note: C_c and C_s represent force in concrete and steel, respectively.
C_{sc} and C_{st} represent force in steel in compression and tension, respectively.

One important point that needs to be mentioned here is that the design philosophy of columns and walls is quite similar. The difference is that at a column, biaxial bending may act and the column is to be designed considering biaxial bending, but in the case of a shear wall only uniaxial bending is considered as out-of-plane bending is negligible in comparison to in-plane bending.

To design a shear wall, the stress blocks of steel both in tension and compression are required along with the stress block of concrete. In IS:456 the parameters of stress block for concrete are given but no such parameters for stress block of steel are present. Hence, before designing a shear wall the stress blocks for steel in both tension and compression are to be developed.

19.9.1 Under Pure Axial Load Only

Under this condition the wall is subjected to a constant compressive strain of 0.002. Hence the stress block for concrete will be a rectangular one having a constant stress of $0.446f_{ck}$, and for steel also it will be a rectangular one with the stress corresponding to the grade of steel for a strain of 0.002. The axial load capacity of the section can be determined from the stress block for concrete and steel. Refer to Figure 19.5.

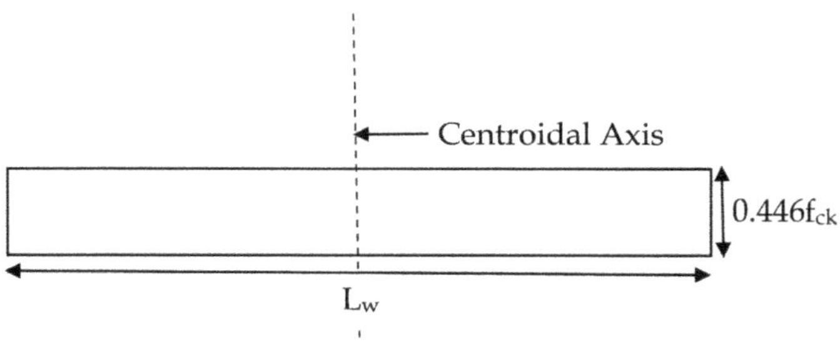

(a) Stress block of concrete in pure axial compression

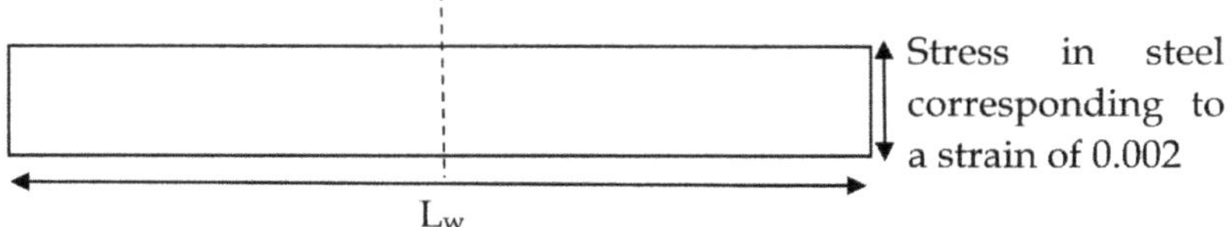

(b) Stress block of steel in pure axial compression

FIGURE 19.5 Stress blocks of concrete and steel under pure axial compression.

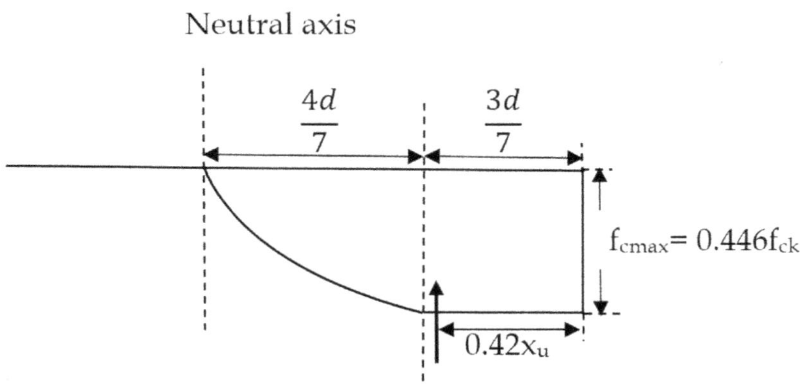

FIGURE 19.6 Stress block for concrete.

19.9.2 Stress blocks for concrete and steel in balanced condition

19.9.2.1 Stress block for concrete

The stress block for concrete is valid for failure under balanced, over-reinforced and under-reinforced sections (Figure 19.6). The parameters of stress block are similar and only the depth of the neutral axis varies depending upon the type of failure.

Area of the stress block = $0.36f_{ck}.x_u$

Distance of the centroid from the outermost fibre in compression = $0.42x_u$

These are valid for the neutral axis within the section. For positions of neutral axis outside the section, the maximum stress at the outermost fibre will vary from 0.0035 to 0.002 depending upon the position of the neutral axis and the portion of the stress–strain diagram of concrete falling within the depth of the section will constitute the stress block. This will be discussed in detail while developing the safety profiles. Refer to Figure 19.6.

19.9.2.2 Stress blocks for steel

The stress blocks act on each lamina of concrete or steel and by multiplying these areas with the width of the concrete section or steel plate the corresponding forces in concrete and steel are obtained. The stress–strain diagrams of HYSD steels are specified in SP:16 (stress–strain relationship of HYSD bars has been discussed in detail in Chapter 11). Beyond the elastic range, the total strains, i.e., sum of elastic and inelastic strains, are related to the corresponding stress levels. The area of the stress–strain diagram from the strain levels of 0 to limiting value, i.e., $\frac{f_y}{1.15E_s}+0.002$, consists of one triangle and five trapeziums which form the stress block on the tension side, whereas for the compression side the shape of the stress block is similar, but the strain values range from 0 to 0.0035. For Fe600, the stress block in compression will be different as the strain of 0.0035 will be reached before the stress reaches $0.975f_{yd}$. The areas of the stress blocks in tension and compression and their centroidal distance from the outermost fibre in tension or compression can be easily calculated. The axial and moment of resistance capacities of the shear wall sections can be determined for all grades of structural concrete up to M 60 (as per IS:456-2000) and HYSD bars of grades Fe415, Fe500, Fe550 and Fe600 with any percentage of vertical longitudinal reinforcement. For flexure design as per IS:456, the stress–strain diagram of HYSD bars is not purely bilinear but consists of an intermediate non-linear portion between the elastic and plastic regions. As per SP:16, for cold worked bars, the stress is proportional to strain up to a stress of $0.8f_y$. The nature of the curve from $0.8f_y$ to $1.0f_y$ is defined by six values of inelastic strains corresponding to stresses at $0.8f_y$, $0.85f_y$, $0.9f_y$, $0.95f_y$, $0.975f_y$ and $1.0f_y$. Though the points are joined by a smooth curve, linear interpolation has been allowed for intermediate values. Hence the relationship between stress and strain may be obtained by drawing five straight lines between the six cardinal points. After the steel yields the curve becomes horizontal. For strain compatibility, strain in steel must be equal to the strain in the surrounding concrete. These are shown in Figures 19.7–19.11.

19.9.2.3 Stress blocks for HYSD steel

19.9.2.3.1 Calculation of stress blocks of HYSD steel in tension

Area of stress block of steel in tension zone

For all grades of HYSD steel the area of the stress blocks in tension is calculated as per the following expression (Refer to Figure 19.7):

$$\begin{aligned}A_T = {} & \frac{1}{2}\,k_1 f_{yd} y_1 (L_w - x_u) + \frac{1}{2}(k_1 f_{yd} + k_2 f_{yd})\,\{y_2(L_w - x_u) - y_1(L_w - x_u)\} \\ & + \frac{1}{2}(k_2 f_{yd} + k_3 f_{yd})\,\{y_3(L_w - x_u) - y_2(L_w - x_u)\} \\ & + \frac{1}{2}(k_3 f_{yd} + k_4 f_{yd})\,\{y_4(L_w - x_u) - y_3(L_w - x_u)\} \\ & + \frac{1}{2}(k_4 f_{yd} + k_5 f_{yd})\,\{y_5(L_w - x_u) - y_4(L_w - x_u)\} \\ & + \frac{1}{2}(k_5 f_{yd} + k_6 f_{yd})\,\{(L_w - x_u) - y_5(L_w - x_u)\}\end{aligned}$$

$$\begin{aligned}\text{Hence, } A_T = {} & \frac{1}{2}(L_w - x_u) f_{yd}\,[y_1 k_1 + (k_1 + k_2)(y_2 - y_1) + (k_2 + k_3)(y_3 - y_2) \\ & + (k_3 + k_4)(y_4 - y_3) + (k_4 + k_5)(y_5 - y_4) + (k_5 + k_6)(1 - y_5)]\end{aligned}$$

Where, $k_1 = 0.8$, $k_2 = 0.85$, $k_3 = 0.9$, $k_4 = 0.95$, $k_5 = 0.975$, $k_6 = 1$

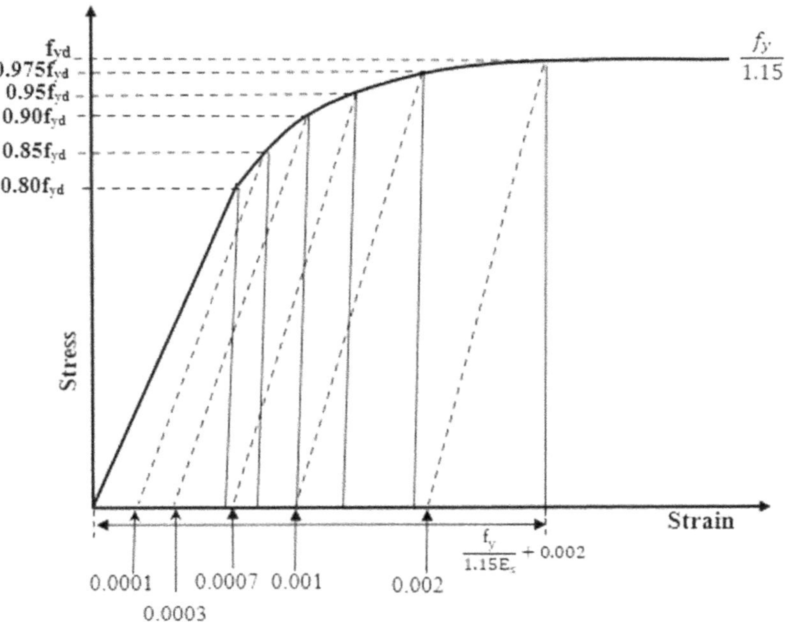

(a) Stress strain diagram

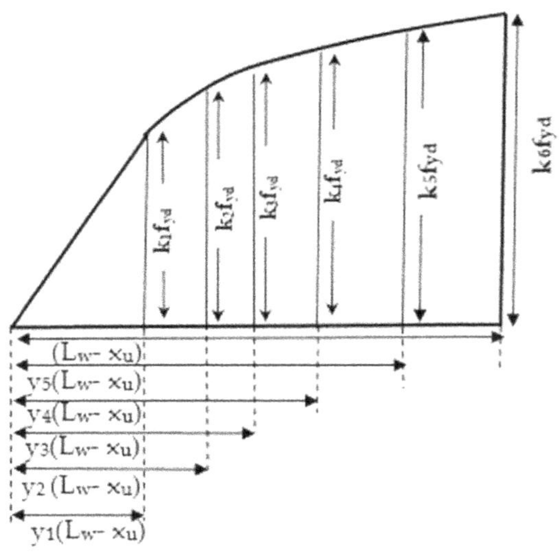

(b) Stress block

FIGURE 19.7 Stress–strain diagram and stress block for Fe415, Fe500, Fe550 and Fe600 grade steel in tension obtained from the stress–strain diagram ranging from strain values of 0 to $\frac{f_y}{1.15E_s} + 0.002$ (Balanced Condition).

Note: *k*: fraction of stress with respect to design stress for different segments of the stress block.
y: Fraction of distances of the ends of the segment from the neutral axis in terms of overall length of the stress block.

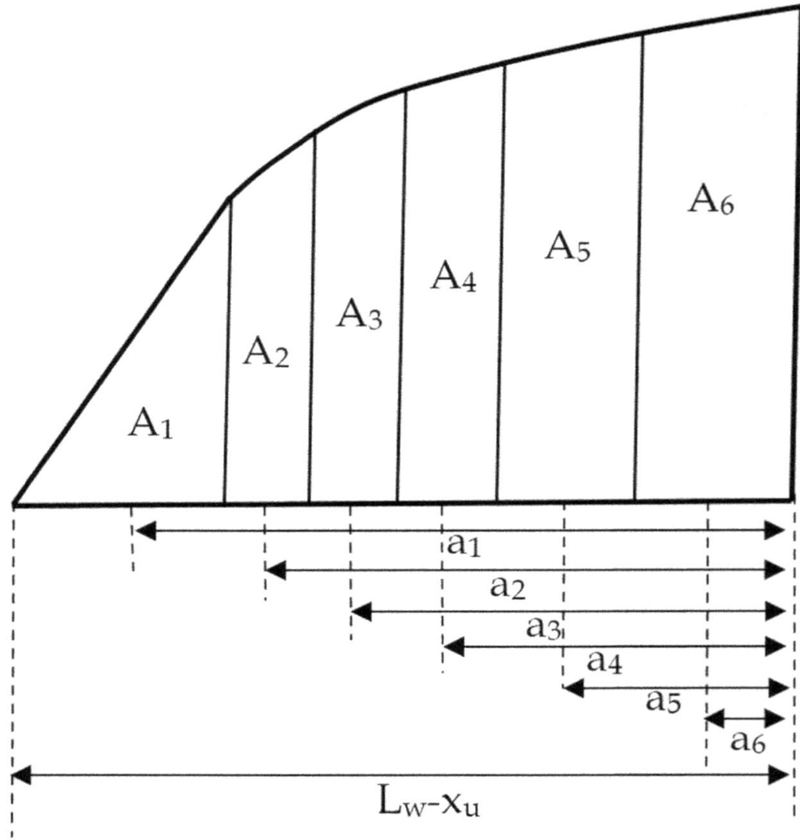

FIGURE 19.8 Different segments of stress blocks showing the areas and centroidal distances.

$$y_1 = (\frac{0.8f_y}{1.15E_s} / y), \qquad y_2 = [(\frac{0.85f_y}{1.15E_s} + 0.0001) / y)]$$

$$y_3 = [(\frac{0.9f_y}{1.15E_s} + 0.0003) / y)], \; y_4 = [(\frac{0.95f_y}{1.15E_s} + 0.0007) / y)]$$

$$y_5 = [(\frac{0.975f_y}{1.15E_s} + 0.001) / y)]$$

Where, $y = \frac{f_y}{1.15E_s} + 0.002$

Calculation of centroidal distance

For Fe415, Fe500, Fe550 and Fe600 grades of steel, stress blocks of tension zone consist of one triangular and five trapezoidal segments. Let the area of triangular segment be A_1, and the areas of the trapezoidal segments be A_2, A_3, A_4, A_5 and A_6, respectively. Let the distance of the centroid of the triangular segment from extreme fibre be a_1 and centroidal distances of the trapezoidal segments from extreme fibre be a_2, a_3, a_4, a_5 and a_6, respectively. Refer to Figure 19.8.

Distance of centroidal axis of stress block from the extreme fibre,

$$x = \frac{A_1 a_1 + A_2 a_2 + A_3 a_3 + A_4 a_4 + A_5 a_5 + A_6 a_6}{A_1 + A_2 + A_3 + A_4 + A_5 + A_6}$$

where, $a_1 = (L_w - x_u) - \frac{2}{3} y_1 (L_w - x_u)$

$$a_2 = (L_w - x_u) \{(\frac{2k_1 + k_2}{k_1 + k_2})(\frac{y_2 - y_1}{3}) + (1 - y_2)\}$$

$$a_3 = (L_w - x_u) \{(\frac{2k_2 + k_3}{k_2 + k_3})(\frac{y_3 - y_2}{3}) + (1 - y_3)\}$$

$$a_4 = (L_w - x_u) \{(\frac{2k_3 + k_4}{k_3 + k_4})(\frac{y_4 - y_3}{3}) + (1 - y_4)\}$$

$$a_5 = (L_w - x_u) \{(\frac{2k_4 + k_5}{k_4 + k_5})(\frac{y_5 - y_4}{3}) + (1 - y_5)\}$$

$$a_6 = (L_w - x_u) (\frac{2k_5 + k_6}{k_5 + k_6})(\frac{1 - y_5}{3})$$

The parameters of the stress block in tension derived for HYSD bars are valid for all grades of steel, namely Fe415, Fe500, Fe550 and Fe600. Hence, using the above expressions, the areas of stress blocks and their centroidal distances from the outermost fibre can be determined for any grade of steel. For failure under balanced condition the areas of the stress blocks in tension and their corresponding centroidal distances are shown in Tables 19.1–19.4.

19.9.2.3.2 Area of stress block for HYSD steel in compression

Due to strain compatibility, the maximum compressive strain at the outermost steel fibre in compression should be equal to the failure strain of concrete in flexural compression, i.e., 0.0035. The

TABLE 19.1
Typical calculations for tension stress block for Fe415 steel as per IS:456-2000

Calculation of tension stress block for Fe415						
Grade of steel (f_y)	415 N/mm^2					
Modulus of elasticity (E_s)	200,000 N/mm^2					
Design stress (f_{yd})	f_y/1.15 N/mm^2					
Coefficient of design stress (k)	0.8	0.85	0.9	0.95	0.975	1
Design stress value (N/mm^2)	288.696	306.739	324.783	342.826	351.848	360.870
Corresponding strain	0.001443	0.001634	0.001924	0.002414	0.002759	0.003804
Fraction of tension zone ($L_w - x_u$)	y_1	y_2	y_3	y_4	y_5	y_6
(y_i)	0.379	0.429	0.506	0.635	0.725	1
Segments of stress block	A_1	A_2	A_3	A_4	A_5	A_6
Area of individual segment	0.132	0.036	0.058	0.104	0.076	0.236
Area of stress block	0.641 ($L_w - x_u$)f_y					
Distance of centroid of individual	a_1	a_2	a_3	a_4	a_5	a_6
segment from extreme fibre	0.747	0.595	0.532	0.429	0.322	0.130
Distance of centroid of stress block from extreme fibre			0.391 ($L_w - x_u$)			

TABLE 19.2
Typical calculations for tension stress block for Fe500 steel as per IS:456-2000

Calculation of tension stress block for Fe500

Grade of steel (f_y)	500 N/mm^2					
Design stress (f_{yd})	f_y/1.15 N/mm^2					
Coefficient of design stress	0.8	0.85	0.9	0.95	0.975	1
Design stress value (N/mm^2)	347.826	369.565	391.304	413.043	423.913	434.783
Corresponding strain	0.001739	0.00194	0.00225	0.00276	0.00312	0.004174
Fraction of tension zone	y_1	y_2	y_3	y_4	y_5	y_6
$(L_w - x_u)$ (y_i)	0.417	0.467	0.541	0.663	0.747	1
Segments of stress block	A_1	A_2	A_3	A_4	A_5	A_6
Area of individual segment	0.145	0.036	0.056	0.098	0.071	0.217
Area of stress block	0.623 $(L_w - x_u)f_y$					
Distance of centroid of individual	a_1	a_2	a_3	a_4	a_5	a_6
segment from extreme fibre	0.722	0.558	0.496	0.398	0.297	0.120
Distance of centroid of stress block from extreme fibre				0.383 $(L_w - x_u)$		

TABLE 19.3
Typical calculations for tension stress block for Fe550 steel as per IS:456-2000

Calculation of tension stress block for Fe550

Grade of steel (f_y)	550 N/mm^2					
Modulus of elasticity (E_s)	200,000 N/mm^2					
Design stress (f_{yd})	f_y/1.15 N/mm^2					
Coefficient of design stress	0.8	0.85	0.9	0.95	0.975	1
Design stress value (N/mm^2)	382.609	406.522	430.435	454.348	466.304	478.261
Corresponding strain	0.001913	0.00213	0.002452	0.002972	0.003333	0.004391
Fraction of tension	y_1	y_2	y_3	y_4	y_5	y_6
zone $(L_w - x_u)$ (y_i)	0.436	0.486	0.558	0.677	0.759	1
Segments of stress block	A_1	A_2	A_3	A_4	A_5	A_6
area of individual segment	0.152	0.036	0.055	0.095	0.069	0.207
Area of stress block	0.614 $(L_w - x_u)f_y$					
Distance of centroid of individual segment from extreme fibre	a_1	a_2	a_3	a_4	a_5	a_6
Area of stress block	0.710	0.539	0.478	0.382	0.285	0.115
Distance of centroid of stress block from extreme fibre	0.380 $(L_w - x_u)$					

TABLE 19.4
Typical calculations for tension stress block for Fe600 steel as per IS:456-2000

Calculation of tension stress block for Fe600						
Grade of steel (f_y)	**600 N/mm²**					
Modulus of elasticity (E_s)	200,000 N/mm²					
Design stress (f_{yd})	f_y/1.15 N/mm²					
Coefficient of design stress	0.8	0.85	0.9	0.95	0.975	1
Design stress value (N/mm²)	417.391	443.478	469.565	495.652	508.696	521.739
Corresponding strain	0.002087	0.002317	0.002648	0.003178	0.003543	0.004609
Fraction of tension zone	y_1	y_2	y_3	y_4	y_5	y_6
$(L_w - x_u)$ (y_i)	0.453	0.503	0.575	0.690	0.769	1
Segments of stress block	A_1	A_2	A_3	A_4	A_5	A_6
Area of individual segment	0.158	0.036	0.055	0.093	0.066	0.198
Area of stress block	$0.605\ (L_w - x_u)f_y$					
Distance of centroid of individual	a_1	a_2	a_3	a_4	a_5	a_6
segment from extreme fibre	0.698	0.522	0.461	0.367	0.273	0.111
Distance of centroid of stress block from extreme fibre	$0.376\ (L_w - x_u)$					

parameters of the stress blocks are to be derived accordingly. For steel grades up to Fe550 the strain value of 0.0035 lies between the strain corresponding to $0.975f_{yd}$ and f_{yd}, however for grade Fe600 this value lies between the strains corresponding to $0.95f_{yd}$ and $0.975f_{yd}$. Accordingly, the areas of the stress blocks and their centroidal distances are to be calculated. Refer to Figure 19.9

For Fe415, Fe500 and Fe550, areas of stress blocks in compression are to be calculated as per the following expression:

$$A_C = \frac{1}{2}y_1 x_u k_1 f_{yd} + \frac{1}{2}(k_1 f_{yd} + k_2 f_{yd})(y_2 x_u - y_1 x_u) + \frac{1}{2}(k_2 f_{yd} + k_3 f_{yd})(y_3 x_u - y_2 x_u) + \frac{1}{2}(k_3 f_{yd} + k_4 f_{yd})(y_4 x_u - y_3 x_u) + \frac{1}{2}(k_4 f_{yd} + k_5 f_{yd})(y_5 x_u - y_4 x_u) + \frac{1}{2}(k_5 f_{yd} + k_6 f_{yd})(x_u - y_5 x_u)$$

$$A_C = \frac{1}{2}x_u f_{yd}\,[y_1 k_1 + (k_1 + k_2)(y_2 - y_1) + (k_2 + k_3)(y_3 - y_2) + (k_3 + k_4)(y_4 - y_3) + (k_4 + k_5)(y_5 - y_4) + (k_5 + k_6)(1 - y_5)]$$

Where, $k_1 = 0.8$, $k_2 = 0.85$, $k_3 = 0.9$, $k_4 = 0.95$, $k_5 = 0.975$ and k_6 is the fraction of design stress f_{yd} corresponding to the strain value of 0.0035. The values of k_6 are 0.9925, 0.9844 and 0.9787 for Fe415, Fe500, and Fe550 respectively.

Calculation of centroidal distances

For Fe415, Fe500 and Fe550 grades of steel, stress blocks of compression zone consist of one triangular and five trapezoidal segments. Let the area of triangular segment be A_1 and the areas of the trapezoidal segments be A_2, A_3, A_4, A_5 and A_6, respectively. Let the distance of centroid of the triangular segment from extreme fibre be a_1 and centroidal distances of the trapezoidal segments from extreme fibre be a_2, a_3, a_4, a_5 and a_6, respectively. Refer to Figure 19.10

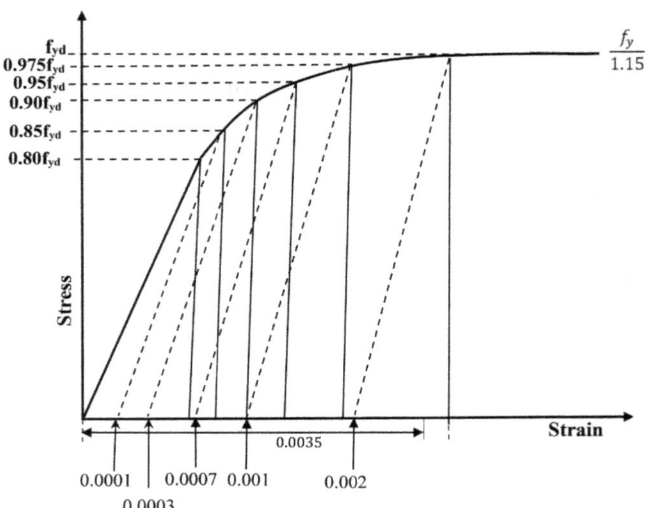

Stress strain diagram

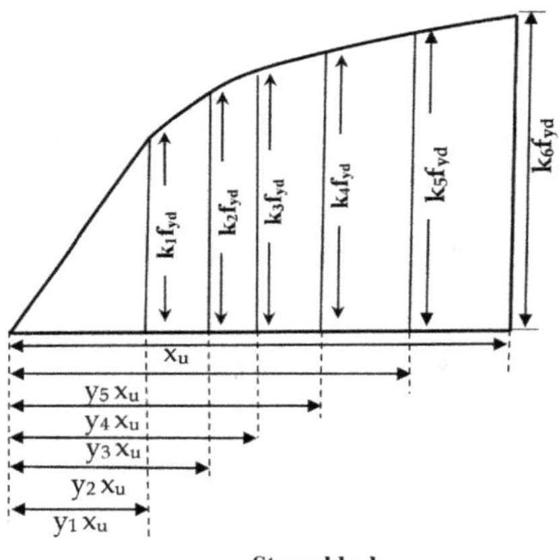

Stress block

FIGURE 19.9 Stress–strain diagram and stress blocks for Fe415, Fe500 and Fe550 grade steels in compression obtained from the stress–strain diagram ranging from strain values of 0 to 0.0035.

Note: *k*: Fraction of stress with respect to design stress for different segments of the stress block. *y*: Fraction of distances of the ends of the segment from the neutral axis in terms of overall length of the stress block.

The distance of centroid of the stress block from the extreme fibre is calculated as per the following expression:

$$\overline{x} = \frac{A_1 a_1 + A_2 a_2 + A_3 a_3 + A_4 a_4 + A_5 a_5 + A_6 a_6}{A_1 + A_2 + A_3 + A_4 + A_5 + A_6}$$

$$\text{where, } a_1 = (x_u) - \frac{2}{3} y_1 x_u$$

$$a_2 = (x_u) \{ (\frac{2k_1 + k_2}{k_1 + k_2})(\frac{y_2 - y_1}{3}) + (1 - y_2) \}$$

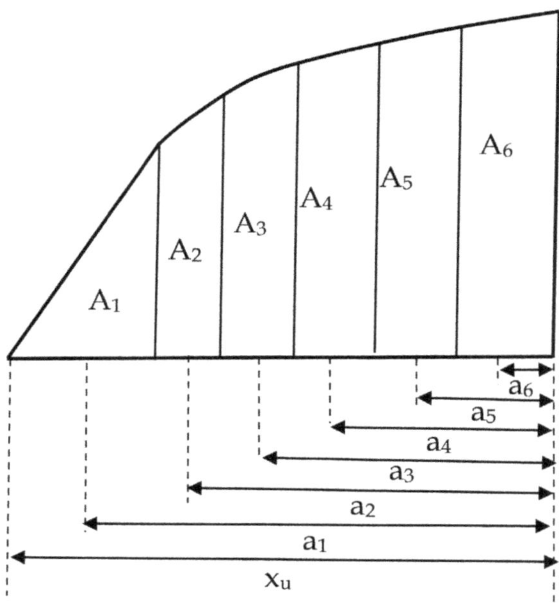

FIGURE 19.10 Different segments of compression stress blocks showing areas and centroidal distances.

$$a_3 = (x_u)\ \{(\frac{2k_2 + k_3}{k_2 + k_3})(\frac{y_3 - y_2}{3}) + (1 - y_3)\}$$

$$a_4 = (x_u)\ \{(\frac{2k_3 + k_4}{k_3 + k_4})(\frac{y_4 - y_3}{3}) + (1 - y_4)\}$$

$$a_5 = (x_u)\ \{(\frac{2k_4 + k_5}{k_4 + k_5})(\frac{y_5 - y_4}{3}) + (1 - y_5)\}$$

$$a_6 = (x_u)\ (\frac{2k_5 + k_6}{k_5 + k_6})(\frac{1 - y_5}{3})$$

The parameters of the stress block in compression derived for HYSD bars are valid for grades of steel Fe415, Fe500 and Fe550. Hence, using the above expressions, the areas of stress blocks and their centroidal distances from the outermost fibre can be determined for these grades of steel. For failure under balanced condition the areas of the stress blocks in compression and their corresponding centroidal distances are shown in Tables 19.5–19.7.

Area of stress block of steel in compression zone for Fe600

In the case of compression, maximum strain at failure is restricted to 0.0035. For Fe600 grade of steel the stress corresponding to this strain is 508.163 N/mm^2. This is less than the value of design stress $0.975f_{yd.}$ Thus, for Fe600 grades of steel only the first five segments of stress block have to be considered in calculation of area and centroidal distances. Refer to Figure 19.11.

TABLE 19.5
Typical calculations for compression stress block for Fe415 steel as per IS:456-2000

Calculation of compression stress block for Fe415						
Grade of steel (f_y)	415 N/mm²					
Modulus of elasticity (E_s)	200,000 N/mm²					
Design stress (f_{yd})	f_y/1.15 N/mm²					
Coefficient of design stress (k_i)	0.8	0.85	0.9	0.95	0.975	0.993
Design stress value (N/mm²)	288.696	306.739	324.783	342.826	351.848	358.242
Corresponding strain	0.001443	0.001634	0.001924	0.002414	0.002759	0.0035
Fraction of x_u (y_i)	0.412	0.467	0.550	0.690	0.788	1
Coefficient of characteristic stress	0.696	0.739	0.783	0.826	0.848	0.863
Segments of stress block	A_1	A_2	A_3	A_4	A_5	A_6
Area of individual segment	0.143	0.039	0.063	0.113	0.083	0.181
Area of stress block				$0.622x_uf_y$		
Distance of centroid of individual segment from extreme fibre in terms of x_u	a_1 0.725	a_2 0.560	a_3 0.491	a_4 0.380	a_5 0.263	a_6 0.102
Distance of centroid of stress block from extreme fibre				0.386 x_u		

TABLE 19.6
Typical calculations for compression stress block for Fe500 steel as per IS:456-2000

Calculation of compression stress block for Fe500						
Grade of steel (f_y)	500 N/mm²					
Modulus of elasticity (E_s)	200,000 N/mm²					
Design stress (f_{yd})	f_y/1.15 N/mm²					
Coefficient of design stress (k)	0.8	0.85	0.9	0.95	0.975	0.984
Design stress value (N/mm²)	347.826	369.565	391.304	413.043	423.913	427.835
Corresponding strain	0.001739	0.001948	0.002257	0.002765	0.003120	0.0035
Fraction of x_u (y_i)	0.497	0.557	0.645	0.790	0.891	1
Coefficient of characteristic stress	0.696	0.739	0.783	0.826	0.848	0.856
Segments of stress block	A_1	A_2	A_3	A_4	A_5	A_6
Area of individual segment	0.173	0.043	0.067	0.117	0.085	0.093
Area of stress block				0.577 x_uf_y		
Distance of centroid of individual segment from extreme fibre in terms of x_u	a_1 0.669	a_2 0.473	a_3 0.399	a_4 0.282	a_5 0.162	a_6 0.053
Distance of centroid of stress block from extreme fibre				0.371 x_u		

For Fe600 grade of steel, area of stress block in compression zone

$$A_C = \frac{1}{2}y_1x_uk_1f_{yd} + \frac{1}{2}(k_1f_{yd} + k_2f_{yd})(y_2x_u - y_1x_u)$$
$$+ \frac{1}{2}(k_2f_{yd} + k_3f_{yd})(y_3x_u - y_2x_u) + \frac{1}{2}(k_3f_{yd} + k_4f_{yd})(y_4x_u - y_3x_u)$$
$$+ \frac{1}{2}(k_4f_{yd} + k_5f_{yd})(x_u - y_4x_u)$$

$$A_C = \frac{1}{2}x_uf_{yd}\,[y_1k_1 + (k_1 + k_2)(y_2 - y_1) + (k_2 + k_3)(y_3 - y_2)$$
$$+ (k_3 + k_4)(y_4 - y_3) + (k_4 + k_5)(1 - y_4)]$$

TABLE 19.7
Typical calculations for compression stress block for Fe550 steel as per IS:456-2000

Calculation of compression stress block for Fe550						
Grade of steel (f_y)	550 N/mm^2					
Modulus of elasticity (E_s)	200,000 N/mm^2					
Design stress (f_{yd})	f_y/1.15 N/mm^2					
Coefficient of design stress (k)	0.8	0.85	0.9	0.95	0.975	0.979
Design stress value (N/mm^2)	382.609	406.522	430.435	454.348	466.304	468.205
Corresponding strain	0.001913	0.002133	0.002452	0.002972	0.003332	0.0035
Fraction of x_u (y_i)	0.547	0.609	0.701	0.849	0.952	1
Coefficient of characteristic stress	0.696	0.739	0.783	0.826	0.848	0.851
Segments of stress block	A_1	A_2	A_3	A_4	A_5	A_6
Area of individual segment	0.190	0.045	0.069	0.119	0.086	0.041
Area of stress block				0.551 $x_u f_y$		
Distance of centroid of individual segment from extreme fibre in terms of x_u	a_1 0.636	a_2 0.422	a_3 0.345	a_4 0.224	a_5 0.102	a_6 0.024
Distance of centroid of stress block from extreme fibre				0.364 x_u		

Where, $k_1 = 0.8, k_2 = 0.85$, $k_3 = 0.9$, $k_4 = 0.95$

k_5 is the coefficient of stress $k_5 f_y$ where $k_5 f_y$ is the design stress corresponding to a strain value of 0.0035. The value of k_5 is 0.845.

$$y_1 = \left(\frac{0.8f_y}{1.15E_s}/y\right), \qquad y_2 = \left[\left(\frac{0.85f_y}{1.15E_s}+0.0001\right)/y\right)\right]$$

$$y_3 = \left[\left(\frac{0.9f_y}{1.15E_s}+0.0003\right)/y\right)\right], \qquad y_4 = \left[\left(\frac{0.95f_y}{1.15E_s}+0.0007\right)/y\right)\right]$$

Where, y = 0.0035

Calculation of centroidal distance for Fe600 from the extreme fibre

For Fe600 grade of steel, stress block of compression zone consist of one triangular and four trapezoidal segments (Figure 19.12). Let the area of the triangular segment be A_1 and the areas of the trapezoidal segments be A_2, A_3, A_4 and A_5, respectively. Let the distance of centroid of the triangular segment from extreme fibre be a_1 and centroidal distances of the trapezoidal segments from extreme fibre be a_2, a_3, a_4, and a_5, respectively.

Distance of centroid of stress block from extreme fibre,

$$\bar{x} = \frac{A_1 a_1 + A_2 a_2 + A_3 a_3 + A_4 a_4 + A_5 a_5}{A_1 + A_2 + A_3 + A_4 + A_5}$$

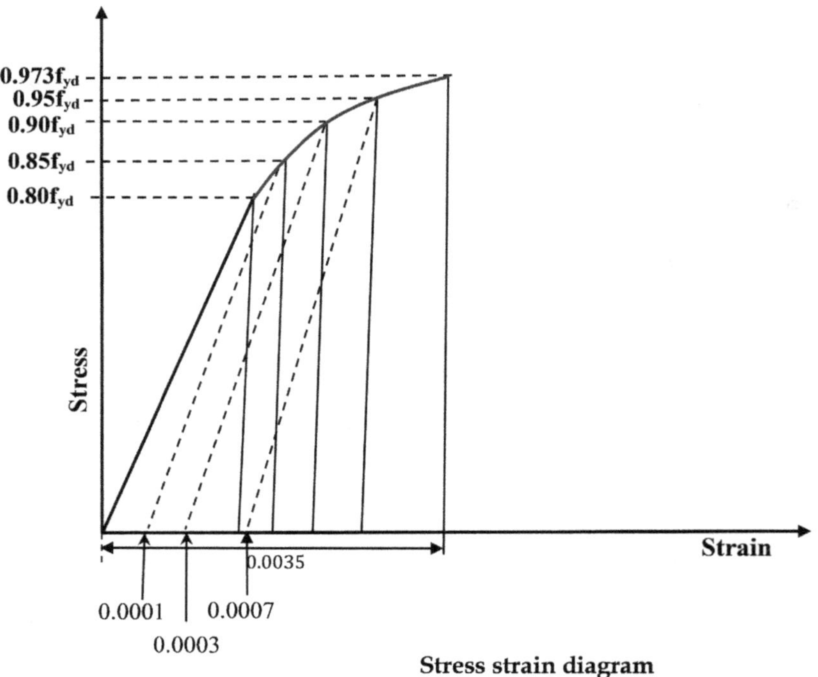

Stress strain diagram

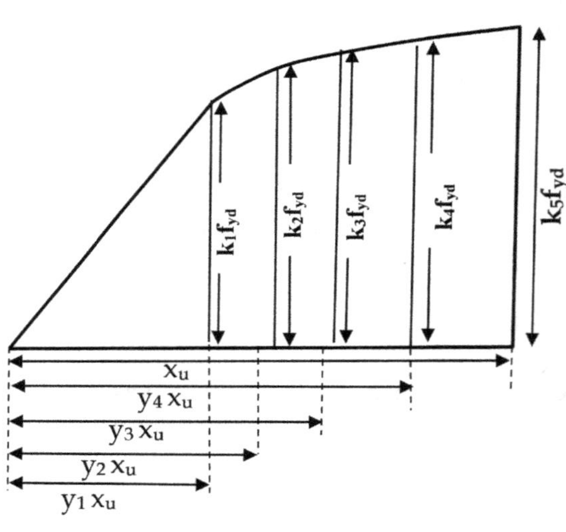

Stress block

FIGURE 19.11 Stress–strain diagram and stress block for Fe600 grade steel in compression obtained from the stress strain diagram ranging from strain values of 0 to 0.0035.

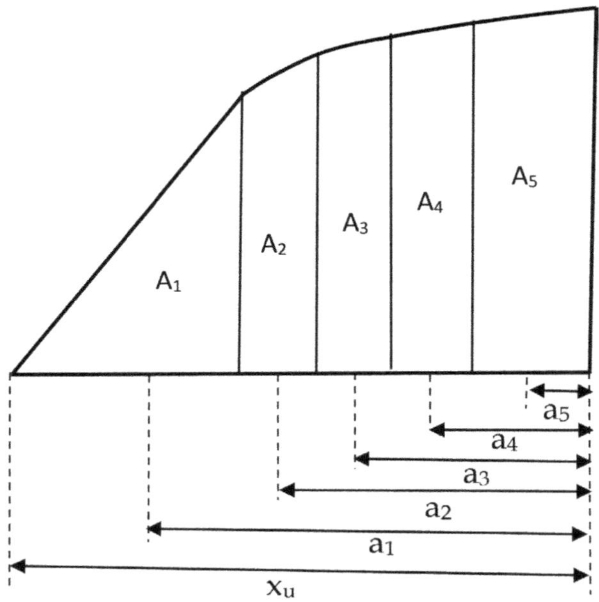

FIGURE 19.12 Different segments of stress blocks showing areas and centroidal distances.

TABLE 19.8
Typical calculations for compression stress block for Fe600 steel as per IS:456-2000

Calculations for compression stress block for Fe600					
Grade of steel (f_y)	600 N/mm²				
Modulus of elasticity (E_s)	200,000 N/mm²				
Design stress (f_{yd})	f_y/1.15 N/mm²				
Coefficient of design stress (k)	0.8	0.85	0.9	0.95	0.973
Design stress value (N/mm²)	417.391	443.478	469.565	495.652	507.539
Corresponding strain	0.002087	0.002317	0.002648	0.003178	0.0035
Fraction of x_u (y_i)	y_1 0.596	y_2 0.662	y_3 0.757	y_4 0.908	y_5 1.000
Segments of stress block	A_1	A_2	A_3	A_4	A_5
Area of individual segment	0.207	0.047	0.072	0.122	0.077
Area of stress block	0.525 $x_u f_y$				
Distance of centroid of individual segment from extreme fibre	a_1 0.602	a_2 0.370	a_3 0.290	a_4 0.167	a_5 0.049
Distance of centroid of stress block from extreme fibre			0.357x_u		

where,

$$a_1 = (x_u) - \frac{2}{3} y_1 x_u$$

$$a_2 = (x_u) \{(\frac{2k_1 + k_2}{k_1 + k_2})(\frac{y_2 - y_1}{3}) + (1 - y_2)\}$$

$$a_3 = (x_u) \{(\frac{2k_2 + k_3}{k_2 + k_3})(\frac{y_3 - y_2}{3}) + (1 - y_3)\}$$

$$a_4 = (x_u) \{(\frac{2k_3 + k_4}{k_3 + k_4})(\frac{y_4 - y_3}{3}) + (1 - y_4)\}$$

$$a_5 = (x_u) (\frac{2k_4 + k_5}{k_4 + k_5})(\frac{1 - y_4}{3})$$

The area of stress block in compression and centroidal distance for Fe600 grade of steel are shown in Table 19.8.

19.9.3 For failure in over-reinforced condition

19.9.3.1 Area of stress block in concrete

Details about the stress block of concrete for over-reinforced condition with neutral axis within or beyond the section has already been discussed in Chapter 11 for columns. These are discussed in detail later in this chapter.

19.9.3.2 Stress block of steel in compression

The parameters of the stress block will be similar to those calculated for a balanced section for the neutral axis lying within the depth of the section. For the neutral axis beyond the depth of the section the portion of the stress–strain diagram of steel falling within the section will constitute the stress block of steel in compression. Hence the stress block will consist of a cropped triangle that is a trapezium plus five more trapeziums up to Fe550 grade and four more trapeziums for Fe600 grade.

19.9.3.3 Stress block of steel in tension

Depending upon the strain at the outermost steel fibre, the stress block of steel in tension is developed. The stress block can correspond only to the elastic part or may go to the inelastic part depending upon the strain levels considered. This is shown in Figures 19.13 and 19.14.

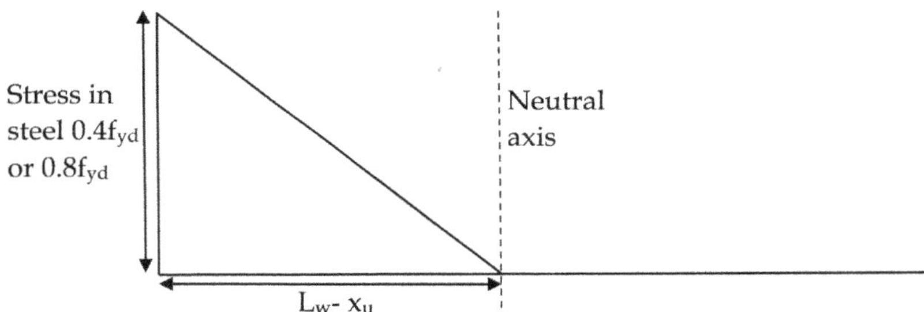

FIGURE 19.13 Stress block of steel in tension for failure in over-reinforced condition for stress at the outermost steel fibre is $0.4f_{yd}$ or $0.8f_{yd}$.

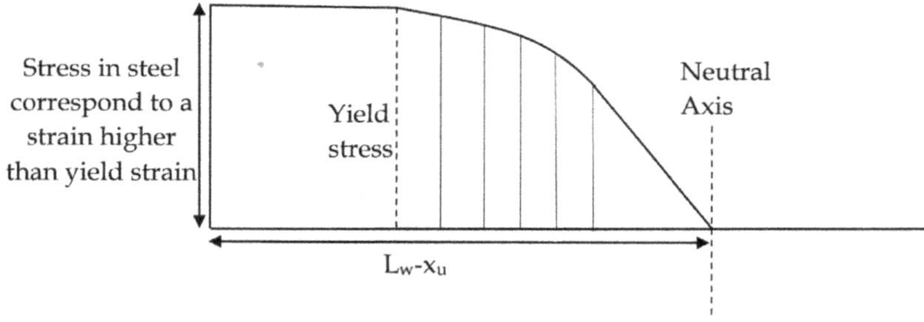

FIGURE 19.14 Stress block of steel in tension for failure in under-reinforced condition.

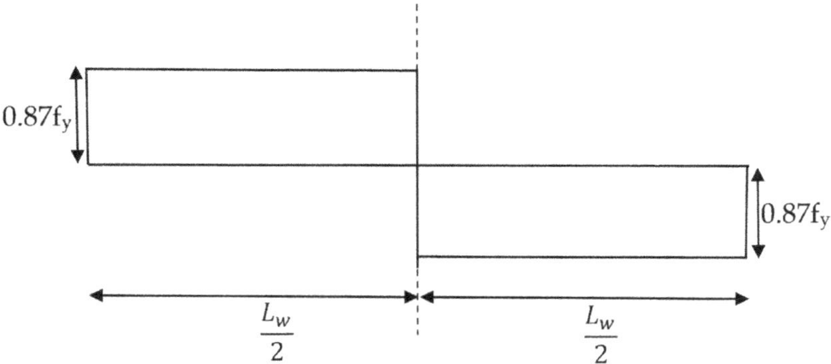

FIGURE 19.15 Stress block of steel for failure under pure flexure only.

19.9.4 For failure in under-reinforced condition

The stress block of concrete will remain similar to that discussed above. However, the stress block of steel will depend upon the strain value at the outermost steel fibre in tension, hence it will consist of the entire stress block under balanced condition plus an additional portion consisting of a rectangular part beyond the yield point. Stress block of steel in compression however will remain similar.

19.9.5 Failure under pure flexure only when the wall acting as a steel beam

Under this condition the compression concrete will be considered ineffective as it can be crushed due to the strain value exceeding the limiting strain value in concrete in flexural compression and the wall section will act as a steel beam. Since the wall has symmetrical longitudinal steel on both sides of the centroidal axis, the wall will resist moment by developing tensile and compressive stresses in the steel on the either side of the neutral axis. The maximum moment capacity will be reached when all the reinforcements yield and reach their limiting stresses. Hence axial force capacity will be zero due to equal values of tensile and compressive forces. The moment of resistance capacity of the section can be determined under this condition. This is shown in Figure 19.15.

19.10 DEVELOPMENT OF FORCE AND MOMENT EQUILIBRIUM EQUATION

For a section subjected to flexure and axial load, the basic principles of equilibrium, compatibility of strains and stress–strain relationships are used to evaluate the position of the neutral axis. The direct solutions of the equations of equilibrium can be very laborious and impractical and hence P–M interaction charts are necessary. The P–M interaction charts can be plotted as a series of lines, although higher sets of values of P and M will result in a more accurate curve.

It needs to be mentioned that while developing the safety profile for a section with a specified percentage of reinforcement the strain distribution diagram is obtained by assuming some stress or strains in concrete or reinforcement based on which the axial and moment capacities of the section are obtained. One needs to understand that the forces and moments are not based on stress resultants acting on the section, rather they are the capacities of the section based on the strain distribution and the corresponding stresses in concrete and steel.

Similarly to columns, wall sections are simultaneously subjected to axial forces along with bending. To ensure the safety of the section, the axial load- and moment-carrying capacities must be greater than or equal to the axial force or moment acting on the section. Safety of the sections needs to be ensured for all probable worst combinations of loads.

Similar to columns, for walls the equations of equilibrium may be stated as,

$$P_u = C_c + C_s$$

$$M_u = C_c \times \text{(appropriate lever arm)} + C_s \times \text{(appropriate lever arm)}$$

where P_u and M_u are the axial loads and moments acting on the wall section, C_c is the force in compression concrete and C_s is the force in steel (may be compression or tension depending on the position of neutral axis). Moments are calculated about the centroidal axis of the wall. The strain distribution across the section will be a function of the moment and axial force acting on the section. The forces in concrete and steel are calculated from the corresponding stress blocks. Once the forces are calculated the moments about the centroidal axis can easily be evaluated.

19.11 SOLUTIONS OF THE EQUATIONS FOR ARBITRARY DEPTHS OF NEUTRAL AXIS

Solutions of the force and moment equilibrium equations for all the worst load combinations is a complex, cumbersome and herculean task. Hence, instead of solving these two equations for the external load conditions, a safety profile of the section with a specified percentage of vertical longitudinal reinforcement is plotted over all the probable positions of the neutral axis from pure axial compression to steel beam conditions. For each assumed condition the force and moment capacities of the section are plotted on a P–M curve and joined by a smooth curve to arrive at the safety profile of the section. The necessary steps are discussed in the following sections.

In the case of construction of safety profiles for shear walls, the different positions of the neutral axis have been considered in a manner similar to columns.

I. When neutral axis is at infinity, i.e., when the wall is under pure axial load.
II. When depth of neutral axis is $kL_w = 1.1L_w$.
III. When the neutral axis is at the face of the section ($kL_w = 1L_w$) and the tensile stress at the longitudinal steel plate at the outermost fibre is zero.
IV. When the tensile stress at the outermost fibre is $0.4f_{yd} = 0.4 \times (0.87f_y)$.
V. When the tensile stress at the outermost fibre is $0.8f_{yd} = 0.8 \times (0.87f_y)$.
VI. When the tensile stress at the outermost fibre is $f_{yd} = 0.87f_y$, and strain in steel $= \dfrac{0.87f_y}{E_s} + 0.002$, i.e., balanced condition.
VII. When depth of the neutral axis $kL_w = 0.42\,L_w$ (under reinforced condition).
VIII. When the wall section behaves as a steel beam.

In the following section, the methodology of development of the safety profile for Fe500 grade steel is demonstrated, for grade of concrete denoted by f_{ck} and percentage of vertical reinforcement as p.

Case I When the wall is under pure axial load

Force in steel $C_s = \dfrac{pt_wL_w}{100}(f_{sc})$ (19.1)

Force in concrete $C_c = 0.446 f_{ck}\,(t_w\text{-}t_s)\,L_w$ (19.2)

Where t_s = thickness of steel plate

From force equilibrium, the axial capacity of the section may be stated as $P_u = C_c + C_s$

$$\text{Or, } P_u = 0.446 f_{ck}\,(t_w - t_s)\,L_w + \frac{pt_wL_w}{100}\,(f_{sc})$$

For Fe500 grade of steel, at a constant strain of 0.002, f_{sc} = 373.1 N/mm².

$$\therefore \quad \frac{P_u}{f_{ck}\, t_w L_w} = 0.446(1 - \frac{p}{100}) + \frac{p}{100 f_{ck}}(f_{sc})$$

$$\frac{P_u}{f_{ck}\, t_w L_w} = 0.446(1 - \frac{p}{100}) + \frac{3.73\, p}{f_{ck}} \tag{19.3}$$

Case II When depth of neutral axis $kL_W = 1.1L_W$ (refer to Figure 19.16)
In this case, the neutral axis lies outside the section and its depth is $1.1L_w$, where L_w is the length of shear wall. The whole section of wall is in compression for this condition. The stress block of steel is determined considering the stress–strain diagram of CTD bars and assumptions for limit state of collapse in compression as per IS:456-2000. The area of stress block and its centroidal distance from the outermost compression fibre of concrete is similar to those developed for columns and have been discussed in Chapter 11.

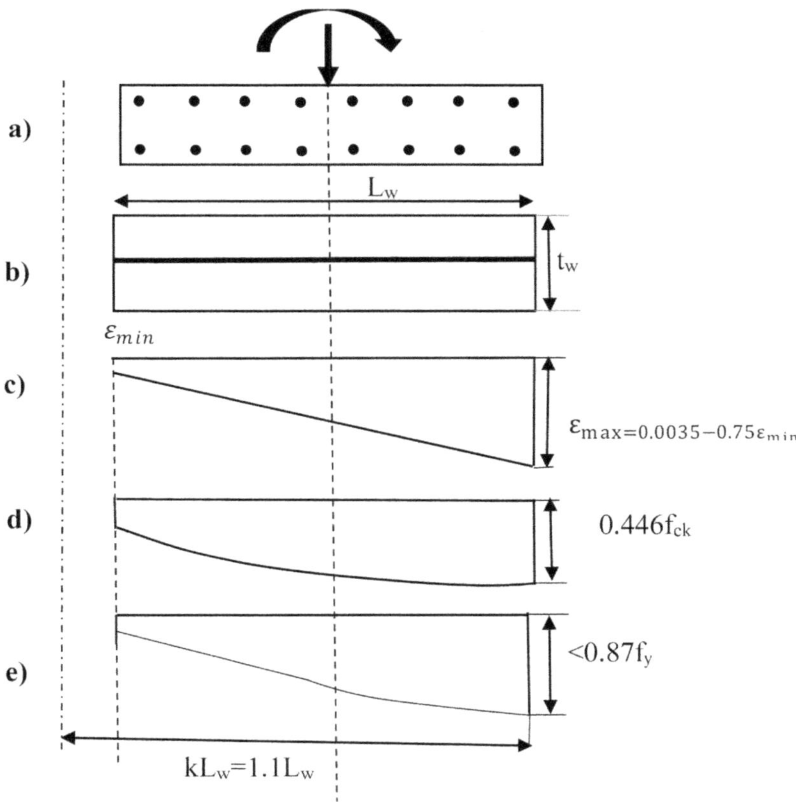

FIGURE 19.16 Strain distributions and stress blocks for shear wall when the neutral axis lies outside the section: (a) cross-section of a rectangular shear wall; (b) cross-section showing vertical reinforcement as an equivalent plate; (c) strain distribution diagram of concrete; (d) stress block for concrete in compression; (e) stress block of steel in compression.

Note: Thickness of steel plate = t_s.

Sample area calculation of stress block of steel in the compression zone of a shear wall for Fe500 at $k = 1.1$. Calculations are performed in a manner similar to Section 19.9.3.2.

For $k = 1.1$, the stress block parameters for concrete are as follows:

$$C_1 = 0.384 \text{ and } C_2 = 0.443.$$

Therefore, area of stress block of concrete = $0.384f_{ck}.(t_w - t_s)\, L_w$ and the distance of its centroid from the maximum compressed edge is = $0.443\, L_w$

Strain at the least compressed edge = $\dfrac{0.002}{4L_w/7 + 0.1L_w}$ x $0.1\, L_w = 0.0003$

Strain at maximum compressed edge = 0.0035 – (0.75 × 0.0003) = 0.003275

Now for Fe500,

Area of stress block of concrete = $0.384\, f_{ck}\,(t_w - t_s)\, L_w$

Area of stress block of steel = $0.608\dfrac{p}{100}\, t_w.L_w f_y$

Therefore,

$$\frac{P_u}{f_{ck} t_w L_w} = 0.384\left(1 - \frac{p}{100}\right) + 0.608\, p \frac{f_y}{100 f_{ck}} \tag{19.4}$$

Distance of centroid of stress block of steel from maximum compressed edge = $0.399L_w$ (refer to Table 19.9).

Distance of centroid of stress block of concrete from right edge = $0.443\, L_w$

$$\frac{M_u}{f_{ck}\, t_w L_w{}^2} = 0.384\left(1 - \frac{p}{100}\right)(0.5 - 0.443) + 0.608\,\frac{p}{f_{ck}}\,\frac{f_y}{100}(0.5 - 0.399) \tag{19.5}$$

TABLE 19.9

Area and centroidal distance for compression stress block for Fe500 steel when depth of neutral axis is at $1.1L_w$

When $k = 1.1$						
Grade of st340eel (f_y) = 500 N/mm²						
Modulus of elasticity (E_s)	200,000 N/mm²					
Design stress (f_{yd})	f_y/1.15 N/mm²					
Coefficient of design stress	0.8	0.85	0.9	0.95	0.975	0.979
Design stress value (N/mm²)	347.82	369.565	391.304	413.043	423.913	425.515
Corresponding strain	0.00173	0.00194	0.00225	0.00276	0.00312	0.00327
Fraction of x_u (y_i)	0.531	0.595	0.689	0.844	0.953	1
Neutral axis depth factor (k)		1.1				
Fraction of L_w	0.584	0.654	0.758	0.929	1.048	1.1
Coefficient of characteristic stress	0.696	0.739	0.783	0.826	0.848	0.851
Segments of stress block	A_1	A_2	A_3	A_4	A_5	A_6
Area of individual segment	0.197	0.050	0.079	0.137	0.100	0.044
Area of stress block				$0.608\, L_w f_y$		
Distance of centroid of individual segment from extreme fibre in terms of x_u	a_1 0.701	a_2 0.480	a_3 0.393	a_4 0.256	a_5 0.115	a_6 0.026
Distance of centroid of stress block from extreme fibre				$0.399\, L_w$		

For Fe500 steel, $$\frac{M_u}{f_{ck}\, t_w L_w^{\,2}} = 0.022\left(1-\frac{p}{100}\right)+30.704\frac{p}{100 f_{ck}} \quad (19.6)$$

Case III When depth of neutral axis $kL_W = 1L_W$ (refer to Figure 19.17)
In this case, the neutral axis lies at the periphery of the section and its depth is $1L_w$ where L_w is the length of the shear wall. The whole section of wall is in compression for this condition. The stress block of steel has been calculated considering the stress–strain diagram of CTD bars as per IS:456-2000. The area of stress block of concrete section can be calculated in a manner similar to columns.

Strain at minimum compressed edge = 0
Strain at maximum compressed edge = 0.0035
For $k = 1$, stress block parameters for concrete are $C_1 = 0.361$ and $C_2 = 0.416$
Therefore, the area of stress block of concrete = $0.361\, f_{ck}.(t_w - t_s).L_w$ and distance of its centroid from maximum compressed edge is = $0.416\, L_w$

The area of the stress block and centroidal distance from maximum compressed fibre for Fe500 grade steel are as follows : for $k = 1$ (Table 19.6), $x_u = L_w$.
Area of stress block = $0.577\, x_u f_y$
Centroidal distance from extreme fibre ($\overline{x}$) =$0.371\, x_u$

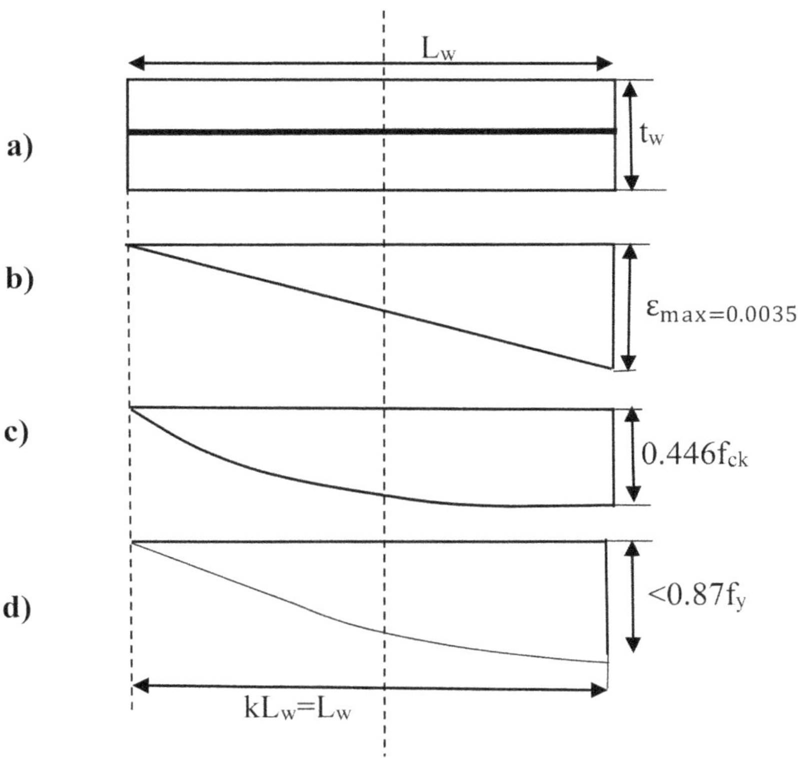

FIGURE 19.17 Strain distribution and stress block for shear wall with neutral axis at face of section: (a) cross-section showing vertical reinforcement as an equivalent plate; (b) strain distribution diagram for concrete; (c) stress block for concrete in compression; (d) stress block of steel in compression.

$$\frac{P_u}{f_{ck}\,t_w L_w} = 0.361\left(1-\frac{p}{100}\right)+\frac{0.577 f_y p}{100 f_{ck}} \tag{19.7}$$

For Fe500 steel,

$$\frac{P_u}{f_{ck}\,t_w L_w} = 0.361\left(1-\frac{p}{100}\right)+\frac{2.885\text{p}}{f_{ck}}$$

$$\frac{M_u}{f_{ck} t_w L_w^{\,2}} = 0.361\left(1-\frac{p}{100}\right)(0.5\text{-}0.416)+\frac{0.577 f_y p}{100 f_{ck}}(0.5-0.371) \tag{19.8}$$

For Fe500 steel,

$$\frac{\text{M}_u}{f_{ck} t_w L_w^{\,2}} = 0.0302\left(1-\frac{p}{100}\right)+37.22\frac{\text{p}}{100 f_{ck}}$$

Case IV When the tensile stress at the outermost steel fibre is $0.4f_{yd}$ = $0.4(0.87f_y)$ (refer to Figure 19.18)

Here the depth of the neutral axis is situated within the section in such a way that the tensile stress at the outermost tension fibre is $0.4\,f_{yd}$.

For compression zone strain at outermost steel fibre = 0.0035

For tension zone strain at outermost tension fibre = $0.4 \times (0.87f_y)/E_s$

$$= \frac{0.4\times 0.87\times 500}{200000} = 0.00087$$

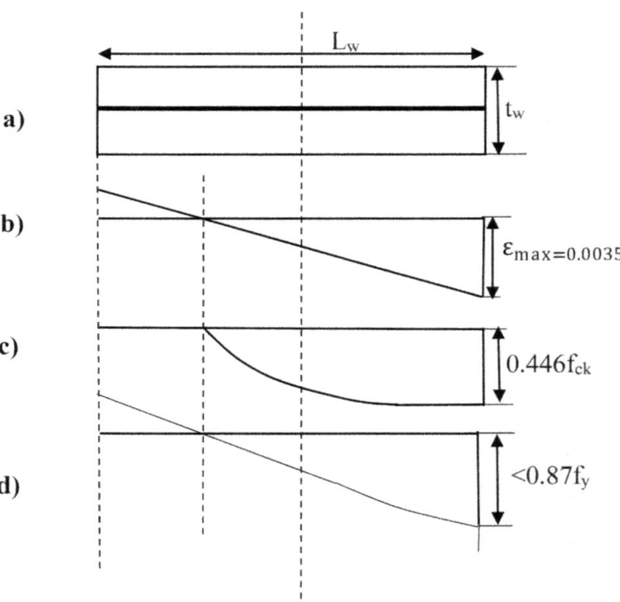

FIGURE 19.18 (a) Cross-section showing vertical reinforcement as an equivalent plate; (b) strain distribution diagram when stress at the outermost tension fiber in steel is $0.4f_{yd}$ or $0.8f_{yd}$; (c) stress block for concrete; (d) stress block for steel.

Distance of neutral axis from the maximum compressed edge is

$$x_u = \frac{0.0035}{0.0035+0.00087} L_w = 0.8009 L_w$$

Now force in concrete $C_c = 0.36 f_{ck} \times 0.8009 \times L_w (t_w - t_s) = 0.288 f_{ck}\, t_w (1 - \frac{p}{100}) L_w$ (refer to Table 19.6)

Force in compression steel $= 0.577\, x_u f_y \times A_{st} = 0.577\, x_u \times \frac{p}{100} f_y \times L_w t_w$

$$= 0.577 \times 0.8009 \times 500 \times \frac{p}{100} L_w t_w = 231.06 \frac{p}{100} L_w t_w$$

Up to a stress of $0.8 f_y$ the stress–strain relationship of HYSD is linear as per IS:456.

Force in tension steel $= 0.5 \times 0.4 \times 0.87 f_y \times (1 - x_u) \times \frac{p}{100} L_w t_w$

$$= 0.5 \times 0.4 \times 0.87 \times 500 \times 0.1991 \frac{p}{100} L_w t_w$$

$$= 17.32 \frac{p}{100} L_w t_w$$

Force in steel $C_s = (231.06 - 17.32) \frac{p}{100} L_w t_w$

$$= 213.74 \frac{p}{100} L_w t_w$$

$$\text{Hence, } \frac{P_u}{f_{ck} t_w L_w} = 0.288\left(1 - \frac{p}{100}\right) + 213.74 \frac{p}{100 f_{ck}} \tag{19.9}$$

$$\frac{M_u}{f_{ck} t_w L_w^2} = 0.288\left(1 - \frac{p}{100}\right)(0.5 - 0.8009 \times 0.42)$$
$$+ 231.06 \frac{p}{100 f_{ck}} \times (0.5 - 0.371 \times 0.8009)$$
$$+ 17.32 \frac{p}{100 f_{ck}} \left[(0.5 - 0.1991) + (0.666 \times 0.1991)\right]$$

$$\frac{M_u}{f_{ck} t_w L_w^2} = 0.0471\left(1 - \frac{p}{100}\right) + 54.38 \frac{p}{100 f_{ck}} \tag{19.10}$$

Case V When the tensile stress at outermost steel fibre is $0.8 f_{yd} = 0.8(0.87 f_y)$ (refer to Figure 19.18)

Here the depth of the neutral axis is situated within the section in such a way that the tensile stress at the outermost tension fibre is $0.8 f_{yd.}$

For Fe500 grade of steel,

Strain at outermost tension fibre in steel $= 0.8(0.87 f_y)/E_s$
$= 0.8 \times 0.87 \times 500/200000$
$= 0.00174$

Distance of neutral axis from the extreme compressed fibre is
$= 0.0035/(0.0035 + 0.00174) = 0.6679 L_w$

Now force in concrete $C_c = 0.36\, f_{ck} \times 0.6679 \times L_w (t_w - t_s) = 0.24 f_{ck}\, t_w (1 - \frac{p}{100}) L_w$

Force in compression steel = $0.577\, x_u \times f_y \times A_{st} = 0.577\, x_u \times f_y \dfrac{p}{100} L_w\, t_w$

$$= 0.577 \times 0.6679 \times 500 \frac{p}{100} L_w\, t_w = 192.689 \frac{p}{100} L_w\, t_w$$

Force in tension steel = $0.5 \times 0.8 \times 0.87 f_y \times (1 - x_u) \times \dfrac{p}{100} L_w\, t_w$

$$= 0.5 \times 0.8 \times 0.87 \times 500 \times 0.3321 \frac{p}{100} L_w\, t_w$$

$$= 57.78 \frac{p}{100} L_w\, t_w$$

Force in steel $C_s = (192.689 - 57.78) \dfrac{p}{100} L_w\, t_w$

$$= 134.91 \frac{p}{100} L_w\, t_w$$

Therefore,

$$\frac{P_u}{f_{ck}\, t_w L_w} = 0.24\left(1 - \frac{p}{100}\right) + 134.91 \frac{p}{100 f_{ck}} \tag{19.11}$$

$$\frac{M_u}{f_{ck} t_w L_w^{\,2}} = 0.24\left(1 - \frac{p}{100}\right)(0.5 - 0.6679 \times 0.42)$$

$$+ 192.689 \frac{p}{100 f_{ck}} \times (0.5 - 0.371 \times 0.6679)$$

$$+ 57.78 \frac{p}{100 f_{ck}} \left[(0.5 - 0.3321) + (0.666 \times 0.3321)\right]$$

$$\frac{M_u}{f_{ck} t_w L_w^{\,2}} = 0.0527\left(1 - \frac{p}{100}\right) + 71.08 \frac{p}{100 f_{ck}} \tag{19.12}$$

Case VI When the neutral axis is at a balanced position (refer to Figure 19.4)

Under balanced condition the strain at the outermost fibre in compression reaches the ultimate or limiting strain in concrete (ε_{cu} = 0.0035) and the outermost fibre of steel in tension reaches the limiting strain in steel ($\varepsilon_s = f_y/1.15E_s + 0.002$) simultaneously.

Limiting strain in tensile zone = 0.0042

Limiting strain in compression zone = 0.0035

Therefore,

$$\frac{x_u}{L_w - x_u} = \frac{0.0035}{0.0042}$$

or, $x_u = 0.4545\, L_w$

Therefore, $k = 0.4545$ and $(L_w - x_u) = 0.5455\, L_w$

In the case of balanced condition, for Fe500

The area of stress block for steel in tension is $0.623(L_w - x_u)f_y$

The area of stress block for steel in compression is $0.577\, x_u f_y$

Therefore, axial force capacity of the section

$$P_u = 0.36 f_{ck} x_u \left(t_w - t_s\right) + 0.577 x_u f_y t_s - 0.623\left(L_w - x_u\right) f_y t_s$$

$$= 0.36 f_{ck} x_u \left(1 - \frac{p}{100}\right) t_w + 0.577 x_u f_y \left(\frac{p t_w}{100}\right) - 0.623\left(0.5455 L_w\right) f_y \left(\frac{p t_w}{100}\right)$$

$$= 0.36f_{ck}\left(0.4545L_w\right)\left(1-\frac{p}{100}\right)t_w + 0.577\left(0.4545L_w\right)f_y\left(\frac{pt_w}{100}\right) - 0.623\left(0.5455L_w\right)f_y\left(\frac{pt_w}{100}\right)$$

$$\frac{P_u}{f_{ck}t_wL_w} = 0.164\left(1-\frac{p}{100}\right) - 38.8\frac{p}{100f_{ck}} \tag{19.13}$$

The distance of centroid of stress block for concrete from maximum compressed fibre is $0.42x_u$.

For Fe500 grade of steel, in the case of balanced condition: the centroidal distance of stress block from maximum tension edge for steel in tension is $0.383(L_w - x_u)$. The centroidal distance of stress block from the maximum compressed edge for steel in compression is $0.371x_u$ (refer to Table 19.2).

Therefore, for the moment capacity of the section

$$M_u = 0.36.f_{ck}x_u\left(1-\frac{p}{100}\right)t_w\left(0.5L_w - 0.42x_u\right) + 0.577x_uf_y\left(\frac{pt_w}{100}\right)\left(0.5L_w - 0.371x_u\right)$$
$$+0.623\left(0.5455L_w\right)f_y\left(\frac{pt_w}{100}\right)\left\{0.5L_w - 0.383\left(L_w - x_u\right)\right\}$$
$$= 0.36f_{ck}\left(0.4545L_w\right)\left(1-\frac{p}{100}\right)t_w\left\{0.5L_w - 0.42\left(0.4545L_w\right)\right\}$$
$$+0.577\left(0.4545L_w\right)f_y\left(\frac{pt_w}{100}\right)\left\{0.5L_w - 0.371\left(0.4545L_w\right)\right\}$$
$$+0.623\left(0.5455L_w\right)f_y\left(\frac{pt_w}{100}\right)\left\{0.5L_w - 0.383\left(0.5455L_w\right)\right\}$$

$$\frac{M_u}{f_{ck}t_wL_w^2} = 0.051\left(1-\frac{p}{100}\right) + 92.91\frac{p}{100f_{ck}} \tag{19.14}$$

Case VII When the neutral axis is at $0.42L_W$

In this case, the depth of the actual neutral axis is less than the balanced depth (Figure 19.19). Hence failure occurs in an under-reinforced condition. Since the strain distribution diagram is linear strain at the outermost tension fibre in steel

$$= \frac{0.0035 \times 0.58L_w}{0.42L_w} = 0.0048$$

Now for Fe500 grade of steel:

Strain of steel at a stress level of $0.87f_y = 0.87f_y/E_s + 0.002$
$= 0.87 \times 500/200000 + 0.002 = 0.0042$

The distance of the point corresponding to the yield strain of 0.0042 from the neutral axis on the tension side = $(0.0042/0.0035) \times 0.42\,L_w = 0.504\,L_w$

Force in concrete $C_c = 0.36f_{ck} \times 0.42\,L_w \times (t_w - t_s) = 0.1512\,(1-\frac{p}{100})f_{ck}\,t_w\,L_w$

Force in compression steel, $C_{sc} = (0.577 \times 0.42\,L_w \times f_y) \times t_s = 121.17\,L_w t_w\frac{p}{100}$

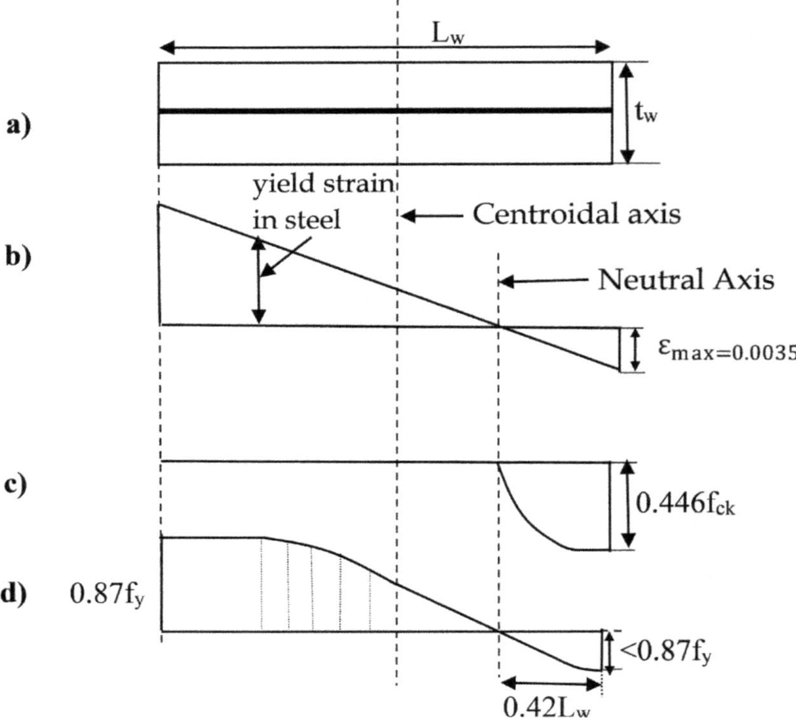

FIGURE 19.19 (a) Cross-section showing vertical reinforcement as an equivalent plate; (b) strain distribution diagram when N.A. is at 0.42 L_w; (c) stress block for concrete; (d) stress blocks for steel.

Force in tension steel, $C_{st} = 0.623 \times 0.504\, L_w \times f_y \times t_w \times \frac{p}{100} + 0.87 f_y \times (0.58 - 0.504)\, t_w \times L_w \times \frac{p}{100}$

$$= 156.996\, L_w t_w \frac{p}{100} + 33.06\, L_w t_w \frac{p}{100}$$

$$\frac{P_u}{f_{ck} t_w L_w} = 0.1512\left(1 - \frac{p}{100}\right) - 68.88\, \frac{p}{100 f_{ck}} \tag{19.15}$$

$$\frac{M_u}{f_{ck} t_w L_w^2} = 0.1512\left(1 - \frac{p}{100}\right) \times (0.5 - 0.42 \times 0.42) + [121.17 \times (0.5 - 0.371 \times 0.42)] \times \frac{p}{100 f_{ck}}$$

$$+ [156.996 \times \{0.5 - (0.504 \times 0.383) + (0.58 - 0.504)\} \times \frac{p}{100 f_{ck}}]$$

$$+ [33.06 \times \frac{p}{100 f_{ck}} \times \{0.5 - (\frac{0.58 - 0.504}{2})\}]$$

$$\frac{M_u}{f_{ck} t_w L_w^2} = 0.0489\left(1 - \frac{p}{100}\right) + 116.794 \times \frac{p}{100 f_{ck}} \tag{19.16}$$

Case VIII When the wall section is acting as a steel beam (refer to Figure 19.15)

Under this condition, the wall section acts as a steel beam. In this case, concrete is considered to be totally ineffective and the reinforcement in the section yields completely. Since the wall has symmetrical vertical reinforcement on both sides of the centroidal axis, the reinforcement will resist pure moment by yielding of both tensile and compressive steel bars with stresses equal to the design stress that is $0.87f_y$. The moment of resistance capacity of the section can be determined as follows:

$$M_u = \left[\frac{0.5pt_wL_w}{100} \times 0.25\,L_w + \frac{0.5pt_wL_w}{100} \times 0.25\,L_w\right] \times 0.87\,f_y$$

$$\text{or, } M_u = \frac{0.25pt_wL_w^2}{100} \times 0.87\,f_y$$

$$\text{or, } M_u = \frac{0.2175f_y pt_wL_w^2}{100}$$

$$\text{or, } \frac{M_u}{f_{ck}t_wL_w^2} = \frac{0.2175f_y p}{100f_{ck}}$$

$$\text{or, } \frac{M_u}{f_{ck}t_wL_w^2} = \frac{108.75\,p}{100f_{ck}}$$

In the above section, the method of calculating coordinates of eight points expressed in the format $\frac{P_u}{f_{ck}t_wL_w}$ and $\frac{M_u}{f_{ck}t_wL_{w^2}}$ for a typical percentage of vertical reinforcement p and grade of steel Fe500 are presented. For any grade of structural concrete and any numerical value of p the values of these coordinates can be easily obtained. The safety profile of a wall section having thickness t_w and length L_w with that specific percentage of reinforcement and grade of concrete with Fe500 steel can be obtained by joining these co-ordinates by a smooth curve to obtain a force–moment relationship. Using different percentages of reinforcement, a number of safety profiles for the wall section with the specified grades of concrete and steel in the form of a P–M interaction chart for the shear wall can be developed.

19.12 ILLUSTRATIVE EXAMPLE

Problem 19.1

Prepare a safety profile for rectangular shear wall with length L_w and thickness as t_w with the grade of structural concrete as M40, grade of steel as Fe550 and percentage of steel is 3%.

Solution

Given data:

Grade of structural concrete as M40, i.e., $f_{ck} = 40$ N/mm²

Grade of steel as Fe550, i.e., $f_y = 550$ N/mm²

Percentage of steel is 3%, i.e., $p = 3$

1st Case: Calculation when the neutral axis is at infinity

Force in concrete $C_c = 0.446f_{ck}\,(t_w - t_s)\,L_w$

Where t_s = thickness of steel plate

From force equilibrium axial capacity of the section $P_u = C_c + C_s$

$$\text{or } P_u = 0.446 f_{ck} (t_w - t_s) L_w + \frac{pt_w L_w}{100}(f_{sc})$$

$$\text{or } \frac{P_u}{f_{ck} t_w L_w} = 0.446(1 - \frac{p}{100}) + \frac{p}{100 f_{ck}}(f_{sc}).$$

Stress and strain value of Fe550

Stress Level at	Strain	Stress
0.8 × (0.87*fy*)	0.001914	382.8
0.85 × (0.87*fy*)	0.00203	406.72
0.90 × (0.87*fy*)	0.002453	430.65
0.95 × (0.87*fy*)	0.00297	454
0.975 × (0.87*fy*)	0.00333	466.304
0.87*fy*	0.00439	478.5

For strain value of 0.002, stress = 400.53 MPa

$$\frac{P_u}{f_{ck} t_w L_w} = 0.7330$$

2nd Case: When depth of neutral axis $kL_w = 1.1L_w$

For $k = 1.1$, the stress block parameters for concrete are

$C_1 = 0.384$ and $C_2 = 0.443$.

Therefore, area of stress block of concrete = $0.384 f_{ck} t_w L_w$ and the distance of its centroid from maximum compressed edge is = $0.443 L_w$.

Strain at the least compressed edge $= \dfrac{0.002}{4L_w / 7 + 0.1L_w} \times 0.1L_w = 0.0003$

Strain at maximum compressed edge = 0.0035 – (0.75 × 0.0003) = 0.003275

Now for Fe550, at $k = 1.1$

Area of stress block of steel = $0.578p\, L_w f_y / 100$

Distance of centroid of stress block of steel from right edge = $0.389L_w$

Area of stress block of concrete = $0.384 f_{ck} (t_w - t_s) L_w$

So, $$\frac{P_u}{f_{ck} t_w L_w} = 0.384\left(1 - \frac{p}{100}\right) + 0.578p\frac{f_y}{100 f_{ck}}$$

$$\frac{P_u}{f_{ck} t_w L_w} = 0.6109$$

Distance of centroid of stress block of concrete from right edge = $0.443 L_w$

$$\frac{M_u}{f_{ck} t_w L_w^2} = 0.384\left(1 - \frac{p}{100}\right)(0.5 - 0.443) + 0.578\frac{p}{f_{ck}}\frac{f_y}{100}(0.5 - 0.389)$$

$$\frac{M_u}{f_{ck} t_w L_w^2} = 0.04769$$

3rd Case: When depth of neutral axis $kL_w = 1L_w$

Area of stress block = 0.551 $x_u f_y$

Centroidal distance from extreme fibre ($\bar{x}$) =0.364 x_u (refer to Table 19.7)

$$\frac{P_u}{f_{ck}\,t_w L_w} = 0.361\left(1-\frac{p}{100}\right)+\frac{0.551 f_y p}{100 f_{ck}}$$

$$\frac{P_u}{f_{ck}\,t_w L_w} = 0.577$$

$$\frac{M_u}{f_{ck} t_w L_w^2} = 0.361\left(1-\frac{p}{100}\right)(0.5\text{-}0.416) + \frac{0.551 f_y p}{100 f_{ck}}(0.5-0.364)$$

$$\frac{\mathrm{M}_u}{f_{ck} t_w L_w^2} = 0.060$$

4th Case: When the tensile stress at the outermost steel fibre is $0.4f_{yd} = 0.4(0.87f_y)$

For compression zone strain at the outermost compression reinforcement = 0.0035

For tension zone strain at outermost tension fibre = 0.4 × $(0.87f_y)/E_s$

$$= \frac{0.4\times 0.87\times 550}{200000} = 0.000957$$

Distance of neutral axis from the maximum compressed edge is

$$\mathrm{x_u} = \frac{0.0035}{0.0035+0.000957} L_w = 0.782 L_w$$

Now force in concrete, $C_c = 0.36 f_{ck} \times 0.782 \times L_w (t_w - t_s) = 0.2815 f_{ck} t_w \left(1-\frac{p}{100}\right) L_w$

Force in compression steel = $0.551\, x_u \times f_y \times A_{st} = 0.551\, x_u \times \frac{p}{100} f_y \times L_w t_w$ (refer to Table 19.7)

$$= 0.551 \times 0.782 \times 550 \frac{p}{100} L_w t_w = 236.98 \frac{p}{100} L_w t_w$$

Up to a stress of $0.8f_y$ stress–strain relationship of HYSD is linear as per IS:456.

Force in tension steel = $0.5 \times 0.4 \times 0.87 f_y \times (1 - x_u) \times \frac{p}{100} L_w t_w$

$$= 0.5 \times 0.4 \times 0.87 \times 550 \times 0.218 \frac{p}{100} L_w t_w$$

$$= 20.86 \frac{p}{100} L_w t_w$$

Force in steel, $C_s = (236.98 - 20.86) \frac{p}{100} L_w t_w$

$$= 216.12 \frac{p}{100} L_w t_w$$

Hence, $$\frac{P_u}{f_{ck}\,t_w L_w} = 0.2815\left(1-\frac{p}{100}\right)+216.12\frac{p}{100 f_{ck}}$$

$$\frac{P_u}{f_{ck}\,t_w L_w} = 0.435$$

$$\frac{M_u}{f_{ck}t_wL_w^2} = 0.2815\left(1-\frac{p}{100}\right)(0.5-0.782\times0.42)$$
$$+236.12\frac{p}{100f_{ck}}\times(0.5-0.364\times0.782)$$
$$+20.86\frac{p}{100f_{ck}}[(0.5-0.218)+(0.666\times0.218)]$$

$$\frac{M_u}{f_{ck}t_wL_w^2} = 0.093$$

5th Case: When the tensile stress at outermost steel fibre is $0.8f_{yd} = 0.8(0.87f_y)$

Strain at outermost tension fibre in steel $= 0.8(0.87f_y)/E_s$
$= 0.8 \times 0.87 \times 550/200000$
$= 0.001914$

Distance of neutral axis from the extreme compressed fibre is

$$0.0035/(0.0035 + 0.001914) = 0.6465L_w$$

Now force in concrete, $C_c = 0.36 f_{ck} \times 0.6465 \times L_w (t_w - t_s) = 0.233 f_{ck} t_w (1 - \frac{p}{100}) L_w$

Force in compression steel $= 0.551\, x_u \times f_y \times A_{st} = 0.551\, x_u \times f_y \frac{p}{100} L_w t_w$

$= 0.551 \times 0.6465 \times 550 \frac{p}{100} L_w t_w = 195.92 \frac{p}{100} L_w t_w$

Force in tension steel $= 0.5 \times 0.8 \times 0.87 f_y \times (1 - x_u) \times \frac{p}{100} L_w t_w$

$= 0.5 \times 0.8 \times 0.87 \times 550 \times 0.3535 \frac{p}{100} L_w t_w$

$= 67.66 \frac{p}{100} L_w t_w$

Force in steel, $C_s = (195.92 - 67.66) \frac{p}{100} L_w t_w$

$= 128.26 \frac{p}{100} L_w t_w$

$$\text{So, } \frac{P_u}{f_{ck}\, t_w L_w} = 0.233\left(1-\frac{p}{100}\right) + 128.26\frac{p}{100f_{ck}}$$

$$\frac{P_u}{f_{ck}\, t_w L_w} = 0.322$$

$$\frac{M_u}{f_{ck}t_wL_w^2} = 0.2331\left(1-\frac{p}{100}\right)(0.5-0.6465\times0.42)$$
$$+195.92\frac{p}{100f_{ck}}\times(0.5-0.364\times0.6465)$$
$$+67.66\frac{p}{100f_{ck}}[(0.5-0.3535)+(0.666\times0.3535)]$$

$$\frac{M_u}{f_{ck}t_wL_w^{\ 2}}=0.1099$$

6th Case: When the neutral axis is at balanced position

Limiting strain in steel at the outermost tension fibre = 0.0044

Limiting strain at the outermost compression fibre = 0.0035

Therefore, $\dfrac{x_u}{L_w - x_u} = \dfrac{0.0035}{0.0044}$

or, $x_u = 0.443\ L_w$

Therefore, $k = 0.443$ and $(L_w - x_u) = 0.557\ L_w$

The area of stress block for steel in tension is $0.614(L_w - x_u)f_y$ (refer to Table 19.3)

The area of stress block for steel in compression is $0.551\ x_uf_y$.

Therefore, force capacity of the section

$$P_u = 0.36f_{ck}x_u\left(t_w - t_s\right) + 0.551x_uf_yt_s - 0.614\left(L_w - x_u\right)f_yt_s$$

$$= 0.36f_{ck}x_u\left(1 - \frac{p}{100}\right)t_w + 0.551x_uf_y\left(\frac{pt_w}{100}\right) - 0.614\left(0.557L_w\right)f_y\left(\frac{pt_w}{100}\right)$$

$$= 0.36f_{ck}\left(0.443L_w\right)\left(1 - \frac{p}{100}\right)t_w + 0.551\left(0.443L_w\right)f_y\left(\frac{pt_w}{100}\right) - 0.614\left(0.577L_w\right)f_y\left(\frac{pt_w}{100}\right)$$

$$\therefore \frac{P_u}{f_{ck}t_wL_w} = 0.1092$$

Distance of centroid of stress block for concrete from maximum compressed fibre is $0.42x_u$.

The centroidal distance of stress block from the maximum tension edge for steel in tension is $0.380(L_w - x_u)$. The centroidal distance of stress block from the maximum compressed edge for steel in compression $0.364\ x_u$ (refer to Table 19.3).

Therefore, the moment capacity of the section, M_u

$$\begin{aligned}
&= 0.36f_{ck}x_u\left(1 - \frac{p}{100}\right)t_w\left(0.5L_w - 0.42x_u\right) \\
&+ 0.551x_uf_y\left(\frac{pt_w}{100}\right)\left(0.5L_w - 0.364x_u\right) \\
&+ 0.614\left(0.557L_w\right)f_y\left(\frac{pt_w}{100}\right)\left\{0.5L_w - 0.380\left(L_w - x_u\right)\right\}
\end{aligned}$$

$$\begin{aligned}
&- 0.36f_{ck}\left(0.443L_w\right)\left(1 - \frac{p}{100}\right)t_w\left\{0.5L_w - 0.42\left(0.443L_w\right)\right\} \\
&+ 0.551\left(0.443L_w\right)f_y\left(\frac{pt_w}{100}\right)\left\{0.5L_w - 0.364\left(0.443L_w\right)\right\} \\
&+ 0.614\left(0.443L_w\right)f_y\left(\frac{pt_w}{100}\right)\left\{0.5L_w - 0.380\left(0.557L_w\right)\right\}
\end{aligned}$$

$$\frac{M_u}{f_{ck}t_w L_w^2} = 0.115$$

7th Case: When the neutral axis is at $0.42L_w$

The outermost tension fibre in steel

$$= \frac{0.0035 \times 0.58L_w}{0.42L_w} = 0.0048$$

Strain of steel at a stress level of $0.87\,f_y = 0.0044$

Distance of the point corresponding to the yield strain of 0.0044 from the neutral axis on the tension side = (0.0044/0.0035) × 0.42 L_w = 0.528 L_w

Force in concrete, $C_c = 0.36\,f_{ck} \times 0.42\,L_w \times (t_w - t_s) = 0.1512\,(1 - \frac{p}{100}) f_{ck}\, t_w\, L_w$

Force in compression steel, $C_{sc} = (0.551 \times 0.42\,L_w \times f_y) \times t_s = 127.28\,L_w t_w \frac{p}{100}$

Force in tension steel, $C_{st} = 0.614 \times 0.528\,L_w \times f_y \times t_w \times \frac{p}{100} + 0.87 f_y \times (0.58 - 0.528)\,t_w \times L_w \times \frac{p}{100}$

$$= 178.305\ \mathrm{L_w t_w}\frac{p}{100} + 24.88\ \mathrm{L_w t_w}\frac{p}{100}$$

$$\frac{P_u}{f_{ck}t_w L_w} = 0.1512(1 - \frac{p}{100}) - 75.63\frac{p}{100 f_{ck}}$$

$$\frac{P_u}{f_{ck}t_w L_w} = 0.0899$$

$$\frac{M_u}{f_{ck}t_w L_w^2} = 0.1512(1 - \frac{p}{100}) \times (0.5 - 0.42 \times 0.42) + 127.28 \times (0.5 - 0.364 \times 0.42)\frac{p}{100 f_{ck}}$$

$$+ 178.305 \times [0.5 - \{(0.524 \times 0.380) + (0.58 - 0.524)\} \times \frac{p}{100 f_{ck}}]$$

$$+ [24.88\frac{p}{100 f_{ck}} \times \{0.5 - (0.58 - 0.524)/2\}]$$

$$\frac{M_u}{f_{ck}t_w L_w^2} = 0.049(1 - \frac{p}{100}) + 119.55\frac{p}{100 f_{ck}}$$

$$\frac{M_u}{f_{ck}t_w L_w^2} = 0.116$$

8th Case: When the wall section is acting as a steel beam

$$\frac{\mathrm{M}_u}{f_{\mathrm{ck}}t_{\mathrm{w}}L_{\mathrm{w}}^{\;2}} = \frac{0.2175 f_y \mathrm{p}}{100 f_{ck}}$$

$$\frac{\mathrm{M}_u}{f_{\mathrm{ck}}t_{\mathrm{w}}L_{\mathrm{w}}^{\;2}} = 0.089$$

Cases	$\frac{P_u}{f_{ck}t_wL_w}$	$\frac{M_u}{f_{ck}t_wL_w^2}$
Case 1	0.733	0
Case 2	0.6109	0.0476
Case 3	0.577	0.06
Case 4	0.435	0.093
Case 5	0.322	0.1099
Case 6	0.1092	0.115
Case 7	0.0899	0.116
Case 8	0	0.0897

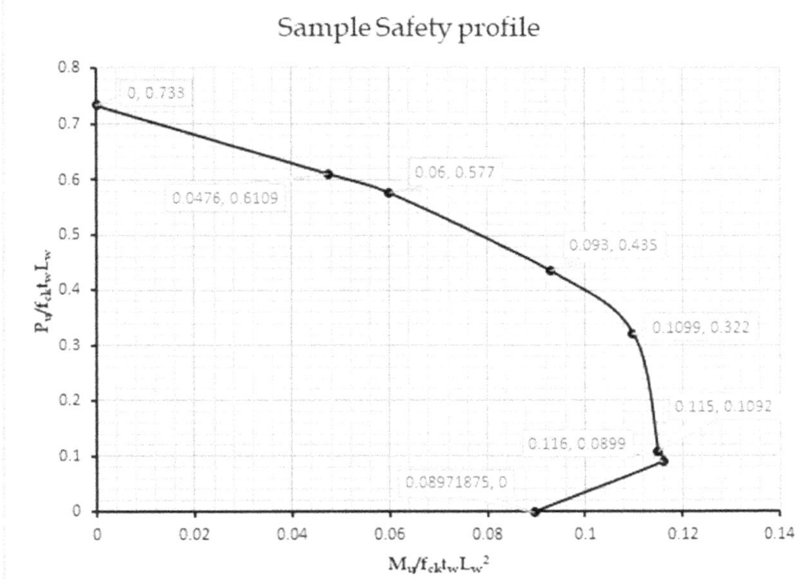

FIGURE 19.20 Safety profile of a rectangular shear wall ($L_w \times t_w$) with three percentage vertical reinforcement using Fe550 grade of steel and M40 grade concrete (f_{ck} = 40 MPa).

The coordinates have been tabulated in Table 19.10 and have been plotted to obtain a safety profile as shown in Figure 19.20.

19.13 DEVELOPMENT OF P–M INTERACTION DIAGRAMS FOR SHEAR WALL FOR NINE GRADES OF CONCRETE (M20 TO M60) AND FOUR GRADES OF STEEL (FE415 TO FE600)

The interaction charts can be developed for structural concrete of M20 to M60 grade using four grades of steel (Fe415, Fe500, Fe550 and Fe600). A total number of 36 (9 × 4) charts for nine grades of concrete and four grades of steel following the requirements of IS:456-2000 can be developed in the manner described above. The interaction charts provide the entire safety profile of a wall section with different percentages of vertical reinforcements. The level of tensile stress at the outermost fibre of steel reinforcement is also depicted in the charts by the equi-stress lines. The interaction charts have been drawn for a wide percentage of vertical longitudinal reinforcements ranging from $0.0025A_g$ to $0.04A_g$ where A_g is the gross cross-sectional area. As per IS:13920-2016, the minimum percentage of vertical reinforcement in a wall is 0.25%. The maximum percentage of vertical reinforcement is not specified but according to Clause 9.6.2 of Eurocode 2, the minimum and maximum amounts of reinforcement required for a reinforced concrete wall are $0.002A_c$ and $0.04A_c$ outside lap locations, respectively. In order to reduce the number of interaction charts without compromising the accuracy of the results, interaction relationships have been plotted for four different grades of steel with M40 grade concrete. It has been observed that using interaction charts with M40 grade of concrete and any grade of steel the results obtained are almost identical with those prepared for M20 and M60 grade concrete using the corresponding grade of steel. The variation lies between –0.1 and 1.1%. Since the variation is negligible, only four interaction charts developed with M40 grade of concrete and Fe415, Fe500, Fe550 and Fe600 grades of steel are considered to be valid for all structural concrete grades from M20 to M60. Results for all grades and percentages of steel are very similar. Hence instead of 36 charts, four charts can be used without compromising the accuracy of the results. The percentages of reinforcement have been expressed as $\frac{p}{f_{ck}}$, whose values range from 0 to 0.2 at intervals of 0.0125. Lower values of neutral axes depths result in negative values of axial force capacity, which indicate that the section is in tension with bending. Since the interaction charts have been developed for compression with bending, the points corresponding to bending with tension have been ignored (refer to Charts 19.1–19.4).

19.13.1 Inferences from the Proposed Charts as per IS:456-2000

The equi-stress lines joining the points corresponding to the balanced conditions have a slope towards the moment axis. Points on the interaction charts (Charts 19.1–19.4) below the balanced condition represent failure in under-reinforced condition and those above the balanced condition represent failure in over-reinforced condition. A close inspection of all the curves indicates the portion representing the under-reinforced condition is very small with respect to the portion representing the over-reinforced region. Hence it is obvious that although failure in under-reinforced condition is desirable, designing under this condition only will restrict the domain of design significantly, leaving very limited options for designers. (For columns refer to Section 18.3)

19.14 DESIGN PROVISIONS AS PER INDIAN STANDARD IS:13920-2016

The basic specifications for shear wall as per the standard are listed below. As per IS:13920-2016, the minimum thickness of a special shear wall shall not be less than 150 mm and 300 mm for buildings with coupled shear walls. This minimum thickness criteria should also satisfy the fire resistance requirements of IS:456.

As per the standard, a special shear wall shall be provided with uniformly spaced reinforcement in its cross-section along vertical and horizontal directions. The minimum reinforcement percentage to be applied in special shear wall has also been specified by this standard.

The design moment of resistance, M_u, of a wall subjected to combined bending moment and compressive axial load shall be estimated in accordance with the requirements of the limit state design method given in IS:456 by using the principles of mechanics involving equilibrium equations, strain compatibility conditions and constitutive laws.

The moment of resistance of a slender rectangular structural wall section with uniformly distributed vertical reinforcement may be estimated using expressions given in Annex A. These

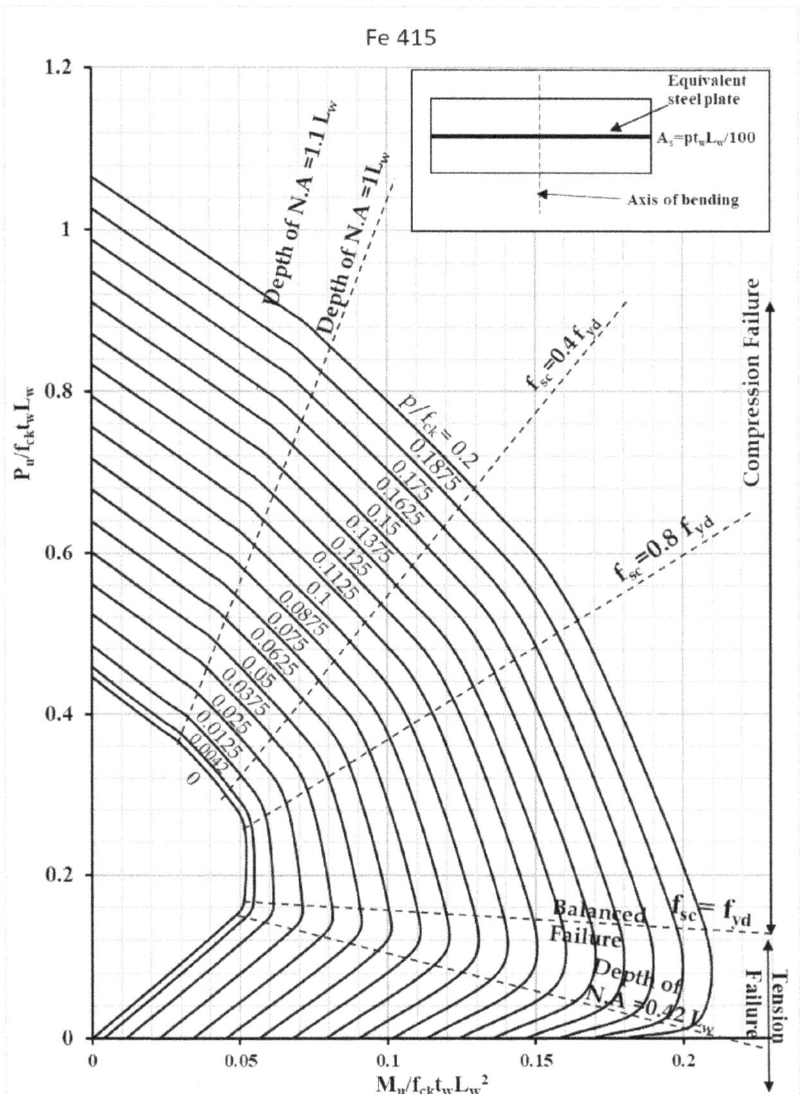

CHART 19.1 Interaction chart for rectangular shear walls subjected to compression with bending as per IS:456-2000.

expressions of Annex A are not applicable to walls with boundary elements. Boundary elements are the portions along the edges of a shear wall that are strengthened by longitudinal and transverse reinforcement. They may have the same thickness as that of the wall web.

As per the standard in a rectangular shear wall, two types of flexural failure may take place in the section. One is flexural tension failure and the other one is flexural compression failure. A flexural tension failure may occur when the tension steel yields before crushing of concrete in the extreme compression fibre. The flexural compression failure takes place when the concrete crushes at the extreme compression fibre before the tensile steel yields. The two sets of closed-form equations provided in the standard deal with these two types of failure only. Failures under balanced condition and when the neutral axis is outside the section or under steel beam condition are not covered in the standard.

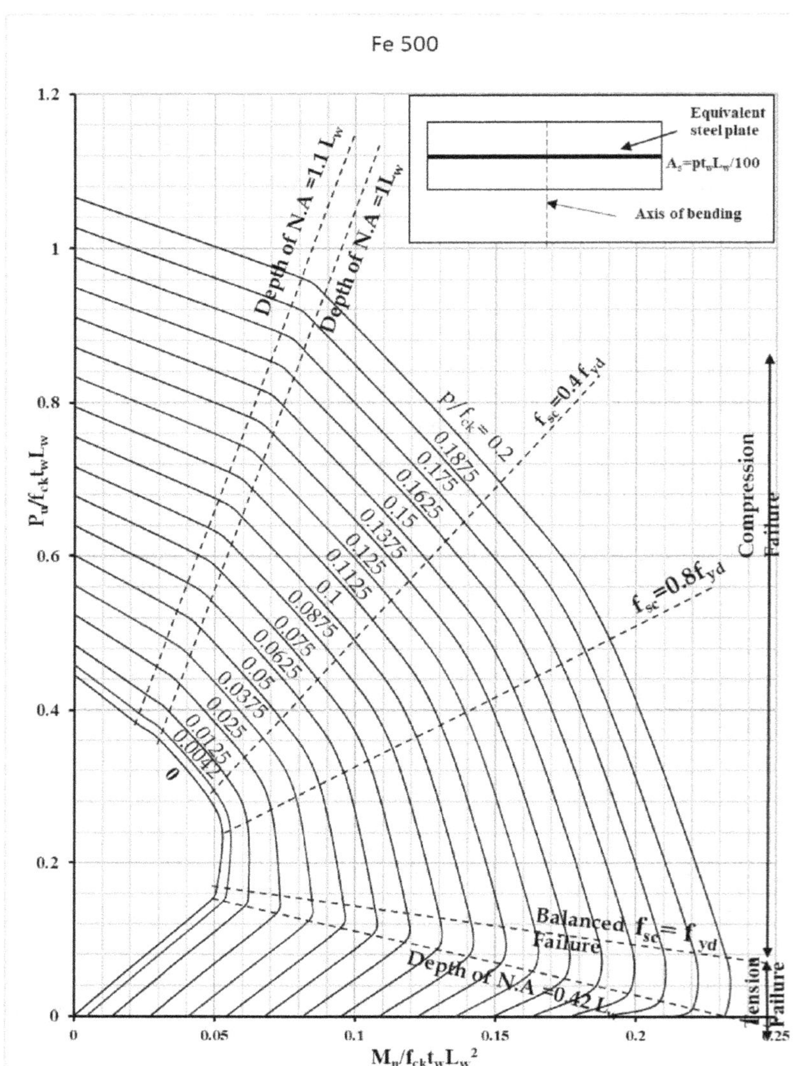

CHART 19.2 Interaction chart for rectangular shear walls subjected to compression with bending as per IS:456-2000.

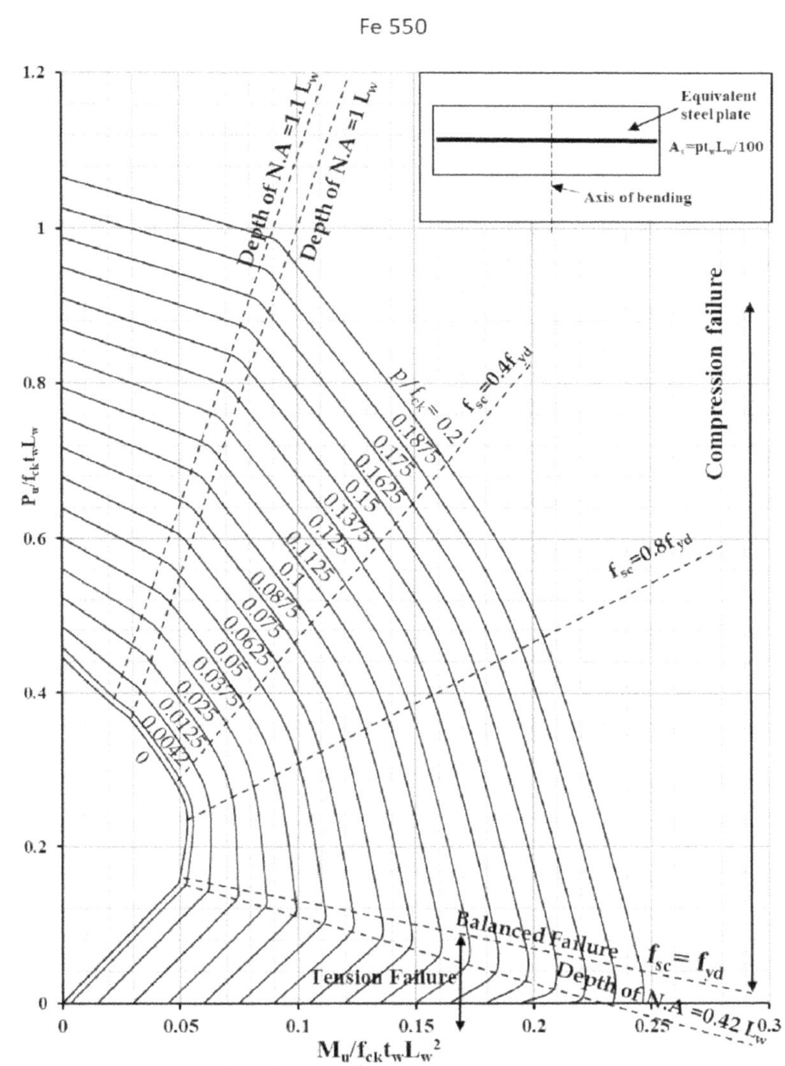

CHART 19.3 Interaction chart for rectangular shear walls subjected to compression with bending as per IS:456-2000.

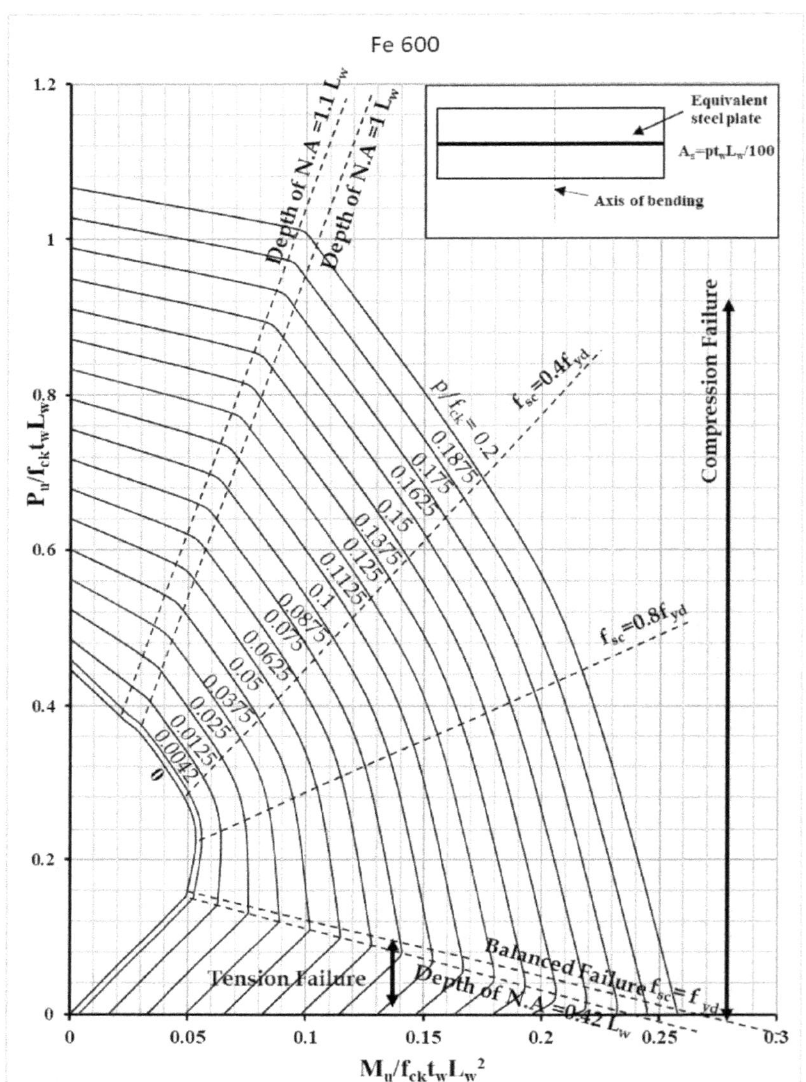

CHART 19.4 Interaction chart for rectangular shear walls subjected to compression with bending as per IS:456-2000.

The minimum vertical and horizontal reinforcements for different types of RC walls along with their minimum spacings are specified in the standard. Requirements for shear design and boundary elements for shear walls are also specified.

19.15 LIMITATIONS OF THE STANDARD

A couple of closed-form expressions have been included in IS:13920-1993 to calculate the moment of resistance of rectangular shear walls with uniformly distributed vertical reinforcement without boundary elements. In those expressions, the stress–strain relationship of steel has been considered as bilinear. Two conditions have been considered, when the depth of the neutral axis is less than the balanced depth and when the depth of the neutral axis is more than the balanced depth of the neutral axis. The depth of the neutral axis is limited to the depth of the wall.

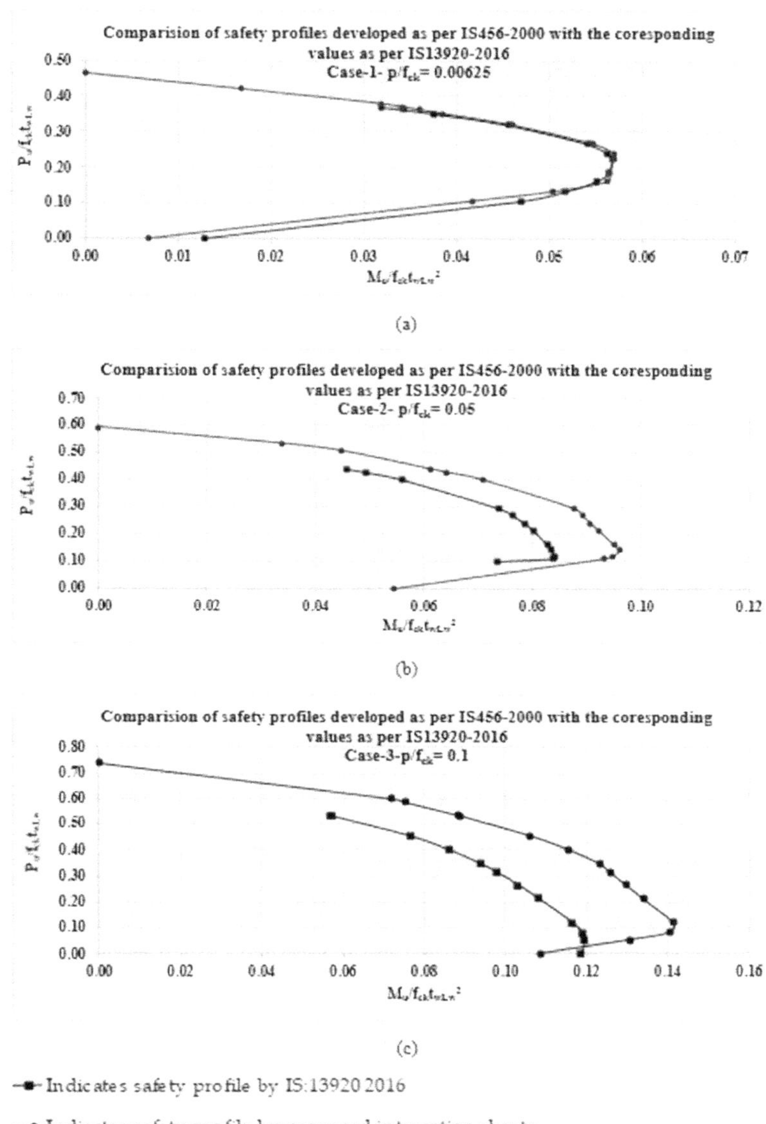

FIGURE 19.21 Comparison of moment of resistance (MOR) values between IS:13920-2016 and proposed interaction charts as per IS:456-2000 for M40 grade concrete and Fe500 steel with t_w = 250mm and L_w = 3750 mm: (a) Case 1, p/f_{ck} = 0.00625; (b) Case 2, p/f_{ck} = 0.05; (c) Case 3, p/f_{ck} = 0.1.

For HYSD bars made by CTD process, as per IS:456-2000, the stress–strain relationship is not bilinear as these steels do not have a definite yield point. When calculating the strength and deformation of reinforced concrete sections, an essential condition which must be satisfied is the compatibility of strains in both concrete and reinforcement, not only at all points across sections but also throughout the lengths of members. When steel and concrete are bonded together, it is assumed that they both have the same strain. The stress–strain relationships of steel in compression and tension are similar. The standard includes these expressions without providing any background information and derivation of the expressions. The expressions do not provide the moment of resistance values

TABLE 19.10
Typical moment of resistance (MOR) values as per IS:13920-2016 vs. those from proposed curves for Fe500, at constant p/f_{ck} = 0.00625 for f_{ck} = 40, t_w = 250 mm and L_w = 3750 mm [Case 1 from Figure 19.21(a)]

$P_u/f_{ck}t_wL_W$	MOR values as per IS: 13920-2016 (kN-m)	MOR values from the proposed chart as per IS:456-2000 (kN-m)	Percentage difference in MOR values with respect to values obtained from IS:13920-2016
0.000	1800	956	46.88
0.107	6623	5850	11.68
0.133	7284	7073	2.90
0.161	7734	7917	-2.36
0.187	7945	7945	0.00
0.228	7988	8002	–0.18
0.240	7917	8016	–1.24
0.267	7622	7706	–1.11
0.320	6413	6469	–0.88
0.352	5273	5414	–2.67
0.363	4823	5063	–4.96
0.371	4472	4795	–7.23
0.380	N.A.	4472	N.A.
0.420	N.A.	2377	N.A.
0.464	N.A.	0	N.A.

TABLE 19.11
Typical moment of resistance (MOR) values as per IS:13920-2016 vs. those from proposed curves for Fe500, at constant p/f_{ck} = 0.05 for f_{ck} = 40, t_w = 250 mm and L_w = 3750 mm [Case 2 from Figure 19.21(b)]

$P_u/f_{ck}t_wL_W$	MOR values as per IS: 13920-2016 (kN-m)	MOR values from the proposed chart as per IS:456-2000 (kN-m)	Percentage difference in MOR values with respect to values obtained from IS:13920-2016
0.100	103359	7650	25.99
0.110	117984	13120	–11.20
0.120	118125	13345	–12.98
0.142	117563	13514	–14.95
0.160	116438	13373	–14.86
0.213	112781	12966	–14.96
0.240	110391	12755	–15.54
0.269	107297	12544	–16.91
0.293	103922	12347	–18.81
0.400	78891	9956	–26.20
0.427	69469	9042	–30.16
0.440	64125	8592	–33.99
0.507	N.A.	6286	N.A.
0.533	N.A.	4753	N.A.
0.592	N.A.	0	N.A.

TABLE 19.12
Typical moment of resistance (MOR) values as per IS:13920-2016 vs. those from proposed curves for Fe500, at constant p/f_{ck} = 0.1 for f_{ck} = 40, t_w = 250 mm and L_w = 3750 mm [Case 3 from Figure 19.21(c)]

$P_u/f_{ck}t_wL_W$	MOR values as per IS: 13920-2016 (kN-m)	MOR values from the proposed chart as per IS: 456-2000 (kN-m)	Percentage difference in MOR values with respect to values obtained from IS:13920-2016
0.000	16678	15300	8.26
0.053	16805	18408	–9.80
0.080	16706	19744	–18.47
0.120	16383	19898	–21.75
0.213	15216	18844	–24.14
0.267	14498	18239	–26.10
0.315	13767	17691	–28.81
0.347	13219	17339	–31.49
0.400	12122	16270	–34.54
0.453	10772	14948	–39.11
0.533	8044	12516	–55.97
0.535	7988	12459	–56.36
0.587	N.A.	10631	N.A.
0.600	N.A.	10167	N.A.
0.738	N.A.	0	N.A.

Note: Negative sign indicates conservative values of moment of resistance as per IS:13920-2016 with respect to the proposed charts.

A positive sign indicates unsafe values of moment of resistance as per IS:13920-2016 with respect to the proposed charts

N.A. means IS:13920-2016 fails to provide any result.

for neutral axis depth higher than the depth of wall, for balanced condition and for steel beam condition. Hence the complete P–M interaction relationship is not available from these expressions.

The interaction charts proposed in this chapter are based on the fundamental principles of the limit state method complying with the requirements of IS:456-2000 and cater to all possible positions of the neutral axis. The results from the proposed charts have been compared with those obtained from IS:13920-2016 using different levels of axial loads, different reinforcement percentages (0.25–4%), different lengths of shear walls (ranging between 600–3750 mm) and thickness (ranging between 150–250 mm) for different grades of concrete (M20–M60) and steel (Fe415–Fe600). It has been observed that IS:13920-2016 yields conservative values for higher values of axial load, higher percentages of reinforcement ($\frac{p}{f_{ck}}$ values ranging from 0.025 to 0.2). However, the code results in unsafe design for lower values of axial load, lower percentages of reinforcements ($\frac{p}{f_{ck}}$ value ranging from 0.00625 to 0.0125). Some typical comparisons along with the percentage variation of results from IS:13920-2016 with respect to the values obtained from the proposed charts are presented in Figure 19.21 and Tables 19.11–19.13. It has been observed that out of a total of 180 typical sample results checked, for 14.4% cases IS:13920-2016 provides unsafe values, for 60.6% cases it provides over-safe values compared to values obtained from the proposed charts as per IS:456-2000 and for 25% of cases it fails to provide any result.

19.16 DEVELOPMENT OF INTERACTION CHARTS FOR SHEAR WALLS AS PER IRC STANDARD/EUROCODE REQUIREMENTS

The design philosophy of RC sections as per IRC:112-2020 is quite similar to the provisions of Eurocode 2. Modifications in the design consideration with respect to IS:456 have been discussed in Chapter 15. The stress–strain relationships of HYSD bars can be considered as one of the most important modifications, which will affect the design of sections. In the subsequent sections, the design of shear wall has been performed as per the requirements of IRC:112-2020.

19.16.1 Flexure design as per IRC:112-2020

EN 1992-1-1 Eurocode 2: Design of concrete structures – Part 1-1: General rules and rules for buildings has provided two stress–strain diagrams for reinforcing steel, namely idealized and design stress strain diagrams as shown in Figure 19.22. IRC:112-2020 has provided similar relationships for steel and defines the diagram with horizontal plastic branch as simplified bilinear and the diagram with inclined plastic branch as idealized bilinear. In the first case the maximum design stress is greater than f_{yd} but the maximum strain is limited to ε_{ud} taken as equal to 0.9 ε_{uk}. An even more simplified option is to limit the maximum stress to f_{yd} with no limit on the maximum strain. For simplicity, the second option is preferred in all common design situations. The nature of the calculation of the forces in steel from stress blocks is similar to the calculations performed for IS:456-2000. The stress block of steel will consist of one triangle and one rectangle, as a result of which the calculations will become significantly simpler. Typical strain distribution diagrams and stress blocks as per IRC:112-2020 for balanced condition are shown in Figure 19.23. The stress–strain relationship of concrete is similar to IS:456-2000 for both IRC:112-2020 and Eurocode 2. Hence the stress blocks for concrete will be identical up to a concrete grade of M60.

Sample calculations for axial force and moment capacities of a rectangular shear wall for failure under balanced condition are shown below.

Under balanced condition the strain at the outermost fibre in compression reaches the ultimate or limiting strain in concrete (ε_{cu} = 0.0035), and the outermost fibre of steel in tension reaches the limiting strain in steel ($\varepsilon_s = f_y/1.15E_s$) simultaneously.

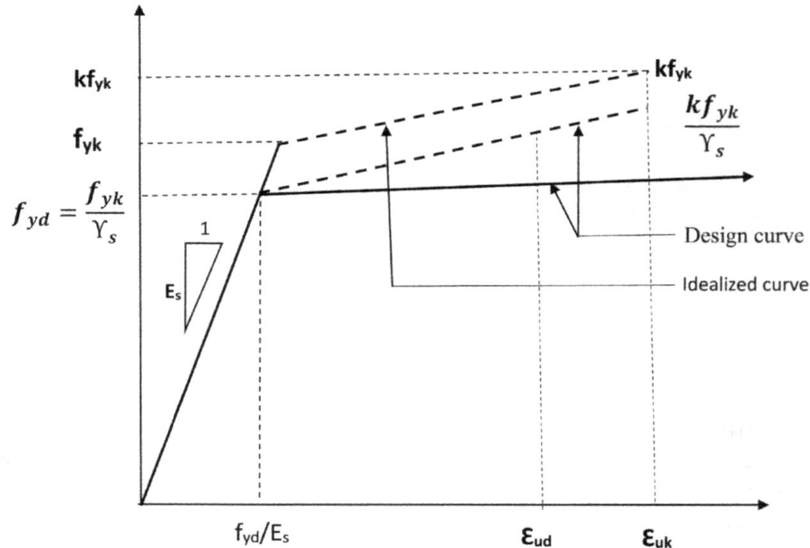

FIGURE 19.22 Stress–strain diagram of steel as per Eurocode 2.

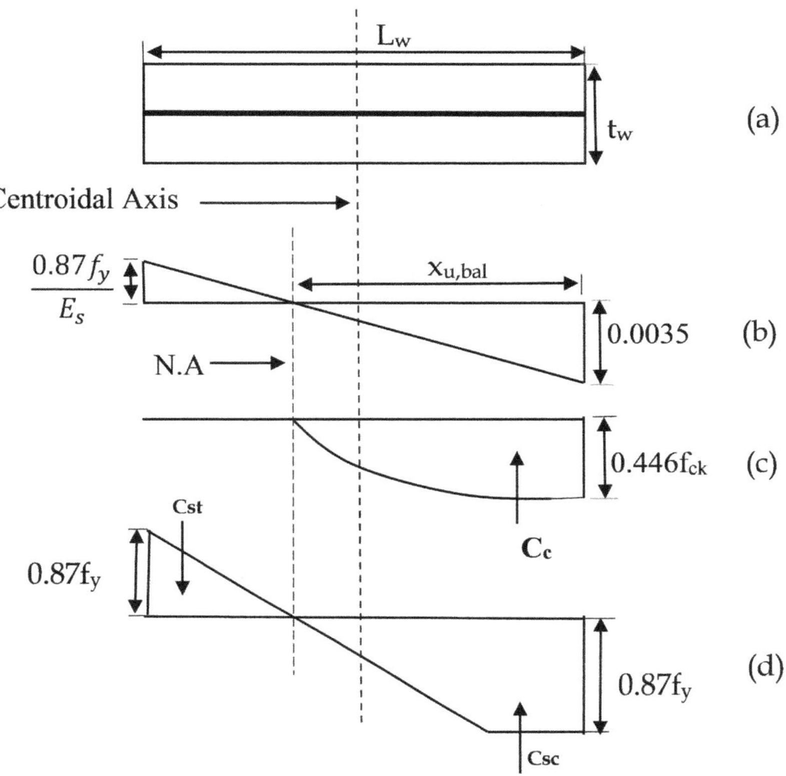

FIGURE 19.23 (a) Cross-section showing vertical reinforcement as an equivalent plate; (b) strain distribution diagram as per IRC:112-2020 in balanced condition; (c) stress block for concrete; (d) stress block for steel.

Let L_w, t_w and t_s be the length of shear wall, width of shear wall and thickness of steel plate, respectively.

For Fe500 grade of steel:

Limiting strain in tension steel $= \dfrac{f_y}{1.15E_s}$

$$= \frac{500}{1.15 \times 200000}$$

$$= 0.0022$$

Limiting strain in compression zone = 0.0035

$$\text{Therefore,} \frac{x_u}{L_w - x_u} = \frac{0.0035}{0.0022}$$

or, $x_u = 0.614\ L_w$

Therefore, $k = 0.614$ and $(L_w - x_u) = 0.386\ L_w$

In the case of balanced condition, for Fe500

The stress block in compression consists of two segments, i.e., segment 1 is a triangle and segment 2 is a rectangle (refer to Figure 19.23).

The strain value at the end of the triangle is $\dfrac{0.87f_y}{E_s} = 0.0022$

The strain value at the end of the rectangle is 0.0035

The fraction of $(L_w - x_u)$ in segment 1 is $\dfrac{0.0022}{0.0035} = 0.449$

Areas of segment 1 and segment 2 are $0.270x_u f_y$ and $0.330x_u f_y$, respectively.
Hence, the total area of stress block for steel in compression is $0.600(x_u)f_y$.
The area of stress block for steel in tension is $0.435\,(L_w - x_u)f_y$.
Therefore, the force capacity of the section

$$P_u = 0.36 f_{ck} x_u \left(t_w - t_s\right) + 0.600 x_u f_y t_s - 0.435\left(L_w - x_u\right) f_y t_s$$

$$= 0.36 f_{ck} x_u \left(1 - \frac{p}{100}\right) t_w + 0.600 x_u f_y \left(\frac{pt_w}{100}\right) - 0.435\left(0.386 L_w\right) f_y \left(\frac{pt_w}{100}\right)$$

$$= 0.36 f_{ck}\left(0.614 L_w\right)\left(1 - \frac{p}{100}\right) t_w + 0.600\left(0.614 L_w\right) f_y \left(\frac{pt_w}{100}\right) - 0.435\left(0.386 L_w\right) f_y \left(\frac{pt_w}{100}\right)$$

$$\frac{P_u}{f_{ck} t_w L_w} = 0.221\left(1 - \frac{p}{100}\right) + 100.245 \frac{p}{100 f_{ck}}$$

Distance of centroid of stress block for concrete from maximum compressed fibre is $0.42x_u$.
For Fe500 grade of steel, in the case of balanced condition:
The centroidal distance of stress block from maximum tension edge for steel in tension is $0.333(L_w - x_u)$.
The centroidal distance of stress block from maximum compressed edge for steel in compression $0.368\,x_u$.
Therefore, the moment capacity of the section:

$$\begin{aligned} M_u = {} & 0.36 f_{ck} x_u \left(1 - \frac{p}{100}\right) t_w \left(0.5 L_w - 0.42 x_u\right) \\ & + 0.600 x_u f_y \left(\frac{pt_w}{100}\right)\left(0.5 L_w - 0.368 x_u\right) \\ & + 0.435\left(0.386 L_w\right) f_y \left(\frac{pt_w}{100}\right)\left\{0.5 L_w - 0.333\left(L_w - x_u\right)\right\} \end{aligned}$$

$$\begin{aligned} = {} & 0.36 f_{ck}\left(0.614\right)\left(1 - \frac{p}{100}\right) t_w \left\{0.5 L_w - 0.42\left(0.614 L_w\right)\right\} \\ & + 0.600\left(0.614 L_w\right) f_y \left(\frac{pt_w}{100}\right)\left\{0.5 L_w - 0.368\left(0.614 L_w\right)\right\} \\ & + 0.435\left(0.386 L_w\right) f_y \left(\frac{pt_w}{100}\right)\left\{0.5 L_w - 0.333\left(0.386 L_w\right)\right\} \end{aligned}$$

$$\frac{M_u}{f_{ck} t_w L_w^2} = 0.0535\left(1 - \frac{p}{100}\right) + 81.67 \frac{p}{100 f_{ck}}$$

Negative signs have been used for tension forces and distances to the tension side from the centroidal axis.

19.16.2 Development of interaction diagrams as per IRC:112-2020

As per IRC:112-2020, P–M interaction charts for rectangular shear walls with uniform distribution of vertical reinforcement can be developed for any grade of structural concrete up to M60. Sample calculations for balanced condition are shown in the previous section. The lines corresponding to the failure under balanced condition have much higher slopes than those obtained from IS:456-2000. In order to properly predict the moment of resistance capacity in the under-reinforced region, two additional points (i.e., when the depth of neutral axis is at $0.52L_w$ and $0.47L_w$) have been considered for the development of the safety profile in this region. The P–M interaction diagrams are shown in Charts 19.5–19.8 for steel grades Fe415 to Fe600, which are valid for all concrete grades up to M60.

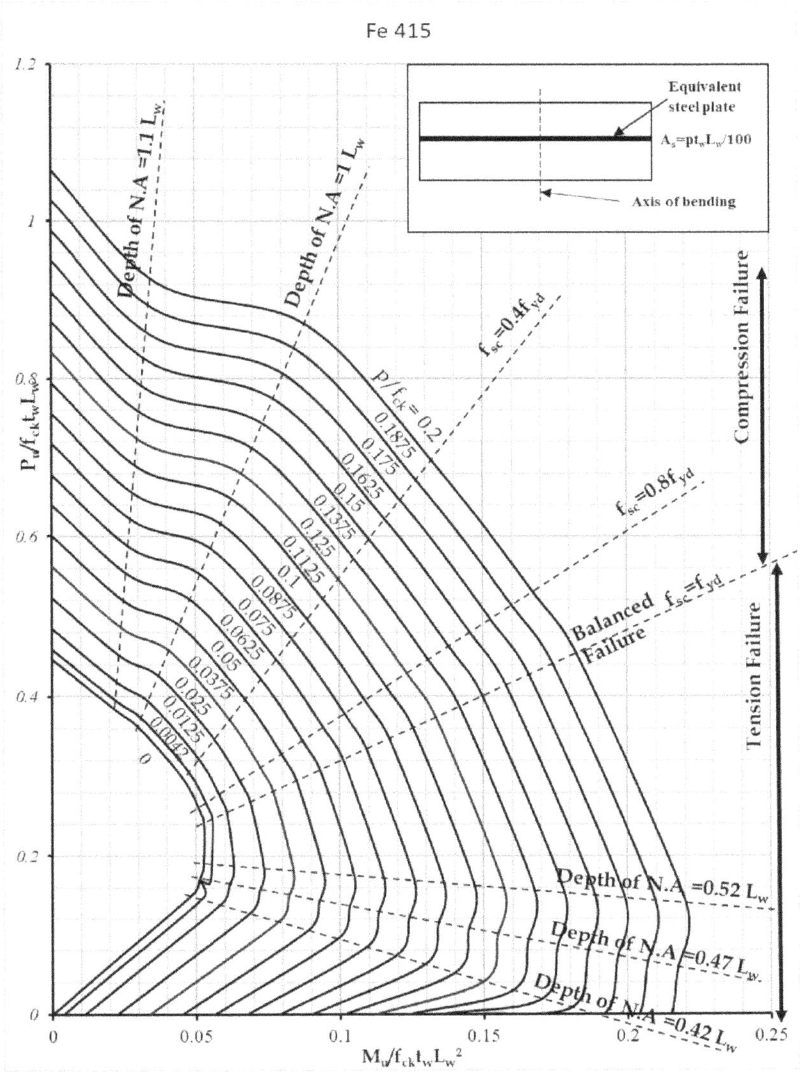

CHART 19.5 Interaction chart for rectangular shear walls subjected to compression with bending as per IRC:112-2020.

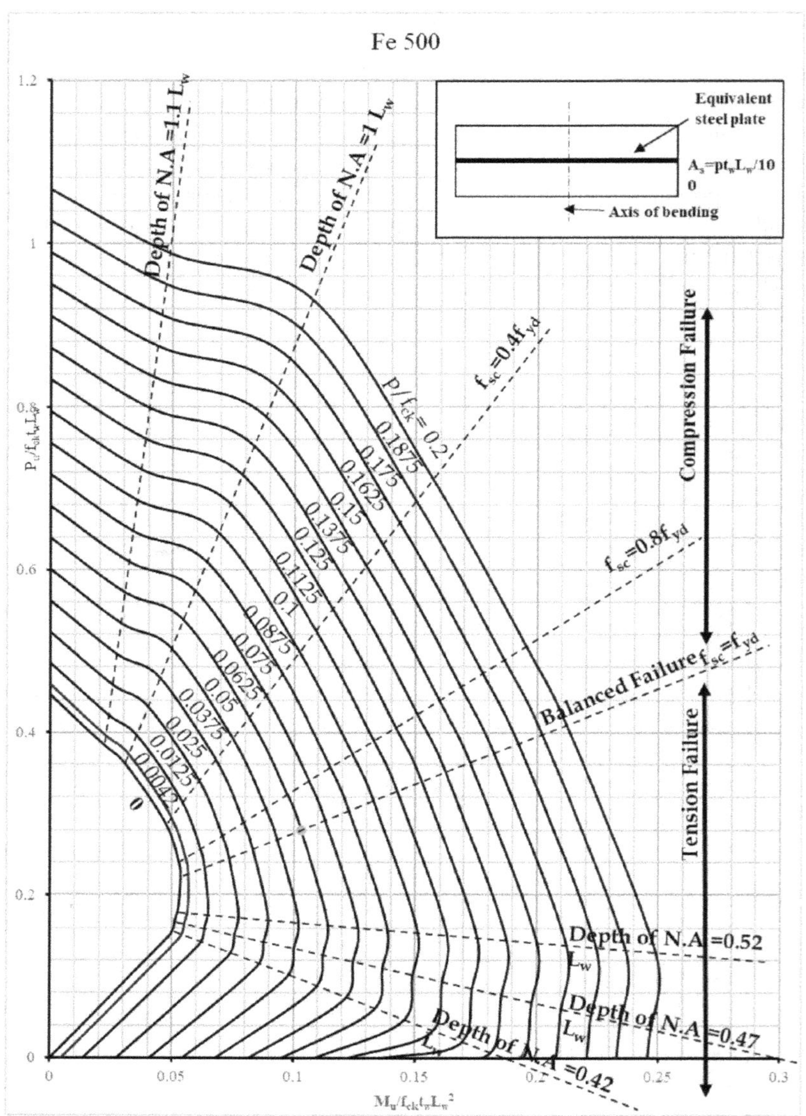

CHART 19.6 Interaction chart for rectangular shear walls subjected to compression with bending as per IRC:112-2020.

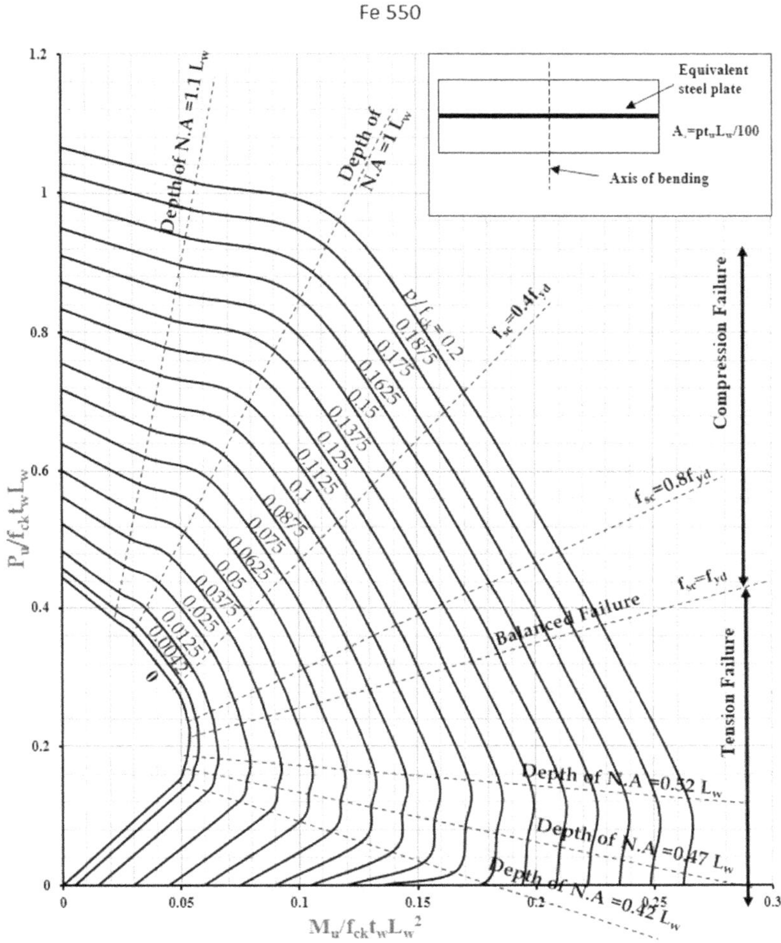

CHART 19.7 Interaction chart for rectangular shear walls subjected to compression with bending as per IRC:112-2020.

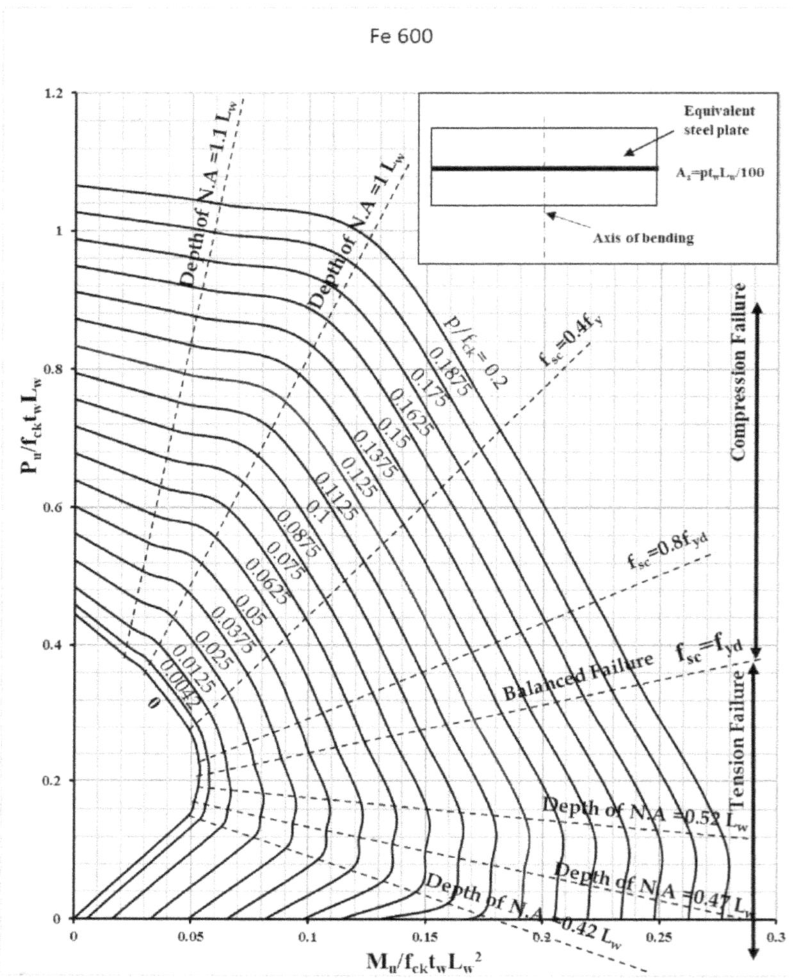

CHART 19.8 Interaction chart for rectangular shear walls subjected to compression with bending as per IRC:112-2020.

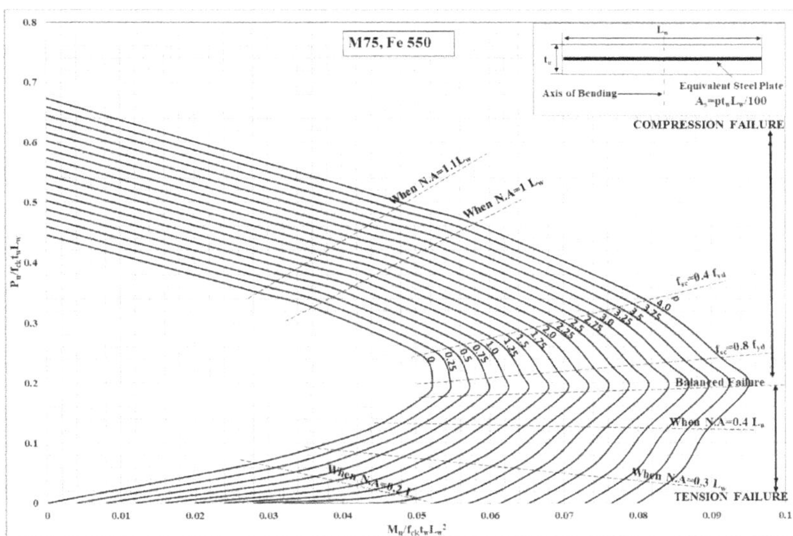

CHART 19.9 Interaction chart for rectangular shear walls subjected to compression with bending as per IRC:112-2020 for M75 concrete and Fe 550 steel.

Stress–strain relationships for higher grade of concrete beyond M60 are available in IRC:112-2020. Using similar methodology interaction charts can be developed for higher grades of concrete with HYSD steel grades Fe415 to Fe600. A typical interaction chart for rectangular shear wall subjected to compression with bending for M75 grade of concrete and Fe550 grade of steel has been presented in Chart 19.9.

19.17 SUMMARY

In this chapter, instead of developing closed-form expressions as formulated in IS: 13920-2016 generalized P–M interaction relationships have been developed for all grades of structural concrete ranging from M20 to M60, which are permitted by IS:456-2000 and commonly used HYSD steel bars from Fe415 to Fe600. A detailed comparison has been performed between the moment of resistance (MOR) values calculated by solving closed-form equations available at Annex A of IS:13920-2016 and the moment of resistance values obtained from the proposed P–M interaction diagrams. It has been observed that the closed-form expressions of IS:13920-2016 for the evaluation of the moment of resistance (MOR) of shear walls do not provide any results when the neutral axis lies outside the section, for balanced condition and steel beam condition. Moreover, the moment of resistance values of the codal expressions give conservative results for the majority of cases with higher reinforcement percentages and axial loads and exhibit unsafe results with lower percentages of reinforcement. The stress–strain relationships for concrete (up to M60 grade) and for HYSD steel (bilinear diagram with horizontal plastic branch) as per IRC:112-2020 are similar to Eurocode 2. In order to meet the requirements of international standards, these relationships have been used to develop P–M interaction charts in a similar manner as performed for IS:456-2000. No significant difference has been observed in the nature of the proposed charts developed as per IS:456-2000 and IRC:112-2020 but it has been observed that there is increased scope for under-reinforced design by using charts prepared by IRC:112-2020 and the later results in economical design. The mother code for concrete in India, IS:456-2000, is on the verge of revision whereby significant changes are likely to be introduced. It is most likely that the modified standard will be in line with the RC design provisions of IRC:112-2020, which have similar specifications as are found in the international standards. The proposed eight interaction charts (four as per IS:456-2000 and four as per IRC:112-2020), which are valid for all grades of structural concrete (up to M60) and commonly used HYSD steel, will significantly simplify the process of design of shear walls and are expected to be useful even with the revised edition of the Indian mother code on concrete.

BIBLIOGRAPHY

Bandyopadhyay, J. N. (2011).*Design of Concrete Structures*, PHI Learning, New Delhi.

Bhatt, Prab, McGinley, Thomas J. and Choo, Ban Seng. (2006). *Reinforced Concrete, Design Theory and Examples*, Third Edition, London, New York, E & FN Spon, Taylor and Francis.

Dasgupta K M., Murty, C. V. R., Shallesh, K. and Agrawal, S. K. (2003). 'Seismic shear design of RC structural walls-Part II: Proposed improvements in IS 13920: 1993 Provisions.' *Indian Concrete Journal*, Vol. 77, No. 11, pp. 1459–1468.

Dasgupta, Kaustubh, Murty, C. V. R., Shailesh, K. and Agrawal, S. K. (2003). 'Seismic shear design of RC structural walls-Part I: Review of code provisions.' *Indian Concrete Journal*, Vol. 77, No. 11, pp. 1423–1430.

DHH. R, Jaiswal, A. K. and Murty, C.V.R. (2013). 'Expressions for moment of resistance of RC walls. *The Indian Concrete Journal* Vol. 87, No. 10, pp. 48–62.

Fintel, Mark. (1986). *Handbook of Concrete Engineering*, Second Edition, CBS Publishers and Distributors, New Delhi.

EN 1992-1-1. (2004). *Eurocode 2: Design of Concrete Structures Part 1-1: General Rules and Rules for Buildings*, European Committee For Standardization, Brussels.

Ferguson, Phil M., Green, John E. and Jirsa, James O. (1988). *Reinforced Concrete Fundamentals*, Fifth Edition, John Wiley, Hoboken, United States.

Indian Standard IS: 456-2000. *Plain and Reinforced Concrete – Code of Practice*, Fourth Revision, Tenth Reprint April 2007, Bureau of Indian Standards, New Delhi.

Indian Standard IS: 13920-2016. (July 2016). *Ductile Design and Detailing of Reinforced Concrete Structures Subjected to Seismic Forces – Code of Practice*, First Revision, Bureau of Indian Standards, New Delhi.

Indian Standard IS: 13920:1993. (November 1996). *Ductile Detailing of Reinforced Concrete Structures Subjected to Seismic Forces -Code of Practice*, New Delhi.

IRC:112-2020. (2020). *Code of Practice For Concrete Road Bridges*, First Revision, Indian Road Congress, New Delhi.

Medhekar, M. S. and Jain, S.K. (1993). 'Seismic behaviour, design and detailing of RC shear walls, Part I: Behaviour and strength', *Indian Concrete Journal*, Vol. 67, No. 7, pp. 311–318.

Medhekar, M. S. and Sudhir K. Jain (1993). 'Design and detailing of RC shear walls, Part II: Design and detailing.' *Indian Concrete Journal* Vol. 67, No. 8, pp. 451–457.

Mondal, A. and Bhanja, S. (2023). 'Modifications proposed in IS: 13920 (2016) regarding shear wall design following the limit state design philosophy of national and international standards', *Indian Concrete Journal*, Vol. 97, No. 6, pp. 9–26.

Mondal, A. and Bhanja, S. (2023). 'Flexure design of high strength concrete slender shear walls using the limit state design philosophy for reinforced concrete sections as per IRC: 112 (2020)', *Indian Concrete Journal*, Vol. 97, No. 10, pp. 40–55.

Mosley, W H., Bungey, J. H. and Hulse, R. (1999). *Reinforced Concrete Design*, Fifth Edition, Palgrave, London.

Park, R. and Paulay, T. (1975). *Reinforced Concrete Structures*, John Wiley, New York.

Prab Bhatt, Thomas J. MacGinley, Ban Seng Choo (2014). *Reinforced Concrete Design to Eurocodes, Design Theory and Examples*, Fourth Edition, CRC Press, Hoboken.

Rajbanshi, S., Kumar, A. and Dasgupta, K. (2020). 'A comparative study of axial force—Bending moment interaction curve for reinforced concrete slender shear wall with enlarged boundary element.' *Proceedings of SECON'19. SECON 2019. Lecture Notes in Civil Engineering*, Vol. 46. Springer, Cham.

SP: 16-1980. *Design Aids for Reinforced Concrete to IS: 456 – 1978*, Fourth Reprint March 1988, Bureau of Indian Standards, New Delhi.

Varghese, P. C. (2005). *Advanced Reinforced Concrete Design*, Second Edition, Prentice-Hall, New Delhi.

20 Design of Staircases

20.1 INTRODUCTION

A staircase is a means of vertical movement from one level to another and consists of an inclined structural slab connecting different floor levels. It generally consists of horizontal and inclined slab portions. The stiffness of this portion of the floor significantly increases with respect to the remaining portion of the floor. Hence, they need to be placed at locations whereby neither significant torsion is introduced in the floor nor the functionality and aesthetic consideration of the staircase are compromised. It is always better to include the stair slab in the analysis of the building so that its effect on the structural system is taken into consideration.

20.2 TYPES OF STAIRCASES

20.2.1 Based on the Geometrical Configurations

- Straight or single flight stair
- Dog-legged stair
- Open-well stair
- Spiral stair
- Helicoidal stair

The type of stair to be selected will depend on the architectural, aesthetic, functional and structural considerations. In this chapter only dog-legged staircases will be discussed in detail.

The details of a typical going of a staircase are shown in the Figure 20.1 along with necessary details.

20.2.2 Structurally, Staircases Can Be Classified into Two Groups Depending upon the Direction of Load Transfer

a) Stair slab spanning longitudinally

Load is transferred predominantly in the direction of the length of the slab. One or more supports can be provided parallel to the riser. Some of the typical supporting systems of a staircase are shown in Figures 20.2 and 20.3.

DOI: 10.1201/9781003415398-20

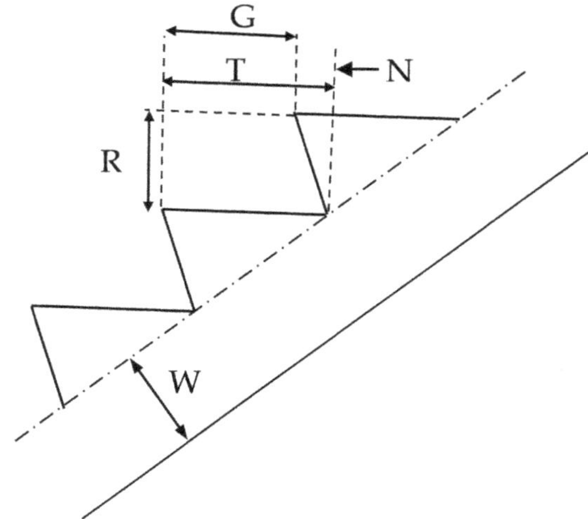

FIGURE 20.1 Details of a waist slab.

Note: W = waist slab thickness;
T = Tread;
R = Rise;
G = Going;
N = Nosing.

If beams are provided at the end of the landings, the span of the stair slab increases, resulting in increased slab thickness. Beams provided at the ends of the going will reduce the slab thickness but these supporting beams, if provided at the half landing level, will require beams parallel to the span direction between the columns, which will increase the stiffness of the stair block portion even more than the other portion of the floor. This will invite the additional tortional moments and hence is undesirable.

The entire load is transferred in one direction, i.e., along the length of the slab. The slab is designed as a beam of 1 m width (unit width) placed on the supporting beams.

b) Stair slab spanning transversely

Here the slab is supported on the sides by two side walls/stringer beams, cantilevering out from a side beam, or doubly cantilevering out from a central spanning beam. Here the load is transferred in the transverse direction.

For this type of stairs, the thickness of the stair slab required is significantly less as the span is greatly reduced.

20.3 IS CODAL REQUIREMENTS

As per IS:456-2000, the effective span of stairs is specified depending on the position of the supporting beams and the direction of load transfer of the landing slab. The distribution of loading on the open well staircase has been specified in the standard. The depth of the section has been specified as the minimum thickness perpendicular to the soffit of the staircase.

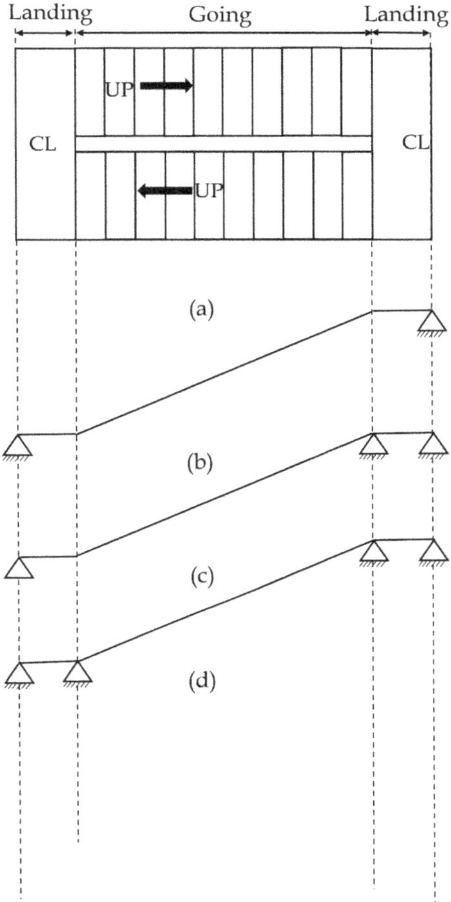

FIGURE 20.2 Staircase spanning in the longitudinal direction: (a) plan; (b), (c) and (d) line elevations showing different positions of supporting beams.

20.4 STRUCTURAL ANALYSIS AND DESIGN CONSIDERATION

Staircases are analysed and designed in a manner similar to floor slabs.

For a dog-legged stair supported by beams at the end there is an inclined portion connecting two horizontal portions at different levels. Such a stair can be analysed as a beam of unit width spanning between the supporting beams. Sometimes the slab is considered as clamped or fixed or hinged at the supports. A fixed support of a beam means all the six degrees of freedom (three translational and three rotational) are fully restrained and a hinged support means that only the three translational degrees of freedom are restrained. The actual condition is, however, intermediate between these two limiting cases. The problem is statically indeterminate and rigorous analysis can be performed using computer software. Moreover, the waist slab being inclined its vertical gravity load has two components, one along the axis of the slab, i.e., tangential component, $w_t = wsin\theta$ and the other being the normal component $w_n = wcos\theta$, where w is the weight of the slab and Θ is the slope of the waist slab with respect to the horizontal. The normal component causes bending of the waist slab and the tangential component results in axial thrust in the form of horizontal force on the supporting beams (Figures 20.4 and 20.5). This thrust is generally ignored in design as the stair slab is analysed considering the projected length or span of the stair on a horizontal plane.

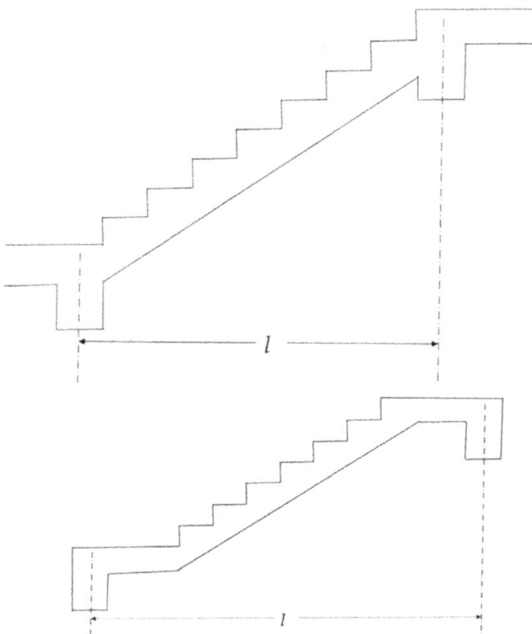

FIGURE 20.3 Typical supports for longitudinally spanning stairs.

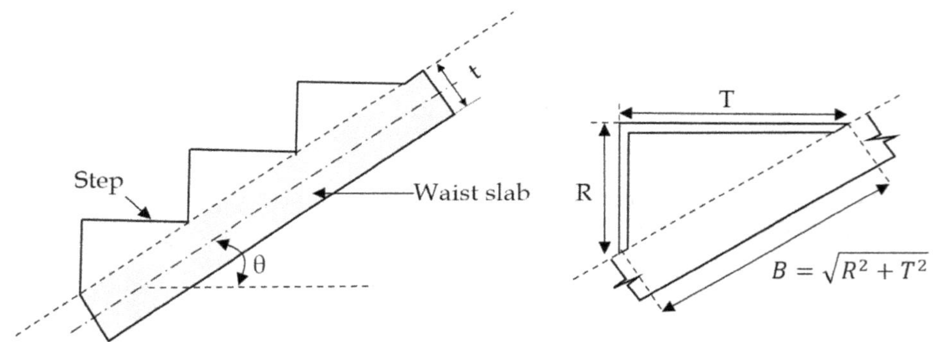

FIGURE 20.4 Details of waist slab.

However, for all practical purposes, simplified analysis of the stair slab is performed using some assumptions. The stairs are assumed to span between the supporting beams at the top and bottom of the flight. The span of the stair is taken as the horizontal distance between the supports, i.e., horizontal projection of the distance between the supports. To determine the weight of the waist slab on plan the weight of the inclined going portion is increased by the ratio, $\frac{\sqrt{R^2 + G^2}}{G}$. The self-weight of the inclined waist slab along with the weight of plaster is increased using this ratio. Steps can be made of concrete or bricks, whose weight act downwards. The self-weight of the bricks is calculated in terms of the weight per metre length of the slab. The stair slab is considered as a beam of unit width. In this manner the loads on the landing and the going portions are considered as load/m length of the beam in plan. For conservative design the stair slab is considered to be simply supported at

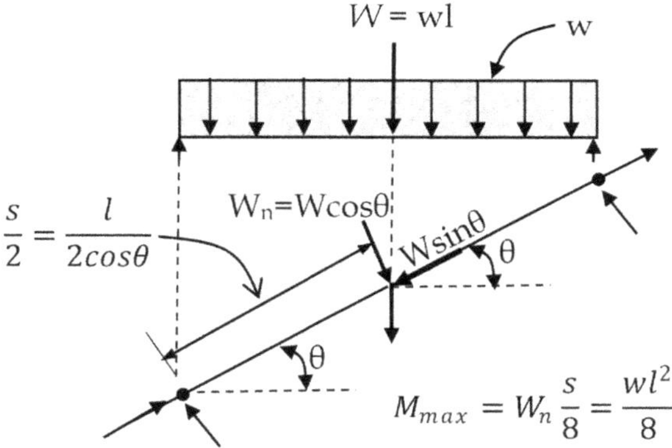

FIGURE 20.5 Load transfer in waist slab.

the supports. Since the slabs and beams are monolithically cast some negative support moment will develop at the ends. Fifty percent of the span reinforcements are provided at the support top.

Deflection check

A flight of stairs is stiffer than a horizontal slab of equivalent thickness. As per British standards, where the stair flight occupies at least 60% of the span, the allowable span/effective depth ratio can be increased by 15%. This is applicable to staircases spanning in the direction of the flight and without stringer beams.

20.5 ILLUSTRATIVE EXAMPLE

Problem 20.1

Design a dog-legged staircase for a residential building using the following data (Figure 20.6):
Floor height = 3 m, width of landing = 1.25 m, imposed load = 3 kN/m^2
Beams provided at the ends have 250 mm width. Use M25 grade of concrete and Fe500 steel. Consider mild exposure condition.

Solution

Consider 10 steps in each flight so that the height of the riser becomes $\frac{3000}{20}$ mm = 150 mm. Consider a tread of 250 mm.

R = 150 mm, T = 250mm

Effective span = 2.25 + (1.25 × 2) + 0.25 = 5 m

Assume a waist slab thickness of $\frac{L}{25} = \frac{5000}{25} = 200\,\text{mm}$

Adopt a thickness of 225 mm for the going and 200 mm for the landing portions.

Effective depth, $d = 225 - \left(20 + \frac{12}{2}\right) = 199\,mm$ (considering the diameter of the main reinforcement as 12 mm and nominal cover as 20 mm for mild exposure condition)

Load analysis:

Going

(i) Self-weight of waist slab in plan $= (0.225 \times 25) \times \frac{\sqrt{250^2 + 150^2}}{250}$ = 6.56 kN / m^2

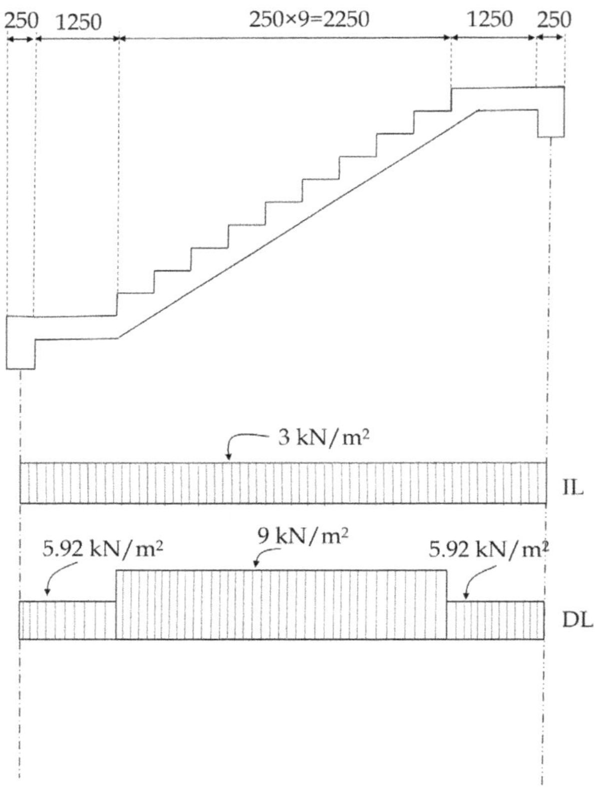

FIGURE 20.6 Details of stair and its loading for Problem 20.1.

(ii) Weight of ceiling plaster 6 mm thick of unit weight 20 kN/m^3 in plan,

$$= (0.006 \times 20) \times \frac{291.5}{250} = 0.14 \text{ kN / m}^2$$

(iii) Weight of step

Number of steps per meter length in plan = $\frac{1000}{250} = 4$

Weight of each step = $\frac{1}{2} \times 0.25 \times 0.15 \times 1 \times 20 = 0.375$ kN / m width (considering unit weight of masonry as 20 kN/m^3)
Weight of steps on the horizontal projection of slab length = 0.375 × 4 = 1.5 kN/m^2

(iv) Weight of floor finish @ 40 mm having unit weight of 20 kN/m^3
=0.04 × 20 = 0.8 kN/m^2

(v) Imposed load = 3 kN/m^2

Total dead load = 6.56 + 0.14 + 1.5 + 0.8 = 9 kN / m^2
Imposed load = 3 kN/m^2
Hence total load = 9 + 3 = 12 kN / m^2

Landing

(i) Self-weight of slab = 0.2 × 25 = 5 kN/m²
(ii) Weight of floor finish = 0.04 × 20 = 0.8 kN/m²
(iii) Weight of ceiling plaster 6 mm thick of unit weight 20 kN/m³ in plan,

$$= (0.006 \times 20) = 0.12\,\text{kN/m}^2$$

(iv) Imposed load = 3kN/m²

Dead load = 5 + 0.8 + 0.12 = 5.92 kN/m². Total load = 5.92 + 3 = 8.92 kN/m²

Support reaction, $R = \left[(8.92 \times 1.375) + \left(12 \times \frac{2.25}{2}\right)\right] = 25.765\,\text{kN/m}$

Maximum moment at midspan=

$$= 25.765 \times 2.5 - 8.92 \times 1.375 \times \left(2.5 - \frac{1.375}{2}\right) - 12 \times \frac{(2.5 - 1.375)^2}{2}$$
$$= 64.41 - 22.23 - 7.59$$
$$= 34.59\ \text{kNm/m}$$

$\frac{M_{ulim}}{bd^2} = 3.44$ for M25 concrete and Fe500 grade of steel.

$$\therefore d_{required} = \sqrt{\frac{34.59 \times 1.5 \times 10^6}{3.44 \times 1000}}$$
$$= 122.81\ \text{mm}$$

$$d_{provided} = 199\ \text{mm} > d_{required}$$

Hence the section is under-reinforced.

From Annex-G of IS:456-2000:

$$M_u = 0.87 f_y . A_{st} . d\left(1 - \frac{A_{st} f_y}{bdf_{ck}}\right)$$

$$\therefore 34.59 \times 1.5 \times 10^6 = 0.87 \times 500 \times A_{st} \times 199 \times \left(1 - \frac{A_{st} \times 500}{1000 \times 199 \times 25}\right)$$

$$\therefore A_{st}\left(1 - \frac{A_{st}}{9950}\right) = 599.4$$

$$\therefore 9950 A_{st} - (A_{st})^2 = 5963793.1$$

$$\therefore (A_{st})^2 - 9950 A_{st} + 5963793.1 = 0$$

$$\therefore A_{st} = 640.62\text{mm}^2$$

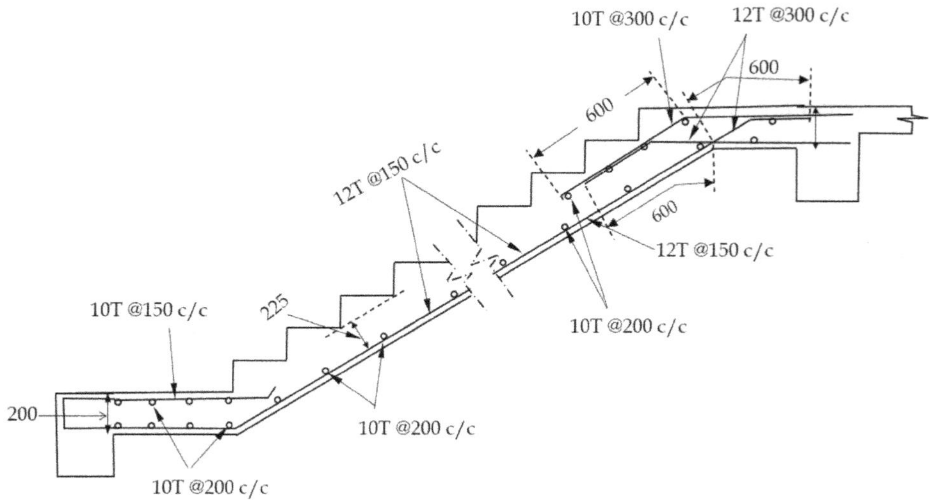

FIGURE 20.7 Reinforcement details of the stair slab.

Required spacing for 12 Tor bars $= \dfrac{113 \times 10^3}{640.62} = 176.4$ mm

Provide 12 Tor @150 c/c (area= 754 mm^2)
Distribution reinforcement $= A_{st,required} = 0.0012\ bd = 0.0012 \times 10^3 \times 199 = 238.8$ mm^2
Provide 10 Tor @200 c/c as distribution reinforcement (area = 393 mm^2).

Reinforcements are to be provided at support due to continuity at the floor level and monolithic construction with beams at landing levels. Necessary reinforcement from floor slab needs to be provided at the top of the landing. The reinforcements at the top of supports for the staircase slab may be provided in the form of 10 Tor @150 c/c (area =524 mm^2). Necessary reinforcement details are shown in Figure 20.7.

Deflection check

For the present problem, the length of the going part is less than 60% of the span. Hence no further enhancement factor has been used for the coefficient. However, the stair slab is much stiffer than a horizontal slab.

Area of steel required at span = 649.18 mm^2
Area of steel actually provided at span = 754 mm^2
Hence, steel stress at service load $= f_s$

$$= 0.58 f_y \frac{\text{Area of cross} - \text{section of steel required}}{\text{Area of cross} - \text{section of steel provided}}$$

$$= 0.58 \times 500 \times \frac{649.18}{754} = 249.68$$

Percentage of actual tension reinforcement provided, $p_t = \dfrac{100 \times 754}{1000 \times 199} = 0.38$

As per clause 23.2, IS:456-2000, modification factor for tension reinforcement = 1.38.

Since the stair slab is continuous at one end and resting on the half landing beam at the other end, the basic value of the modification factor may be taken as 23 (average of that for simply supported and continuous conditions) .

$$d_{required} = \frac{5000}{23 \times 1.38} = 157.53 \text{ mm}$$

$$D_{required} = 157.53 + \left(20 + \frac{12}{2}\right) = 183.53 \text{ mm}$$

$$D_{provided} = 225 \text{ mm} > D_{required}, \text{hence OK.}$$

20.6 SUMMARY

This chapter lists the different types of staircases commonly used based on geometrical and structural considerations. The different components of a typical stair have been clearly depicted. Although the stair slab is monolithically cast with beams for RC frame buildings, the analysis methodology is significantly simplified in design calculations. The inclined portion of the stair flight, i.e., the going part, will result in axial thrust on the supporting beams in addition to causing flexure on the waist slab. This axial thrust is generally ignored as analysis is performed using simplified assumptions. The stair flight is generally designed for a simply supported span moment along with tension reinforcement at the support top to resist any negative moment developing at the support. A typical example of a dog legged staircase has been illustrated with necessary details.

BIBLIOGRAPHY

Allen, H. (1988). *Reinforced Concrete Design to BS 8110, Simply Explained*, E & F. N. Spon, London.

Bandyopadhyay, J. N. (2011). *Design of Concrete Structures*, PHI Learning, New Delhi.

Indian Standard IS: 456-2000, *Plain and Reinforced Concrete – Code of Practice*, Fourth Revision, Tenth Reprint April 2007, Bureau of Indian Standards, New Delhi.

Pillai, S. U. and Menon, D. (2012). *Reinforced Concrete Design*, Third Edition, Reprint 2012, Tata McGraw Hill Education, New Delhi.

Sinha, S. N. (2002). *Reinforced Concrete Design*, Second revised edition, Tata McGraw-Hill, New Delhi.

SP 34: 1987. (1999). *Handbook on Concrete Reinforcement and Design*, Bureau of Indian Standards, New Delhi.

Varghese, P. C. (2005). *Limit State Design of Reinforced Concrete*, Second Edition, Prentice-Hall, New Delhi.

21 Design of Foundations

21.1 INTRODUCTION

A structure has two parts – the superstructure and substructure or foundation. The foundation is that part of the structure which provides support to the superstructure and transfers all actions safely from the superstructure to the supporting soil. Foundation failure can cause the destruction of the entire structure and hence its design requires special care and attention. The present trend worldwide is to allow inelastic deformation in part of the superstructure so that energy dissipation can take place, and this concept plays a very vital role in aseismic design. However, foundation design is generally performed using the conventional concept of strong conservative design.

Foundation design for any structure should satisfy the following design criteria:

I. There should be an adequate factor of safety against bearing capacity failure.
II. Settlement of the foundation must be within permissible limits.

In fact, foundation design requires the combined efforts of geotechnical and structural engineers. The bearing capacity of soil and its settlement characteristics are specified by geotechnical engineers, based on which the foundation design is to be performed. Foundation design consists of two distinct parts:

1) Proportioning, i.e., determining the area required to transmit the actions coming on it properly to supporting soil.
2) Structural design of the foundation.

Since this part of the structure remains embedded within the soil, whose properties can be highly variable, certain additional precautions are required for foundation design. Some of these are discussed below.

A. Generally, the superstructure and substructure are separately treated for analysis and design. The most important point which needs to be addressed by the designer is the location of the supporting systems along with the boundary conditions of the superstructure. Generally, the connection between the structural elements of the moment-resisting frame, which is the most common structural system for RC buildings, is monolithically cast and should be considered as rigid joints. This means that the angle between structural elements meeting at that joint are the same before and after deformation. These joints are often misunderstood as supports, but they should not be treated as supports.

DOI: 10.1201/9781003415398-21

A support may be defined as a node where the degrees of freedom are defined or known. Depending upon the restraints to these degrees of the freedom, supports are classified into different types such as fixed, hinged, roller, etc. All connecting nodes between the beam–column elements are joints, but all joints are not supports. The degrees of freedom at the beam–column joints of the superstructure are not defined or known and hence they should not be considered as supports. The substructure which is embedded within the ground provides necessary supporting systems to the superstructure. The superstructure is analysed for all forms of external and internal actions and their combinations. The stress resultants obtained from the analysis at the supporting systems are thereafter used for the analysis and design of the foundation. The methodology and principles of foundation design are abundantly available, but detailed considerations of the supporting system provided by the substructure to the superstructure are not as abundant.

Concrete structures and the structural elements are robust, massive and quite rigid. Hence how the support systems will behave needs to be properly understood. It is a general custom to consider the columns coming down from the superstructures as fixed at the footing levels. A fixed support means that all three translational and three rotational degrees of freedom are restrained. Generally, the footing plate which is placed within the soil is not expected to translate or rotate unless the soil is very poor. The thickness of the base is generally obtained as sufficiently thick to resist the bending moments and shear forces, and in general it acts as a rigid member. Therefore, the consideration that the footing can provide a fixed support to the column seems to be quite justified. Provision of a pedestal in footings not only makes the footing design economical but also helps in providing proper development length in compression of column bars. In aseismic design, overloading can be a routine phenomenon and hence provision of full development length is always desirable. The moments of inertia of pedestals due to their larger cross-sections are higher than those of columns. The heights of pedestals are significantly less than those of columns. Hence the stiffness of the pedestals is very high and columns may be considered to be fixed at the points where they meet with the pedestals. Tie beams may also be provided at those levels and when the top of the tie beams are flushed with the top of the pedestals the system becomes even stiffer. In fact, the above discussions are mostly generalized in nature and the actual situation needs to be handled by the designer as per his or her discretion.

B. For foundation design, as mentioned earlier, two distinct steps, proportioning and structural design, are to be performed. There is a considerable amount of confusion amongst designers about the loading to be considered for these two conditions. For structural design of RC footing sections, ultimate loads are to be used. However, it needs to be clearly understood that serviceability and strength design are specified by the code for the design of reinforced concrete sections but proportioning of footings deals with the bearing capacity of the soil based on its strength and settlement criteria and has no relation with concrete. As per National Building Code of India 2016 (NBC-2016) – Part 6, section 2, Soils and Foundations, the load combinations for design are specified. The foundation shall be proportioned for the following combination of loads:

a) Dead load + Imposed load; and
b) Dead load + Imposed load + Wind load or seismic loads, whichever is critical.

However, the detailed load combinations are provided in Part 6 of the National Building Code (NBC-2016) structural design Section 1 Loads, Forces and Effects, which states:

The following loading combinations, whichever combination produces the most unfavourable effect in the building, foundation or structural member concerned may be adopted (as a general guidance). It should also be recognized in load combinations that the simultaneous occurrence of maximum values of wind, earthquake, imposed and snow loads is not likely:

a. DL
b. DL and IL
c. DL and WL
d. DL and EL
e. DL and TL
f. DL, IL and WL
g. DL, IL and EL
h. DL, IL and TL
i. DL, WL and TL
j. DL, EL and TL
k. DL, IL, WL and TL
l. DL, IL, EL and TL

(DL = dead load, IL = imposed load, WL = wind load, EL = earthquake load and TL = temperature load.)

As per IS:456-2000, in ordinary buildings, such as low-rise dwellings whose lateral dimensions do not exceed 45 m, the effects due to temperature fluctuations and shrinkage and creep can be ignored in design calculations. However, for structures other than buildings, other relevant loads and their combinations, as applicable, should be considered.

21.2 EFFECT OF WIND FORCE AS PER NBC-2016

Where the bearing pressure due to wind is less than 25 percent of that due to dead and imposed loads, it may be neglected in design. Where this exceeds 25 percent, foundations may be so proportioned that the pressure due to combined dead, imposed and wind loads does not exceed the allowable bearing pressure by more than 25 percent.

21.3 EFFECT OF EARTHQUAKE FORCE AS PER IS:1893 – PART 1, 2016

When earthquake forces are included, net bearing pressure in soil can be increased as per Table 21.1, depending on type of foundation and type of soil. Details about the types of soil are available in the standard. In soft soils, no increase shall be applied in bearing pressure, because settlements cannot be restricted by increasing bearing pressure.

21.4 TYPES OF FOOTING

Foundations can be classified as shallow or deep, depending on the depth where the foundation transfers the loads to the soil. The different types of foundation are listed in Figure 21.1.

TABLE 21.1
Permissible increase in bearing capacity of soil

Soil type	Percentage increase allowable
Type A Rock or hard soil	50
Type B Medium or stiff soil	25
Type C Soft soil	0

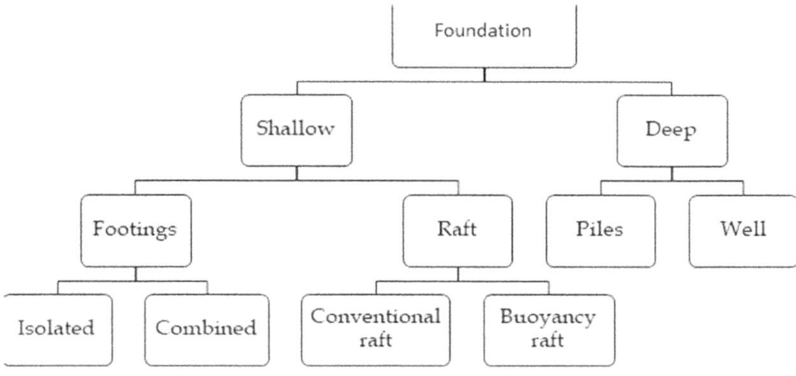

FIGURE 21.1 Types of foundations.

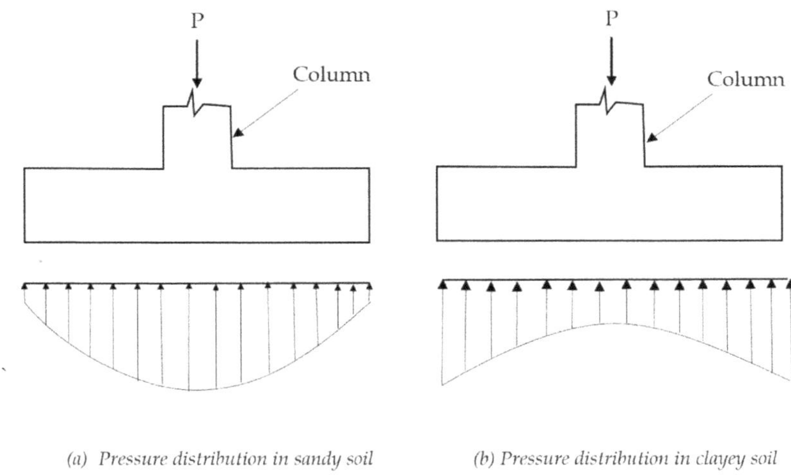

FIGURE 21.2 Distribution of soil pressure below foundations.

21.5 DISTRIBUTION OF SOIL PRESSURE AND BEHAVIOUR OF ISOLATED FOOTING

The foundation is generally assumed to act as a rigid body. The thickness is relatively high and the system is embedded within the soil. Hence it may be considered to act as a rigid member and the general assumption made as a linear distribution of base pressure holds good. The distribution of base pressure is different for different types of soil. Typical distributions in sandy and clayey soils are shown in Figure 21.2.

However, for all practical purposes, footing is considered to be a perfectly rigid body and hence the pressure distribution below its base is considered as linear. For concentrically loaded footings, the pressure distribution will be uniform. However, in real-life situations the supporting nodes of the columns or structural walls are subjected to support reactions in the form of vertical and horizontal loads and bending moments. Thus, the footings are subjected to axial loads, shear forces and moments about both the axes, which are obtained as a sum of the column moments at the base + the corresponding shear force at the column base multiplied by the depth of the footing. As a result, the pressure distribution below the columns become non-uniform. In this chapter, discussion will be made only for isolated footings.

In the following section the distribution of base pressure will be explained for footings subjected to axial loads only, i.e., concentrically loaded footings and footings subjected to axial loads, moments and shear forces which can be considered as eccentrically loaded footings.

(a) Footing subjected to axial load only, i.e., concentrically loaded footings

Footings are placed at some depths below the soil and are filled up or backfilled to the ground level. The net force to which the soil is subjected to at the base is equal to load coming from the superstructure + difference between self-weight of the footing and weight of soil occupying the footing volume. This load divided by the area of the footing gives the net pressure at the base of the footing, which should be less than the net safe bearing capacity of the soil.

(b) Footing subjected to axial load and moment

Figure 21.3 shows a rectangular footing subjected to axial load and moments. In real-life situations the superstructure is analysed using software, and from the analysis results the support reactions at the base of the column (the points where they are considered to be fixed) are obtained. Figure 21.3 shows a rectangular isolated footing subjected to the different stress resultants. The loads acting at the base of the footing can be listed as follows:

Total vertical load, $P_T = P_1$ + difference in weight of footing and soil occupying the footing volume

Total moment about x-axis $M_{x,T} = M_x + S_y.D$

Total moment about y-axis $M_{y,T} = M_y + S_x.D$

where, P_1 = vertical load coming from the superstructure

M_x and M_y = moments about x and y axes acting at the column base

S_x and S_y = shear forces along the x and y axes coming from the superstructure

D = depth of the footing from the top of the pedestal

B = width of the footing

L = length of the footing

b and d = dimensions of the pedestal

For the rectangular footing, the areas and moment of inertias are calculated as

Area, $A = B \times L$

Moment of inertia about y-axis, $I_y = \dfrac{BL^3}{12}$

Moment of inertia about x-axis, $I_x = \dfrac{LB^3}{12}$

Hence the base pressure can be calculated as follows,

$$q = \frac{P_T}{A} \pm \frac{M_{x,T}(y)}{I_x} \pm \frac{M_{y,T}(x)}{I_y} \tag{21.1}$$

where, x and y are the distances of points under consideration from the centroidal axis of the footing.

Using the above equation, the maximum base pressures at the four corner points are determined. The maximum value should not exceed the net safe bearing capacity of the soil and the minimum value should not be negative because soil cannot sustain tension. If the stress resultants from the

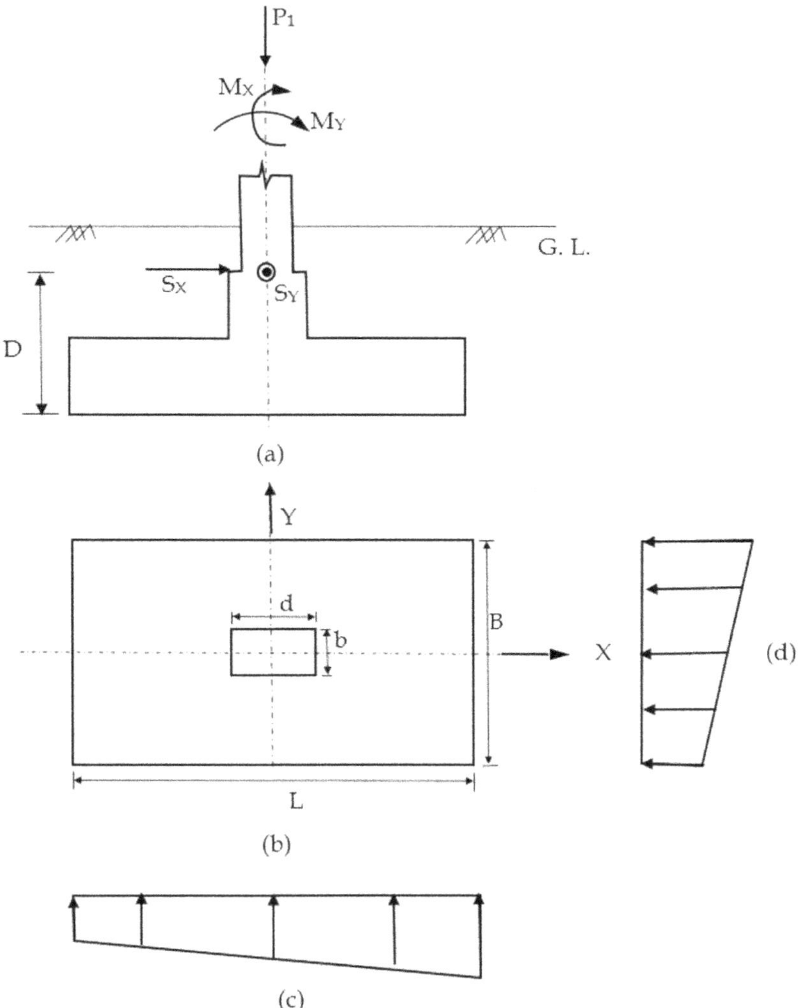

FIGURE 21.3 Isolated footing subjected to axial load and bending moments: (a) elevation; (b) plan; (c) pressure distribution along the X-direction; (d) pressure distribution along the Y-direction.

superstructure involve wind loads or earthquake loads, a necessary increase in bearing pressure as permitted by the standards should be used in the calculations.

For designing along the length or width of the footing, the corresponding moments are to be taken into consideration for calculating the base pressure as shown in Figure 21.3(c) and (d).

Moment acting on a footing can be considered equivalent to the vertical load acting at an eccentricity, i.e., $M = P.e$. Thus, the base pressure due to the action of the vertical load and moments can be expressed as

$$q = \frac{P}{A} \pm \frac{p.e}{I_y}.x \tag{21.2}$$

Considering $A = BL$, $I_y = \frac{BL^3}{12}$, and $x = \frac{L}{2}$

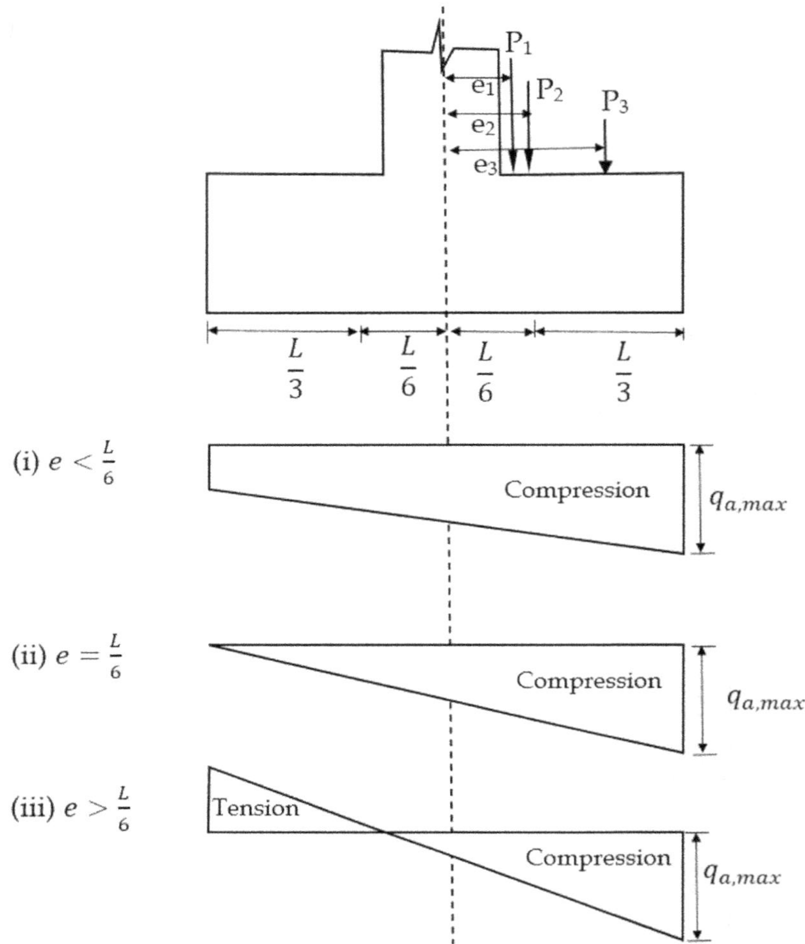

FIGURE 21.4 Footing subjected to different eccentric loadings.

pressure at point P of Figure 21.4 is:

$$q_P = \frac{P}{BL}\left[1-\left(\frac{6e}{L}\right)\right] \tag{21.3}$$

From the above equation, three possible cases can result:

I. When e is less than $\frac{L}{6}$, q_p is positive,

II. When $e=\frac{L}{6}$, $q_P = 0$

III. When $e>\frac{L}{6}$, q_p is negative, i.e., soil is subjected to tension

However, soil cannot sustain tension, and hence a modified stress distribution below the footing is to be determined based on force and moment equilibrium as described below.

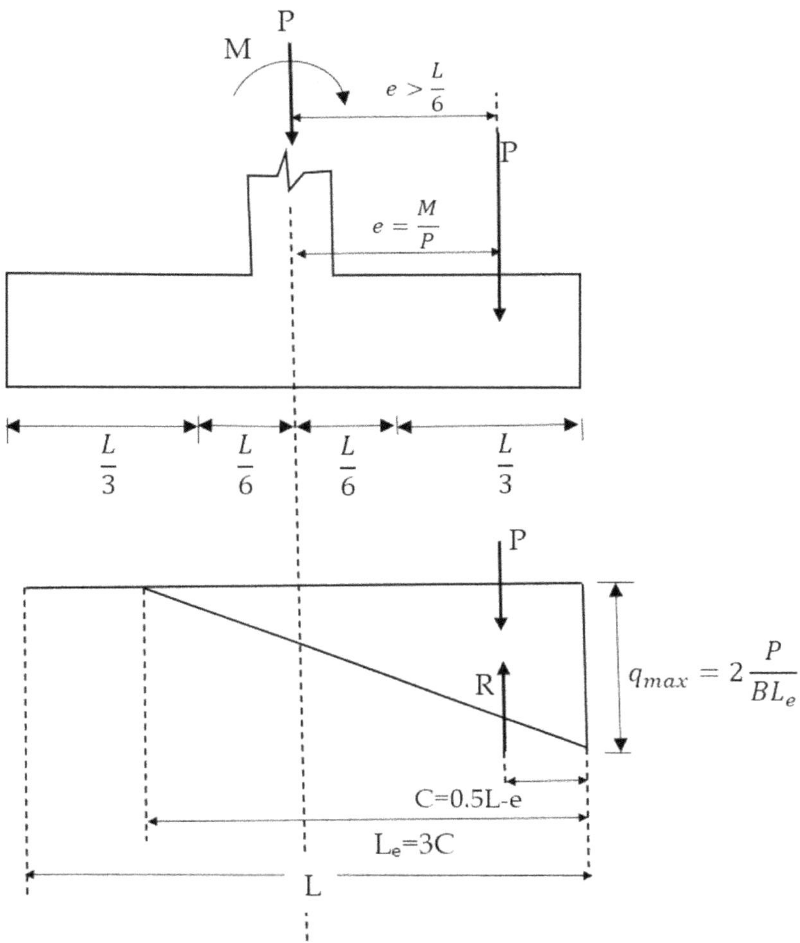

FIGURE 21.5 Eccentrically loaded isolated footing ($e > L/6$).

For arriving at the new pressure diagram two unknowns are to be solved: length of the new pressure diagram and ordinate of the pressure diagram. The two conditions mentioned below can be used to solve the two unknowns:

a) The vertical load must be equal to the area of the pressure diagram below the footing.
b) The resultant force of the pressure diagram acting below should be coincident with the point of application of the resultant vertical force on the footing.

This condition is shown in Figure 21.4. From simple statics, the values obtained for a footing of dimensions $L \times B$ subjected to an eccentric load P are shown in Figure 21.5.

21.6 GENERAL DESIGN CONSIDERATIONS AND IS CODAL REQUIREMENTS

As per IS:456-2000 for concrete, in contact or buried under non-aggressive soil/ground water should be considered as subjected to moderate exposure condition. As per the code for moderate exposure condition, the minimum structural grade of concrete should be M25. Thus, it can be inferred that substructure should always be made with at least M25 grade of concrete.

As per IS:456-2000, the minimum nominal cover for footings should be 50 mm.

21.6.1 Thickness at the Edge of the Footing

As per IS:456-2000 in reinforced and plain concrete footings, the thickness at the edge shall not be less than 150 mm for footings on soil nor less than 300 mm above the top of the piles for footing on piles.

21.6.2 Bending Moment

As per IS:456-2000, the bending moment at any section shall be determined by passing through the section a vertical plane which extends completely across the footing and computing the moment of the forces acting over the entire area of the footing on one side of the said plane. The greatest bending moment to be used in the design of an isolated concrete footing which supports a concrete/masonry column, pedestal or wall shall be calculated at the face of the column, pedestal or wall for footings supporting a concrete column, pedestal or wall.

21.6.3 Shear and Bond

The shear strength of footings is governed by the more severe of the following two conditions:

(i) The footing acting essentially as a wide beam, with a potential diagonal crack extending in a plane across the entire width. The critical section for this condition shall be assumed to be a vertical section located from the face of the column, pedestal or wall at a distance equal to the effective depth of footing in the case of footing on soils and at a distance equal to half the effective depth of footing for footing on piles.

(ii) Two-way action of the footing, with potential diagonal cracking along the surface of a truncated cone or pyramid around the concentrated load. In this case, the footing shall be designed for shear in accordance with the requirements of flat slab design.

21.6.4 Critical Section

The critical section for checking the development length in a footing shall be assumed at the same plane as those described for bending moment and also at all other vertical planes where abrupt changes in section thickness occur. If reinforcement is curtailed, the anchorage requirements shall be checked in accordance with the requirements of curtailment of tension reinforcement in flexural members.

21.6.5 Tensile Reinforcements

The total tensile reinforcement at any section shall provide a moment of resistance at least equal to the bending moment on the section. Total tensile reinforcement shall be distributed across the corresponding resisting section as described below.

(a) In one-way reinforced footing the reinforcement shall be distributed uniformly across the full width of the footing

(b) In two-way reinforced square footing, the reinforcement extending in each direction shall be distributed uniformly across the full width of the footing

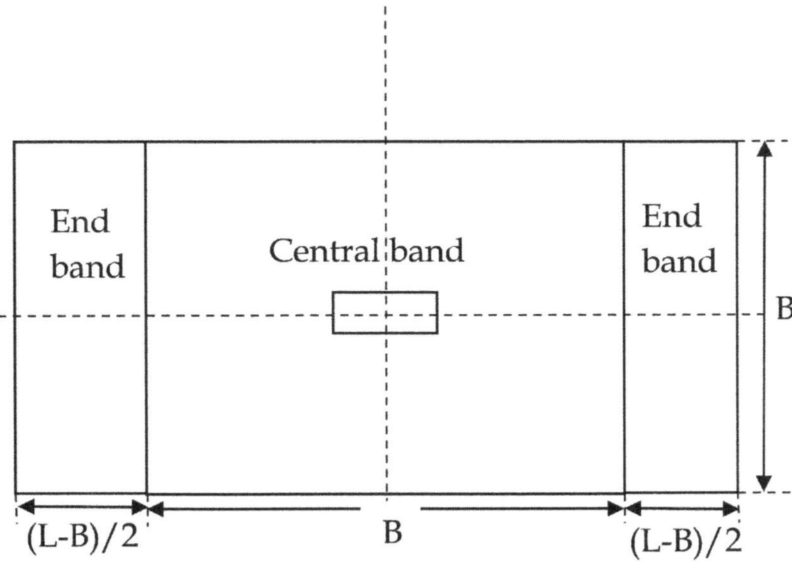

FIGURE 21.6 Distribution of reinforcement in rectangular footings.

(c) In two-way reinforced rectangular footing the reinforcement in the long direction shall be distributed uniformly across the full width of the footing. For reinforcement in the short direction a central band equal to the width of the footing shall be marked along the length of the footing and will contain a portion of the reinforcement as per the following expression:

$$\frac{\text{Reinforcement in central band width}}{\text{Total reinforcement in short direction}} = \frac{2}{\beta + 1}$$

where β is the ratio of the long side to the short side of the footing. The remainder of the reinforcement shall be uniformly distributed in the outer portions of the footing. (Refer to Figure 21.6.)

21.6.6 Transfer of Load at the Base of Column

The compressive stress in concrete at the base of a column or the pedestal shall be considered to be transferred by bearing to the top of the supporting pedestal or footing. The bearing pressure on the loaded area shall not exceed the permissible bearing stress in direct compression, multiplied by a value $= \sqrt{\frac{A_1}{A_2}}$ but not greater than 2, where A_1 = supporting area for bearing of footing which in sloped or stepped footing may be taken as the area of the lower base of the largest frustum of a pyramid or cone contained wholly within the footing and having for its upper base the area actually loaded and having side slopes of one vertical to two horizontal. A_2 = loaded area of the column base.

For the limit state method of design the permissible bearing stress on the full area of concrete shall be taken as $0.45 f_{ck}$. If the permissible bearing stress in the supporting or supported member is exceeded, reinforcement shall be provided for developing the excess force either by extending the longitudinal bars into the supporting member or by dowels, satisfying the necessary requirements of development length and relevant codal clauses.

21.6.7 Nominal reinforcement

The minimum reinforcement and spacing shall be as per the requirements of a solid slab. The nominal reinforcement for concrete sections of thickness greater than 1 metre shall be 360 mm^2 per metre length in each direction on each face with due consideration of minimum tension reinforcements based on the depth of the section.

21.7 PROPORTIONING OF FOOTINGS

The first step of footing design is to determine the area required to transmit the external actions coming at the base of columns or walls safely to the soil, without exceeding its bearing capacity and keeping the displacement within permissible limits. The footing area should be calculated on the basis of worst load combinations. For isolated footings, either a square, rectangular or circular footing is generally adopted. While proportioning footings, geotechnical considerations should also be given due importance and every attempt should be made to ensure that pressure bulbs from surrounding footings do not overlap. The designers must ensure that the maximum pressure below the footings under all possible load combinations does not exceed the safe bearing capacity of the soil after allowing for a necessary increase, if permissible. The minimum pressures should never be tensile and in the case that calculations lead to negative pressures, the pressure distribution diagram should be modified as soil cannot resist tension.

21.8 STRUCTURAL DESIGN OF FOOTINGS

Isolated footings behave as a plate supported fully at the bottom with a concentrated load of the column generally along with moments, which consists of the superstructure load. Hence when loaded at the centre the plate will try to bend up from the sides resulting in tension at the bottom face and compression at the top. Hence the footing can be designed as a cantilever slab fixed at the column face. Shear force for these cantilever elements needs to be checked in a manner similar to beam shear.

If the thickness of the plate is low and the column load is high, the column can punch through the plate. This phenomenon is seen in punching paper using a punching machine. Hence the thickness has to be such that the column cannot punch through.

21.9 ILLUSTRATIVE EXAMPLE

Problem 21.1

Design an isolated square footing using the following information

Bearing capacity of soil 120 kN/m^2, size of pedestal = (500×500) mm^2, load from superstructure = 1200 kN (in dead load + live load condition), grade of concrete = M25 and grade of steel = Fe500

Solution

Load transferred to soil = load from superstructure + difference between the self-weight of footing and weight of soil occupying the footing volume.

For all practical purposes for simplifying the calculation an additional load of 10% of superstructure is considered

Total load= $1200 \times 1.1 = 1320$ kN

$$\text{Footing area required} = \frac{1320}{120} = 11\ m^2$$

Each side of the footing required, considering a square footing = 3.31 m

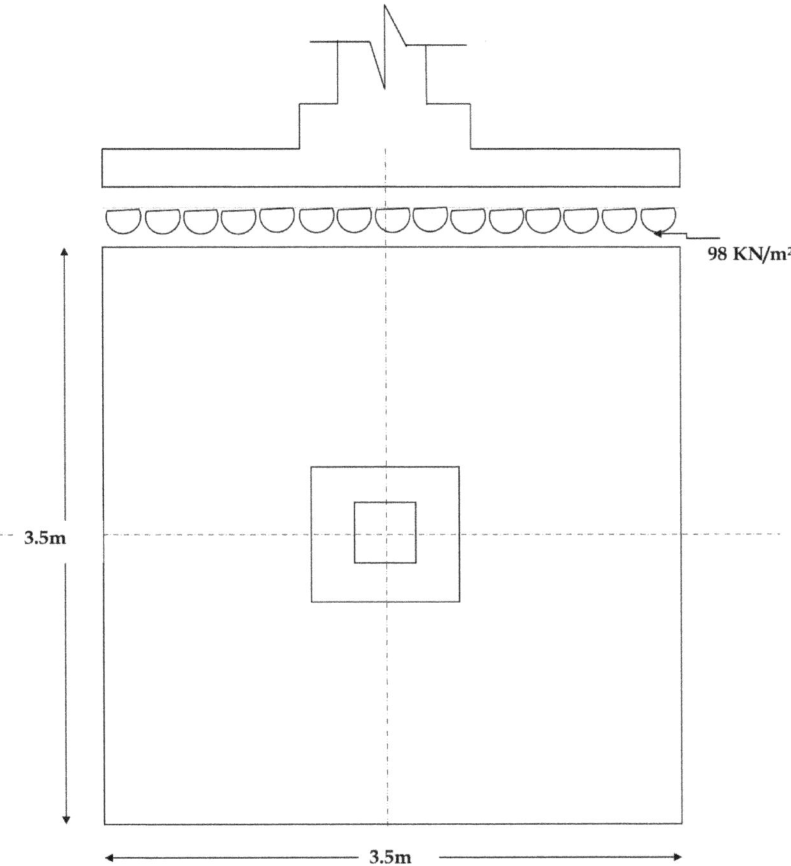

FIGURE 21.7 Plan and elevation of footing showing pressure distribution.

Adopt 3.5 m × 3.5 m square footing

Footing self-weight is transmitted to the soil directly by bearing and hence the superstructure load is responsible for bending in footing (Figure 21.7).

$$\text{Effective pressure on footing} = \frac{1200}{3.5 \times 3.5} = 98\,\text{kN}/\text{m}^2$$

<u>Design for flexure</u>

Bending moment (at face of pedestal) $= 98 \times \dfrac{1.5^2}{2}$

(Cantilever length from column face = 1.5 m)

= 110.25 kN-m/m width of slab.

For M25 grade of concrete and Fe500 the limiting moment of resistance factor value is 3.444

$$d_{required} = \sqrt{\frac{110.25 \times 1.5 \times 10^6}{3.444 \times 1000}} = 219.13\,\text{mm}\,(\text{using a load factor of } 1.5)$$

$$D_{required} = 219.13 + 50 + 10 = 279.13\,\text{mm}$$

(Assuming 50 mm clear cover and 20 mm bars to be used.)

Footing design is governed by shear and not bending moment. Hence a depth greater than that required in flexure is to be generally provided.

Let D = 400 mm.

Therefore, d = 340 mm.

Area of reinforcement required by the limit state method is 12.206 cm^2/m width

This area of reinforcement will be compared with that area required for one-way shear and the higher of the two will be provided.

Design for shear

(i) One-way shear

Critical section located at a distance of d from the face of pedestal
Shear force = 98 × (1.5 – 0.34) =113.68 kN/m width
Hence, shear stress, τ_v =113.68 × 1.5 × 10^3/(1000 × 340) = 0.501 MPa
$\tau_{c,max}$ for M25 grade concrete is 3.1 N/mm^2 which is more than τ_v
Hence, this is OK.
The percentage of reinforcement required corresponding to τ_v (as per IS:456-2000) is:
$p_t = 0.5437\%$
A_{st} = 18.48 cm^2/m
This area is greater than that required for flexure.
Hence provide 20 mm Tor steel bar @ 150 mm c/c (area 20.94 cm^2/m width)

(ii) Punching shear (as per IS:456).

Critical section is at a distance of $d/2$ from the face of the pedestal

$$\text{Shear stress} = \frac{V}{b_0 d}$$

b_0 = perimeter of the critical section
= 4(500 + 340) = 3360 mm
Punching force = total downward force – reactive force offered by punching area

$$= 98\,[3.5^2 - 0.84^2] = 1131.4 \text{ kN}$$

$$\text{Punching shear stress } (\tau_s) = \frac{1131.4\times10^3\times1.5}{3360\times340} = 1.49\,\text{MPa}$$

Permissible shear stress (Cl. 31.6.3) = $k_s.\tau_c$
Where k_s= 0.5 + β_c not greater than 1
β_c = ratio of short side to long side of column/capital
For the present problem, β_c = 1

Hence, $\tau_c = 0.25\sqrt{f_{ck}} = 0.25\sqrt{25}$ = 1.25 MPa, which is greater than the punching shear developed in the section.

Hence the section is unsafe.

Increase the overall depth to 500 mm. Performing calculations in a similar manner the punching shear stress is calculated as 1 MPa which is less than the permissible value. Hence the section is safe. Longitudinal reinforcements of 16 mm Tor provided at a spacing of 150 mm c/c satisfy the requirement of flexure and shear.

The details of the reinforcement are shown in Figure 21.8.

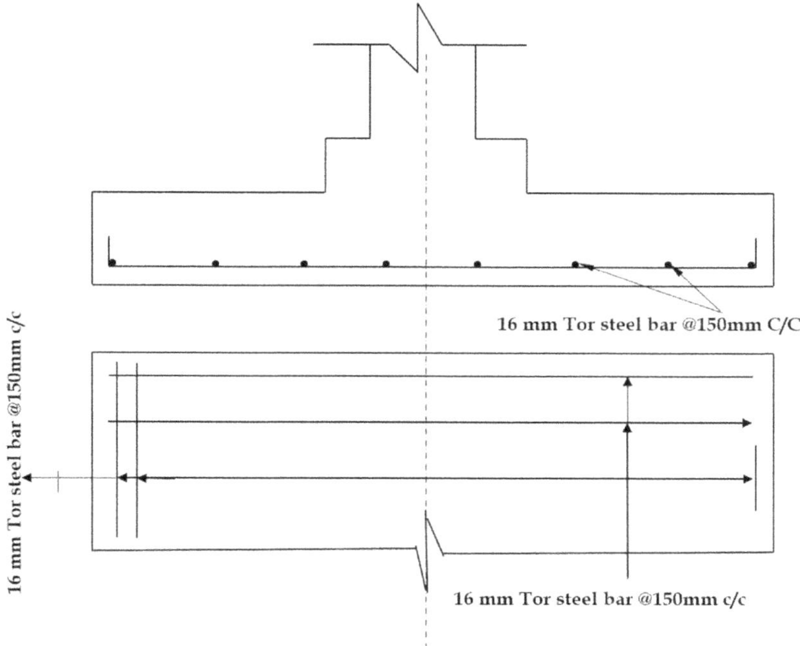

FIGURE 21.8 Reinforcement details of the footing as per Problem 21.1.

21.10 SUMMARY

In this chapter the basic requirements of foundation design have been described. The design of the foundation is performed in two distinctly different steps: step 1 – proportioning of the footing, i.e., finding out the dimensions of the footing based on the external loads and bearing capacity of the soil, and step 2 – structural design of footing which involves determining the depth and reinforcement required depending upon the grades of concrete and steel. The pressure distribution below the foundation will depend upon the type of supporting soil and stress resultants acting on the footing. In this chapter pressure distribution due to axial column loads and axial loads with moments have been discussed in detail for isolated footings. The requirements of footing design as per IS:456-2000 also have been highlighted in this chapter. A typical numerical example demonstrating checks for flexure, one-way shear and punching shear have been presented.

BIBLIOGRAPHY

Bandyopadhyay, J. N. (2011). *Design of Concrete Structures*, PHI Learning, New Delhi.

Bhatt, Prab, McGinley, Thomas J. and Choo, Ban Seng. (2006). *Reinforced Concrete, Design Theory and Examples*, Third Edition, E & FN Spon, Taylor and Francis, London.

Ferguson, Phil M., Green, John E. and Jirsa, James O. (1988). *Reinforced Concrete Fundamentals*, Fifth Edition, John Wiley, Hoboken.

Indian Standard IS: 456-2000, *Plain and Reinforced Concrete – Code of Practice*, Fourth Revision, Tenth Reprint April 2007, Bureau of Indian Standards, New Delhi.

Indian Standard IS: 1893 (Part-1):2016, *Criteria for Earthquake Resistant Design of Structures*, Part 1 General Provisions and Buildings, Sixth Revision, Bureau of Indian Standards, New Delhi.

Mallick, S. K. and Gupta, A. P. (1989). *Reinforced Concrete*, Fifth Revision, Oxford & IBH, New Delhi.
Pillai, S. U. and Menon, D. (2012). *Reinforced Concrete Design*, Third Edition, Reprint 2012, Tata McGraw-Hill, New Delhi.
Sinha, S. N. (2002). *Reinforced Concrete Design*, Second revised edition, Tata McGraw-Hill, New Delhi.
SP 7: 2016, National Building Code of India 2016, Third Revision, Bureau of Indian Standards, New Delhi.

Index

A

Abrams, Duff, 11
Abram's law, 11–12
Admixtures, 9–10, 13
  benefits of, 9
  chemical, 20, 24
  defined, 9
  IS:9103 (Specifications for Admixture for Concrete), 23
  mineral, 20
  reasons for using, 9
  superplasticizers (SP), 10
  types of, 9–10
  water-reducing, 10
Aggregates
  grading of, 8
  influence on the properties of concrete, 8
American Concrete Institute (ACI), 18
Anchor, failure modes for, 197
Anchorage (development) bond, 189, 190, 193–194
Anchorage length, 106, 191
Aspdin, Joseph, 58
ATC40, 275, 286, 302
Atmospheric cooling, 30
Axial compression, 62, 75, 78, 90, 114, 117, 122, 180, 218, 220, 274, 308–309, 317, 334
  effect on curvature ductility, 300–301
Axial compressive stress, 309–310
Axial deformations, of beams and columns, 44
Axial load capacity, 117, 126, 144, 266, 318
Axially loaded columns, design of, 117

B

Balanced failure, 86, 119
Balanced reinforcement, 67, 295, 302
Balanced section stress, 88
Bearing capacity of soil, permissible increase in, 376, 378, 386
Beeby, A. W., 208
Bending
  classical theory of, 202
  stress, 181, 279
Bends, 191
Bernoulli–Euler bending equation, 279
Bernoulli's hypothesis, 84–85
Bernoulli's principle, 61, 84, 91, 117, 281, 316–317
Bilinear stress–strain diagram, of reinforcing steel, 38
Blast furnace, 2, 18, 26, 29
Bogue, R. H., 6
Bogue's compounds, 6
Bond stress, in reinforced concrete
  anchorage (development bond), 189, 190, 193–194
  determination of, 191
  development of, 191
  flexural bond stresses, 189, 194–195
Bonding between concrete and reinforcement, concept of, 199
Braced walls, 313–314
Breaking strength of concrete
  of anchors in tension, 199
  calculation of, 196–199
  concrete breakout, 196–199
  meaning of, 196
  pull-out, 196
  steel failure, 196
Brittle failure, 30, 78, 86, 277, 285
Buckling, 50, 54, 75, 114–116, 201
Bureau of Indian Standards, 147, 310

C

Calcium chloride, 9
Calcium silicates, 5–7
Calcium sulphate, 5
Cement, 5–7
  chemical composition of
    compound composition, 6
    determination of, 6–7
    oxide composition, 6
  fineness of, 7
  hydration of, 7
  hydraulic, 5
  manufacture of, 5–6
    raw materials used in, 6
  Portland, 5, 7
  structure of hydrated cement paste, 7
Cement and Concrete Association, 208
Cement concrete, 1, 26, 58
Civil engineering project, 50
Coarse aggregates (CA), 8, 17, 19
Codes for reinforcement in concrete
  Eurocode 2, 27–28, 30
  IRC:112-2020, 26–30
  IS:383, 23
  IS:456-2000, 22–25, 28
  IS:3812, 26
  IS:9103, 23
  IS:12089, 26
  IS:13920-2016, 25
  IS:15388, 26
Codes of practice, 16, 23, 32, 42–43, 48–51, 60, 72, 75, 77–78, 80, 98, 101, 115, 121, 209, 212, 220, 225, 271, 274–275, 277, 299, 306
Coefficients of thermal expansion, of steel and concrete, 54
Cold twisted deformed (CTD) process, 34, 35, 82
Collapse
  in concrete, 85
  limit states of, 75
Collapse prevention (CP), 275, 302

Columns
under axial load with
biaxial bending, 180
uniaxial bending, 117, 234
braced and unbraced, 114–115
circular, 115
classification and definitions, 114
confinement of concrete by circular and square hoops, 116
design of
axially loaded columns, 117
charts for, 116–117
compression design and shortcomings of SP16, 146–147
equilibrium equations for, 119–120
methodology of, 114
need for P–M interaction charts for, 120–121
desirable choices of section, 115–116
development of strain distribution diagram and stress blocks
when neutral axis is outside the section, 121–123
when the neutral axis is within the section, 123–124
evaluation of ductility for, 300
interactions charts for rectangular section with reinforcement
distributed on two sides, 148–163
equally distributed on four sides, 164–179
longitudinal reinforcement, 115
modes of failure in, 114, 117–119
under balanced conditions, 118, 119
over-reinforced conditions (compression failure), 118, 119
under-reinforced conditions (tension failure), 118, 119
P–M interaction diagrams
need for, 120–121
problem example, 129–146
steps for preparation of, 128–129
when neutral axis is at infinity, 130–131
when neutral axis is outside the sections, 131–133
rectangular section,
interaction charts for, 147–163
safety profile of
determination of stress and strain values, 124
different arbitrary positions of the neutral axis, 125–128
longitudinal reinforcements, 124–125
when the column section behaves like a steel beam, 127–128
when the neutral axis is at infinity, 125–126
when the neutral axis is outside the sections, 126–127
when the neutral axis is within the sections, 127
transverse reinforcement, 115
Compaction, degree of, 11
Compression design, sample design aids for, 249–251
Compression members, design for, 233
Compression reinforcement, 107, 302
in load design, 101, 107
Compressive force, 93, 116
in concrete, 222
line of action of, 223–224, 256–266
Compressive strength, 2, 11–12, 14–15, 24, 26, 54, 82, 116, 207, 309
of the concrete, 15–16, 18–19, 199
Compressive stress block, 56, 61, 84, 93, 254–256
Compressive stress distribution, 55–56, 84, 85
Concrete breakout capacity
calculation of, 196
of a single anchor in tension, 196
Concrete capacity design (CCD), 196
Concrete mix design
approach to, 15
basics of
economy, 14
strength and durability, 14
workability of, 14
concept of trial mixes, 16–17
methods of, 15
of ordinary and standard grades, 15–16
parameters considered in, 15
process of, 14
purpose/objectives of, 15
quality control, 14
Concrete road bridges, code of practice for, 26–27, 214
Concrete technology, 17, 19, 214
Confined concrete, compressive strength of, 309
Confinement of concrete, effect of stirrup in, 187
Crack control, guidelines to provide, 211–212
Crack width, calculations of, 209–211
Cracking
limits as per Indian standards, 212
phenomenon of, 209
Creep
coefficient, 207
deflection due to, 207
Curvature ductility, 285
for balanced section, 286–291
determination of $X_{U,Max}$ or $X_{U,BALANCED}$, 289–291
determination of X_y, 288–289
for different grades of concrete and steel, 291
definition of, 87, 300
of doubly reinforced sections, 295–298
calculation of depth of neutral axis at ultimate stage, 296–298
calculation of neutral axis at yield, 295–296
effect of axial compressive force on, 300–301
of maximum under-reinforced sections, 291–295
determination of Xu,min, 291–293
determination of Xy,min, 293–295
prescriptive design as per Indian standards, 298–299
of under-reinforced sections, 300
values of reinforced concrete sections, 299–300

D

Dead loads, 41, 42, 51
Deflection, of reinforced concrete members
calculation of
deflection behaviour of beams, 202
by elastic theory, 204

flexural rigidity, 204, 205–207
tension stiffening effect and, 205, 210
control of, 201, 208
effective moment of inertia formulation, 207
excessive, 201
limits/requirements as per Indian standards, 208
long-term deflection, 201
due to creep, 207
due to shrinkage, 207
short-term deflection, 204
Deflocculation, 10
Deformation load, 32
Design charts, needed for column design, 116
Design failures
criterion of, 271
displacement-based, 273
load-based, 272–273
Design loads, 43
behaviour of structural elements beyond limit state, 271
criterion of, 271
displacement-based, 273
ductility design, 273–274
load-based design failures, 272–273
load–deformation behaviour, 271
overloading beyond, 270
Design for compression members,
for balanced condition, 221–225
balanced/limiting moment of resistance
moment of resistance from compression concrete and tension steel, 225
for columns under axial load and uniaxial bending, 234
for compression members, 233
beyond M60 grade of concrete, 268–269
for doubly reinforced sections, 233
for maximum percentage of tension reinforcement, 229–233
for maximum under-reinforced condition, 225–232
numerical example for columns
concrete grade beyond M60, 251–267
concrete grade up to M60, 235–251
preparation of graphical chart for columns, 267
strain distribution diagrams and stress blocks for, 234–236, 317
Design for flexural members,
balanced percentage of tension reinforcement, 95
balanced section, 65, 70, 71, 85, 86, 88, 89, 92, 93, 99, 106, 281, 284–287, 289, 290
limiting moment of resistance, 96, 99, 105, 224, 225, 387
limiting values of neutral axis depth, 94–95
maximum under-reinforced section, 96–98, 103, 220, 221, 226, 227, 233
Design shear forces
calculation of, 187
due to the formation of plastic hinges at beam ends, 187
Development length
concept of, 190
of column and beam reinforcement, 193
of column bars at
footing level, 192
pedestal level, 193
critical sections for checking, 192–193
equation of, 193–194
lapping length, 192
as per IS:456-2000, 193
at point of inflection, 195
values for fully stressed (HYSD) bars in
compression, 194
tension, 194
Displacement ductility, 285, 299
Displacement-based design, 273, 275, 276
Displacement-based systems, 51, 273
Ductile failures, 50, 78, 86, 270, 278, 285, 309
Ductile ferrite–pearlite structure, 30
Ductile flexural failure, 107
Ductility,
of balanced sections, 278, 286, 291, 295, 299, 300, 304
of maximum under-reinforced sections, 278, 281, 282, 284, 291–295, 298, 299, 302–304
for columns, 300
concept of, 58, 73, 274, 278, 307
curvature, 82, 86, 87, 285, 291, 294, 295, 298, 299, 300, 302, 304, 308, 309
determination of, 278
displacement, 285, 299
effect of axial force on, 301
measures of, 285–286
provisions for ductility design and detailing as per IS:13920-2016, 309–310
role in design, 278–279
rotational, 286
significance of, 278
strain, 282, 283
of structural concrete members, 309
determination of yield and ultimate curvature, 279–281
elements which need, 307–309
need for, 277–278
Ductility-based design (DBD), 306, 309
Durability
of concrete, 9, 14

E

Earthquake ground motion, 275, 302
Earthquake loads, 41, 43, 62, 273, 381
Earthquake-resistant design, of RC structures, 77, 86, 101, 273, 276, 277, 306–310
inelastic behaviour of structures, 306
philosophy of, 307
for safety against earthquakes, 306
Element stiffnesses, 44
EN codes, on concrete structures, 274
Engineering stress–strain curve, determination of, 32
Eurocode 2 (Design of Concrete Structures), 26–30, 35, 39, 54, 76, 84, 90, 91, 214–219, 232–234, 269
basic groups of limit states, 90–91
design methodology, 218
stress–strain relationship as per, 35–39
Eurocode 8 (design of structures for earthquake resistance), 99, 309

F

Failure, in the limit state of collapse
  balanced failure, 86, 119
  under design load, 270
  modes of, 3, 77, 79, 80, 90, 117, 186
  multi stage failure, 78, 79, 112
    Stage I, 270, 274, 300
    Stage II, 270, 274, 300
  under-reinforced failure, 271
Failure, in working stress method of design
  Single stage, 66, 72, 73, 78, 79, 86, 274
FEMA356, 275
Fibre-reinforced polymers, 1
Fine aggregates (FA), 8, 16–17, 19, 20, 23, 26
Fire insulation, 312
Flexural bond stresses, 189
Flexural compression, 61–63, 78, 90–92, 109, 117, 121, 215, 218, 316–317, 323, 333, 351
Flexural design, 92, 103, 387–389
Flexural ductility, 299
Flexural failure, 3, 56, 78, 86, 101, 107, 186, 188, 278, 307, 310, 316, 351
Flexural member, deformation of, 280
Flexural reinforcement, yielding of, 316
Flexural rigidity, 202, 204, 205–207, 279
Flexural strength, of concrete elements, 91
Flexural strength ratio, between columns and beams, 101
Flexure–shear failure, 316
Fly ash, 2, 10, 17, 18, 26
Footing, types 378–379
Force equilibrium equation, 99, 103–104, 109, 110, 112, 221, 229, 233, 293, 334
Foundation design
  and behaviour of isolated footing, 379–383
  bending moment, 384
  codal requirements, 383–386
  critical sections, 384
  design criteria, 376
  distribution of soil pressure, 379–383
  effect of
    earthquake force as per IS:1893, 378
    wind force as per NBC-2016, 378
  footing subjected to axial load and moment, 380–383
  footing subjected to axial load only, 380
  general design considerations, 383–386
  nominal reinforcement, 386
  proportioning of, 386, 389
  shear and bond, 384
  structural design, 386, 389
  tensile reinforcement, 384–385
  thickness at the edge of footing, 384
  transfer of load at the base of column, 385
  types of footing, 378–379
Fresh concrete, properties of, 10–11

G

Geo-polymer concrete, 1
Gravity loading, 44
Gravity loads, 44, 272, 312, 315, 369
Ground granulated blast furnace slag (GGBFS), 18, 26
Gypsum, 7

H

Hardened concrete, properties of, 11–14
High-performance concrete (HPC), 10, 17–20, 26
  admixtures, 19, 20
  coarse/fine ratio, 20
  concept of, 17
  main ingredients of, 19–20
  phases of, 19
  principle of, 19
  requirements for constituent materials and mix proportioning, 19
  selection of materials for, 19–20
High-strength deformed steel (HSD), 3, 23, 26, 28, 29, 76
  mechanical properties of, 30–31
  process for making
    CTD process, 28, 29, 34, 82, 86, 90, 214, 355
    thermo-mechanical treatment, 29
    TMT process, 30
  stress blocks for
    area of stress block for HYSD steel in compression, 323–332
    calculation of stress blocks of HYSD steel in tension, 320–323
  stress–strain relationship, 30–32
    engineering stress–strain curves, 32–33
Hooke's law, 33, 71
Hook and bend, 191
  anchorage value of, 195
Horizontal loading, 44
Hydraulic calcium silicates, 5
Hydraulic cement, 5
Hydrostatic fluid pressure, 55
HYSD steel, 3, 23, 28, 61, 69, 82, 90, 91, 110, 124, 146, 147, 180, 199, 214–216, 233, 308, 312, 320, 323, 365, 366
  grades of HYSD reinforcement, 125
  stress–strain diagram of, 28, 90, 110, 215–216, 218, 220, 221, 231–233, 235, 320, 345, 358, 365–366
  used in earthquake zones, 77

I

Igneous rocks, 8
Immediate occupancy (IO), 275, 302, 303
Imposed loads, 41–42, 51, 62, 76, 186, 272, 371, 378
Indian Standard for High-Strength Deformed Steel Bars and Wires for Concrete Reinforcements, 3
Inelastic deformations, 34, 54, 229, 274, 277, 306
Inelastic strain, 33, 34, 82, 89–90, 110, 119, 124, 221, 235, 289
Inflection point, 195
Ingredients of concrete
  admixtures, 9–10
  aggregates, 8
  cement, 5–7
  water, 9
Interface shear, 189

Intermediate walls, 313
IRC:112-2020 (Code of Practice for Concrete Road Bridges), 26–29, 90, 214, 366
  assumptions in ultimate limit state for design of members, 215–220
  for columns under axial load and uniaxial bending, 234
  for design for compression members, 233
  for design for flexural members
    singly reinforced sections, 220–221
  for development of strain distribution diagrams and stress blocks, 234–235
  for doubly reinforced sections, 233
  stress–strain relationships as per, 35, 214–215
IS:383 (Specifications for Coarse and Fine Aggregates from Natural Sources and Concrete), 23
IS:456-2000 code, for concrete in India, 22, 28–30, 54, 82, 84, 191, 272, 281, 317, 366
  on deflection limits, 208–209
  limit state method of design, 117, 180
  provisions for plain concrete and reinforcements
    inspection and testing, 25
    materials and general considerations, 23–24
    workmanship, 24–25
  provisions for structural concrete, 25
  provisions for working stress method, 60–61
  stress–strain diagram of steel for HYSD bars, 90
  stress–strain relationships, 34–35
IS:9103 (Specifications for Admixture for Concrete), 23
IS:13920-2016 (Code of Practice for Ductile Design and Detailing of Reinforced Concrete Structure Subjected to Seismic Forces), 25, 76, 90, 99, 103, 106, 107, 186, 187, 226, 278, 281–285, 310, 312, 314, 316

J

Japan Code, on concrete structures, 274

L

Lapping length, 192
Lateral load-resisting system, in multi-storeyed buildings, 312
Life safety (LS), 275, 302–304
Limit state method (LSM), for design of RC structures, 22, 59, 303
  analysis/check type problems, 103–104
  assumptions as per, 99
    Indian standards, 90
    international guidelines, 90–91
  based on
    capacity, 101–102
    loads, 101
  characteristic and design values of loads and material strengths, 76–77
  computation of design parameters for flexural members
    balanced percentage of tension reinforcement, 95–96
    balanced section, 92–94
    limiting values of neutral axis depth, 94–95
    maximum under-reinforced section, 96–98
  computation of limiting moment of resistance, 96
  concepts of, 74, 75
  design stress in reinforcement, 186
  in doubly reinforced sections, 98
    balanced in nature, 99–100
    introduction and need for, 98–99
  earthquake-resistant, 101
  illustrative examples of, 102–104
  maximum under-reinforced sections, 96–98, 281–282
  mode of failure of, 78–79
  multistage failure of, 78–79
  numerical examples of, 104–112
  objectives of, 74
  provisions of IS:456-2000 for, 77–78
  for real-life beams, 100–101
  of serviceability, 75, 78
  in singly reinforced sections, 92
    theoretical condition, 98
  types and classification of, 75–76
  under-reinforced doubly reinforced sections, 100
Limit state of collapse, 75, 78
  codal requirements in, 187–188
  in shear, 187–188
  stress–strain diagram for
    concrete, 272
    steel, 271
Limiting moment of resistance, 224–225
  computation of, 96, 105
  for different grades of steel, 96
Linear elastic theory, 44, 71, 77, 285
Live loads (imposed loads), 41, 62
Load combinations, for RC structures, 43, 309, 334, 377, 378
Load design, 78, 100, 101, 106–108, 111, 186, 277
  compression reinforcement in, 101, 107
Load-based design failures, 272–273
Load–deformation curve, 32
Longitudinal reinforcement in beams, termination of, 192

M

Mechanical anchorages, 191, 195
Metallurgy, of steel, 85
Metamorphic rocks, 8
Mild steel (MS), 23, 26, 28–30, 34, 35, 37, 61, 76
Modular ratio, of concrete, 61
Modulus of elasticity, of concrete, 13, 207
Moment equilibrium equation, development of, 333–334
Moment of inertia
  calculation of
    cracked section, 206–207
    gross cross-section, 205
    uncracked transformed section, 205–206
  effective formulation of, 207
  of pedestals, 377
Moment of resistance (MOR), 365
  calculation of, 111–112
  of shear walls, 365
  values as per IS:13920-2016 *vs.* those from proposed curves for Fe500, 356
Moment–axial force interaction relationship, 312

Moment–curvature relationship, 302
in RC members, 279
for under-reinforced beams, 203
Moment-resisting frame, beam–column joint of, 107

N

National Building Code of India 2016 (NBC-2016), 377
Neutral axis
calculation of
at ultimate stage, 296–298
at yield, 295–296
limiting values of depth, 94–95
minimum depth at maximum strain in steel, 98
solutions of the equations for arbitrary depths, 334–343
Newton's laws of motion, 42, 273

O

Operational (O), 275
Over-reinforced, 66, 84, 86, 87, 219, 271, 281

P

Partial safety factors γ_f for loads, values of, 76
Performance objectives, definition of, 275
Performance-based design (PBD), 204, 274–275, 279, 300, 302–304
goal of, 275
Permissible stresses, in concrete, 62, 66, 67, 71
Plain concrete, 1–2, 22
provisions of IRC:112-2020 on, 26–27
Plastic deformation, 32
Plastic hinge, in reinforced concrete sections, 286
Plastic rotations
of balanced section, 279, 284, 286–291
for different grades of concrete and steel, 290
of doubly reinforced section, 297
of flexural member, 302
prescriptive design as per Indian standards, 298–299
of structural elements at different states, 302
of under-reinforced sections, 278–279, 291–295
Plastic zone, 33
P–M curve, 334
P–M interaction charts, 315, 333
for column design, 120–121
Poisson's ratio, 53
Portland cement, 5, 7, 17–18
discovery of, 58
Post-cracking stress, 204
Potable water, 9, 23
Pozzolanic materials, 58
Pull-out strength, 196
Pure bending states, theory of, 279

Q

Quality control and workmanship, 26
Quenching, 29–30

R

RC structures
analysis of
considerations for, 43–44
deformations, 44
element stiffnesses, 44
guidelines for, 44
methods of, 44
beam–column connections, 47
behaviour under loads, 49
characteristic and design loads of, 43
combinations of loads, 43
design process, 49
aims and objectives of, 50
code-based methodology, 51
general requirements, 50
performance-based, 51
prescriptive method as per standards, 51
earthquake force, 42, 378
effects of cracking in, 44
loading, 41–42, 44
plastic hinge in, 286
snow loads, 42
special loads, 42
strength and serviceability design based on loads, 51
stress resultants of structural elements of, 45–48
stress–strain relationships of, 49
Real-life beams
design of
based on capacity, 109–112
span section, 106
support section, 105
determination of
effective depth, 104
limiting moment of resistance, 105
doubly reinforced, 100–101
Reinforced concrete
advantages of, 2
brittle failure, 30
definition of, 1
disadvantages of, 2
ductility of, 30
flexure failure in, 102
invention of, 2
moment-resisting frames, 47, 104, 107, 307, 310, 313, 388
as principal construction material, 2
properties of
role of concrete and reinforcement on, 2–3
Reinforced concrete (RC) buildings
earthquake-resistant, 314
lateral load-resisting structural elements for, 306, 307, 312, 314
Reinforced concrete (RC) design, 317
aims and objectives of, 50
basic assumptions of, 54
code-based methodology for, 51
general requirements for, 50
historical development of, 58

limit state method (LSM), 59
performance-based, 51
prescriptive method as per standards, 51
strain distribution diagrams, 56
stress–strain diagrams of concrete
development of stress blocks from, 56–57
strain distribution, 56
unconfined and confined, 54–56
ultimate load method (ULM), 58–59
working stress method (WSM) for, 58
Reinforcing steel
longitudinal reinforcements, 98
method for increasing yield strength of, 29
strains in, 283
stress–strain diagram of, 26
transverse reinforcement, 98
Roman concrete structures, 58
Rotational ductility, 286

S

Sample design aids, for compression members beyond M60 grade of concrete, 268–269
Sedimentary rocks, 8
Seismic base isolation (SBI), 309
Seismic forces, 3, 23, 25, 42–43, 51, 90, 185, 186, 273, 275, 277–278, 307, 312
codes dealing with, 273, 277
Seismic loads, 51, 76–77, 186, 277, 309, 313, 377
Seismic performance, definition of, 274–275, 302
Seismic vulnerability, 41, 77, 192
Seismic-resistant design, requirements of, 50, 73, 98, 310
Self-tempering, 30
Serviceability design, of commonly used structural elements, 201–202
Serviceability limit states (SLS), 27, 43, 75, 90, 201, 212, 217
Servo-hydraulic Universal Testing Machine, 32
Shear deformation, 44, 46, 314
Shear design, concept of, 186
Shear failure, 101, 107, 182, 186, 316
in reinforced concrete, 181
Shear forces, 56, 98, 181, 183–187, 189, 190, 195, 312, 315–316, 377, 379–380, 386, 388
Shear loading, 197
Shear reinforcements, 107
action of, 183–186
in bends, hooks and mechanical anchorages, 191
Shear strength
of RC sections, 182–183
of shear reinforcement, 184
Shear stress, 181, 189
in beams, 182
calculation of, 181–182
Shear walls, 312
advantages of, 313
behaviour of a cantilever shear wall, 313
definition and classifications of, 312–313
design of, 315
subjected to axial load and uniaxial bending, 316–317
design provisions as per Indian standard IS:13920-2016, 350–354
development of interaction charts for
flexure design as per IRC:112-2020, 358–360
interaction diagrams as per IRC:112-2020, 361–365
development of P–M interaction diagrams for, 349–350
inferences from the proposed charts as per IS:456-2000, 350
development of strain distribution diagrams and stress blocks, 317–333
for failure in over-reinforced condition, 332–333
failure under pure flexure only when the wall acting as a steel beam, 333
or failure in under-reinforced condition, 333
under pure axial load only, 318–319
stress block for concrete, 319
stress block for HYSD steel, 320–332
interaction chart for rectangular shear walls, 351
limitations of the standard, 354–357
low-rise, 315
modes of failure in, 315–316
moment of resistance (MOR) of, 365
Slender cantilever walls, failure modes for, 316
Slender walls, 313, 316
stiffness of, 314–315
lateral, 314
stress resultants in, 315
Squat walls, 313
Shrinkage, deflection due to, 207–208
Silica fume, 10, 17–18, 26
Snapping of steel, 78, 90, 274, 315
Snow loads, 42, 43, 51, 76, 378
Sound insulation, 312
Special loads, 42
Staircases
based on the geometrical configurations, 367
classification of, 367
codal requirements, 368–369
deflection check, 371, 374
determination of the weight of the waist slab, 370
dog-legged, 371
load transfer of the landing slab, 368
reinforcement details of, 374
for residential building, 371–375
stair slab spanning
longitudinally, 367–368
transversely, 368
structural analysis and design consideration, 369–371
types of, 367–368
Steel concrete interface, 189, 194
Steel failure, 196, 197
Steel reinforcements, 1
effect on properties of reinforced concrete, 2–3
Steel truss, 184
Steel under stress, behaviour of, 33–34
Strain distribution, 27, 78
determination of, 100
Strain distribution diagram, 61, 99, 103, 218, 229, 256, 341
for balanced condition and with a minimum amount of tension reinforcement, 284

for a balanced section, 93, 287
of columns
when neutral axis is outside the section, 121–123
when the neutral axis is within the section, 123–124
development of, 92, 317–332
linear, 84, 86
for maximum under-reinforced section, 97
of singly reinforced sections, 92
and stress blocks, 84, 93
of under-reinforced doubly reinforced section, 101
Strain ductility, 229, 282
calculation and interpretation of, 282–284
Strain hardening, 32, 186
of steel, 232
zone, 34
Strength of concrete
effect of porosity on, 11
factors affecting, 12–13
Stress blocks
for concrete, 61–62, 89, 215, 317, 319
determination of, 100
development of, 317–332
for HYSD steel, 320–332
of under-reinforced doubly reinforced section, 101
Stress distribution diagram, 57, 66–67, 121, 210, 292
Stress–strain curve, 32, 34, 56, 84
for cold worked steel, 36
of concrete, 80, 229
as per IS:456-2000, 81
factor of safety on, 85
for mild steel, 35
for steel, 82
for CTD 415, 83
with definite yield point, 82
for TMT 415, 83
Stress–strain diagram, 59
for concrete, 13, 54–56, 61, 72, 84, 319
showing the condition corresponding to the limit state of collapse, 272
development of stress blocks from, 56–57
of HYSD steel, 110, 320
of reinforcing steel, 26, 33, 103
showing failure initiation stage in steel, 271
of steel with inclined plastic branch
computation of design parameters for concrete and steel, 231–233
for TMT bars, 36
of typical reinforcing steel, 38
of unconfined and confined concrete, 115
of untensioned steel, 37
Stress–strain relationship, 13, 22, 80, 333
of concrete, 49, 71, 80–82, 187, 214–215, 222, 284
as per IS:456-2000, 91
for different types of steel reinforcements, 34
of high-strength deformed bars and wires, 30
for hot rolled and cold worked steel, 35–39
for HYSD bars, 28–29, 215, 233
for monotonic loading of confined and unconfined concrete, 55
as per Eurocode 2, 35–39
as per IRC:112, 35
as per IS:456, 34–35
of reinforcing steel, 82–83, 214–215
of steel in compression and tension, 355
of unconfined concrete, 26
for uniaxial compression, 81
Structural concrete, 1–2
grades of, 3
provisions of
IS:456-2000, 25
IS:13920-2016, 25
Structural concrete members, ductility of, 309
Superplasticizers (SP), 10, 12, 17, 19

T

Tensile force, 32, 84, 93, 103, 115, 120, 144, 185, 196, 224, 296
Tensile loading, 197
Tensile reinforcement, 67, 71, 88, 93, 294, 384
Tensile strength, of concrete, 85, 91, 196
Tensile stresses, 2, 64, 182
in concrete, 61
Tension reinforcement, 107–108, 284, 302, 308
area of, 208
balanced percentage of, 95–96
modification factor for, 209
Tension stiffening effect, 204–205, 210
Thermal expansion, coefficient of, 1
Thermo-mechanically treated bars (TMT), 35
Translational degrees of freedom, 45, 369
Trial mixes
cement–water ratio, 17
concept of, 16
constituents of, 16
Truss, 183–184

U

Ultimate limit states (ULS), 27, 90–91, 214, 217, 218, 231, 234, 281
Ultimate strength, 13, 34, 75, 196
Unbraced walls, 313–314
Under-reinforced doubly reinforced section, forces in, 100–101, 112
Under-reinforced failure, 87, 271
Uniaxial tensile force, 32

V

Vertical reinforcement, 309, 315–318, 343, 349–350, 354, 361
Very high wind loads, 313
Vibrations, in the superstructure, 42

W

Water–cement ratio, 11–12, 16, 201
relation with compressive strength for trial mixes, 17
Wind loads, 41–42, 51, 76, 272, 273, 275, 313, 378, 381
Wind-induced oscillations, 42

Working stress method (WSM), for design of RC structures, 58
  assumptions regarding, 61
  for balanced, under-reinforced and over-reinforced, 66
  of compression members, 71
  design philosophy of, 60
  for doubly reinforced sections, 67
  flexural members, 63–65
  of member subjected to direct tension, 71
  mode of failure in, 66–67
  permissible stresses, 62
  problems regarding
    analysis, 67–70
    design, 68–71
  provisions as per IS:456-2000, 60–61
  reasons for getting outdated throughout the globe, 72–73
  serviceability requirements, 71
  shortcomings of, 71–72
  single-stage failure, 66–67
  for singly reinforced section
    balanced section, 65
    over-reinforced section, 66
    under-reinforced section, 65–66
  strength and serviceability design, 60
  stress blocks in concrete, 61–62

Y

Yield deformation, 72, 274
Yield point, 28, 32, 34–35, 49, 72, 78–79, 82, 86, 90, 119, 140, 214, 218, 221, 229, 231–232, 263, 274, 286, 289, 310, 333, 355
Yield strain, 34, 78, 88, 90, 92, 121, 219, 231–232, 235, 271, 274, 281–282, 289, 291, 317, 341, 348
Yield stress, 82, 192, 235
  for CTD bars, 86
Yielding, concept of, 274
Young's modulus of elasticity, 279

For Product Safety Concerns and Information please contact our EU representative GPSR@taylorandfrancis.com
Taylor & Francis Verlag GmbH, Kaufingerstraße 24, 80331 München, Germany

www.ingramcontent.com/pod-product-compliance
Lightning Source LLC
Chambersburg PA
CBHW080653220326
41598CB00033B/5197

* 9 7 8 1 0 3 2 5 4 1 4 5 7 *